O
TA
418.22   Creep of engineering
.C73   materials and of the
          earth.

O
TA
418.22   Creep of engineering
.C73   materials and of the
          earth.

CREEP OF ENGINEERING MATERIALS
AND OF THE EARTH

CREEP OF ENGINEERING MATERIALS AND OF THE EARTH

A ROYAL SOCIETY DISCUSSION
ORGANIZED BY
A. KELLY, F.R.S., ALAN H. COOK, F.R.S.,
AND G. W. GREENWOOD

HELD ON 24 AND 25 FEBRUARY 1977

LONDON
THE ROYAL SOCIETY
1978

Printed in Great Britain for the Royal Society
at the
University Press, Cambridge

ISBN 0 85403 099 9

First published in *Philosophical Transactions of the Royal Society of London*,
series A, volume 288 (no. 1350), pages 1–236

Published by the Royal Society
6 Carlton House Terrace, London SW1Y 5AG

PREFACE

This meeting was organized in order to bring together scientists concerned with the behaviour of engineering materials and those concerned with understanding geological and geophysical phenomena in terms of materials behaviour. The topic chosen for consideration was the slow deformation and fracture at low rates of strain of metals, and of rocks. The hope was to cover the laws of flow and the development of textures in rocks, minerals and metals, so that scientists in engineering, geology, geophysics and mineralogy might stimulate one another by learning about each other's methods, problems and progress.

The deduction of the creep laws for the mantle of the Earth was discussed by Professor J. Weertman and this was followed by a review by Professor F. A. Leckie of how constitutive equations governing the flow of metals can be set up and how the accumulation of creep damage during the life of a component may be used to predict when unsafe behaviour or rupture will occur. A detailed account was given by Professor A. Nicolas of how stress estimates may be made from the examination of dislocation or subgrain structures in mantle peridotites, and this was followed by a comprehensive review of our knowledge of the flow and fracture of materials under high temperatures and moderate hydrostatic pressures. Professor M. F. Ashby in his presentation showed how to produce deformation maps indicating the stress temperature and strain rate régimes, within which certain mechanisms of flow are likely to predominate. All save one of the papers presented at the meeting are reproduced here; the only one omitted is that on the flow of quartz and quartzite by hydrolytic weakening. A detailed account of creep in olivine was given by Professor C. Goetze. The recrystallization of metals during hot deformation and the recrystallization textures produced in metals and quartz were covered in detail by Dr C. M. Sellars and by Professor R. W. Cahn respectively. Dr J. H. Gittus gave an account of the high temperature deformation of two phase structures.

Lastly, fracture was covered in a discussion by Dr W. B. Beeré of the stresses and deformation at grain boundaries in metals and this was followed by an account of similar phenomena at geologic faults by Dr G. C. P. King. Finally, an account of fracture during creep was given by Professor G. W. Greenwood.

At the start of the meeting, apparent similarities between metallurgical and geological processes were emphasized and, in the summary given at the end of the meeting, the significant divergences in the scale, and how the similarities may arise, were commented upon.

The discussions were most stimulating and have suggested many interconnections between the disciplines of metallurgy, mechanical engineering, geology and geophysics.

A. KELLY
G. W. GREENWOOD
A. H. COOK

CONTENTS

Phil. Trans. R. Soc. Lond. A. **288**, 3–8 (1978)   [3]
Printed in Great Britain

Introduction

By A. Kelly, F.R.S.
Vice-Chancellor's Office, University of Surrey, Guildford

The basis of this conference, which was something of an experiment, was the idea of bringing together two groups of people concerned with a similar problem, namely the mechanisms of slow deformation of solids at elevated temperatures, but who are differently motivated and so follow different disciplines in their study of the phenomena.

Both the structural geologist and the geophysicist clearly have an interest in the hot deformation of solids, since the increase in temperature within the Earth is considerable, even in quiet areas (i.e. those away from volcanic activity), ranging from 10 to 50 °C/km. This alone, if continued, would lead to temperatures of between 100 and 500 °C at the bottom of the crust under ocean floors. The variation in temperature gradient arises usually from variation of thermal conductivity, the product of the two being remarkably constant at a value around $60 \, \mathrm{mW/m^2}$ ($1.5 \times 10^{-6} \, \mathrm{cal \, cm^{-2} \, s^{-1}}$) (see, for example, Horai & Simmons 1969), leading to an observed outflow of heat from the Earth of some $2.5 \times 10^{13} \, \mathrm{W}$. Movement of material within the mantle and crust occurs slowly except during earthquakes and these movements correspond to rates of strain and estimated stresses, which lead to values of the viscosity of the material of $10^{21} \, \mathrm{Pa \, s}$ ($10^{22} \, \mathrm{P}$) (see Anderson 1966).

The normal definition of a solid, used by the materials scientist based on experience with glass, is that it possesses a viscosity greater than $10^{14} \, \mathrm{Pa \, s}$ ($10^{15} \, \mathrm{P}$), so that the above rates of creep correspond to flow within the solid state.† A creep-resistant nickel alloy showing 2 % elongation in 1000 h under a stress of 61 MPa shows a viscosity of $3.7 \times 10^{15} \, \mathrm{Pa \, s}$. Creep tests on engineering materials only very rarely correspond to viscosities greater than $10^{18} \, \mathrm{Pa \, s}$ ($10^{19} \, \mathrm{P}$) because the stresses used are those to produce strain rates of $3 \times 10^{-10} \, \mathrm{s^{-1}}$ ($10^{-6} \, \mathrm{h^{-1}}$). However, components are designed for service lives of 20 years and if strains are to be less than say 0.1 % in this time under an operating stress of 75 MPa, then the engineer and materials scientist are interested in materials with nominal viscosities deduced from the equation

$$\sigma = 3\eta\dot{\epsilon}, \tag{1}$$

where σ is the tensile stress, η the coefficient of viscosity and $\dot{\epsilon}$ the tensile strain rate, of $1.5 \times 10^{19} \, \mathrm{Pa \, s}$ ($1.5 \times 10^{20} \, \mathrm{P}$).

The laws of creep deduced by the materials scientist are usually produced in the hope that they may help the engineer in the process of design, either to predict the rate of creep of engineering components, or else to defer the failure of these components. He knows the history of the synthesis of the material which is is studying.

The situation for the geologist is quite different. He attempts to explain physical features of the Earth and to account for motions in the Earth's crust and upper mantle. The history is, to a large extent, unknown. For instance, we know that the material at the top of the mantle (the

† The strain point, a definition used in glass manufacture, corresponds to that temperature at which internal elastic strains in a piece of glass are removed in four hours in a conventional article and to an estimated coefficient of Newtonian viscosity of $10^{14.5}$ P.

mantle being defined as that below the Moho (Mohorovičić discontinuity)) is spreading laterally across the Earth's surface with an upswelling beneath the ocean ridges and subduction under the continental masses (Hess 1962). This is accompanied by fracture of the crust in many cases, with the attendant production of transcurrent faulting in the crust above the mid-ocean ridges (see, for example, Vacquier 1965), and with a pattern of cracking and faulting elsewhere (see references for example in the Royal Society Symposium on Continental Drift (Blackett, Bullard & Runcorn (eds) 1965)). The flow of the material of the upper mantle appears to be more akin to that of a creeping solid than of a Newtonian fluid (even though the motion may be driven by convection) because postulated viscosities become of somewhat obscure meaning with values of 10^{20} Pa s being necessary (Anderson 1966), and, in addition, there is evidence of the existence of differences between principal stresses of as much as $(1–2) \times 10^7$ Pa (Munk & MacDonald 1960).

We also know from geodetic, tidal data, and tilted and raised beaches, that where ice sheets have melted, what is called isostatic rebound is occurring, decaying at a rate approximately exponential with time (Farrand 1962), so that crustal features are being elevated due to the release of the pressure due to the ice sheets. Superficial measurements may be made of the vertical movements in this case, which are of the order of centimetres per year, and of the horizontal movements in the case of continental drift. The rates in the latter case are again of the order of 1 cm/a and may be up to 10 cm/a in parts of the Pacific Ocean. Creep is clearly involved, but the question of what material is undergoing these rates of slow creep is not, of course, entirely clear. If an input can be obtained, either from the science of materials or from other sciences, concerning possible lower and upper limits to the rates of creep for known materials, then some rational deductions may be made concerning the type of material involved in the creep processes in the mantle. The constitution of the mantle rocks is not subject to direct experimental verification, although very plausible arguments suggest that they may be identified close to the mid-ocean ridges and when overriding occurs during mountain building (Jacobs, Russell & Wilson 1974, pp. 87 and 469).

There has, apparently, been some success achieved in interpreting rates of creep involved in geological processes in terms of laws deduced from materials science (see, for example, Gordon 1965). In addition, Orowan (1965) has considered the whole question of continental drift from the point of view of the science of materials and theory of plasticity.

The theories which have been used to understand creep of the upper mantle of the earth have nearly all, as far as I am aware, been concerned with so called 'steady-state' creep, in which the strain varies linearly with time. However, other forms of creep – often referred to as transient creep (see, for example, Garofalo 1965) may be important over a geological time scale (see Murrell (1976); Weertman, this volume, p. 9).

The type of creep involved in the flow of the mantle has been postulated to be either that called Nabarro-Herring creep (Nabarro 1948), Coble (1963) creep, or Nabarro's modification of the former for coarse grained materials in which dislocations form the principal sources and sinks for vacancies (Nabarro 1967). The general theory for volume conserving diffusion creep in a polycrystal of a one component solid (Raj & Ashby 1971) yields for the tensile strain rate an equation of the form

$$\dot{\epsilon} = \frac{14\sigma\Omega}{kT}\frac{1}{d^2}D_\mathrm{V}\left(1+\frac{\pi\delta}{d}\frac{D_\mathrm{B}}{D_\mathrm{V}}\right), \tag{2}$$

where σ is the stress, Ω the atomic volume, k and T have their usual meanings, d is the grain diameter and δ the effective cross section of a grain boundary. D is the diffusion coefficient with subscript B for boundary and V for volume diffusion. There is a linear dependence of strain rate upon stress and so an effective Newtonian viscosity can be identified with an effective temperature dependence of the form $(1/T)\exp(-Q/kT)$. Nabarro's volume diffusion predominates at very large grain sizes and at very high temperatures (where D_B/D_V becomes reduced). When diffusion creep in compounds is considered, attention must be paid to the nature of the diffusing unit and to the effects of departures from stoichiometry, as well as the effects of specific impurities upon the value of the diffusion coefficient. The simple use of tracer diffusion coefficients is inappropriate (Pascoe & Hay 1973; Stocker & Ashby 1973).

When dislocations become involved as sources and sinks for vacancies, then the strain rate depends upon a higher power of the stress and this is so for all processes of dislocation creep. For instance, for many coarse-grained face-centred cubic and body-centred cubic metals, the secondary creep rates are given by an equation of the form

$$\dot{\epsilon} = KDGb(\sigma/G)^n/kT, \tag{3}$$

where G is the shear modulus, b the Burgers vector and K is a constant. The exponent n varies between a value of 3 and 7, consistently with shear modulus and stacking fault energy for the face centred cubic metals. The constant K can vary widely over a factor of 10^6 depending upon the exact dislocation process involved (Bird, Mukherjee & Dorn 1969).

A dependence of strain rate upon stress to a high power is generally true for many materials (see Finnie & Heller 1959; Lubahn & Felgar 1961). We have, as yet, no fundamental theories of creep under other than uniaxial stress.

There are clear phenomena in the deformation of rocks and minerals which are not dealt with in the conventional discipline called Materials Science. In the régime of cold deformation, we have the process of cataclasis, where the material is effectively fragmented along a plane and relative movement of the fracture faces occurs. At elevated temperatures there is also the recently recognized importance of the phenomenon of pressure solution (Ramsay 1967; Rutter 1976). In this, diffusion is assisted by the presence of water at grain boundaries and it has some geometrical similarities to the diffusion creep described above.

If, as a result of this meeting, the materials scientist recognizes some additional modes of deformation, or if superplasticity (see Gittus (1975) for a review) be seen to be important in geological processes, the organizers will feel gratified. The constitutive equation describing superplastic flow shows a dependence of strain rate on stress to a higher power than in equation (2), and the microstructural characteristics of a material undergoing superplastic flow, characterized by a fine and stable grain size, are different from those of a material deforming by Nabarro creep or dislocation motion.

Another point worth making at the outset is that the materials scientist and engineer is usually able to distinguish quite clearly between structures produced by the process of slow deformation at elevated temperatures, whether or not accompanied by fracture, and those produced during recrystallization in which the microstructure is totally reconstituted. Creep under conditions of periodic recrystallization has been studied by Gilbert & Munson (1965). Of course, low melting point metals are well known to recrystallize during deformation at room temperature or slightly above, so that the microstructure progressively changes with deformation and then shows a discontinuous change when recrystallization occurs (see, for example, Gay & Kelly 1953). With the hot creep of engineering materials or of ceramics, whether or not recrystallization has intervened, is usually known from X-ray or microscopical observation. In the case of deformation of subsurface rocks, of course, it will not be known how far recrystallization may have altered the microstructures and how far not. When metals are subject to heavy deformation and subsequently allowed to recrystallize, sharp textures are produced and these sharp textures can be related to the preceding texture produced by the deformation (R. W. Cahn, this volume, p. 159). I do not know whether or not discussion of this topic could illuminate some riddles of geological structure.

The engineer, when considering creep, usually likes to ignore the microstructure and to deduce constitutive equations describing the material's behaviour. F. A. Leckie, in the second paper, calls this the 'black box' approach. It often fails, and, when a material shows wide deviations in creep ductility, leads to a very conservative design. It leaves out of account methods of examining a component part-way through its life in order to assess the damage accompanying creep. However, it has the great advantage of considering multiaxial stress states.

At a meeting such as this where creep is to be considered under widely different conditions, the stress system is very important. In materials science it is clearly recognized that there is a difference between the effects of tensile and shear stresses, and the different effects on the microstructure can be identified. For instance, shear strains produce voids at grain boundaries, but a tensile stress is needed in order for these to grow (Dyson, Loveday & Rodgers 1976). The terrestrial creep, which we shall be considering at this meeting, occurs under conditions of moderate pressure, e.g. at a depth of 7 km under the ocean, the pressure at the top of the mantle is about 6×10^7 Pa, and at 50 km directly below the stable platforms of a continent some 15×10^8 Pa. If we consider that flow can occur in the soft upper mantle – the asthenosphere ('$\alpha\sigma\theta\epsilon\nu\epsilon\iota\alpha$, lack of strength) – at depths of 100 km, then the pressures attain 3×10^9 Pa, or a small percentage of the bulk modulus at atmospheric pressure.

Creep and fast slip also occur at what geologists call transform faults, and there appears to be some similarity in this to grain boundary sliding. The differences in scale are enormous, e.g. that between 10 μm and 100 km, or a factor of 10^{10}. However, the geometrical similarities are so striking that I would like to believe they cannot be ignored and may be helpful in arriving at possible modes of relative motion.

In an alloy at temperatures above one half of the melting temperature, the grain boundaries deform easily and the grains themselves are relatively, but not completely, rigid. The grains slide past one another and there is some deformation within the grains. This deformation may be accomplished either by diffusional flow or by some type of dislocation motion; the dislocation motion often results in kinks and folds. The driving force is the ability of the applied stress to do work upon the specimen. The analogy is between this process and that in which relatively rigid lithospheric plates are pushed apart by upswelling of magma at an inconformity between the plates – compare Figures 1 and 2 of Bhattacharji & Koide (1975) describing motions in the Red Sea area (see also figure 1) and the illustration of a fold near a triple-point grain junction in Chang & Grant's (1956) work on the slow deformation of an aluminium–zinc alloy at a temperature of 260 °C, where the grain boundaries are weak and their sliding produces a deformation band within the grains (see figure 2).

The similarity between the work of the materials scientist and that of the structural geologist or geophysicist is that each has been concerned with a deduction of properties, and in the geologist's case also a deduction of history, from a knowledge of crystal structure and microstructure. The engineer uses this knowledge to derive constitutive equations, indicating rates of flow under prescribed conditions of temperature, pressure, strain rate, stress state, etc. Now that crustal blocks of the Earth are seen to be in constant motion, the science of geology may be able to predict

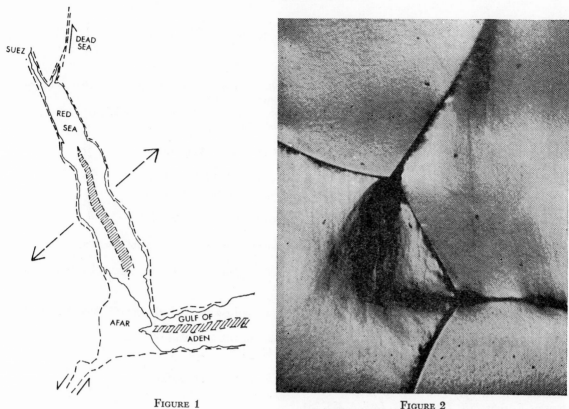

FIGURE 1   FIGURE 2

FIGURE 1. The formation of the Red Sea by extension (scale 1:28 600 000). (Reproduced from Girdler 1965.)

FIGURE 2. Fold formation in an aluminium–zinc alloy at 260 °C and 15.9 MPa (magn. × 92). (Reproduced by permission of the copyright owner, American Institute of Mining, Metallurgical and Petroleum Engineers, from Chang & Grant (1956).)

triumphantly the expected changes in geography and knowledge of the constitutive equations will be necessary for this. In figure 3 I have illustrated the conventional flow of information, which has usually been vertical, between the materials scientist and engineer and between the geologist and geophysicist. By placing one above the other, I am *not* emphasizing a hierarchy.

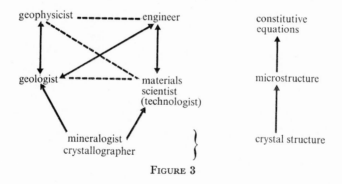

FIGURE 3

There has been some contact, of course, between the geologist and the engineer, via soil mechanics and structural phenomena. The aim of this conference is to bring about a better horizontal flow of information and also a better diagonal flow of information, e.g. between the geophysicist and materials scientist.

REFERENCES (Kelly)

Anderson, D. L. 1966 *Science, N.Y.* **151**, 321.

Bhattacharji, S. & Koide, H. 1975 *Nature, Lond.* **255**, 21.

Bird, J. E., Mukherjee, A. K. & Dorn, J. E. 1969 In *Quantitative relation between properties and microstructure* (eds D. G. Brandon & A. Rosen), p. 255. Jerusalem: Israel Universities Press.

Blackett, P. M. S., Bullard, E. C. & Runcorn, S. K. (eds) 1965 *Phil. Trans. R. Soc. Lond.* A **258**, 1.

Chang, H. C. & Grant, N. J. 1956 *Trans. Am. Inst. Min. metall. Petrol. Engrs.* **206**, 544.

Coble, R. L. 1963 *J. appl. Phys.* **34**, 1679.

Dyson, B. F., Loveday, M. S. & Rodgers, M. J. 1976 *Proc. R. Soc. Lond.* A **349**, 245.

Farrand, W. R. 1962 *Am. J. Sci.* **260**, 181.

Finnie, I. & Heller, W. R. 1959 *Creep of engineering materials.* New York: McGraw-Hill.

Garofalo, F. 1965 *Fundamentals of creep and creep rupture in metals.* New York: Macmillan.

Gay, P. & Kelly, A. 1953 *Acta. crystallogr.* **6**, 172.

Gilbert, E. R. & Munson, D. E. 1965 *Trans. Am. Inst. Min. metall. Petrol. Engrs* **233**, 429.

Girdler, R. W. 1965 *Phil. Trans. R. Soc. Lond.* A **258**, 123.

Gittus, J. 1975 *Creep viscoelasticity and creep fracture in solids.* London: Applied Science Publishers Ltd.

Gordon, R. B. 1965 *J. geophys Res.* **70**, 2413.

Hess, H. H. 1962 In *Petrological studies* (eds A. E. J. Engel, H. L. James & B. F. Leonard). Geological Society of America.

Horai, K. & Simmons, G. 1969 *Earth & planet. Sci. Lett.* **6**, 386.

Jacobs, J. A., Russell, R. D. & Tuzo Wilson, J. 1974 *Physics and geology*, 2nd edn, pp. 87, 469. New York: McGraw-Hill.

Lubahn, J. D. & Felgar, R. P. 1961 *Plasticity and creep of metals*, p. 175. New York: J. Wiley.

Munk, W. H. & MacDonald, G. J. F. 1960 *J. geophys. Res.* **65**, 2169.

Murrell, S. F. 1976 *Tectonophysics* **36**, 5.

Nabarro, F. R. N. 1948 *Bristol conf. on strength of solids*, p. 75. London: Physics Society.

Nabarro, F. R. N. 1967 *Phil. Mag.* **16**, 231.

Orowan, E. 1965 *Phil. Trans. R. Soc. Lond.* A **258**, 284.

Pascoe, R. T. & Hay, K. A. 1973 *Phil. Mag.* **27**, 897.

Raj, R. & Ashby, M. F. 1971 *Met. Trans. Am. Soc. Metals* **2**, 1113.

Ramsay, J. G. 1967 *Folding and fracturing of rocks.* New York: McGraw-Hill.

Rutter, E. H. 1976 *Phil. Trans. R. Soc. Lond.* A **283**, 203.

Stocker, R. L. & Ashby, M. F. 1973 *Rev. Geophys. Space Phys.* **11**, 391.

Vacquier, V. 1965 *Phil. Trans. R. Soc. Lond.* A **258**, 77.

Phil. Trans. R. Soc. Lond. A. **288**, 9–26 (1978)   [9]

Printed in Great Britain

Creep laws for the mantle of the Earth

By J. Weertman

Departments of Materials Science and Engineering and Geological
Sciences, Northwestern University, Evanston, Illinois 60201, U.S.A.

The analyses of glacial rebound data by Cathles and by Peltier and Andrews have led them to the conclusion that the flow law of the mantle of the Earth is Newtonian and that the viscosity is essentially a constant (10^{22} P) throughout the mantle. In this paper it is concluded that no large strain, steady-state creep process in mantle rock can account for a Newtonian, constant viscosity mantle. It is suggested that small strain, transient creep and not steady-state creep is involved in the isostatic rebound phenomenon. Since convective motion in the mantle involves large creep strains, conclusions about the effective viscosity of mantle rock undergoing such flow that is based on isostatic rebound data are likely to be wrong. If Post's and Carter & Mercier's laboratory results of the stress dependence of the grain size in a mantle type rock are representative of the actual grain sizes in the mantle, power law creep is almost certainly the creep law that governs convective flow in the mantle.

Introduction

The temperature of the rock in the Earth's crust and mantle ranges from a small fraction of the melting (solidus) temperature, at shallow depths near the Earth's surface, up to the melting (solidus) temperature at depths within the low-velocity, low strength zone from which magmas of many volcanoes originate. Rocks in the Earth's crust and mantle also are subjected to a wide range of stresses. At teleseismic distances from an earthquake source the stresses are much smaller than those required to activate the Frank–Read sources necessary for plastic deformation by dislocation movement. At the other end of the spectrum explosive events, natural and man-made, may cause the theoretical shear strength of the rock to be exceeded.

The type of response of rock to the stress that deforms it depends both on the magnitude of that stress and the temperature of the rock. The stress and temperature ranges of the different types of responses are conveniently indicated on a diagram in which one axis is a normalized stress σ/μ, where σ is the stress and μ is the shear modulus of rock, and the other axis is the normalized temperature T/T_{m}, where T is the temperature and T_{m} is the melting (solidus) temperature.

Figure 1 shows such a diagram, which we have called a creep diagram when applied to creep phenomena (Weertman & Weertman 1965, 1975), for crystalline material like rock minerals, whose dislocations cannot move easily at the lower temperatures because of the existence of a high Peierls stress. The effective Peierls stress at the higher temperatures is reduced because of thermally produced stress fluctuations in the thermal sound waves within the solid. It should be pointed out that most mechanisms that give rise to anelastic deformation, and thus produce the damping of seismic waves, also can operate at stresses above the Peierls stress-dislocation source stress boundary in figure 1. Moreover deformation mechanisms, such as the Nabarro–Herring mass transport of point defects between grains boundaries by diffusion, can operate below the Peierls stress-dislocation source stress boundary. In this region Nabarro–Herring creep can produce large plastic deformation in fine-grained material although the effect would be negligible in large crystals. Rock within the Earth's mantle also is subjected to high hydrostatic pressure. Since hydrostatic pressure raises the melting temperature of most materials the effect of pressure can be taken into account in a diagram such as figure 1 through its influence on the melting temperature.

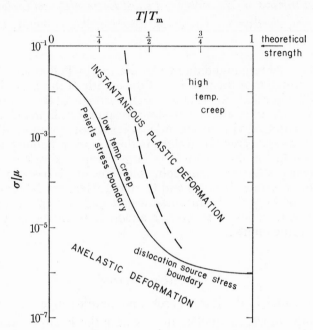

FIGURE 1. Schematic diagram of normalized temperature and normalized stress showing fields where different types of deformation and creep occur for crystalline material with a large Peierls stress.

In the case of steady-state creep processes a qualitative diagram like figure 1 can be made quantitative by plotting on it lines of constant creep rate. This is what Ashby and his co-workers (Ashby 1972; Stocker & Ashby 1973) have done and the resultant plots have been called by him deformation maps. The constant creep rate curves can be obtained over part of the diagram from experimental data, if it is available. In regions where no experimental data exist, theoretical curves can be calculated from theories that give steady-state creep rates based on the creep mechanism judged most likely to be dominant in the particular region. The deformation maps have not included transient creep data up to now but transient creep maps could be made by plotting, say, fixed times to reach some standard creep strain.

A geophysicist who works on a problem such as the process of continental drift or isostatic rebound from glacial loads wants to know the creep behaviour of mantle rock at the stresses, temperatures and pressures that exist within the Earth's mantle. He needs a deformation map which contains plots of constant creep rate in the area that corresponds to mantle conditions, but this area cannot be reached in laboratory experiments. So he must try to get there, by hook or crook, from extrapolation of experimental data, from geophysical field data, and from theories. Review articles on creep of rock and creep of the mantle (Gordon 1965, 1967; McKenzie 1968; Weertman 1970; Stocker & Ashby 1975; Weertman & Weertman 1975; Kirby & Raleigh 1973; Kohlstedt, Goetze & Durham 1976; Tsukahara 1974, 1976; Meissner & Vetter 1976; Heard 1976; Murrell 1976; Lorimer 1976; Nicolas 1976; Nicolas & Poirier 1976; Carter 1976), of which Carter's and Nicolas & Poirier's are the most complete and up to date, have summarized the experimental data and theories used in attempts to fill in the mantle area of the deformation map.

In recent years there has been controversy over what is the most likely creep law and creep mechanism for rock which is deforming under the stress, temperature, and pressure conditions found within the mantle. This paper is addressed to this controversy. The main point in dispute is the following: Is the appropriate creep law a linear law (Newtonian flow) or a power law with a power significantly larger than one?

Stress range – creep rate range

The order of magnitude of the creep rates and stresses of interest to flow processes within the mantle can be estimated with little difficulty. Plate motion is known to occur with velocities of the order of 1–$10 \, \mathrm{cm} \, \mathrm{a}^{-1}$. The convection within the mantle that gives rise to the motion of plates must take place over distances which are not much smaller than the thickness of a plate (about $100 \, \mathrm{km}$), and no larger than the thickness of the mantle ($3000 \, \mathrm{km}$). Hence the creep rates for convective motion are within the range of $10^{-16} \, \mathrm{s}^{-1}$ to $3 \times 10^{-14} \, \mathrm{s}^{-1}$. The isostatic rebound that follows the removal of a glacial load from the Earth's surface produces a maximum vertical displacement of the crust of about one third the thickness of the ice (*ca.* $3 \, \mathrm{km}$) that is removed. Since the horizontal scale of a large ice sheet is several thousand kilometres, rebound probably involves the mantle down to depths of comparable distances. Thus the total strain is of the order of 10^{-3}. Walcott (1972) has estimated uplift velocities over the last $6000 \, \mathrm{a}$ of the order of $2 \, \mathrm{cm} \, \mathrm{a}^{-1}$ near the centres of the former ice age ice sheets. This velocity would correspond to creep rates of the order of $10^{-15} \, \mathrm{s}^{-1}$, a value similar to that estimated from the convective motion.

A creep rate of the order of $10^{-15} \, \mathrm{s}^{-1}$ is well beyond a meaningful measurement in a laboratory experiment. At that creep rate in a test of one year's duration the total creep strain is only 3×10^{-8}. To produce such a strain, each dislocation in a sample with a dislocation density of $10^{12} \, \mathrm{m}^{-2}$ ($10^{8} \, \mathrm{cm}^{-2}$) would have to move, on the average, about one half a lattice spacing. In the case of creep produced by dislocation motion, to reach steady-state creep each dislocation source must produce many dislocation loops and each loop must move through distances orders of magnitude larger than the interatomic distance.

An estimate can be made of the order of magnitude of the stress needed to produce a creep rate of about $10^{-15} \, \mathrm{s}^{-1}$ in the mantle. An ice sheet with an average thickness of 2–$3 \, \mathrm{km}$ pushes the Earth's crust downwards with a pressure of the order of 20–$30 \, \mathrm{MPa}$ (200–$300 \, \mathrm{bar}$). The non-hydrostatic stresses that this applied load sets up within the mantle before isostatic adjustments, take place may be $\frac{1}{2}$–$\frac{1}{4}$ this amount; that is, about 5–$10 \, \mathrm{MPa}$. Removing the ice load after equilibrium has been reached would produce stresses of the same magnitude but of the opposite sign. Walcott (1972) estimates that about $300 \, \mathrm{m}$ of uplift still will take place in Canada. This present depression of the Earth's crust corresponds to a non-hydrostatic stress component of about 2–$5 \, \mathrm{MPa}$ (20–$50 \, \mathrm{bars}$).

For convective flow in the mantle a rough estimate of the order of magnitude of the stress can be made. The stresses should be some fraction of the quantity $\rho g \alpha_t \Delta T L$, where ρ is the density of rock, g is the gravitational acceleration, α_t is the coefficient of thermal expansion of the rock, ΔT is the average temperature difference that drives the convection, and L is a characteristic distance over which the convection is taking place. The amount of heat transported per unit time by convection to the Earth's surface is equal to $\mathscr{A}Q$, where $\mathscr{A}$ is the area of the Earth and Q is the average heat flux at the Earth's surface. The heat transport also is of the order of magnitude of $\mathscr{A}vc\Delta T$, where v is the average velocity of rock within the mantle and c is the specific heat of rock. Thus the magnitude of the stresses is some fraction of the expression $\rho g\alpha_t LQ/vc$. For $\rho = 5 \times 10^3 \, \mathrm{kg\,m^{-3}}$, $g = 9.8 \, \mathrm{m\,s^{-2}}$, $\alpha_t = 2 \times 10^{-5} \, \mathrm{K^{-1}}$, $L = 1000 \, \mathrm{km}$, $Q = 0.052 \, \mathrm{W\,m^{-2}}$, $v = 1\text{--}10 \, \mathrm{cm\,a^{-1}}$, and $c = 5 \, \mathrm{MJ\,m^{-3}\,K^{-1}}$, the stresses are some fraction of 3–30 MPa (30–300 bar).

It seems not unreasonable to expect that the stresses that produce flow in the mantle are of the order of magnitude of 5 MPa (50 bar) plus or minus half an order of magnitude.

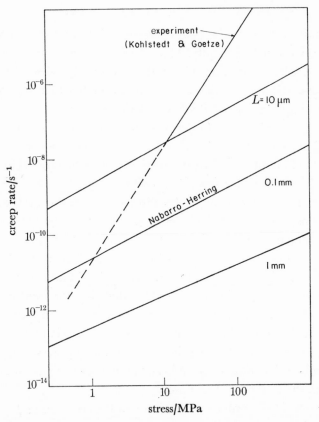

FIGURE 2. Plot of creep rate against stress for olivine, normalized to a temperature of 1400 °C. Solid part of line is in region where data were obtained by Kohlstedt & Geotze (1974). Also shown are creep rates predicted from the Nabarro–Herring creep equation for various grain sizes. Data used to obtain the plots are given in the text.

CROSS-OVER STRESS FOR DISLOCATION CREEP AND NABARRO–HERRING CREEP

There has been a persistent controversy whether the dominant type of creep in the mantle is that produced by dislocation motion or is that produced by the diffusional mass transport of atoms between grain boundaries: the Nabarro–Herring creep mechanism. The relative importance of these two mechanisms in the mantle depends upon the stress level and the grain size. This point can be understood most easily through figure 2. This figure shows for olivine

$$((\mathrm{Mg_{0.92}Fe_{0.08}})_2\mathrm{SiO_4})$$

at 1400 °C and atmospheric pressure logarithmic plots of creep rate $\dot{\epsilon}$ against stress calculated for various grain sizes from the Nabarro–Herring creep equation

$$\dot{\epsilon} = \alpha(D/L^2)\,(\sigma\Omega/kT), \tag{1}$$

where α is a dimensionless constant ($\alpha \approx 5$), L is the grain size, k is Boltzmann's constant, T is the temperature, Ω is the atomic volume, and D is the diffusion coefficient. The diffusion coefficient is given by the equation

$$D = D_0\exp\left(-Q/kT\right)\exp\left(-P\Delta V/kT\right), \tag{2}$$

where D_0 is a constant, Q is the activation energy of diffusion, ΔV is the activation volume of diffusion, and P is the hydrostatic pressure. (In our earlier papers (Weertman 1970; Weertman & Weertman 1975) we have assumed that the diffusion coefficient is also given by

$$D = D_0\exp\left(-gT_{\mathrm{m}}/T\right), \tag{3}$$

where g is a dimensionless constant ($g \approx 29$) and T_{m} is the melting temperature at the hydrostatic pressure P.) The Nabarro–Herring plots in figure 2 were calculated using

$$\Omega = 1.71 \times 10^{-28}\,\mathrm{m}^3$$

and the diffusion data estimated by Goetze & Kohlsteds (1973) from climb of dislocation loops ($D_0 = 3\,\mathrm{m}^2\,\mathrm{s}^{-1}$ and $Q = 565\,\mathrm{kJ\,mol}^{-1}$ ($135\,\mathrm{kcal\,mol}^{-1}$).)

Figure 2 also shows a plot of the steady-state creep rate of olivine single crystals measured by Kohlstedt & Goetze (1974) and normalized to a temperature of 1400 °C using the measured activation energy of creep of $524\,\mathrm{kJ\,mol}^{-1}$ ($125\,\mathrm{kcal\,mol}^{-1}$). The magnitude of the creep rate of these data seems to be representative of that reported in the literature for dunite under dry conditions. Only data of Griggs & Post (1973) and Post (1973) have substantially smaller creep rates. The dashed part of the line shown in figure 2 is an extrapolation from the stress range where the experimental data was obtained. The experimental data obey the following power law creep equation with a stress exponent $n = 3$:

$$\dot{\epsilon} = C\sigma^3\exp\left(-Q/kT\right), \tag{4}$$

where C is a constant ($C = 4.2 \times 10^{-13}\,\mathrm{Pa}^{-3}\,\mathrm{s}^{-1}$) and Q is the activation energy of creep. (The two activation energies used in figure 2 are within experimental error of each other.) Equation (4) is of the form of theoretical creep equations that are based on a dislocation mechanism. A third power dislocation creep equation (stress exponent $n = 3$) can be expressed as

$$\dot{\epsilon} = \gamma(D/b^2)\,(\mu\Omega/kT)\,(\sigma/\mu)^3, \tag{5}$$

where γ is a dimensionless constant whose value depends upon the particular mechanism that controls the dislocation motion, b is the length of the Burgers vector of the dislocation, and μ is the shear modulus. (If $C = 4.2 \times 10^{-13}\,\mathrm{Pa}^{-3}\,\mathrm{s}^{-1}$, $\mu = 79.1\,\mathrm{GPa}$, and $b = 0.698\,\mathrm{nm}$, then $\gamma \approx 0.058$.) Combining equations (1) and (5) gives the following equation for the stress at which the Nabarro–Herring creep equation and the dislocation creep equation predict the same creep rate:

$$\sigma = \mu(\alpha/\gamma)^{\frac{1}{2}}\,(b/L). \tag{6}$$

The cross-over stress that is given by equation (6) is independent of temperature and hydrostatic pressure except as μ varies with pressure. Equation (6) can be rearranged to give the grain size at which equations (1) and (5) predict the same creep rate for a constant stress. The equation is

$$L = b(\alpha/\gamma)^{\frac{1}{2}} (\mu/\sigma). \tag{7}$$

From figure 2 and equation (6) it is seen that the cross-over stress for olivine is 10 MPa (100 bar) for a grain size of 10 μm, 0.1 MPa (1 bar) for 1 mm, and 10 kbar (0.1 bar) for 1 cm. These grain sizes, particularly the larger ones, are all within the range of possible ones for mantle rock. The cross-over stresses they give bracket the rough estimates for stresses existing in the mantle that is given in the previous section.

Post (1973) has found in high-temperature creep experiments carried out on dunite that contains 98 % olivine in the deviatoric stress range of 100 MPa–1 GPa (1–10 kbar) that recrystallization occurred during the creep runs and that the recrystallized grain size is given by the equation

$$L/L_0 \approx C'^{\frac{3}{2}}(\sigma_0/\sigma)^{\frac{3}{2}}, \tag{8}$$

where the constants $L_0 = 1$ μm, $\sigma_0 = 100$ MPa, and $C' = 19$. Equation (8) applies whether the creep runs are made under 'wet' or 'dry' conditions. Carter and Mercier (reported in Carter 1976) also measured the grain size as a function of stress on a dunite from the same source as used by Post and on a synthetic dunite. They also found that equation (8) gives the grain size, and values of the constants L_0, σ_0, and C' that they determined are essentially the same as those found by Post.

If the grain size L that is given by equation (8) is larger than that given by equation (7) power law creep is dominant. Because equation (7) gives $L \sim 1/\sigma^{\frac{3}{2}}$ there is a critical stress *below* which the dislocation creep mechanism (power law creep) is dominant and *above* which the Nabarro–Herring creep mechanism is dominant (assuming that equation (8) can be extrapolated outside the stress range for which it is experimentally established). This critical stress is found by combining equations (7) and (8) to give

$$\sigma = C'^3(\gamma/\alpha) (\sigma_0^3/\mu^2) (L_0/b)^2. \tag{9}$$

For $\alpha/\gamma = 86$ and the values of C', L_0 and σ_0 quoted above the critical stress given by equation (9) is equal to $\sigma = 26$ GPa (260 kbar). Thus if equation (8) offers a reasonable estimate of the grain size of mantle rock it would be very unlikely that Nabarro–Herring creep is the important creep mechanism in the mantle for steady-state creep. (It should be noted that for a stress of $\sigma = 5$ MPa the grain size predicted by equation (8) is $L = 7.4$ mm, a not unreasonable value at such a low stress level. From figure 2 it is clear that Nabarro–Herring will not be the rate controlling creep mechanism if the dislocation creep mechanism does indeed operate at 5 MPa.)

At lower temperatures and for finer grain sizes grain boundary diffusion can increase significantly the Nabarro–Herring creep rate (Coble 1963). The Nabarro–Herring–Coble creep equation is the following modification of equation (1):

$$\dot{\epsilon} = \alpha(D/L^2) (\sigma\Omega/kT)[1 + (\pi\delta/l) (D_{\mathrm{B}}/D)], \tag{10}$$

where D is again the bulk or lattice diffusion coefficient, D_{B} is the grain boundary diffusion coefficient, and δ is the effective thickness of a grain boundary as a fast diffusion path ($\delta \approx b$).

If it is assumed that the grain boundary activation energy is approximately two thirds of that of the bulk diffusion activation energy then for olivine with a value of $Q = 565$ kJ/mol (135 kcal/mol), the ratio $D_{\mathrm{B}}/D = 7.9 \times 10^5$ at 1400 °C and $D_{\mathrm{B}}/D = 3.7 \times 10^4$ at 1890 °C (the melting temperature of an olivine that contains almost no iron (Yoder 1976)). Thus the Coble creep rate is faster than the Nabarro–Herring creep rate in olivine for grain sizes smaller than 1.7 mm at 1400 °C (and 8 μm at the melting temperature of 1890 °C). In figure 2 the creep rate for a grain size of $L = 1$ mm is increased by a factor 2.7, the curve with $L = 0.1$ mm by a factor 17, and the curve of $L = 10$ μm by a factor of 170 if grain boundary diffusion is taken into account.

If equations (5) and (10) are combined the cross-over stress is found to be equal to

$$\sigma = \mu(\alpha/\gamma)^{\frac{1}{2}} (b/L) [1 + (\pi\delta/L) (D_B/D)]^{\frac{1}{2}}. \tag{11}$$

If equations (8) and (11) are combined, the following equation is found for the stress *above* which Nabarro–Herring–Coble creep is dominant and *below* which dislocation creep is the important creep mechanism

$$\sigma = C'^3 \frac{\gamma}{\alpha} \frac{\sigma_0^3}{\mu^2} \left(\frac{L_0}{b}\right)^2 \bigg/ \left[1 + \frac{\pi\delta}{L_0} \frac{D_B}{D} \left(\frac{\sigma}{\sigma_0}\right)^{\frac{3}{2}} C'^{-\frac{3}{2}}\right]. \tag{12}$$

At temperatures so low that the ratio D_B/D is a very large number, equation (12) reduces to

$$\sigma = \sigma_0^{\frac{3}{5}} \sigma^{*\frac{2}{5}} C'^{\frac{3}{5}}(L_0/\pi\delta)^{\frac{2}{5}} (D/D_B)^{\frac{2}{5}}, \tag{13}$$

where $\sigma^* = C'^3(\gamma/\alpha) (\sigma_0^3/\mu^2) (L_0/b)^2$. With the values of the terms α, γ, etc., used previously at one half of the melting temperature of magnesium rich olivine the stress σ given by equation (13) is $\sigma = 14$ MPa. Thus even at this rather low temperature it is doubtful that a Nabarro–Herring–Coble creep mechanism can control the creep rate in the mantle.

It is known that the power law creep rate of rock is increased substantially when water is present (the hydrolytic weakening effect discovered by Blacic & Griggs: see the discussion in Nicolas & Poirier 1976). If hydrolytic weakening only increased the power law creep rate and not the Nabarro–Herring–Coble creep rate the presence of water in mantle rock would make it even more probable that the Nabarro–Herring–Coble creep mechanism does not control the creep rate of the mantle. Of course, it is possible that the hydrolytic weakening effect occurs because it reduces the activation energy of diffusion. If such is the case the presence of water will increase the creep rates of power law creep and Nabarro–Herring–Coble creep by equal amounts.

SUPERPLASTIC CREEP

Gueguen & Boullier (1976) have seen evidence of superplastic flow in the deformation textures of some mantle peridotite nodules they have examined. Twiss (1976) has proposed, in fact, that at least in the upper mantle superplastic creep may be the dominant creep mechanism. Superplastic creep occurs at a faster rate than Nabarro–Herring–Coble creep and it has a stress exponent n that is smaller than that found in dislocation produced creep. Generally, the exponent is of the order of $n = 1.5$–2 but values as low as $n = 1$ are reported. (A review of superplastic creep is given in the article of Edington, Melton & Cutler (1976). It also is discussed by Nicolas & Poirier (1976) as applied to rocks.) Thus if superplastic creep is the dominant creep mechanism in the mantle, the mantle would be Newtonian or near Newtonian in its flow properties.

Twiss based his quantitative calculations on the probability of superplastic flow on the following superplastic creep equation that was derived by Ashby & Verrall (1973):

$$\dot{\epsilon} = \alpha^*(D/L^2) (\sigma\Omega/kT)[1 + (\pi\delta/L) (D_B/D)], \tag{14}$$

where the dimensionless constant $\alpha^* \approx 100$. Equation (14) is essentially the Nabarro–Herring–Coble creep equation but with a creep rate that is one order of magnitude larger because the constant α^* is an order of magnitude larger than the constant α of equations (1) and (10).

Recently Nix (1976, private communication) has objected (and I believe his objection is well founded) to the model used by Ashby & Verrall (1973) in the derivation of their equation (14). The reason that the Ashby–Verrall creep equation is an order of magnitude faster than the Nabarro- Herring–Coble creep equation is that the main diffusional flow paths in their model

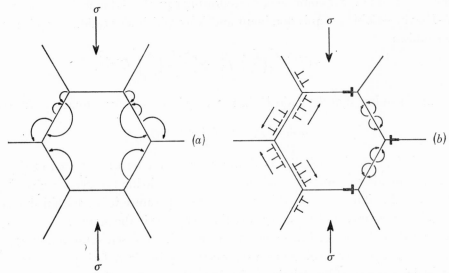

FIGURE 3. (a) Mass transport from grain boundaries as shown in fig. 7 of Ashby & Verrall (1973). (b) On the left hand side of the figure is shown the equivalent dislocations at grain boundaries that dislocation sliding would produce. These dislocations are roughly equivalent to the 'super' dislocations shown on the right hand side of the figure. The stress fields of the dislocations would cause mass transport as indicated by the arrows.

do not start at one grain boundary and end at a different grain boundary. Instead, as shown in figure 3a, the diffusion paths generally start and terminate at the same grain boundary. The paths are shorter and the amount of material transported needed to attain the same creep strain is smaller than in the Nabarro–Herring–Coble model. Hence, the predicted creep rate is faster. However, Nix points out that the grain boundary sliding that is part of the Ashby–Verrall model produces an effective dislocation distribution at the grain boundaries that is shown on the left hand side of figure 3b. The stress fields of these dislocations are approximately the same as those of the 'superdislocations' shown on the right hand side of figure 3b. These latter dislocations have stress fields that will cause diffusional mass transport along paths also indicated on the right hand side of figure 3b. This mass transport is symmetrical and is not the same as that used to obtain equation (14) and will not produce an increase in the creep rate.

Whether the derivation of equation (14) is correct or not Ashby and Verrall have shown that it does predict the right creep rate for at least one superplastic alloy (Zn–0.2 mass % Al) of 3.5 μm grain size. Moreover, for a number of superplastic alloys the grain size dependence of the creep rate does agree with equation (14). That is $\dot{\epsilon} \propto 1/L^m$ where $m = 2$–3 (Edington et al. 1976). It seems not unreasonable to use their equation, as Twiss has done, in considering the possibility of superplastic flow in the mantle. The equations for cross-over stress, etc., discussed in the previous section are unaltered if the Ashby–Verrall equation is used. It is only necessary to increase the value of the constant α in those equations by one order of magnitude.

Cathles–Pelter–Andrews constant viscosity Newtonian mantle

Cathles (1975) has analysed the post glacial rebound data from N. America in a very ambitious attempt to determine the viscosity of the Earth's mantle from shallow to the deepest depths. He assumed that the flow law of the mantle was that of a Newtonian solid and found that the field data was consistent with this assumption and that, moreover, the coefficient of viscosity was essentially a constant and independent of depth within the mantle. The viscosity coefficient has the value of 10^{22} P (10^{21} Pa s). Cathles's results rule out the existence of a thick (*ca.* 350 km) low viscosity channel. However, he does conclude that a thin (*ca.* 75 km) low viscosity channel does exist in the upper mantle which has a viscosity (4×10^{20} P) smaller by a factor of 25 than for the rest of the Earth. Peltier & Andrews (Peltier & Andrews 1976; Peltier 1974, 1976) have reached essentially the same conclusions as did Cathles using a different mathematical analysis of the rebound data.

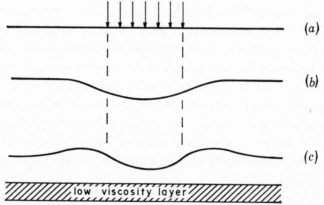

FIGURE 4. (*a*) Stress applied to a limited area on a half space. (*b*) Deformation produced by stress of (*a*) of surface of half space made of material that obeys a linear flow law with constant viscosity (or constant elastic constants). (*c*) Deformation of surface when material obeys a linear flow law but a layer of significant thickness exists below the surface within which the viscosity is very much lower than elsewhere in the half space. The deformation of the half space surface shown in (*c*) also occurs, whether the low viscosity layer exists or not, if the flow law of the half space material obeys a power law equation with an exponent significantly larger than unity.

Cathles's and Peltier & Andrews's conclusions depend not only on numerical results, but also on the qualitative character of the results. The general type of qualitative results used can be explained with the aid of figure 4. In figure 4*a* is shown a stress, that might correspond to a glacial load, pushing down over a limited area on the surface of a half space. If the material of the half space obeys the linear flow law with a constant viscosity everywhere, the deformation produced by the stress in (*a*) will be that shown schematically in figure 4*b*. The result to note is that the surface is lowered adjacent to the area subjected to the stress. If a low-viscosity channel of appreciable thickness exists below the free surface at a depth that is comparable with the horizontal extent of the surface area that is stressed, as is shown in figure 4*c*, the surface deformation will be that shown schematically in figure 4c. The surface adjacent to the stress area is raised rather than lowered. If the direction of the stress is reversed, as occurs during unloading by the melting away of an ice sheet, these deformations are also reversed. (The actual surface displacement histories actually are more complex than indicated by the schematic figure 4.)

The areas covered by Lake Bonneville and the Fennoscandia Ice Sheet were considerably smaller than that covered by the Laurentide Ice Sheet. Cathles found that the Lake Bonneville and the Fennoscandia rebound data corresponded essentially to the model of figure 4b and the Laurentide rebound data to figure 4c. Since the Laurentide rebound data sample the mantle deformation behaviour to deeper depths than do the rebound data of Lake Bonneville and Fennoscandia, Cathles and Peltier & Andrews were able to reach the conclusion that the Earth's mantle has constant viscosity except for a 75 km thick channel where the viscosity is reduced by a factor of 25.

Cathles ruled out a mantle governed by power law creep with an exponent n significantly larger than 1. A power law creep equation will produce a surface deformation that is shown in figure 4c. The qualitative behaviour is the same as found in channel flow. Thus the qualitative nature of Cathles's and Peltier & Andrews's results appear to be evidence against a power law creep being important for the mantle *if it is assumed that steady-state creep equations apply to the isostatic rebound phenomenon.*

The Fennoscandia rebound data have been used as evidence for a mantle that obeys a power law creep equation (Griggs & Post 1973; Post 1973). Because both channel flow and power law flow give the same qualitative rebound response the surface deformation history cannot be used as evidence for power law creep against Newtonian creep. Griggs and Post based their conclusion about power law creep on the curve of uplift at the centre of the rebounding area versus time. They found that the observed rebound curve fitted the one predicted using a power law creep equation (with $n = 3.21$) and not the one using a Newtonian creep equation ($n = 1$). However, as Cathles has pointed out, the shape of the uplift versus time curve depends upon the value estimated of the final uplift after infinite time. Cathles disputes the final uplift estimate used by Griggs and Post.

Cathles's and Peltier & Andrews's conclusions about the value of the viscosity of the upper and lower mantle are not in major conflict with those of others (see the review article of Walcott 1973; O'Connell 1971) except for McConnell's (1968) conclusion that the rebound data indicates that the viscosity of the lower mantle increases significantly with depth. However, it is very difficult to account for the findings that the viscosity of the lower mantle is low in magnitude and is a constant and that the flow of the lower mantle is Newtonian.

The most popular mechanism that has been invoked to give a Newtonian law for the mantle is the Nabarro–Herring one. Nabarro himself proposed this mechanism for creep in the Earth in his original paper (Nabarro 1948). The first quantitative calculations that used this mechanism to determine the viscosity of the mantle were made by Gordon (1965), who showed that the viscosity should vary over many orders of magnitude. The reason that the variation is so large is that the ratio T/T_m, where T is the temperature and T_m is the melting (solidus) temperature, at a given depth in the mantle presumably varies from a value near 1 in the low viscosity channel to a value of the order of $\frac{1}{2}$ at the core–mantle boundary. The takeover of Coble creep at the lower values T/T_m reduces the variation of the predicted viscosity with depth but the variation is still in violent disagreement with the conclusion of a constant viscosity mantle. (From data in fig. 7 of Weertman & Weertman 1975, the variation of viscosity over a temperature range of

$\frac{1}{2} \leqslant T/T_{m} \leqslant 1$ for Nabarro–Herring–Coble creep for a grain size of 1 mm is a millionfold.) If superplastic creep exists in the mantle the predicted viscosity variation is just as large as that for Nabarro–Herring–Coble creep.

If a power law creep equation governs mantle flow the variation of 'effective' viscosity with depth is greatly reduced (Weertman 1970; Weertman & Weertman 1975). But this reduction is no help in explaining Cathles' and Peltier & Andrews' results because they conflict qualitatively with the surface displacements predicted with a power law creep equation. In fact, any diffusion controlled steady-state creep mechanism has great difficulty in giving an account of a Newtonian Earth with essentially a constant viscosity. Only in the very unlikely event that the ratio T/T_{m} is almost constant through the mantle is it possible to do so.

It would appear that we are at an impasse. Cathles's and Peltier & Andrews's conclusions about the Earth's viscosity and predictions made using steady-state creep equations just do not agree. A way out of the dilemma is to abandon, at least for the lower mantle, the steady-state creep equations.

The creep that occurs during glacial rebound is not necessarily a steady-state type of creep. It has already been mentioned that the total strain that occurs during rebound is of the order of 10^{-3}. The rebound data that Cathles used to obtain information about the viscosity of the lower mantle involve strains that actually are considerably smaller than 10^{-3}. For example, the measured uplift curves from regions far south of the edge of the Laurentide Ice Sheet (see Cathles 1975, pp. 226–228) have total displacements no larger than 10 m. For these data to be a sample of deformation of the lower mantle at depths of the order of 2000–3000 km implies that the total strain involved is only of the order of $10\,m/2500\,km = 4 \times 10^{-6}$. Strains of such small magnitude can hardly be considered to be produced in ordinary steady-state creep.

It seems reasonable to conclude that the viscosity of the Earth (at least of the lower mantle) as measured from glacial rebound is a measure of the transient creep properties of mantle rock, not a measure of the large strain, steady-state creep properties of mantle rock that determine the convective flow in the mantle and the motion of the plates.

It is possible for transient creep, over a limited range of strain, to be Newtonian. In Chalmers's original experiments on microcreep (Chalmers 1936) it was observed that at very small creep strains that the creep rate (of tin) is proportional to the stress. Thus microcreep is Newtonian. Moreover, the creep rate is much larger than the true steady-state creep rate measured at large creep strains. I have argued that other cases of Newtonian behaviour seen at small creep strain, including observations in ice, is Chalmers's microcreep phenomenon (Weertman 1967, 1969). (The explanation of microcreep is that, because the creep strains are so small, the dislocation density remains essentially constant during creep. If the dislocation velocity is proportional to the stress and the dislocation density remains constant the creep rate produced by the dislocation motion is proportional to the stress.)

Transient, Newtonian creep behaviour also may occur over a rather large range in strain. Harper & Dorn (1957) discovered in high-purity polycrystalline (3 mm grain size) aluminum that at low stresses ($\sigma \approx 2 \times 10^6 \mu$) the creep strain was proportional to time and the creep rate proportional to the stress for creep runs carried out to strains as large as 0.01. The creep rates were 3 orders of magnitude larger than that predicted from the Nabarro–Herring creep equation. Subsequent work (Harper, Shepard & Dorn 1958; Barrett, Muehleisen & Nix 1972; Muehleisen 1969; Mohamed, Murty & Morris 1975) has shown that Harper–Dorn creep occurs in single crystals of aluminium and in other metals and that it is not likely to be produced by

Nabarro–Herring subgrain creep. The status of our understanding of Harper–Dorn creep was reviewed recently by Mohamed *et al.* (1975).

It seems reasonably clear (Mohamed *et al.* 1975) that in Harper–Dorn creep the dislocation density is independent of stress. Application of a low stress will not immediately cause the dislocation density to change through dislocation multiplication. (In fact, the attainment of equilibrium dislocation density appropriate to the applied stress level might even require the annihilation of dislocations. For example, in unstressed crystal the equilibrium dislocation density is no dislocations at all. Yet dislocations exist in almost all crystals, regardless of how long they have been annealed at a high temperature.) A creep strain as large as 0.01, however, cannot be reached before appreciable dislocation multiplication is needed. The creep strain is equal to the product of the Burgers vector, the average distance a dislocation moves, and the total length of dislocation line per unit volume that moves during creep. For a dislocation density of $10^8 \, \text{m/m}^3$ ($10^4 \, \text{cm/cm}^3$) that have been reported for Harper–Dorn creep specimens and for an average dislocation displacement equal to the average spacing between dislocations (*ca.* 170 μm) the total creep strain would be only of the order of 1×10^{-5}. Thus dislocation multiplication is needed or much larger dislocation displacements are required to obtain a strain of 10^{-2}. Therefore, Harper–Dorn creep and Chalmers microcreep must be different. It has yet to be made clear in any paper so far published how the Newtonian behaviour of Harper–Dorn creep comes about.

Can Harper–Dorn creep – if it exists in rock – be important in the mantle? The activation energy of Harper–Dorn creep is that of self-diffusion. Therefore, the same objection can be raised against it as against Nabarro–Herring–Coble creep: if Harper–Dorn creep controls glacial rebound then the viscosity of the mantle should vary over many orders of magnitude. Since Chalmers's microcreep mechanism also involves diffusion mechanisms the objection raised against Harper–Dorn creep and Nabarro–Herring–Coble creep also can be raised against it.

In normal transient creep the creep rate decreases as the creep strain increases. Thus the effective viscosity in transient creep increases with increasing strain. Since the creep strain involved in the isostatic rebound phenomenon must decrease with depth in the mantle, this strain effect would cause a decrease in the effective viscosity with increasing depth in the lower mantle. This decrease might, therefore, compensate for the increase in effective viscosity at a fixed strain that occurs because the temperature ratio T/T_{m} decreases with depth in the lower mantle. Therefore, it is possible to see the possibility of an explanation of why the mantle has a constant viscosity if normal transient creep occurs during isostatic rebound.

DISCUSSION

'The resistance arising from want of lubricity in parts of a fluid, is, other things being equal, proportional to the velocity with which the paths of the fluid are separated from one another' (Newton 1947).

Up to now the discussions about whether the mantle of the Earth lacks lubricity and is or is not Newtonian have not distinguished between large strain creep deformation (which occurs in convective motion and which requires at least one of many possible steady-state creep mechanisms to produce it) and small strain creep deformation such as occur during glacial rebound. It appears to be very important that this distinction be made. Cathles and Peltier & Andrews have shown that glacial rebound data can only be explained with a Newtonian mantle of essentially constant viscosity. It is impossible for steady-state creep mechanisms to account for this result. Since glacial rebound requires only relative small creep strains the viscosity values determined from rebound data undoubtedly are a measure of transient creep that occurs in the mantle rather than steady-state creep. Unfortunately transient creep has not been studied, either experimentally or theoretically, as extensively as has steady-state creep and we are not as well equipped to make comparisons between rebound data and experimental data and theories; but perhaps it will prove possible in the future to decide if the viscosity of the mantle as measured with rebound data can be explained with transient creep phenomenon.

The value of the viscosity determined in rebound measurements most probably sets a lower limit to the value of the effective viscosity of large strain flow. The larger the transient creep strain, and hence the shallower the depth in the mantle, the closer this limit will be to the true effective viscosity of large strain flow.

Whether the large strain, steady-state creep flow that occurs during convection is or is not Newtonian depends upon the value of the grain size of mantle rock. If the experimental results of Post (1973) and Carter and Mercier (Carter 1976) on the stress dependence of the grain size in dunite are at all representative of the stress dependence of the grain size of rock in the mantle it is extremely unlikely that the mantle is Newtonian at stress levels that are likely to occur during convective motion. When power law creep operates the effective viscosity of the mantle, as calculated under conditions of either constant stress or constant strain rate, varies over about 4–6 orders of magnitude (Weertman 1970; Weertman & Weertman 1975; Stocker & Ashby 1973). The effective viscosity in the lower mantle is likely to be several orders of magnitude larger than the glacial rebound value of the viscosity, but it still may be low enough to permit convection there.

This work was supported by the National Science Foundation under Grant number AER 75-00187.

References (Weertman)

Ashby, M. F. 1972 *Acta metall.* **20**, 887.

Ashby, M. F. & Verrall, R. A. 1973 *Acta metall.* **21**, 149.

Barrett, C. R., Muehleisen, E. C. & Nix, W. D. 1972 *Mater. Sci. & Eng.* **10**, 33.

Carter, N. L. 1976 *Rev. Geophys. Space Phys.* **14**, 301.

Cathles, L. M., III 1975 *The viscosity of the Earth's mantle.* Princeton: Princeton University Press.

Chalmers, B. 1936 *Proc. R. Soc. Lond.* A **156**, 427.

Coble, R. L. 1963 *J. appl. Phys.* **34**, 1679.

Edington, J. W., Melton, K. N. & Cutler, C. P. 1976 *Prog. Mater. Sci.* **21**, 63.

Geotze, C. & Kohlstedt, D. L. 1973 *J. geophys. Res.* **78**, 5961.

Gordon, R. B. 1965 *J. geophys. Res.* **70**, 2413.

Gordon, R. B. 1967 *Geophys. J. R. astr. Soc.* **14**, 33.

Griggs, D. T. & Post, R. L., Jr. 1973 *Science, N.Y.* **181**, 1242.

Guegen, Y. & Boullier, A. M. 1976 *Physics and chemistry of minerals and rocks* (ed. R. G. J. Strens), p. 19. New York: John Wiley.

Harper, F. G. & Dorn, J. E. 1957 *Acta metall.* **5**, 654.

Harper, J. G., Shepard, L. A. & Dorn, J. E. 1958 *Acta metall.* **6**, 507.

Heard, H. C. 1976 *Phil. Trans. R. Soc. Lond.* A **283**, 173.

Kirby, S. H. & Raleigh, C. B. 1973 *Tectonophysics* **19**, 165.

Kohlstedt, D. L. & Geotze, C. 1974 *J. geophys. Res.* **79**, 2045.

Kohlstedt, D. L., Geotze, C. & Durham, W. B. 1976 *The physics and chemistry of minerals and rocks* (ed. R. G. J. Strens), p. 35. New York: John Wiley.

Lorimer, G. W. 1976 *The physics and chemistry of minerals and rocks* (ed. R. J. G. Strens), p. 3. New York: John Wiley.

McConnell, R. K. 1968 *J. geophys. Res.* **72**, 7089.

McKenzie, D. P. 1968 *The history of the Earth's crust* (ed. R. A. Phinney), p. 28. Princeton: Princeton University Press.

Meissner, R. O. & Vetter, U. R. 1976 *Tectonophysics* **35**, 137.

Mohamed, F. A., Murty, K. L. & Morris, J. W., Jr 1975 *Rate processes in plastic deformation of materials* (eds J. C. M. Li & A. K. Mukherjee), p. 459. Metals Park: American Society for Metals.

Muehleisen, E. C. 1969 The role of structure in Newtonian deformation of metals. Ph.D. Thesis, Stanford University.

Murrell, S. A. F. 1976 *Tectonophysics* **36**, 5.

Nabarro, F. R. N. 1948 *Strength of solids*, p. 75. London: The Physical Society.

Newton, I. 1947 *Mathematical principles* (ed. F. Cajori), book II, section IX, p. 385. Berkeley: University of California.

Nicolas, A. 1976 *Tectonophysics* **32**, 93.

Nicolas, A. & Poirier, J. 1976 *Crystalline plasticity and solid state flow in metamorphic rocks*. New York: John Wiley.

O'Connell, R. J. 1971 *Geophys. J. R. astr. Soc.* **23**, 299.

Peltier, W. R. 1974 *Rev. Geophys. Space Phys.* **12**, 649.

Peltier, W. R. 1976 *Geophys. J. R. astr. Soc.* **46**, 669.

Peltier, W. R. & Andrews, J. T. 1976 *Geophys. J. R. astr. Soc.* **46**, 605.

Post, R. L., Jr 1973 The flow laws of Mt. Burnett dunite. Ph.D. Thesis, University of California, Los Angeles.

Stocker, R. L. & Ashby, M. F. 1973 *Rev. Geophys. Space Phys.* **11**, 391.

Tsukahara, H. 1974 *J. Phys. Earth* **22**, 345.

Tsukahara, H. 1976 *J. Phys. Earth* **24**, 89.

Twiss, F. J. 1976 *Earth planet. Sci. Lett.* **33**, 86.

Walcott, R. I. 1972 *Rev. Geophys. Space Phys.* **10**, 849.

Walcott, R. I. 1973 *Annual review of earth and planetary sciences* (eds F. A. Donath, F. G. Stehli & G. W. Wetherill), vol. 1, p. 15. Palo Alto: Annual Reviews Inc.

Weertman, J. 1957 *J. Glaciol.* **3**, 38.

Weertman, J. 1967 *Trans. metall. Soc. A.I.M.E.* **239**, 1989.

Weertman, J. 1969 *J. Glaciol.* **8**, 494.

Weertman, J. 1970 *Rev. Geophys. Space Phys.* **8**, 145.

Weertman, J. & Weertman, J. R. 1965 *Physical metallurgy* (ed. R. W. Cahn), p. 79. Amsterdam: North-Holland.

Weertman, J. & Weertman, J. R. 1975 *Annual review of earth and planetary sciences* (eds F. A. Donath, F. G. Stehli & G. W. Wetherill), vol. 3, p. 293. Palo Alto: Annual Reviews Inc.

Yoder, H. S., Jr 1976 *Generation of basaltic magma*. Washington: National Academy of Sciences.

Discussion

J. WEERTMAN. D. C. Tozer pointed out in the discussion after the presentation of my paper (see also Tozer 1977) that convection in the mantle self-regulates the viscosity value of rock of the mantle. (That is, if the mantle were cold its viscosity would be very high, but radioactive heating in time would raise the temperature of the mantle and reduce the value of the viscosity. On the other hand, if the temperature were very high, the viscosity of the mantle would be very low, but the resultant fast convective motion within the mantle would reduce the mantle temperature and increase the value of the viscosity of the mantle rock.) He proposed that the viscosity of the mantle is a constant throughout because of this self-regulation.

I agree with him that the mantle self-regulates to a large extent the value of its viscosity, and point out that if power law creep is the important creep law for the mantle that the mantle can even self-regulate the value of its effective viscosity by changing the stress level without even changing the mantle temperature. In my paper (Weertman 1970), in which the importance for the mantle of power law creep was stressed, it was pointed out that the variation with depth of effective viscosity is smaller for power law creep than for Newtonian creep. However, self-regulation is no guarantee that the effective viscosity of the mantle will have a constant value everywhere. It is only necessary to note the obvious fact that self-regulation has not kept the viscosity of the outer core of the Earth from being numerous orders of magnitude smaller and the effective viscosity of the Earth's crust from being orders of magnitude larger than that of the mantle.

I am not convinced that Tozer's self-regulation mechanism is the explanation of the constant viscosity values that come out of the glacial rebound analysis. Convective motion within the mantle drives the temperature profile of the mantle towards an adiabatic temperature profile. The pressure gradient of an adiabat in the mantle presumably is smaller than the pressure gradient of a melting temperature curve (Jacobs, Russell & Wilson 1974). (However, in the liquid outer core of the earth Kennedy & Higgins (1973) believe that the adiabat gradient is larger than the melting point gradient.) Thus the temperature profile within the mantle should lie between the melting temperature profile and an adiabat. If there is no zone of appreciable thickness within the mantle in which the actual temperature is above the solidus temperature then the ratio T/T_m (where T_m is the solidus temperature and T is the actual temperature) will be smaller the greater the depth within the mantle. Thus the effective viscosity at a fixed strain rate or stress should increase with depth in the lower mantle.

Even if the actual temperature of the mantle is close to the melting point everywhere, as suggested by Kennedy & Higgins (1972) and by Stacey (1975), there is another reason why the mantle viscosity may increase with depth in the lower mantle. The viscosity is inversely proportional to the diffusion coefficient D. The diffusion coefficient D depends upon the melting temperature of the material by the empirical equation (3), $D = D_0 \exp\left(-gT_m/T\right)$, of the text. (For magnesium-rich olivine–forsterite of a melting temperature 2163 K and a diffusion activation energy of 565 kJ/mol the value of g is 31.5.) In solid solutions the solidus temperature is a reasonable temperature to use for T_m in equation (3) because it is the temperature at which a solid of a given composition begins to melt as the temperature is raised. The liquidus temperature, on the other hand, gives the temperature at which a solid of a *different* composition begins to freeze out of a melt as the temperature is lowered. The only available test of equation (3) for metal alloys which form single-phase solid-solutions below the solidus temperature (Birchenall 1951; Shewmon 1963) shows that the diffusion coefficient varies with change in composition in the same way that the solidus temperature does. (However, in the alloy systems studied the liquidus temperatures were close to the solidus temperatures and hence it is not possible to decide if the solidus or the liquidus temperature is the better temperature to use for T_m.) In the lower mantle it has been suggested that olivine, which has transformed to a spinel structure at intermediate depths, transforms into periclase (MgO) and stishovite (SiO_2) (Kennedy & Higgins 1972). More recently it has been proposed that the transformation is into periclase and a perovskite phase ($MgSiO_3$) (Liu 1976a, b; Ito 1977). Suppose that the creep properties of the lower mantle are determined by the creep properties of the periclase mineral. Suppose further that the actual temperature in the lower mantle is near an eutectic temperature of a two or more phase mixture, as suggested by Kennedy & Higgins (1972). What is the melting temperature T_m to be used in equation (3)? It is not the eutectic temperature. Consider figure 5 which shows a schematic phase diagram for an eutectic system. Suppose periclase is the α phase indicated in the figure. If the actual temperature T is somewhat below the eutectic temperature, as indicated in figure 5, the α phase in a two-phase mixture has a composition that is different from what it would have at the eutectic temperature. The solidus temperature of the α phase of a composition at temperature T in a two-phase mixture is that indicated by the temperature T_m in figure 5. The solidus temperature T_m can be considerably higher than the eutectic temperature. It is the most reasonable temperature to use in equation (3) for the melting temperature of the α phase of a two-phase mixture. It is the melting temperature of the α phase of a composition that is different from the composition the α phase has at the eutectic temperature. In other words, the melting temperature to use in equation (3) for a two-phase mixture will itself depend on the actual temperature because the composition of the phases varies with temperature. Therefore, even if the actual temperature is close to the eutectic temperature it may not be close to the appropriate melting temperature T_m. Only if the actual temperature is exactly equal to the eutectic temperature is T_m equal to the eutectic temperature. Newton, Jayaraman & Newton (1962) have pointed out that an eutectic trough deepens with increasing pressure. That is, the melting temperature of the end members of an eutectic system increases much more rapidly with increasing pressure than does the eutectic temperature. Hence even if the mantle were everywhere slightly below an eutectic temperature the ratio T/T_m could decrease with increasing depth in the mantle.

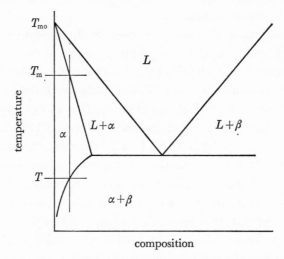

FIGURE 5. Schematic phase diagram for an eutectic system.

It should also have been pointed out in the text that an increase of an order of magnitude in the viscosity will be produced in the lower mantle (for power law creep) because the elastic constants there increase by a factor of 3. The transformation into denser phases also can produce a decrease of one to two orders of magnitude in the diffusion coefficient and thus a comparable increase in the viscosity (Shewmon 1963).

In an earlier paper (Weertman 1970) I had suggested that if the glacial rebound stresses are superimposed on the stresses that are produced during convection in the mantle and if the rebound stresses are small compared with the later one, then the rebound phenomenon will be pseudo-Newtonian although the creep rate is determined by the power law creep equations. I still believe this possibility. However, if the effective viscosity for convection in the lower mantle is, say, 10^{24} P, it would not be possible for the pseudo-Newtonian viscosity for the rebound phenomenon to be lower than this value (in steady-state creep).

It also should have been pointed out in the paper that data on the recrystallized grain size in olivine that was considered by Kohlstedt, Geotze & Durham (1976) also is described by equation (8). Sellars and co-workers (references given in Nicolas & Poirier 1976) have found in metals that the recrystallized grain size also is proportional to $\sigma^{-\frac{3}{2}}$.

References

Birchenall, C. E. 1951 *Atom movements*, p. 112. Cleveland: American Society for Metals.
Ito, E. 1977 *Geophys. Res. Lett.* **4**, 72.
Jacobs, J. A., Russell, R. D. & Wilson, J. T. 1974 *Physics and geology*, 2nd edn. New York: McGraw-Hill.
Kennedy, G. C. & Higgins, G. H. 1972 *The upper mantle* (eds A. R. Ritsema, K. Aki, P. J. Hart & L. Knopoff),
     p. 221. Amsterdam: North-Holland.
Kennedy, G. C. & Higgins, G. H. 1973 *The Moon* **7**, 14.
Liu, L.-G. 1976a *Nature, Lond.* **262**, 770.
Liu, L.-G. 1976b *Earth Planet. Sci. Lett.* **31**, 200.
Newton, R. C., Jayaraman, A. & Kennedy, G. C. 1962 *J. geophys. Res.* **67**, 2559.
Shewmon, P. G. 1963 *Diffusion in solids*. New York: McGraw-Hill.
Stacey, F. D. 1975 *Nature, Lond.* **255**, 44.
Tozer, D. C. 1977 *Sci. Prog., Oxf.* **64**, 1.

S. WHITE (*Department of Geology, Royal School of Mines, Imperial College, London SW7 2BP*). Phase transformations are often overlooked when mantle deformation is considered, although they are among the least speculative of mantle phenomena. There is a growing volume of literature associating deep seated earthquakes with phase transformations (see Sung & Burns 1976). However, these earthquakes are directly related to the volume changes associated with the phase transformation. There is another process associated with transformations, namely transformation enhanced ductility which is termed 'transformation superplasticity' in the metallurgical literature. Greenwood & Johnston (1965) showed that volume changes associated with phase changes in metals can lead to the generation of appreciable stresses across the phase boundary. These stresses lower the applied stress required to induce flow in the weaker phase. In some instances the stresses may exceed the yield stress and consequently may induce flow in the absence of an applied stress. Transformation enhanced ductility has also been observed in a variety of ceramics and minerals (Edington, Melton & Cutler 1976).

Pressure induced phase transformations appear to be common in the Earth's mantle (Akimoto, Matsui & Syona 1975). A major zone of transformation occurs at a depth of 400 km where olivine transforms to a spinel structure. It is reasonable to assume that appreciable stresses, especially when compared to the 10 MPa differential stress thought to be responsible for most mantle deformations, develop in the above zone. Consequently transformation enhanced ductility should be associated with transformation zones and rocks moving through these should exhibit an abnormal weakness. This can lead to homogeneous deformation in the deep mantle.

Normally, only small strains are induced by a transformation in metals, and cycling through the transformation is necessary to obtain large strains. However, the strains associated with mantle phase transformations may be greater because of the very slow, continuous passing of rock through the transformation. Consequently transformation enhanced ductility may make a significant contribution to mantle deformation processes.

References

Akimoto, S., Matsui, Y. & Syona, Y. 1975 High-pressure crystal chemistry of orthosilicates and the formation
     of the mantle transition zone. In *The physics and chemistry of minerals and rocks* (ed. R. G. J. Strens), pp. 327–363.
     London: Wiley.
Edington, J. W., Melton, K. N. & Cutler, C. P. 1976 Superplasticity *Prog. Mater. Sci.* **21**, 63–170.
Greenwood, G. W. & Johnston, R. H. 1965 The deformation of metals under small stresses during phase trans-
     formations. *Proc. R. Soc. Lond.* A **283**, 403–422.
Sung, C. M. & Burns, R. G. 1976 Kinetics of high-pressure phase transformations: Implications to the evolution
     of the olivine-spinel transition in the down-going lithosphere and its consequences on the dynamics of the
     mantle. *Tectonophysics* **31**, 1–32.

J. WEERTMAN. Transformational superplasticity certainly is a possible deformation mode in those parts of the mantle in which phase changes occur. I would doubt that large plastic strains are produced because of the need to cycle back and forth through the phase transformation to obtain this type of superplasticity. However, for the rebound phenomenon, which involves only small plastic strains, it might be an important deformation mechanism.

ALAN H. COOK, F.R.S. (*Department of Physics, Cavendish Laboratory, Madingley Road, Cambridge CB3 0HE*). It is generally agreed that a series of polymorphic changes occurs in minerals in the upper mantle. Olivine, in particular, is known to change to a so-called β-phase, then to a spinel structure and finally, probably, to a mixture of high pressure forms of the constituent oxides. Could Professor Weertman say how such changes will affect creep rates within the upper mantle?

J. WEERTMAN. A phase change to a more close packed phase is known to decrease the diffusion coefficient and the creep rate of metals by one to two orders of magnitude (Shewmon 1963; Sherby & Burke 1967). I would expect that a similar decrease in the magnitude of the creep rate would occur for polymorphic changes into denser phases in the mantle. The effect of the phase change could also be accounted for by a change in the melting temperature T_m in equation (3).

Reference

Sherby, O. D. & Burke, P. M. 1967 *Prog. Mater. Sci.* **13**, 325.

when the strains at rupture are much smaller. In practice the operating stress levels are normally on the low side being below those occurring in the transition region of the rupture curve. Consequently failure is normally associated with internal damage. If t_0 is the rupture time associated with the applied stress σ_0 then the creep rupture time t_R for the applied stress σ is given by

$$t_R/t_0 = (\sigma_0/\sigma)^\nu, \tag{2.1}$$

where ν is a material constant equal in magnitude to the slope of the rupture curve $g\sigma$ against $\lg t_R$.

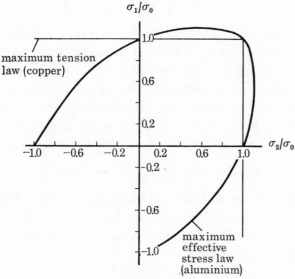

FIGURE 3. Plane stress isochronous rupture loci for copper and aluminium alloys.

In the absence of damage the steady state creep rate $\dot{\epsilon}_s$ can usually be expressed in the form

$$\dot{\epsilon}_s/\dot{\epsilon}_0 = (\sigma/\sigma_0)^n. \tag{2.2}$$

In this equation $\dot{\epsilon}_0$ is the steady state creep rate corresponding to the applied stress σ_0 and n is a material constant. In many materials the value of n is constant except for high values of stress when the value can increase.

Since most engineering components operate under triaxial stress states there is a particular desire to understand how materials behave in these circumstances. The experimental studies of Johnson, Henderson & Khan (1962) show that the magnitudes of the strain rate is dictated by the Mises or effective stress ($\bar{\sigma}$) criterion where

$$\bar{\sigma} = 2^{-\frac{1}{2}}[(\sigma_1 - \sigma_2)^2 + (\sigma_2 - \sigma_3)^2 + (\sigma_3 - \sigma_1)^2]^{\frac{1}{2}}. \tag{2.3}$$

If the effective strain rate $\bar{\dot{\epsilon}}_s$ is defined according to the condition

$$\bar{\dot{\epsilon}}_s = 2^{\frac{1}{2}} \times \tfrac{1}{3}[(\dot{\epsilon}_1 - \dot{\epsilon}_2)^2 + (\dot{\epsilon}_2 - \dot{\epsilon}_3)^2 + (\dot{\epsilon}_3 - \dot{\epsilon}_1)^2]^{\frac{1}{2}}, \tag{2.4}$$

then the general form of equation (2.2) becomes

$$\bar{\dot{\epsilon}}_s/\dot{\epsilon}_0 = (\bar{\sigma}/\sigma_0)^n. \tag{2.5}$$

In the above equations the subscripts 1, 2, 3 refer to the principal stresses and strain rates with σ_1 being the maximum value.

The effect of multiaxial states of stress on rupture are normally expressed in terms of *isochronous surfaces*. These are surfaces in stress space for which rupture times are constant. As a result of experiments on tubes subjected to torsion and tension Johnson *et al.* (1962) concluded that the rupture life of copper tested at 250 °C is controlled by the value of maximum stress σ_1 while that of an aluminium alloy tested at 200 °C is controlled by the value of effective stress $\overline{\sigma}$. These criteria for biaxial states of stress are illustrated in figure 3 and Johnson *et al.* suggested that the two criteria represent two extremes of material behaviour. Hayhurst (1972), after an extensive review of available data, concluded that isochronous surfaces for rupture time t_0 can be represented by the equation

$$\Delta(\sigma_{ij}/\sigma_0) = \left[\alpha\frac{\sigma_1}{\sigma_0} + \beta\frac{\overline{\sigma}}{\sigma_0} + \gamma\frac{J_1}{\sigma_0}\right] = 1. \tag{2.6a}$$

In this expression α, β, γ are constants with $\alpha + \beta + \gamma = 1$, σ_1 is the maximum stress, $\overline{\sigma}$ the effective stress and $J_1(= \sigma_1 + \sigma_2 + \sigma_3)$ is the first stress invariant. When $\beta = \gamma = 0$ the rupture criterion is maximum stress dependent, and when $\alpha = \gamma = 0$ it is effective stress dependent. Both of these rupture criteria pass through the point (1.1) on the plane stress loci shown in figure 3. The constant γ is introduced to describe the reduction in the value of stress sometimes observed when metals are subjected to equal biaxial states of stress. It appears however, that γ is usually small (Hayhurst 1972) and when neglected (2.6a) simplifies to the form suggested by Sdobyrev (1958):

$$\Delta(\sigma_{ij}/\sigma_0) = [\alpha(\sigma_1/\sigma_0) + (1-\alpha)(\overline{\sigma}/\sigma_0)] = 1. \tag{2.6b}$$

A survey of the results of tests designed to investigate the effect of variable stress histories on rupture life has been carried out by Penny & Marriott (1971). The most common test consists of the application of cyclic stress histories and it is then found that the lives are predicted quite satisfactorily by the time summation rule

$$\Sigma t_i/t_R^i = 1, \tag{2.7}$$

where t_R^i is the rupture time for a constant stress test at stress σ_i, and t_i is actual time of application of the stress σ_i. Storåkers (1969) has shown that this result can be less accurate for other forms of stress history and a reliable form of law valid for any stress history remains to be found.

The material information described in this section is the most that the designer can normally expect. A more common experience is that only the results of short-term tests are available and time extrapolation methods must be used. A variety of extrapolation techniques are available but the most scientific and reliable appear to be the so-called 'creep rupture maps' proposed by Ashby & Raj (1975). In these maps the regions in which various creep damage mechanisms are dominant are defined by boundaries of stress and temperature. A given working stress and temperature will fall within a region for which a particular mechanism is dominant and it is important to limit the use of extrapolation techniques to that particular region.

3. Approximate description of component behaviour

On the basis of available information it is now possible to speculate on how different components might behave. The basis of the speculations are:

(*a*) Stress redistribution occurs so that the rupture life of a component is based on average rather than maximum stresses, the supporting evidence being the results of the experiments described in § 1.

(*b*) Creep strains are proportional to the effective stress $\overline{\sigma}$ to some power n.

(*c*) Rupture life of the material is inversely proportional to $\Delta(\sigma_{ij}/\sigma_0)$ (equation 2.6) to some power ν. The equation $\Delta(\sigma_{ij}/\sigma_0) = 1$ defines the isochronous surface for failure time t_0.

After the suggestion of Johnson *et al.* (1962) that the rupture life of copper is maximum stress (σ_1) dependent while that of aluminium is effective stress ($\overline{\sigma}$) dependent, it appears sensible to investigate the properties of different load bearing components made from these different materials. Two simple components have been selected to illustrate the difference in behaviour which might be expected. These are the Andrade disk (figure 4, plate 1) and the Bridgman notch (figure 5).

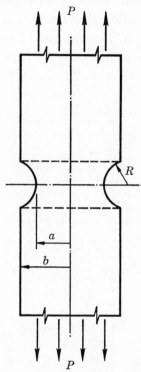

FIGURE 5. The Bridgman notch.

The copper specimens were tested at 250 °C and for copper it is found that in equations (2.1) and (2.5) $n \approx \nu \approx 6$ and that the strain at failure is ϵ_0^R approximately 7 %. For the aluminium alloy HF9 tested at 150 °C the corresponding constants are $n \approx \nu \approx 10$ and the strain at failure ϵ_0^R is 1.5 %. Strain at failure in both of these materials shows small variations with stress, the strain decreasing with decreasing stress. However, it might be noted that there is considerable difference between the failure strains of the copper and aluminium.

(a) *The Andrade & Jolliffe disk* (1952)

This is a circular disk which is profiled so that when the disk is subjected to torque about the axis of symmetry the shear stress τ is everywhere constant (figure 4). For this field the maximum stress is $\sigma_1 = \tau$ and the effective stress $\overline{\sigma} = \tau\sqrt{3}$ so that the ratio $\sigma_1/\overline{\sigma} = 1/\sqrt{3} = 0.577$.

Since the rupture life of copper is maximum stress (σ_1) dependent then the stress required in the disk to give a rupture time t_0 is $\sigma_1 = \tau = \sigma_0$. Compared with the corresponding uniaxial test with rupture life t_0 the effective stress is increased by a factor $\sqrt{3}$ and consequently the strain to failure is $\epsilon_0^R(\sqrt{3})^n = 27.0\,\epsilon_0^R$. Hence the strain at failure of the disk is many times greater than the failure strain in the uniaxial specimen for the same rupture time.

The rupture life of the aluminium alloy is effective stress $(\bar{\sigma})$ dependent and the shear stress required to give a rupture time t_0 is given by the equation $\bar{\sigma} = \tau\sqrt{3} = \sigma_0$ or $\tau = \sigma_0/\sqrt{3}$. Since the effective stress $\bar{\sigma} = \sigma_0$ in the corresponding uniaxial test with rupture time t_0, the strain at failure in the disk is equal to that in the corresponding uniaxial test.

For ease of comparison the shear stresses required to give a rupture time t_0 together with the strains at failure are given in table 1. In this test therefore the copper disk should be considerably stronger than the aluminium disk when expressed in terms of the normalizing uniaxial stress σ_0. If test results of shear stress τ against failure time are plotted on uniaxial rupture data the results should have the form shown in figure 6. The strain at failure in the copper disk should be many times greater than the uniaxial strain while that in the aluminium disk should be the same.

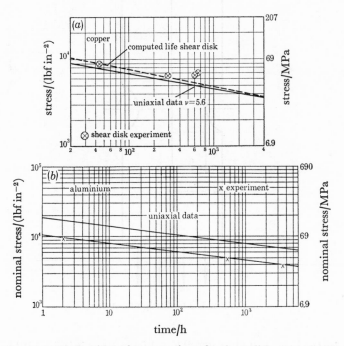

FIGURE 6. Comparison of rupture times for shear disk and uniaxial tension specimens (a) copper and (b) aluminium.

TABLE 1. STRESSES AND STRAINS FOR RUPTURE TIME t_0

component	max. stress (σ_1)	effective stress $(\bar{\sigma})$	$\sigma_1/\bar{\sigma}$	failure stress for rupture time t_0		failure strain	
				copper	aluminium	copper	aluminium
uniaxial	σ_0	σ_0	1.00	σ_0	σ_0	ϵ_0	ϵ_0
Andrade disk	τ	$\tau\sqrt{3}$	0.577	$\tau = \sigma_0$	$\tau = \sigma_0/\sqrt{3}$	$27\epsilon_0$	ϵ_0
Bridgman notch	σ_N	$\sigma_N/1.33$	1.33	$\sigma_N = \sigma_0$	$\sigma_N = 1.33\sigma_0$	$0.18\epsilon_0$	ϵ_0

Experiments have been made on copper and aluminium disks by Hayhurst & Storåkers (1976). The experimental results are in good agreement with the theoretical predictions (figure 6). The large strains observed in the copper disks are illustrated in figure 4 which shows the distorted form of lines which were initially radial.

(b) The Bridgman notch (1952)

The introduction of a notch into a cylindrical specimen of the type analysed by Bridgman for perfectly plastic materials can be used to create multiaxial states of stress. For the geometry shown in figure 5 the average normal stress is $\sigma_N = P/\pi a^2$ where P is the applied load and a the radius of the specimen at the minimum cross section. The average maximum stress is then $\sigma_1 = \sigma_N$ while the effective stress is $\bar{\sigma} = \sigma_N/1.33$. Hence the ratio $\sigma_1/\bar{\sigma} = 1.33$, which is approximately twice the value occurring in the Andrade disk.

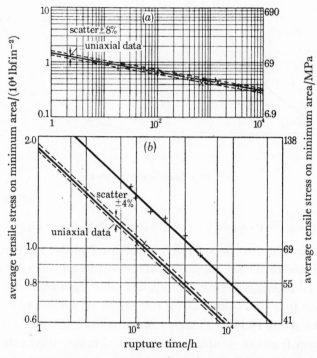

FIGURE 7. Comparison of rupture times for the Bridgman notch and uniaxial tension specimens (a) copper (b) aluminium.

In a copper notch the failure time is achieved with a normal stress $\sigma_N = \sigma_1 = \sigma_0$. The effective stress is $\bar{\sigma} = \sigma_0/1.33$ and consequently the strain at failure should be $\epsilon_0(1/1.33)^6 = 0.18\epsilon_0$. In an aluminium notch the time to failure t_0 should be achieved when $\bar{\sigma} = \sigma_N/1.33 = \sigma_0$, or $\sigma_N = 1.33\sigma_0$. Since the effective stress is equal to that in a uniaxial test the strain at failure should be ϵ_0 (table 1). When the rupture life stress σ_N results are plotted on uniaxial data the results should have the form illustrated in figure 7.

Tests on Bridgman notches have been performed by Leckie & Hayhurst (1974). The results of these tests are in agreement with the predictions of theory. The aluminium specimens show the expected strengthening when compared with uniaxial data, and the failure strains in the copper notches are approximately 2 % which is considerably less than the 7 % observed in uniaxial tests (figure 7).

The predictions and experimental observations illustrate that the behaviour of components is strongly dependent on the combined effect of the multiaxial stress state and the isochronous surface. In the Andrade disk it is the copper which is stronger than aluminium while in the

Bridgman notch the position is reversed. In aluminium the strain at failure is equal to the strain at failure in the uniaxial test. However, in copper the strain to failure can differ greatly in the different components. This result indicates that a commonly held view that the strain at failure is constant may not always be justified.

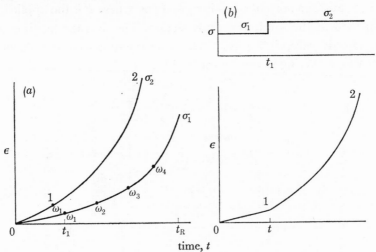

FIGURE 8. State variable tests.

4. CONSTITUTIVE EQUATIONS USING THE STATE VARIABLE DESCRIPTION

The predictions made in the previous section were based on the assumption that the rupture life of a component is based on the average values of the normal and effective stresses. The average values were determined using stress distributions available from plasticity theory. The experimental results indicate that procedures along the suggested lines may be valid, but a precise theoretical justification is still lacking. Furthermore, engineers are anxious to study the growth of strains and of damage fronts in components since local failure might result in leakage, while general failure is not imminent. The engineering approach is to develop constitutive equations which describe the macroscopic behaviour of the material. It is the constitutive equations in conjunction with the laws of continuum mechanics which determines component behaviour. In the case of creep rupture, attempts will be made to calculate the time at which cracks first appear, the growth of damage fronts and the time for complete rupture. In order to describe the macroscopic behaviour of the materials it is necessary to introduce internal state variables which, in some sense, are a measure of the physical state of the material. Since engineers do not attempt to give a precise physical description of the state variable this approach, on the face of it, is at variance with the approach of the metallurgist who wishes to give an accurate description of the mechanisms which take place. It will be suggested in the text, however, that the state variable method and the methods used by metallurgists to describe physical phenomena are, in some circumstances, almost identical. In these circumstances it would appear that the state variable description can be used with confidence.

(a) The state variable description for uniaxial stress

If, for convenience of discussion, the primary portion of the creep curve is neglected then the strain/time curve in a constant stress test will have the form shown in figure 8a. The creep rate increases from the initial steady state value as damage causes an advance into the tertiary portion of the curve. The steady-state creep rate can be expressed as a function of the applied stress σ alone. In order to account for the increase in strain rate during a constant stress test it is necessary to introduce a new variable into the strain equation. Since the increase in strain rate is the result of a damage process the variable ω introduced is referred to as the damage state variable. The strain rate equation then takes the form

$$\mathrm{d}\epsilon/\mathrm{d}t = f(\sigma, \omega), \qquad (4.1a)$$

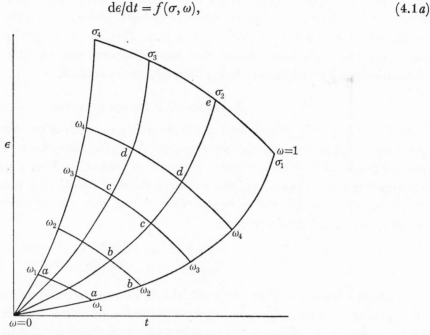

FIGURE 9. Constant damage contours.

where f is a function which has yet to be defined. However, in the undamaged state when $\omega = 0$ the equation must reduce to the form observed for steady state conditions. To define the strain rate it is necessary that the value of ω as well as the stress σ be known and consequently an equation must be introduced which defines the growth of the damage state variable. Assuming that the damage rate depends both on the current state of stress and damage gives the equation

$$\mathrm{d}\omega/\mathrm{d}t = \mathrm{g}(\sigma, \omega). \qquad (4.1b)$$

The problem is how to find the functions f and g in a systematic manner. This may be achieved by following a testing procedure suggested by Leckie & Hayhurst (1975) which makes use of the results of tests with step changes in stress. Neglecting the primary portion of the creep curve of the strain/time graphs for two constant stress tests carried out at stresses σ_1 and σ_2 would have the form shown in figure 8a. Suppose in the constant σ_1 test that ω varies between zero at time $t = 0$ and unity at the rupture time t_R. Select an arbitrary variation of ω along the σ_1 curve (figure 8a). Now perform a test in which the stress σ is applied for time t_1 when the stress is increased to σ_2 and maintained constant at this value to give the creep curve shown in figure 8b. Since a single state variable is sufficient to define the creep curves it follows that section 12 of the curve 012 must be identical in shape with a portion of the constant stress creep curve σ_2. In this way the state ω_1 may be defined on the σ_2 curve. By performing similar tests constant ω contours may be constructed as shown in figure 9. This information can then be used to predict the variation of strains and damage in a result of variable stress histories (Leckie & Hayhurst 1975).

The procedures described above are applicable provided that the state of damage can be described by a single damage parameter. If a single state parameter suffices it might still be necessary to modify the equations (4.1) for loading and unloading structures. This is certainly the case for creep deformations in the primary region when it is found that the growth laws of dislocation density differ for loading and unloading. This point has been discussed by Leckie & Ponter (1974). A systematic programme of step loadings for defining the growth laws of copper in the tertiary region is being performed by D. R. Hayhurst & C. J. Morrison (private communication). The test results indicate that more than one state variable is required with both dislocation density and damage being necessary state variables.

(b) The Rabotnov–Kachanov equations

Unfortunately it is difficult and time consuming to conduct experiments described in the previous section and the variable stress experiments which have been performed are limited to the prediction of rupture life under cyclic conditions of loading. Faced with this difficulty Rabotnov (1969) proposed modifications of constitutive equations first suggested by Kachanov (1958) which described existing experimental results. For uniaxial stress tests the growth laws are assumed to have the simple form

$$\dot{\epsilon}/\dot{\epsilon}_0 = (\sigma/\sigma_0)^n/(1-\omega)^m;$$
$$\dot{\omega}/\dot{\omega}_0 = (\sigma/\sigma_0)^\nu/(1-\omega)^\eta. \qquad (4.2a, b)$$

It is assumed that $\omega = 0$ when the material is in its undamaged state and $\omega = 1$ at rupture. In the equations, n, m, ν, η, $\dot{\epsilon}_0$, $\dot{\omega}_0$ and σ_0 are constants to be defined; η is defined below. It should be noted that in common with the spirit of §4.1 no precise physical definition of the parameter ω is attempted. In fact this point has been specifically made by Rabotnov (1969) but for some reason there has been a persistent misrepresentation in some literature by interpreting the term $(1-\omega)$ as the current cross-sectional area and the term $\sigma/(1-\omega)$ as the effective stress.

When at the beginning of the test and the material is undamaged $\omega = 0$ and equation (4.2a) reduces to

$$\dot{\epsilon}/\dot{\epsilon}_0 = (\sigma/\sigma_0)^n, \qquad (4.3)$$

which is the usual form of the steady state creep equation. In this equation $\dot{\epsilon}_0$ is the steady-state strain rate corresponding to the applied stress σ_0 and the constant n is the stress index.

For constant stress it is easy to integrate the equations (4.2b) to give the time variation of strain and damage. By applying the rupture condition $\omega = 1$ it is also possible to determine the time to rupture t_R. Applying these conditions gives the following results:

$$\epsilon/\epsilon^* = \lambda[1 - \{1 - t/t_R\}^{1/\lambda}];$$
$$\omega = 1 - (1 - t/t_R)^{1/(\lambda+1)}, \qquad (4.4a, b)$$

where
$$t_R = 1/(\eta + 1)\,\dot{\omega}_0(\sigma/\sigma_0)^\nu;$$
$$\epsilon^* = \dot{\epsilon}_0(\sigma/\sigma_0)^n\,t_R = \dot{\epsilon}_0(\sigma/\sigma_0)^{n-\nu}/(\eta+1)\,\dot{\omega}_0;$$
$$\lambda = (\eta + 1)/(\eta + 1 - m). \qquad (4.4c-e)$$

From equation (4.4c) the rupture time t_0 corresponding to the applied stress σ_0 is

$$t_0 = 1/(\eta + 1)\,\dot{\omega}_0,$$

and ϵ^* is equal to the initial strain rate times the time to rupture. At rupture the strain is $\lambda\epsilon^*$ and the strain/time curve has the form shown in figure 10. It should be noted in passing that n is normally greater than ν so that prediction of the theory is that the creep strain at rupture decreases with decreasing stress. This is in accord with experimental observations. The value of λ can be obtained from a measurement of the strain at rupture and it can be seen that the form of the dimensionless strain/time curve is dependent on the value of λ only. From the expression for the time to rupture and the test results it is possible to obtain the values of $(\eta + 1)\,\dot\omega_0$ and ν.

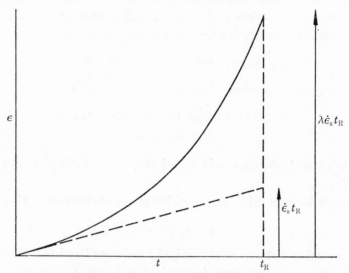

FIGURE 10. Determination of constant λ.

However, the information available is insufficient to pinpoint all the constants, which implies that there is an excess of constants available. If, for convenience, m is selected to be equal to the constant n the equations then become

$$\epsilon/\dot\epsilon_0 = \{\sigma/(1-\omega)\,\sigma_0\}^n;$$

$$\dot\omega = (\lambda-1)\,(\sigma/\sigma_0)^\nu/\lambda n t_0\,(1-\omega)^\eta, \qquad (4.5\,a, b)$$

where
$$\eta = \lambda n/(\lambda-1) - 1.$$

The constants are now in a form which is convenient for comparison with experimental results.

(c) Generalization of the Rabotnov–Kachanov equations

Leckie & Hayhurst (1974) have attempted to generalize the Rabotnov–Kachanov equations to multiaxial states of stress. The form of the generalizations was influenced by the experimental results of Johnson et al. (1962) which indicate that the strain rates are dependent on the effective stress $\bar\sigma$ and that the components of strain rate $\dot\epsilon_{ij}$ are proportional to the deviatoric stress s_{ij}. These results when expressed mathematically take form

$$\dot\epsilon_{ij}/\dot\epsilon_0 = \tfrac{3}{2}(\bar\sigma/\sigma_0)^{n-1} s_{ij}/\sigma_0, \qquad (4.6)$$

with n, $\dot\epsilon_0$ and σ_0 having the same significance as in (4.5). Investigation of the torsion-tension experimental results of Johnson et al. on copper and aluminium show in both cases that, when deterioration is taking place in the tertiary region, the ratio of the strain rate components remains sensibly constant and equal to the value in the steady state condition. Consequently the strain rates in the tertiary region must be represented by constitutive equations with the form

$$\dot{\epsilon}_{ij}/\dot{\epsilon}_0 = \tfrac{3}{2}k(t)\,(\overline{\sigma}/\sigma_0)^{n-1}\,s_{ij}/\sigma_0, \tag{4.7}$$

where $k(t)$ is a scalar quantity increasing monotonically with time. Johnson *et al.* also observed that the growth of k is dependent on the multiaxial form of the stress field. In copper it is maximum stress dependent and in aluminium shear stress dependent. In fact the stress state affecting the growth of k appears to be the same as that dictating the form of the isochronous surface.

Comparing the form of equation (4.7) with the Rabotnov–Kachanov uniaxial relation (4.5 a) suggests the following constitutive equation for the strain rates:

$$\dot{\epsilon}_{ij}/\dot{\epsilon}_0 = \tfrac{3}{2}(\overline{\sigma}/\sigma_0)^{n-1}\,(s_{ij}/\sigma_0)/(1-\omega)^n. \tag{4.8 a}$$

The proposed damage rate equation following (4.5 b) has the form

$$\dot{\omega} = \{(\lambda-1)/\lambda n t_0\}\,\varDelta^\nu(\sigma_{ij}/\sigma_0)/(1-\omega)^\eta, \tag{4.8 b}$$

where

$$\eta = \{\lambda n/(\lambda-1)\}-1,$$

and t_0 and ν have the same significance as in equations (4.5) and $\varDelta(\sigma_{ij}/\sigma_0) = 1$ is the isochronous surface for failure time t_0.

Integrating this equation and using the condition $\omega = 1$ at the rupture time t_R gives

$$t_R/t_0 = 1/\varDelta(\sigma_{ij}/\sigma_0)^\nu.$$

The damage rate equation reduces to the uniaxial form and satisfies the condition that the isochronous rupture surface for rupture time t_0 is $\varDelta(\sigma_{ij}/\sigma_0) = 1$. The value of ω in equation (4.8 b) is also a function of $\varDelta(\sigma_{ij}/\sigma_0)$ thereby satisfying the condition that the growth of the scalar k in equation (4.7) is affected by the same stress state as that for rupture. The proposed constitutive equations appear to satisfy the macroscopic observations and are in a mathematical form which allows further exploitation in continuum mechanics.

5. The materials science approach

(a) Introduction

In this section an attempt is made to present, in a unified form, the studies of materials scientists and to identify features which are common to those of the phenomenological approach described above.

A general formulation for the growth of damage at grain boundaries has been given by Ashby & Raj (1975). The growth of damage is expressed in terms of two mechanisms. One mechanism, referred to as nucleation, gives a measure of the rate at which grain boundary voids are formed. The other mechanism, referred to as growth, gives a measure of the rate of growth of void size.

The nucleation rate is represented by dn/dt and is measured in the number of voids formed per unit time on unit area of grain boundary. Holes apparently nucleate at points of strain concentration which can occur at inclusions contained in the grain boundary or at the extremities of slip bands within the crystal. The growth of holes can apparently be the result of the diffusion of vacancies along the grain boundary or of concentration of creep strains at inclusions.

Since voids are forming and growing continuously it is necessary to use the integral relation derived by Ashby & Raj to obtain a measure of the total damage. Suppose at time τ the nucleation rate is $\dot{n}(\tau)$ so that in time interval $\mathrm{d}\tau$ the number of new voids formed is $\dot{n}(\tau)\,\mathrm{d}\tau$. At a later time, t, these holes will be growing and the rate of increase of cross sectional area of the voids formed at time τ is $\dot{a}(t,\tau)$. During the time interval $(t-\tau)$ the holes formed at time τ will then have a cross sectional area

$$\dot{n}\,\mathrm{d}\tau \int_{\tau}^{t} \dot{a}(t,\tau)\,\mathrm{d}t.$$

The total area of the voids at time t_{f} is then

$$A(t_{\mathrm{f}}) = \int_{\tau=0}^{\tau=t_{\mathrm{f}}} \dot{n}(\tau)\,\mathrm{d}\tau \int_{t=\tau}^{t=t_{\mathrm{f}}} \dot{a}(t,\tau)\,\mathrm{d}t. \tag{5.1}$$

Instead of using A as a measure of damage, some authors prefer to use the volume of the voids. Using the same method, illustrated above, the total volume V of the voids is

$$V(t_{\mathrm{f}}) = \int_{\tau=0}^{\tau=t_{\mathrm{f}}} \dot{n}(\tau)\,\mathrm{d}\tau \int_{t=\tau}^{t=t_{\mathrm{f}}} \dot{v}(t,\tau)\,\mathrm{d}t, \tag{5.2}$$

where $\dot{v}$ is the volume growth rate of the voids.

The formulation is quite general but it implies that to calculate damage it is necessary to keep track of the size of each void as it is formed. This is equivalent to a multistate variable theory in which the number of state variables corresponds to the number of voids and is increasing with time. It is unlikely that the integral can be calculated in closed form except for simple loading histories or for special forms of the rate equations.

Various studies have been made with the objective of formulating the equations which define the nucleation and void growth rates. Two particular examples are now discussed against the background of the phenomenological procedures previously discussed.

(b) The Greenwood growth equations

Greenwood (1973) has studied the growth of creep damage in copper at a temperature of $500\,^{\circ}\mathrm{C}$. The tests were performed under conditions of constant uniaxial stress and measurements made of the number and size of voids. The visual observations indicate that the voids are all approximately the same size which implies that when a void is first formed it grows rapidly in size until it catches up with the voids formed earlier. It is possible to express this observation in a suitable mathematical form (Leckie & Hayhurst 1977) which can be substituted in equation (5.1), but it is easy to see that the total cross-sectional area is simply $A = na$. No attempt was made to study the effect of damage on strain rate although it appears that the tertiary strains were little bigger than those predicted by steady state theory.

Suppose that a uniaxial test is conducted at constant stress σ_0. Let the rupture time be t_0 when the strain is ϵ_0, the hole density n_0 and the average volume of the holes v_0. With reference to these physical values the growth equations for a uniaxial stress σ are

$$\mathrm{d}(n/n_0)/\mathrm{d}t = (\sigma/\sigma_0)^2 \mathrm{d}(\epsilon/\dot{\epsilon}_0)/\mathrm{d}t;$$

$$\mathrm{d}(v/v_0)/\mathrm{d}t = (\sigma/\sigma_0)/t_0;$$

$$\mathrm{d}(\epsilon/\epsilon_0)/\mathrm{d}t = (\sigma/\sigma_0)^5/t_0. \tag{5.3a--c}$$

In the first of these equations the nucleation rate is proportional to the strain rate but is, in addition, proportional to the square of the applied stress. In the second the volume rate of the voids is assumed to be controlled by a diffusion process which is proportional to the stress. The third equation illustrates that the strain rate is proportional to the fifth power of applied stress and the effect of damage on strain rate is neglected.

The damage is defined as

$$\omega = n v^{\frac{2}{3}} / n_0 v_0^{\frac{2}{3}} = A/A_0, \tag{5.3d}$$

and the rupture condition as $\omega = 1$. This is the same condition as proposed by Greenwood (1973) who used the concept of a critical area fraction at failure. For multiaxial states of stress Leckie & Hayhurst (1977) suggested that equations (5.3a–d) take the form

$$\mathrm{d}(n/n_0)/\mathrm{d}t = (\sigma_1/\sigma_0)^2 \, \mathrm{d}(\bar{\epsilon}/\epsilon_0)/\mathrm{d}t;$$

$$\mathrm{d}(v/v_0)/\mathrm{d}t = (\sigma_1/\sigma_0)/t_0;$$

$$\mathrm{d}(\bar{\epsilon}/\epsilon_0)/\mathrm{d}t = (\bar{\sigma}/\sigma_0)^5/t_0;$$

and the damage is
$$\omega = n v^{\frac{2}{3}} / n_0 v_0^{\frac{2}{3}}. \tag{5.4a–d}$$

Integrating the above conditions and applying the rupture condition $\omega = 1$ gives the isochronous surface for rupture time t_0:

$$\Delta(\sigma_{ij}/\sigma_0) = (\sigma_1/\sigma_0)^{\frac{8}{23}} (\bar{\sigma}/\sigma_0)^{\frac{15}{23}} = 1. \tag{5.5a}$$

Leckie & Hayhurst (1977) have shown that for proportional loading (i.e. stress histories for which the ratio of component stresses remain constant) the equations (5.4a–d) can, by introducing certain small modifications, be expressed in the form

$$\mathrm{d}(\epsilon/\epsilon_0)/\mathrm{d}t = (\bar{\sigma}/\sigma_0)^5/t_0,$$

$$\mathrm{d}\omega/\mathrm{d}t = \tfrac{2}{3}\Delta^{\frac{23}{5}} (\sigma_{ij}/\sigma_0)/(1-\omega)^{\frac{1}{2}}, \tag{5.5b, c}$$

with the rupture condition $\omega = 1$. These equations are a special case of the Kachanov–Rabotnov type equations (4.8) in which the damage state variable ω is equal to the normalized form of the damage relation of equation (5.1).

(c) The Dyson–McLean equations

Dyson & McLean (1977) have performed constant uniaxial tension and torsion tests on Nimonic 80A at 700 °C. In addition to measuring tertiary strains and times to rupture, measurements were also made of the number of voids per unit area of grain boundary and of the total volume of voids.

For constant uniaxial stress tests carried out at stress σ_0, the rupture time is t_0 at which time the void density is n_0 and the total volume of the voids is V_0. The constitutive equations proposed by Dyson & McLean (1977) give expressions for the rate of void nucleation and of void growth. Damage is expressed as the total volume of voids and the effect of damage on the creep rate is also included. For multiaxial states of stress the proposed rate equations may be written in the form

$$\mathrm{d}(\bar{\epsilon}/\epsilon_0)/\mathrm{d}t = 3(\bar{\sigma}/\sigma_0)^4 \, (v/v_0)^{\frac{4}{9}}/t_0;$$

$$\mathrm{d}(n/n_0)/\mathrm{d}t = \tfrac{1}{2}(\sigma_1/\bar{\sigma})^2 (\epsilon_0/\bar{\epsilon})^{\frac{1}{2}} \, \mathrm{d}(\epsilon/\epsilon_0)/\mathrm{d}t;$$

$$\mathrm{d}(v/v_0)/\mathrm{d}t = (\sigma_1/\bar{\sigma})^{0\cdot7} \, \mathrm{d}(\epsilon/\epsilon_0)/\mathrm{d}t; \tag{5.6a–c}$$

where
$$v_0 = 3V_0/2n_0.$$

The damage is defined in terms of the total volume of voids which are defined by the integral relation of equation (5.2). Then

$$\omega = \frac{V}{V_0} = \frac{3}{4} \int_{\epsilon_1=0}^{\epsilon_1=\epsilon} (\sigma_1/\overline{\sigma})^2 \, (\epsilon_0/\epsilon_1) \, \mathrm{d}(\epsilon_1/\epsilon_0) \int_{\epsilon_1}^{\epsilon} (\sigma_1/\sigma)^{0.7} \, \mathrm{d}(\epsilon/\epsilon_0).$$

For the case of proportional loading Leckie & Hayhurst (1977) were able to show that the above equations may, with small modification, be reduced to the following growth equations

$$\mathrm{d}(\epsilon/\epsilon_0)/\mathrm{d}t = \{\overline{\sigma}/\sigma_0/(1-\omega)\}^4;$$

$$\mathrm{d}\omega/\mathrm{d}t = \Delta^4(\sigma_{ij}/\sigma_0)/(1-\omega)^4;$$

where
$$\Delta(\sigma_{ij}/\sigma_0) = [\sigma_1^{1.8}\,\overline{\sigma}^{2.2}/\sigma_0^4]^{\frac{1}{4}},$$

with $\Delta(\sigma_{ij}/\sigma_0) = 1$ defining the isochronous surface for rupture time t_0. The rupture condition is again $\omega = 1$.

(d) Some observations

It would appear that for proportional stress[†] histories the phenomenological constitutive equations developed in §4 using a single state damage variable satisfy the macroscopic observations and are also capable of physical interpretation. However, for non-proportional loading it appears necessary to provide growth equations for both nucleation and void growth and in these circumstances at least two state variables will be required. Fortunately, in many practical components, while stresses vary they do remain sensibly proportional. Nevertheless, problems will arise in which the stress fields have large rotations when the sequence of loading is applied. Growth and nucleation will occur at different rates in particular directions and it is to be expected therefore that the description of damage will be tensorial in form. It will also be necessary to understand how and if damage introduces important anisotropic effects. An experimental investigation into this problem has been started by Hayhurst & Morrison (1977). In these experiments plates have been subjected to plane stresses in the ratio 2:1 and after times in excess of half life the stress field is rotated to give the ratio 1:2. If the material deforms according to a Mises law then no deformation should be observed in the direction of the lower stress in a 2:1 ratio test. It is found that this rule is obeyed even after the stress rotation which suggests that it is a scalar form of the damage tensor which affects the value of the strain rate. It is not surprising that this result should hold for aluminium since damage and strain appear to be related, but it is more surprising in the case of copper for which damage occurs in the direction of maximum stress. Once this test programme has been completed it should be possible to formulate more complete constitutive laws which can describe the effect of large rotations of stress.

6. Applications of the constitutive laws

By using the constitutive equations (4.8) in conjunction with the continuum conditions of equilibrium and compatibility it is possible to calculate the time-dependent stress and strain fields in load bearing components. It is also possible to calculate the growth of damage so that the time to first local failure and final rupture may be determined. Apart from problems of especially simple geometry it is normally necessary to resort to the use of the finite element method. An example illustrating the procedure is that solved by Hayhurst, Dimmer & Chernuka (1975) of the plate under tension penetrated by a hole. The results of the calculations indicate

[†] I.e. a stress régime in which the applied forces maintain their direction while varying in magnitude.

that high stresses exist at the edge of the hole during the early part of the component life. However, as time progresses damage develops in the vicinity of the hole with the result that the high stresses are relaxed and equilibrium is maintained by an increase of stress in the less severely loaded regions. This stress redistribution is beneficial in extending the life of the component, but it makes it difficult to estimate the remaining life of a component on the basis of physical measurement of the damage as suggested by Dyson & McLean (1972). Patterns of damage calculated by Hayhurst *et al.* (1975) are plotted in figure 1. and compare well with the damage observed in a uniaxial plate made from copper. In order to maintain numerical stability in the calculations it is necessary to use very small time step increments and, as a result, even for this simple component heavy demands are made on the IBM 370/195 computer. There is therefore a strong motivation to develop approximate methods which give results good enough for design purposes and which avoid the need for demanding and detailed analysis. A complete analysis often provides more information than is required and the calculations often obscure rather than highlight those features which dominate component performance. It will be illustrated how approximate calculations isolate the reference rupture stress as the material parameter which most strongly influences component life.

(a) Upper bound calculations

Upper bound calculations of rupture life can be obtained by making the assumption that the isochronous surface $\Delta(\sigma_{ij}/\sigma_0) = 1$ is convex. The experimental evidence presented by Hayhurst (1972) suggests that in many circumstances this is likely to be a justified assumption. The assumption appears always to be valid when the compressive stress is smaller in magnitude than the tensile stress. In aluminium the isochronous surface is effective stress dependent even when the stresses are compressive so that the surface is always convex. However, in copper the isochronous surface according to equation (5.5a) is concave when the magnitude of the compressive stress is somewhat greater than the tensile stress. More testing is required in the region of compressive stress space before the isochronous surface for copper can be fixed.

Suppose a structure of total volume V is subjected to a set of constant loads P_i. An upper bound t_u on the rupture life t_R of the structure is given by the following expression (Leckie & Wojewódzki 1975)

$$t_R/t_0 < t_u/t_0 = V \bigg/ \int_V \Delta^\nu(\sigma_{ij}^s/\sigma_0) \, \mathrm{d}v. \tag{6.1}$$

In this expression σ_{ij}^s is the stress distribution found by performing the analysis on a structure of identical geometry and load to that under consideration but with a creep law of the form

$$\dot{\epsilon}_{ij}/\dot{\epsilon}_0 = \Delta^{\nu-1}(\sigma_{ij}/\sigma_0) \, \partial\Delta/\partial(\sigma_{ij}/\sigma_0). \tag{6.2}$$

This calculation is equivalent to the standard steady state analysis but with a different creep law. The substitution of the stress field σ_{ij}^s into equation (6.1) is a simple calculation.

It is found convenient in practice to express the result (6.1) in terms of the so-called reference rupture stress. If a uniaxial specimen is subjected to a constant stress σ_u so that the rupture life is t_u then

$$t_u/t_0 = (\sigma_0/\sigma_u)^\nu.$$

Equating this to the time t_u calculated for the component gives an expression for σ_u according to the expression

$$\frac{\sigma_u}{\sigma_0} = \left\{ \frac{1}{V} \int_V \Delta^\nu(\sigma_{ij}^s/\sigma_0) \, \mathrm{d}V \right\}^{1/\nu}. \tag{6.3}$$

If this value σ_u is used in conjunction with the uniaxial creep rupture data then an upper bound is obtained on the time to rupture of the structure. This is a convenient way of relating component performance directly to uniaxial data and avoids data fitting procedures. The stress is known as the *reference rupture stress*.

Another upper bound on rupture time has been developed by Goodall, Cockroft & Chubb (1975). In their calculation the limit load P_0 of a component of identical geometry is determined for a material with a yield stress σ_0 and a yield surface

$$\Delta(\sigma_{ij}/\sigma_0) = 1,$$

which is identical in shape to the isochronous surface. Then the representative rupture stress σ_u is given by

$$\sigma_u/\sigma_0 = (P_i/P_0), \tag{6.4}$$

and once again when used in conjunction with the uniaxial creep rupture data gives an over-estimate of the rupture life of the component. This expression gives a better bound than that of (6.3) but the difference is not great and the selection of the procedure will normally be related to convenience and local computing expertise.

The expression (6.3) illustrates that the reference rupture stress is related to a weighted average stress over the volume which is in accord with the experimental observations discussed in § 1. While the above procedures do give upper and therefore unsafe bounds on rupture life, experiences reported in the quoted references suggest that the bounds are often good enough for practical design purposes. However, more experience about the effectiveness of the bounds is still required.

(b) Lower bound calculations

It has proved difficult to find a reference rupture stress which, when used in conjunction with the uniaxial data, gives a lower bound on rupture life. Some limited progress has been made for components which are kinematically determinate (in contrast to a statically determinate system in which all of the stresses can be expressed in terms of the loads by means of equilibrium considerations, a kinematically determinate structure is one for which all strains can be expressed in terms of a discrete displacement parameter using the conditions of compatibility). For materials for which the isochronous surface is given by the Mises condition $\Delta(\sigma_{ij}/\sigma_0) = (\overline{\sigma}/\sigma_0)$ then the reference rupture stress σ_L found by Leckie & Hayhurst (1974) giving a lower bound on time is

$$\sigma_L/\sigma_0 = \left\{ \int_V (\overline{\sigma}_s/\sigma_0)^{n+1+\nu} dv \Big/ \int_V (\overline{\sigma}_s/\sigma_0)^{n+1} dv \right\}^{1/\nu}, \tag{6.5}$$

where $\overline{\sigma}_s$ is the steady state effective stress distribution.

For components in a state of plane stress the reference rupture stress is

$$\sigma_L/\sigma_0 = \left\{ \int_V (\overline{\sigma}_s/\sigma_0)^{n+1} \Delta^\nu(\sigma_{ij}^s/\sigma_0) \, dv \Big/ \int_V (\overline{\sigma}_s/\sigma_0)^{n+1} dv \right\}^{1/\nu}, \tag{6.6}$$

where σ_{ij}^s is the steady state stress distribution.

Again it can be observed that this reference stress, which gives a lower or conservative estimate of rupture life, is a weighted average of stresses over the component volume, and can be calculated using steady state stress distributions.

(c) Examples

The formulae which have been developed for the reference rupture stress require stress distributions which may be obtained using steady state creep or limit load analysis, both of which are standard engineering calculations. With these stress distributions available, the calculations of the integrals is straightforward and avoids the computationally demanding task of integrating forward in time. The results of the steady state analysis of a plate in tension pierced by a hole were substituted in equation (6.6) (Leckie & Hayhurst 1974) to give a reference rupture stress which is only 1 % greater than the average stress at the minimum cross section. This result is in accord with the experimental result discussed in §1 which indicated that the rupture life of the plate was dependent on the average stress at the minimum cross section. The calculations were very much less demanding than those required for the full analysis reported by Hayhurst *et al.* (1975).

TABLE 2. REFERENCE RUPTURE STRESS FOR CYLINDER WITH INTERNAL PRESSURE

$\sigma_R/2p$	0.742	0.730	0.725
n	3	5	7

Apart from their computational utility perhaps the real advantage of the expressions for reference rupture stress is that they provide physical insight so that certain results can be anticipated.

Provided local stresses are kept to the level expected in practical design (for example elastic stress concentration factors are limited to 2.25 in certain pressure vessel components) it is found in steady state calculations that the value of $(\overline{\sigma}/\sigma_0)^{n+1}$ does not vary greatly throughout the component. Assuming this to be the case (equation (6.5) for example) gives the result that the reference rupture stress is independent of the values of the constants n and ν. In practice the value of $(\overline{\sigma}/\sigma_0)^{n+1}$ does vary so that the reference stress must be dependent on the values of ν and n, but it is to be expected that the dependence is not very strong.

The calculation of the reference stress σ_R for a cylinder with $\Delta(\sigma_{ij}/\sigma_0) = \overline{\sigma}/\sigma_0$ subjected to internal pressure p has been shown (Leckie & Hayhurst 1974) to be

$$\sigma_R/2p = \{[1 - (a/b)^{2(1+\nu)/n}]/(1 + \nu)[1 - (a/b)^{(1+\nu)}]\}^{1/\nu}/n,$$

where a is the internal and b the external radius. If the value of ν is taken to be $0.7n$ (Odqvist 1974) then for the ratio $b:a = 2$ the values of the reference stress for different values of n are those shown in table 2. A practical expression for this component might then be

$$\sigma_R/p = 1.5, \tag{6.7}$$

which is independent of n and ν. The aim of most engineering calculations is to use simple expressions of this type which are particularly helpful in the initial stages of design. This example also illustrates how the reference stress σ_R may be regarded as a material parameter. If the required component life is t_R then σ_R is obtained from the uniaxial creep rupture data and the pressure p which the component can contain for time t_R can then be determined using equation (6.7). The design life of many components operating at temperature is 10^5 h and referring for example to pages 352 and 353 of the *British Steelmakers Creep Committee high temperature data*, the reference rupture stresses for different temperatures of a 1 % Cr $\frac{1}{2}$ %Mo alloy steel are those given in table 3. It is suggested that the representative rupture stress is a convenient material property playing a rôle similar to that of yield stress so that for a given temperature only one value is required and the complex and bulky data appearing in the literature are greatly condensed. In performing the calculations relating component performance and material properties sensible values of ν and n ought to be selected but there seems little advantage to be gained in attempting to find precise values.

The formulae for representative stress have been determined for a number of components and the predictions compared with the results of tests on the components (Leckie & Hayhurst 1974). The comparisons are favourable, and the often dramatic effects resulting from the shape of the isochronous surface are verified. For example, when the formulae are applied to the notched cylindrical bar described in § 2 it is found that the reference stress for copper bars is approximately equal to the average stress at the minimum section, while for aluminium bars the reference stress is considerably smaller than the average stress and accurately predicts the strengthening observed in experiment.

TABLE 3. REFERENCE RUPTURE STRESSES FOR AN ALLOY STEEL

$\sigma_{\mathrm{R}}/(\mathrm{kgf}/\mathrm{mm}^2)$	30	15	5.5
$T/^\circ\mathrm{C}$	450	500	550

In order to complete the calculation for the reference rupture stress it is necessary to know the form of the isochronous surface. This information may not always be available and it is then necessary to assume a shape which gives the most conservative results. For example, there are many engineering components which operate in a state of plane stress and it is then conservative to assume that the isochronous surface is defined by the effective stress, i.e. $\varDelta(\sigma_{ij}/\sigma_0) = (\overline{\sigma}/\sigma_0)$. It is then not too difficult to find a representative stress. If the classical limit load of a component is P_0 for a yield stress σ_0 then the reference representative stress when the applied load is P_i is, from equation (6.4),

$$\sigma_{\mathrm{R}}/\sigma_0 = P_i/P_0 = \lambda. \tag{6.8}$$

This approximation can be used to advantage in the design of reinforcement of pressure vessel intersections. The cylinder/sphere intersection for example is normally designed by providing reinforcement (figure 11, plate 2) so that the limit load of the reinforcing pad is equal to the limit load of the membrane. In this way the pressure vessel is of uniform strength. If the intersection designed according to this procedure operates at temperature then the value of λ in equation (6.8) is the same for the membrane and the reinforcement so that the failure should apparently occur in both simultaneously. However, a further modification must be made to the calculations to account for the effect of geometry change resulting from creep deformation. It is known that as the intersection deforms its limit load increases while that of the membrane decreases. Consequently, the value of λ continuously decreases for the reinforcement and increases for the membrane (i.e. the reference stress decreases for the reinforcement and increases for the membrane). Hence the intersection becomes stronger than the membrane and consequently it is the membrane which should rupture first. The results of experiment verify this prediction (figure 12, plate 2), and the example illustrates that the behaviour of the intersection can be determined without the need of any complex calculations.

7. Conclusions

Some progress has been made towards formulating constitutive equations which can be used in a continuum mechanics approach to determine the growth of material damage in load bearing components operating in the creep range. It is suggested that the single damage parameter used in the phenomenological constitutive equations has a rather precise physical interpretation expressed in terms of an integral of the nucleation and growth of voids. The constitutive equations are limited, however, to proportional loading. For non-proportional loading the physical studies suggest that at least two state variables will be necessary in the constitutive equations.

In principle the constitutive equations can be used together with the continuum equations to calculate precisely the growth of damage in components, but the calculations are very demanding. Another approach has been to develop approximate methods and results are obtained which are much simpler to apply. The formulae indicate that it is average rather than point values which dictate failure life, and this is consistent with experimental observation. When applying the approximate methods it becomes evident that it is the reference rupture stress which is an important material parameter, and that other constants in the constitutive equations play only a minor rôle in influencing performance.

To date, studies have been limited to constant and proportional loading and attempts must now be made to extend the understanding of the effects of non-proportional loading especially those involving stresses induced by fluctuations in temperature. An increase of local temperature is normally accompanied by a decrease of stress and vice versa with a decrease in local temperature. There is limited understanding of this problem and few test results available to provide the necessary guidance. The study of the effectiveness of the procedures described has been largely limited to structures with small values of stress concentration factors. It remains to be seen if the methods are reliable for conditions involving high stress concentrations and in particular when these concentrations appear as the result of cracks and flaws in the component.

References (Leckie)

Andrade, E. N. da C. & Jolliffe, K. H. 1952 Flow in polycrystalline metals under simple shear. *Proc. R. Soc. Lond.* A **213**, 3.

Ashby, M. F. & Raj, M. 1975 Creep fracture. *Proceedings of the Conference on the Mechanics and Physics of Metals.* Cambridge: The Metals Society, Institute of Physics.

Bridgman, P. W. 1952 *Studies in large plastic flow and fracture.* New York: McGraw-Hill.

Dyson, B. F. & McLean, D. 1972 A new method of predicting creep life. *Met. Sci. J.* **6**, 220.

Dyson, B. F. & McLean, D. 1977 Creep in torsion and tension. *Metal Sci.* (In the press.)

Goodall, I. N., Cockroft, R. D. H. & Chubb, E. J. 1975 An approximate description of the creep rupture of structures. *Int. J. Mech. Sci.* **17**, 351.

Greenwood, G. 1973 Creep life and ductility. Int. Congress on Metals, Cambridge. *Microstructure and the design of alloys* **2**, 91–105.

Hayhurst, D. R. 1972 Creep rupture under multiaxial states of stress. *J. Mech. Phys. Solids* **20**, 381.

Hayhurst, D. R., Dimmer, P. R. & Chernuka, M. W. 1975 Estimates of the creep rupture lifetime of structures using the finite element method. *J. Mech. Phys. Solids* **23**, 335–355.

Hayhurst, D. R. & Morrison, C. J. 1977 Private communication.

Hayhurst, D. R. & Storåkers, B. 1976 Creep rupture of the Andrade shear disc. *Proc. R. Soc. Lond.* A **349**, 369.

Johnson, A. E., Henderson, J. & Khan, B. 1962 *Complex stress and creep relaxation and fracture of metallic alloys.* Edinburgh: H.M.S.O.

Kachanov, L. M. 1958 Time of the fracture process under creep conditions. *Izv. Akad. Nauk. SSSR O.T.N. Tekh. Nauk.* **8**, 26.

Leckie, F. A. & Hayhurst, D. R. 1974 Creep rupture of structures. *Proc. R. Soc. Lond.* A **240**, 323.

Leckie, F. A. & Ponter, A. R. S. 1974 On the state variable description of creeping materials. *Ing. Arch.* **43**, 158.

Leckie, F. A. & Hayhurst, D. R. 1975 The damage concept in creep mechanics. *Mech. Res. Commun.* **2**, 1.

Leckie, F. A. & Wojewódzki, W. 1975 Estimates of rupture life – constant load. *Int. J. Solids & Struct.* **11**, 1357.

Leckie, F. A. & Hayhurst, D. R. 1977 Constitutive equations for creep rupture. Leicester University Engineering Department Internal Report 77/1. *Acta metall.* (In the press.)

Odqvist, F. K. G. 1974 *Mathematical theory of creep and creep rupture.* 2nd edn. London: Oxford University Press.

Penny, R. K. & Marriott, D. L. 1971 *Design for creep.* New York: McGraw-Hill.

Rabotnov, Y. N. 1969 *Creep problems in structural members.* Amsterdam: North Holland Publishing Company.

Sdobyrev, V. P. 1958 Long term strength of the alloy E1/437B under complex stress. *Izv. Akad. Nauk. SSSR O.T.N.* **4**, 92.

Storåkers, B. 1969 Finite creep of a circular membrane under hydrostatic pressure. *Acta polytech. scand.* **Me 44**, 1.

Discussion

S. A. F. MURRELL (*Department of Geology, University College London, Gower Street, London, WC1E 6BT*). Professor Leckie has asked whether creep rupture and the growth of voids by creep were processes which could occur in the depths of the Earth, where matter is subjected to high pressure. Geologists and geophysicists would answer that they do envisage processes in the Earth in which voids may play a rôle. At depths down to levels at which hydrous minerals can still survive as stable phases there is the possibility of such minerals decomposing and releasing water into pores in the rock (Murrell & Ismail 1976a). Carbon dioxide may also exist in pores in mantle rocks. In addition, partial melting would create voids containing melt under high pressure (Murrell & Ismail 1976b). Magmas are generated in the upper parts of the mantle, and earthquakes occur at depths down to 700 km. In so far as earthquakes and the movement of magma involve faulting processes it seems likely that voids have a rôle in them.

References

Murrell, S. A. F. & Ismail, I. A. H. 1976a The effect of decomposition of hydrous minerals on the mechanical properties of rocks at high pressures and temperatures. *Tectonophysics* **31**, 207–258.

Murrell, S. A. F. & Ismail, I. A. H. 1976b The effect of temperature on the strength at high confining pressure of granodiorite containing free and chemically-bound water. *Contrib. Mineral. Petrol.* **55**, 317–330.

Phil. Trans. R. Soc. Lond. A. **288**, 49–57 (1978)   [49]

Printed in Great Britain

Stress estimates from structural studies in some mantle peridotites

By A. Nicolas

Laboratoire de Tectonophysique, Nantes, France

[Plates 1 and 2]

Deviatoric stresses acting in rock deformation can be estimated by measuring parameters connected with the dislocation microstructure, after an experimental calibration. In olivine, the available structural geopiezometers are based on dislocation curvature, dislocation density, sub-boundary size and recrystallized grain (neoblast) size. Their application in the case of olivine bearing rocks (peridotites) deformed in mantle conditions is critically assessed. The most reliable geopiezometer is the one based on olivine neoblast size. It yields values in the range of 1 kbar (1 kbar = 10^8 Pa) and over in the case of the sheared nodules in kimberlites, of 0.3–0.5 kbar in the case of basalt nodules and in the Lanzo massif, although locally the stress can be much higher. These values are compared with those ascribed to mantle flow by various independent methods and which tend to indicate a lower deviatoric stress for asthenosphere flow (10–100 bar). In the state of the art, the disagreement between this stress estimate and those in basalt nodules and in massifs, which is nearly one order of magnitude, can be explained by the various uncertainties in the estimate, thus leaving room for the possibility that the flow structures in these peridotites do represent asthenosphere conditions.

Introduction

Direct evidence on the composition and the geodynamic properties of the upper mantle is provided by peridotites either brought up to the surface by basaltic and kimberlitic magmas or intruded as mantle slices in orogenic belts. The structures and minerals preferred orientations have been investigated in these various types of peridotites and the main results concerning the flow mechanisms and kinematics have been recently reviewed (Nicolas 1976).

It is now accepted that these peridotites are largely representative of the upper mantle so far as their geochemical and geophysical properties are concerned; the question addressed in this paper is whether the structures observed in peridotite nodules from kimberlites, basalts and in peridotite massifs are representative of the asthenosphere flow in the upper mantle or, on the contrary, of local and incidental phenomena. In the latter case they could not be considered to deduce the rheological properties of the flowing asthenosphere, because both the flow mechanisms and the deviatoric stresses that they indicate would not be appropriate. By the same token their pyroxenes composition would be of no use to derive any representative geotherm for a large area in the mantle.

The analysis of structures and minerals preferred orientations in the various types of mantle peridotites indicates that the flow mechanism is dislocation creep controlled by recovery (Nicolas, Boudier & Boullier 1973). In mylonites from kimberlite nodules, evidence for structural superplasticity has been presented by Gueguen & Boullier (1976), who ascribe it to diffusion accommodated grain boundary sliding. It must be emphasized that this situation is exceptional. Dislocation creep which is the flow mechanism in natural peridotites is also that commonly accepted for the asthenosphere (Weertman 1970), although different modes of diffusion creep

4

have been recently advocated as an important (Meissner & Vetter 1976) or a dominant (Twiss 1976) creep mechanism in the upper mantle. The flow geometry in nodules from basalt and in massifs which was found to be dominantly rotational and possibly approaching simple shear (Mercier & Nicolas 1975) is also compatible with the rotational shear flow expected for the asthenosphere. Finally, the structures in peridotite massifs from ophiolite complexes can be ascribed to asthenosphere flow as shown by the excellent agreement between the seismic anisotropy calculated for these massifs once restored to their original orientation and that measured at sea (Peselnick & Nicolas, in press). This does not preclude the possibility that some deformation structures could be due to high temperature emplacement (Mercier 1976).

The question of whether the peridotites under consideration represented the asthenosphere flow has recently arisen from considerations on stress and recovery. Goetze (1975a, b) has contended that the deformed structures in Boyd & Nixon's (1972) sheared nodules from kimberlites could not result from the asthenosphere flow as his estimation of the stress responsible for this deformation was two orders of magnitude larger than that expected in the asthenosphere (2 or 3 kbar as compared with 10 or 100 bar). This conclusion has also been attained by Boullier & Nicolas (1975) considering the limited degree of recovery in the sheared nodules compared with that in the other nodules, although the former are equilibrated at the highest temperatures. This suggested that the sheared nodules had been deforming until they were extracted by the kimberlite magmas at strain rates largely higher than that expected in the asthenosphere.

THE POTENTIAL GEOPIEZOMETERS

Empirical relations have been established in metals between the dislocation microstructures and deviatoric stress and their applications in the case of rocks discussed by Nicolas & Poirier (1976, p. 137). The main relations tie the dislocation curvature, the dislocation density and the subgrain size to the applied stress.

Durham, Goetze & Blake (1977) have considered the relation between the minimum radius of curvature (R, measured in micrometres) of dislocation loops and the stress (σ, measured in kilobars) in olivine. They observe a fairly good experimental correlation:

$$\sigma \approx 0.6\,R^{-1},$$

which when compared with the expected relation

$$\sigma = K\mu b R^{-1},$$

where μ is the shear modulus and b the Bergers vector, yields a K value of 1.84×10^{-3} kbar cm.

The relation between the free dislocation density (ρ) and the stress has been experimentally investigated by Kohlstedt & Goetze (1974), by Goetze (1975a, b) and by Durham et al. (1977) who find a good experimental correlation expressed in the general form:

$$\sigma = K'\mu b \rho^{\frac{1}{2}},$$

$$\sigma = 9 \times 10^{-5} \rho^{0.5} \quad \text{(Goetze 1975a, b)},$$

$$\sigma \approx 2 \times 10^{-5} \rho^{0.61} \quad \text{(Durham et al. 1977)}.$$

The relation between the subgrain size (d, measured in micrometres) and the maximum stress is expected to have the following form:

$$\sigma = K'' \mu b d^{-1}.$$

In olivine, the [100] tilt wall spacing is principally measured, due to its predominance over any other substructure and to its smallest spacing when compared to the other subboundary systems. The considered subgrains are only those optically visible and not those which subdivide them at the t.e.m. scale (Green & Radcliffe 1972). They are now observed using a new decoration technique (Kohlstedt, Geotze, Durham & Vander Sande 1976) which, making it possible to observe about ten times as many tilt walls than before, can explain the large discrepancy between Raleigh & Kirby's (1970), Goetze's (1975 a, b), Mercier's (1976) and Durham et al. (1977) relations.

$$\sigma = 17\,d^{-1} \quad \text{(Geotze 1975a, b),}$$

$$\sigma = 115\,d^{-1} \quad \text{(Mercier 1976),}$$

$$\sigma = 10\,d^{-1} \quad \text{(Durham et al. 1977).}$$

The size of dynamically recrystallized olivine grains (neoblasts) has also been considered as a potential geopiezometer for mantle rocks. The following experimental relations have been found:

$$\sigma = 11\,d^{-0.5} \quad \text{(Goetze 1975a, b),}$$

$$\sigma = 19\,d^{-0.67} \quad \text{(Post 1973),}$$

$$\sigma = 40\,d^{-0.81} \quad \text{(Mercier 1976).}$$

Mercier obtains a weak temperature dependence for this relation:

$$d = 1.13\sigma^{-1.4}\exp{(13500/RT)}.$$

Otherwise, as expected from the metallurgical literature, no temperature dependence is found for all these relations.

CRITICAL ASSESSMENT OF THE GEOPIEZOMETERS

Piezometers based on dislocation microstructure

In olivine, as in metals, the dislocation density is very sensitive to further straining or recovery. Durham et al. (1977) have shown experimentally that in this mineral 0.01–0.1 % strain is sufficient to completely obscure the initial dislocation structure, whereas a new steady state substructure is obtained for 2–3 % strain. Many rocks have suffered such moderate strains at lower temperature and higher stress after the major deformation which is investigated. This will result in a dislocation substructure which reflects a higher stress than the one responsible for large creep.

In all types of mantle peridotites, evidence of late high stress deformation can be found. For instance, Gueguen (1977), who has systematically investigated the dislocation substructures in olivine from kimberlite and basalt nodules, describes in some nodules (110) slip bands comprising {110}[001] dislocations which represent a low-temperature slip system (Carter & Ave'Lallemant 1970) (figure 1, plate 1). In contrast with the other dislocation systems found in these nodules, this system cannot be related to the mineral preferred orientation developed during the main flow. These bands are attributed to some late event, possibly occurring during the ascent. In the

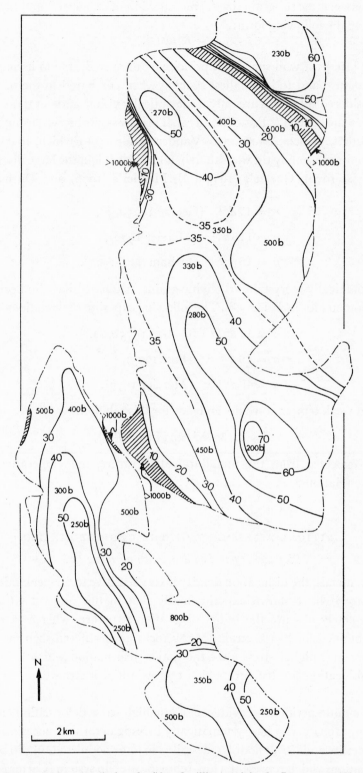

FIGURE 2. Map of olivine neoblast size (in hundredths of millimetres) in the Lanzo massif based on 125 measurements. The hatched pattern corresponds to mylonitic zones. The stress estimates in bars (b) are derived from Mercier's and Post's calibrations (see text).

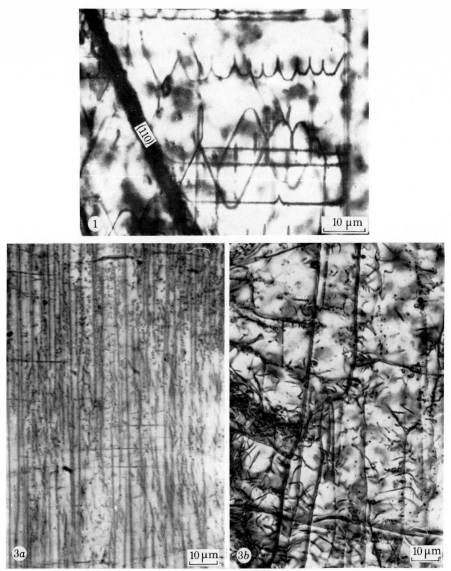

FIGURE 1. (110) slip band superimposed on the low stress dislocation substructure in a decorated olivine crystal from a basalt nodule (picture by Y. Gueguen).

FIGURE 3. This shows d_{100} spacings (vertical lines) in decorated olivine grains: (*a*) starting specimen, a dunite from Norway; (*b*) after 1 h of annealing at 1700 °C, argon atmosphere (from D. Ricoult).

FIGURE 5. Partial recrystallization of an olivine porphyroclast in a peridotite nodule from Lunar Crater basalt. The shaded subgrains grade into neoblasts (slightly darker or lighter) which results from their progressive misorientation. A few smaller and often lighter neoblasts are ascribed to the nucleation–grain boundary migration mechanism.

FIGURE 6. Two generations of olivine neoblasts in a mylonitic peridotite from the Lanzo massif. The centre of the picture is occupied by a porphyroclast whose substructure evolves in a first generation of 100 μm neoblasts. This structure is cut by mylonitic bands at the upper and lower part of the picture in which the olivine grains are only 10 μm.

Lanzo massif, evidence of minor strain under high stresses is provided by the discrepancy between the stress values and pattern deduced from the neoblast size (figure 2 and table 1) and those deduced from dislocation density and d_{100} spacings (table 1) observed with the decoration technique. The higher and more uniform values obtained in the latter case indicate that during the last stage of its intrusion the massif was submitted in a fairly uniform fashion to stresses in the 1 kbar range (in the mylonitic bands it would attain 5 kbar).

TABLE 1. STRESS ESTIMATES IN MANTLE ROCKS

kimberlite nodules

$\rho \approx 2.10^7 \, \mathrm{cm}^{-2}$ $\quad\quad\quad\quad$ $\sigma \approx 0.4 \, \mathrm{kbar}$
$d_{100} \approx 12 \, \mu\mathrm{m}$ $\quad\quad\quad\quad$ $1 \, \mathrm{kbar} < \sigma < 9 \, \mathrm{kbar}$
$d_{\mathrm{n}} \approx 75 \, \mu\mathrm{m}$ $\quad\quad\quad\quad$ $1 \, \mathrm{kbar} < \sigma < 1.3 \, \mathrm{kbar}$

basalt nodules

$\rho \approx 9.10^6 \, \mathrm{cm}^{-2}$ $\quad\quad\quad\quad$ $\sigma < 0.25 \, \mathrm{kbar}$
$d_{100} \approx 50 \, \mu\mathrm{m}$ $\quad\quad\quad\quad$ $0.35 \, \mathrm{kbar} < \sigma < 2.3 \, \mathrm{kbar}$
$d_{\mathrm{n}} \approx 500 \, \mu\mathrm{m}$ $\quad\quad\quad\quad$ $0.25 \, \mathrm{kbar} < \sigma < 0.5 \, \mathrm{kbar}$

Lanzo massif

$\rho \approx 10^7 – 10^8 \, \mathrm{cm}^{-2}$ $\quad\quad\quad\quad$ $0.3 \, \mathrm{kbar} < \sigma < 1 \, \mathrm{kbar}$
$d_{100} \approx 50 \, \mu\mathrm{m}$ $\quad\quad\quad\quad$ $0.35 \, \mathrm{kbar} < \sigma < 2.3 \, \mathrm{kbar}$
$d_{\mathrm{n}} \approx 400 \, \mu\mathrm{m}$ $\quad\quad\quad\quad$ $0.3 \, \mathrm{kbar} < \sigma < 0.55 \, \mathrm{kbar}$

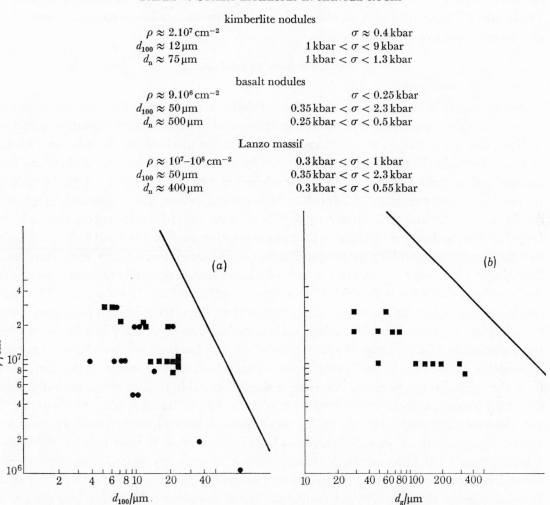

FIGURE 4. (a) Relation between ρ, the dislocation density, and d_{100} in olivine from kimberlite nodules.
(b) Relation between ρ and the size of olivine neoblasts in the same suite of nodules. The solid lines are the experimental relations observed by Goetze (1975 a, b).
Circles: coarse granular and tabular nodules; squares: porphyroclastic nodules (after Gueguen 1977).

Recovery, on the other hand, can lower the dislocation density and increase the d_{100} spacing. This latter possibility is denied by Goetze & Kohlstedt (1973) but Ricoult (figure 3, plate 1) has brought experimental evidence for it. The mean d_{100} spacing observed at a magnification of 1000 with the decoration technique increases from $4 \, \mu\mathrm{m}$ in the starting specimen to $16 \, \mu\mathrm{m}$ after 1 h of annealing at 1700 °C. Meanwhile the optically visible d_{100} tilt walls (misorientation $\geqslant 1°$), when compared to the totality of decorated tilt walls, have a percentage which increases from 4 to 9 (D. Ricoult, personal communication). This suggests that during annealing the highly misoriented walls remain stable and that the increase in mean d_{100} spacing is due to the destruction of the least misorientated walls.

Gueguen (1977) finds in kimberlite nodules very poor correlations between the dislocation density (ρ) and the d_{100} spacing (figure 4) and between ρ and the neoblast size. Moreover, his data fall largely below the experimental line from Goetze (1975 a, b). He concludes that all these nodules have been statically recovered. In the basalt nodules, the correlation between ρ and d_{100} is no better. The dislocation density also varies by a factor of 10–20 in a same grain at a few hundredths of microns distance, an observation that Gueguen ascribes to recovery, some microstructures being more stable than others.

Piezometer based on neoblast size

Recrystallization mechanism

During progressive deformation the olivine porphyroclasts tend to recrystallize into neoblasts. Two mechanisms have been recognized: nucleation and grain boundary migration (n.g.b.m.) and progressive subgrain rotation (p.s.g.r.). In the latter mechanism, the subgrains tend to increase their misorientation during creep until beyond 15° in olivine they evolve into independent grains (Poirier & Nicolas (1975) for olivine; Hobbs (1968), White (1973) for quartz). In peridotites experimentally deformed at high stresses (above a few kilobars), n.g.b.m. is dominant (Ave'Lallemant & Carter 1970; Nicolas *et al.* 1973; Mercier 1976), although Post (1973) ascribes to the other mechanism the recrystallization produced at 5 and 7 kbar, suggesting that the parameter controlling the recrystallization mechanism is temperature, high temperature favouring n.g.b.m. over p.s.g.r. On the basis of observations in naturally deformed peridotites, the hypothesis that stress is the critical parameter is preferred here. In these rocks, it is usually concluded that, whatever the temperatures during flow, the mechanism has been p.s.g.r. even though secondary grain boundary migration is sometimes recorded. Only in a few cases where the neoblast size is in the range of a few tenths of microns has evidence been found of n.g.b.m. A peridotite nodule from Lunar Crater basalts contains both types of neoblasts; the ones derived from the subgrains are 60 µm in diameter, a size comparable to that of the optically visible subgrains (40 µm), and those produced by n.g.b.m. $\leqslant$ 30 µm (figure 5, plate 2). Applying the experimental relations given above, this would set the limit between the two mechanisms around 2 kbar. Above this stress, provided there has been enough strain to store through dislocations a sufficient energy, n.g.b.m. would dominate; below it, the recovery processes (dislocation climb into sub-boundaries) would keep the strain energy at an insufficient level for n.g.b.m. and again, for strains larger than *ca.* 0.3 % the subgrain misorientation would evolve into p.s.g.r. recrystallization.

The experimentally established relation between neoblast size and deviatoric stress corresponds to a recrystallization operating mainly by n.g.b.m. Considering this mechanism, it is not understood why such a relation exists. There is now a need for neoblast size/stress calibration in the stress field below 2 kbar where p.s.g.r. recrystallization is thought to dominate.

This type of piezometer is less sensitive to the history of the rock subsequent to the major flow, that is to annealing and high-stress low-strain deformation. Therefore, once these problems are solved, it will be superior to those based on the dislocation microstructure.

During annealing, however, a grain growth can occur, driven by the surface energy. This would evidently ruin all possibilities of using neoblast grain size as piezometers. This grain growth is documented in experiments (Mercier 1976) and in some peridotites from kimberlites (Harte, Cox & Gurney 1975; Boullier 1975), from basalts (Mercier & Nicolas 1975) and from massifs (Boudier 1976). It is illustrated by olivine growth enclosing other minerals, mainly spinel. In the case of dunites in massifs, exaggerated growth leads to a grain size attaining 50 mm. However, in most cases the olivine does not enclose the small spinel grains, indicating that the grain boundary migration is absent or at least moderate. It is also observed that the neoblast size then compares with that of *optically visible* subgrains (Poirier & Nicolas 1975) (figure 6, plate 2). The correlation between neoblast size and sub-boundary spacing is not valid for the subgrains only visible with the decoration technique and which have misorientations smaller than 1°: only those with misorientations greater than one degree are potentially able to evolve in grain boundaries. Earlier it has been proposed that the latter are also more stable during recovery.

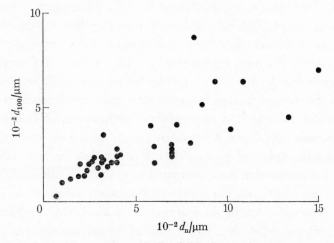

FIGURE 7. Relation between the optically visible d_{100} spacing (misorientation $\geqslant 1°$) and the neoblast size in olivine from basalt nodules.

Strains of the order of 1 % under high stress at low-temperature conditions which are sufficient to alter the dislocation substructure will not change the neoblast size as both recrystallization mechanisms require larger strains. Moreover, in the Lanzo massif where locally large strains have produced mylonitic bands at low temperatures it is still possible to distinguish the neoblasts resulting respectively from the low and high stress deformations (figure 7).

THE PERIDOTITE STRUCTURES AND THE ASTHENOSPHERE FLOW

Stresses in the mantle are estimated by three independent means: stress drop calculated from appropriate earthquakes, data on the external gravitational potential and rheological models of mantle creep.

Stress estimates from earthquakes can reasonably be suspected of being more representative of lithosphere conditions than of asthenosphere ones even for the deep earthquakes which are ascribed to a subducting lithosphere. The stress release measurements may also reflect only a part of the deviatoric stress (Kanamori & Anderson 1975). This is certainly true for shallow earthquakes in which the stress release can be only 0.01–0.1 of the total stress (Wyss & Molnar 1972); it is more dubious for deep earthquakes (H. Berckhemer, personal communication). The shallow earthquakes are also those in which the stress drop is smallest: 10–100 bar compared with 100–1000 bar for the deep earthquakes (Aki 1972). The comprehensive study of Kanamori & Anderson (1975) results in stress drop estimates of 30 bar for interplate earthquakes, 100 bar for intraplate ones and 60 bar as an average value.

Stress estimates bases on gravity data are values obtained for the whole mantle and are possibly more representative of the lower mantle (Jeffreys (1964) and Caputo (1965), cited in Clark & Ringwood 1968). Minimum stresses thus estimated are 30–40 bar. Kaula (1963) reports a stress of 160 bar for the lower mantle.

Finally, stress estimates based on the modelling of the rheology of the asthenosphere concern without ambiguity this part of the upper mantle. They all rely entirely on the rheology of olivine extrapolated from laboratory to asthenosphere conditions. The main uncertainties derive from a poor knowledge of (1) the creep mechanism in the asthenosphere, (2) the pressure dependence in the creep equation of olivine, (3) the geotherms in the asthenosphere, (4) the thickness and therefore the strain rates acceptable for the asthenosphere. For these reasons the stress estimates are considered with some diffidence: Schubert, Froidevaux & Yuen (1976) predict stresses lower than 100 bar, Melosh (1976), of at most a few tens of bars, Carter, Baker & George (1972), between 7 and 16 bar and Meissner & Vetter (1976), locally below one bar. These stress estimates are somewhat lower than the stresses derived from the structural geopiezometers contained in the various types of peridotites (table 1). The stresses for peridotites compare better with the 200–300 bar estimated by McKenzie (1972) and Artyushkov (1973) for the lithosphere.

This discrepancy can be explained by considering either that the stresses estimated for the asthenosphere are too low and/or that those estimated from the peridotites structure are too high, or that the observable peridotites do not represent the steady state asthenosphere flow. This second interpretation is accepted for the sheared nodules in kimberlites and for mylonites which are present in all categories of peridotites. In the state of the art, too many uncertainties remain about stress in the asthenosphere and stress calibration in peridotites to rule out that the other deformed peridotites do represent asthenosphere flow.

The author wishes to thank F. Boudier, M. Darot, D. Ricoult and P. Coisy for providing him with unpublished data, H. Berckhemer and J. P. Poirier for profitable discussions.

REFERENCES (Nicolas)

Aki, K. 1972 Earthquake mechanism. *Tectonophysics* **13**, 423–446.

Artyushkov, E. V. 1973 Stresses in the lithosphere caused by crustal thickness inhomogeneities. *J. geophys. Res.* **78**, 7675–7708.

Ave Lallemant, H. G. & Carter, N. L. 1970 Syntectonic recrystallization of olivine and modes of flow in the upper mantle. *Bull. geol. Soc. Am.* **81**, 2203–2220.

Boudier, F. 1976 Le massif lherzolitique de Lanzo, étude structurale et pétrologique. Thèse Doctorat d'Etat, University of Nantes.

Boullier, A. M. 1975 Structure des peridotites en enclaves dans les kimberlites d'Afrique du Sud. Thèse Doctorat 3e cycle University of Nantes.

Boullier, A. M. & Nicolas, A. 1975 Classification of textures and fabrics of peridotites xenoliths from South African kimberlites. In *Physics and chemistry of the Earth* (ed. L. H. Ahrens), vol. 9, pp. 97–105. London: Pergamon Press.

Boyd, F. R. & Nixon, P. H. 1972 Ultramafic nodules from the Thaba Putsoa Kimberlite pipe. *Yb. Carnegie Instn Wash.* **71**, 362–373.

Carter, N. L. & Ave'Lallemant, H. G. 1970 High temperature flow dunite and peridotite. *Bull. geol. Soc. Am.* **81**, 2181–2202.

Carter, N. L., Baker, D. W. & George, R. P. 1972 Seismic anisotropy, flow and constitution of the upper mantle. Flow and fracture of rocks, the Griggs. *Am. geophys. Un. Geophys. Mon. Ser.* **16**, 167–190.

Clark, S. P. & Ringwood, A. E. 1967 Density, strength and constitution of the mantle. In *The Earth's mantle* (ed. T. F. Gaskell), pp. 111–123. New York: Academic Press.

Durham, W. B., Goetze, C. & Blake, B. 1977 Plastic flow of oriented single crystals of olivine. II. Observations and interpretations of the dislocation structures. *J. geophys. Res.* (In the press.)

Goetze, C. 1975*a* Sheared lherzolites: from the point of view of rock mechanics. *Geology* **3**, 172–173.

Goetze, C. 1975*b* Textura land microstructural systematics in olivine and quartz. *Trans. Am. geophys. Un.* **56**, 455.

Goetze, C. & Kohlstedt, D. L. 1973 Laboratory study of dislocation climb and diffusion in olivine. *J. geophys. Res.* **78**, 5961–5971.

Green, H. W. & Radcliffe, S. V. 1972 Dislocation mechanisms in olivine and flow in the upper mantle. *Earth & planet. Sci. Lett.* **15**, 239–247.

Gueguen, Y. 1977 Dislocations in mantle peridotite nodules. *Tectonophysics* **39**, 231–251.

Gueguen, Y. & Boullier, A. M. 1976 Evidence of superplasticity in mantle peridotites. In *The physics and chemistry of minerals and rocks* (ed. R. J. G. Strens), pp. 19–33. New York: Wiley-Interscience.

Harte, B., Cox, K. G. & Gurney, J. J. 1975 Petrography and geological history of upper mantle xenoliths from the Matsoku kimberlite pipe. In *Physics and chemistry of the Earth* (ed. L. H. Ahrens), vol. 9, pp. 617–646. London: Pergamon.

Hobbs, B. E. 1968 Recrystallization of single crystals of quartz. *Tectonophysics* **6**, 353–401.

Kanamori, H. & Anderson, D. L. 1975 Theoretical basis of some empirical relations in seismology. *Bull. seism. Soc. Am.* **65**, 1073–1095.

Kaula, W. M. 1963 Elastic models of the mantle corresponding to variations in the external gravity field. *J. geophys. Res.* **68**, 4967–4978.

Kohlstedt, D. L. & Goetze, C. 1974 Low stress, high temperature creep in olivine single crystals. *J. geophys. Res.* **79**, 2045–2051.

Kohlstedt, D. L., Goetze, C., Durham, W. B. & Vander Sande, J. 1976 New technique for decorating dislocations in olivine. *Science, N.Y.* **191**, 1045–1046.

McKenzie, D. P. 1972 Plate tectonics. In *The nature of the solid Earth* (ed. Robertson), pp. 323–360. New York: McGraw-Hill.

Meissner, R. O. & Vetter, U. R. 1976 Isostatic and dynamic processes and their relation to viscosity. *Tectonophysics* **35**, 137–148.

Melosh, H. J. 1976 Plate motion and thermal instability in the asthenosphere. *Tectonophysics* **35**, 363–390.

Mercier, J. C. 1976 Natural peridotites: chemical and rheological heterogeneity of the upper mantle. Ph.D. thesis, Earth and Planetary Science Department, University of New York, Stony Brook.

Mercier, J. C. & Nicolas, A. 1975 Textures and fabrics of upper mantle peridotites as illustrated by xenoliths from basalt. *J. Petrol.* **16**, 454–487.

Nicolas, A. 1976 Flow in upper mantle rocks: some geophysical and geodynamic consequences. *Tectonophysics* **32**, 93–106.

Nicolas, A., Boudier, F. & Boullier, A. M. 1973 Mechanism of flow in naturally and experimentally deformed peridotites. *Am. J. Sci.* **273**, 853–876.

Nicolas, A. & Poirier, J. P. 1976 *Crystalline plasticity and solid state flow in metamorphic rocks.* New York: Wiley-Interscience.

Peselnick, L. & Nicolas, A. Seismic anisotropy of an ophiolite peridotite: application to oceanic upper mantle. *J. geophys. Res.* (In the press.)

Poirier, J. P. & Nicolas, A. 1975 Deformation-induced recrystallization due to progressive misorientation of subgrains with special reference to mantle peridotites. *J. Geol.* **83**, 707–720.

Post, R. 1973 The flow laws of MT Burnett dunite. Ph.D. Thesis, Geophysics Department, University of California, Los Angeles.

Raleigh, C. B. & Kirby, S. H. 1970 Creep in the upper mantle. *Mineral Soc. Am. Spec. Paper* **3**, 113–121.

Schubert, G., Froidevaux, C. & Yuen, D. A. 1976 Oceanic lithosphere and asthenosphere: thermal and mechanical structure. *J. geophys. Res.* **81**, 3525–3540.

Twiss, R. J. 1976 Structural superplastic creep and linear viscosity in the Earth's mantle *Earth. & planet. Sci. Lett.* **33**, 86–100.

Weertman, J. 1970 The creep strength of the Earth's mantle. *Rev. Geophys. Space Phys.* **8**, 145–168.

White, S. 1973 Syntectonic recrystallization and texture development in quartz. *Nature, Lond.* **244**, 276–278.

Wyss, M. & Molnar, P. 1972 Efficiency, stress drop, apparent stress, effective stress and frictional stress of Denver, Colorado, earthquakes. *J. geophys. Res.* **77**, 1433–1438.

Phil. Trans. R. Soc. Lond. A. **288**, 59–95 (1977) [59]
Printed in Great Britain

Micromechanisms of flow and fracture, and their relevance to the rheology of the upper mantle

By M. F. Ashby and R. A. Verrall
Cambridge University, Engineering Department,
Trumpington Street, Cambridge CB2 1PZ

Crystalline solids respond to stress by deforming elastically and plastically, and by fracturing. The dominant response of a given material depends on the magnitude of the shear stress (σ_s), on the temperature (T) and on the time (t) of its application. This is because a number of alternative mechanisms exist which permit the solid to flow, and its fracture, too, occurs by one of a number of competing mechanisms. Their rates depend on σ_s, T and t: it is the fastest one which appears as dominant.

In geophysical problems, pressure appears as an additional variable. At pressures corresponding to depths of a few kilometres below the surface of the Earth, the mechanisms of fracture are the most affected; but at depths of a few hundred kilometres, plasticity, too, is influenced in important ways.

This paper outlines the mechanisms of flow and fracture which appear to be relevant in the deformation of materials of interest to the geophysicist, and the way pressure affects them. The results are illustrated and their shortcomings emphasized by using them to calculate the mechanisms of flow and fracture to be expected in the upper mantle of the Earth.

1. Introduction

1.1. *Micromechanisms of flow and fracture*

There exist a number of alternative mechanisms by which a crystalline solid may deform or fracture. At low temperatures, it may fracture by *cleavage* before it yields. It may yield or twin, exhibiting *low-temperature plasticity*, terminated by one of several sorts of *ductile fracture*. As the temperature is raised, the solid may start to flow by *power-law creep*, and even, at sufficiently high temperatures, by *diffusional flow*, until it fails by one of a number of *creep fracture* processes.

Although none of these is completely understood, the underlying physics is sufficiently clear that each can be modelled approximately. The models lead to *constitutive laws*: equations relating the strain increment (or strain-rate) to the stress and temperature applied to the solid, and to the time. These constitutive laws, when fitted to experimental data, give a useful description of the mechanisms of flow or fracture in a form which can be used for solving boundary-value problems.

In materials science and engineering, it is uncommon to regard the hydrostatic pressure as an independent variable: it is almost always determined by the shear strength of the material and seldom exceeds 1 % of its bulk modulus (K). Because of this, the effect of pressure on mechanical strength is less studied than the effect of, say, temperature. However, in many geophysical problems, pressure is as important a variable as temperature, and at great depths below the Earth's surface, the pressures are immense: at a depth of 400 km, for instance, the pressure is about $K/10$.

Even small pressures ($K/100$ or less) have a large effect on mechanisms of fracture. A super-imposed pressure of the same magnitude as the yield strength, for example, suppresses most modes of ductile fracture. Cleavage fracture is an exception: as described in § 3, even a large pressure does not suppress it entirely, though it may make it so difficult that some other mechanism of flow appears instead.

Plasticity and creep are much less affected by pressures. In engineering design, it is normal to assume that it is only the shearing, or *deviatoric*, part of the stress field which causes flow; pressure has no effect whatever. There is some justification for this: neither low-temperature plasticity nor creep are measurably affected by pressures less than $K/100$. But when the pressure exceeds this value, the rates of flow are slowed, and pressure must be regarded, with the temperature and shear stress, as an independent variable.

1.2. *Mechanisms important in geophysical modelling*

In understanding and modelling phenomena such as the flow of the Earth's mantle, or the creep of a large ice body, or formation of a salt dome, it is important to identify the dominant mechanism of deformation, since this determines how the strains (and thus, displacements) are related to the stress and temperature. We shall consider as an example the mechanisms involved in the first of these problems, but the same mechanisms and method could be applied equally well to the other two.

Geophysical and petrologic evidence suggest that the dominant phase in the upper mantle is olivine Fo_{85}–Fo_{95} (Fo_{85} is $(Mg_{0.85}Fe_{0.15})_2 SiO_4$) (Birch 1969; Ringwood 1970), so that, as a first approximation, the upper mantle can be treated rheologically as pure olivine. (There are certain risks in this: in the low velocity zone, a shell of the upper mantle about 150 km below the surface, a basaltic phase in the mantle rock may melt, locally changing the mechanical behaviour. Flow in the presence of a fluid phase cannot yet be modelled adequately, but it is discussed in § 7.)

The mechanisms responsible for steady-state deformation in the upper mantle have been considered by several authors (Gordon 1965, 1967; McKenzie 1968; Weertman 1970; Raleigh & Kirby 1970; Carter & Ave' Lallemant 1970; Stocker & Ashby 1973). This paper extends the earlier work. In particular, it includes cataclastic flow as a mechanism; it incorporates recent developments in the understanding of plastic flow and creep and in the way in which pressure affects them; and it reviews and uses new data on the transport and mechanical properties of olivine.

The mechanisms which appear to be most important in such problems are:

(*a*) *Cataclastic flow* (§ 3): to a materials scientist, this is not flow, but fracture. It is modelled as repeated cleavage fracture together with the rolling of already fragmented particles over each other.

(*b*) *Low-temperature plasticity* (§ 4): plastic deformation involving the gliding motion of dislocations.

(*c*) *Diffusional flow* (§ 5): deformation by the diffusive motion of single ions, possible only at high temperatures, and leading to flow which is, at least approximately, Newtonian viscous.

(*d*) *Power-law creep* (§ 6): a non-linear flow involving both the climb and glide motion of dislocations.

There exist several other mechanisms – twinning, for instance – which we do not understand well enough to model in detail. Two of these are clearly of importance in the deformation of the Earth's crust and mantle, and are the major gaps in our present understanding. They are:

(*e*) *Fluid-phase transport* (§ 7): flow in the presence of an aqueous fluid phase or a partial melt.

(*f*) *Creep with dynamic recrystallization* (§ 7): creep accelerated, or made possible, by continuous recrystallization.

Throughout the paper we are concerned with steady-state flow. When strains are small ($< 10^{-2}$), as they are when the upper mantle deforms in response to surface loads, for instance, transient effects cannot be neglected (Goetze 1971). When the strains are large ($\geqslant 1$), as they are in mantle convection, the contribution of transients depends on how rapidly the stress and temperature change along a stream path; but for all but the most rapidly changing conditions, the assumption of steady-state flow is a good first approximation.

TABLE 1. MATERIAL PROPERTIES FOR OLIVINE

material property	symbol	value	units	remarks
molecular volume per oxygen ion at 1 atm.†	Ω^0	1.23×10^{-29}	m^3	This value used for diffusional flow, equation (5.1)
oxygen ion volume at 1 atm.	Ω_i^0	1.15×10^{-29}	m^3	This value used for activation volumes, equation (5.9).
Burgers vector at 1 atm.	b^0	6.0×10^{-10}	m	An average of several values – § 4
melting point at 1 atm.	T_M	2140	K	
shear modulus at 300 K and 1 atm.	μ^0	8.13×10^{10}	Pa	⎫
T-dependence of shear modulus	$(T_M/\mu^0)(d\mu/dT)$	0.35	—	⎬ See §§ 2.2 and 2.3 for references
P-dependence of shear modulus	$d\mu/dP$	1.8	—	⎭
bulk modulus at 300 K and 1 atm.	K^0	1.27×10^{11}	Pa	⎫
T-dependence of bulk modulus	$(T_M/K^0)(dK/dT)$	0.26	—	⎬ See §§ 2.2 and 2.3 for references
P-dependence of bulk modulus	dK/dP	5.1	—	⎭
pre-exponential, lattice diffusion	D_{ov}	0.1	m^2/s	⎫
activation energy, lattice diffusion	Q_v	522	kJ/mol	⎬ The data are discussed in § 5 and plotted in figure 9
activation volume/oxygen ion volume	V^*/Ω_i	0–1	—	⎭
pre-exponential, boundary diffusion	δD_{OB}	1×10^{-10}	m^3/s	There are no data for olivine.
activation energy, boundary diffusion	Q_B	350	kJ/mol	These are obtained by scaling: $\delta D_{OB} = 10^{-9} D_{ov}$ and $Q_B = \tfrac{2}{3} Q_v$
activation volume/oxygen ion volume	V^*/Ω_i	0–1	—	
first creep exponent	n_1	3	—	⎫ See § 6.4 for the rate equation used for power-law creep. The values refer to creep in shear (equation (6.4))
second creep exponent	n_2	5	—	
first creep constant	A_1	0.45	—	⎭
second creep constant	A_2	5.4×10^4	—	⎫
activation energy for creep	Q_{cr}	522	—	⎬ Q_{cr} and V_{cr}^* are the same as those for lattice diffusion
activation volume/oxygen ion volume	V_{cr}^*/Ω^0	0–1	—	⎭
flow stress at 0 K, (lattice resistance)/modulus	τ_p/μ	3.3×10^{-2}	—	⎫
pre-exponential for lattice resistance	$\dot{\gamma}_p$	10^{11}	s^{-1}	⎬ This determines the flow stress at 0 K and throughout the plasticity field – see also § 4
activation energy for lattice resistance	$\Delta F_p/\mu b^3$	0.05	—	⎭
flow stress at 0 K, (obstacle control)/modulus	$\hat{\tau}_0/\mu$	8×10^{-3}	—	⎫
pre-exponential, obstacle control	$\dot{\gamma}_0$	10^6	s^{-1}	⎬ This determines the plateau in the flow stress separating plasticity from power-law creep – see also § 4
activation energy for obstacle control	$\Delta F_0/\mu b^3$	0.5	—	⎭
cleavage stress in tension modulus	σ_f/μ	5×10^{-3}	—	See § 3 and table 4

† 1 atm $\approx 10^5$ Pa.

The following sections give, with explanation, a simple constitutive law for each mechanism of flow, including the effects of pressure. They have the form

$$\dot{\gamma} = f(\sigma_s, p, T, M_1, M_2, \ldots), \tag{1.1}$$

where $\dot{\gamma}$ is the shear strain-rate, σ_s the shear stress, p the hydrostatic pressure, T the absolute temperature, and M_1, M_2, etc., are material properties. Each section contains a discussion of the material properties for olivine; they are summarized in table 1, which also serves to define the symbols. In a later section (§ 8) the equations and data are used to construct deformation-mechanism maps for olivine, both under laboratory conditions (constant pressure) and under the conditions presumed to obtain in the upper mantle.

It is worth pointing out that, for all the mechanisms except cataclastic flow, pressure enters the laws only through the influence on the material properties (M_1, M_2, etc.). For these mechanisms one may write

$$\dot{\gamma} = f(\sigma_s, T, M_1(p), M_2(p) \ldots).$$

Because pressure enters in this way, the Prandtl–Reuss generalization still applies, and the equation can be written in a form suitable for numerical computation:

$$\mathrm{d}e_{ij}/\mathrm{d}t = f(\overline{\sigma}_s, T, M_1(p), M_2(p) \ldots) S_{ij}/2\overline{\sigma}_s, \tag{1.2}$$

where the function f is the same as before. Here $\mathrm{d}e_{ij}$ are the components of the strain increment, $\mathrm{d}t$ is the increment of time, and $\overline{\sigma}_s$ is an equivalent shear stress:

$$\overline{\sigma}_s^2 = \tfrac{1}{6}[(\sigma_1 - \sigma_2)^2 + (\sigma_2 - \sigma_3)^2 + (\sigma_3 - \sigma_1)^2] = \tfrac{1}{2} S_{ij} S_{ij}; \tag{1.3}$$

p is the hydrostatic pressure:

$$p = -\tfrac{1}{3}(\sigma_1 + \sigma_2 + \sigma_3) = -\tfrac{1}{3}\sigma_{kk}; \tag{1.4}$$

and S_{ij} is the deviatoric part of the stress tensor:

$$S_{ij} = (\sigma_{ij} - \tfrac{1}{3}\delta_{ij}\,\sigma_{kk}). \tag{1.5}$$

The quantities σ_1, σ_2 and σ_3 are the principal stresses, and σ_{ij} is the stress tensor.

It is not permissible to generalize equations for fracture in this way because pressure enters the equations directly, not merely through its influence on a material property.

For later use, it is convenient to define an equivalent shear strain-rate:

$$\overline{\dot{\gamma}}^2 = \tfrac{2}{3}[(\dot{\epsilon}_1 - \dot{\epsilon}_2)^2 + (\dot{\epsilon}_2 - \dot{\epsilon}_3)^2 + (\dot{\epsilon}_3 - \dot{\epsilon}_1)^2] = 2\dot{\epsilon}_{ij}\dot{\epsilon}_{ij}, \tag{1.6}$$

where $\dot{\epsilon}_1$, $\dot{\epsilon}_2$, and $\dot{\epsilon}_3$ are the principal strain-rates, and $\dot{\epsilon}_{ij}$ is the strain-rate tensor.

2. Effect of pressure and temperature on ionic volumes and moduli

2.1. *Effect of pressure on the ionic volume and Burgers vector*

In a linear-elastic solid of bulk modulus K^0, the atomic or ionic volume varies with pressure as

$$\Omega = \Omega^0 \exp\{-(p - p^0)/K^0\}, \tag{2.1}$$

and the lattice parameter a (and the Burgers vector $\boldsymbol{b}$) as

$$a = a^0 \exp\{-(p - p^0)/3K^0\}, \tag{2.2}$$

where Ω^0 and a^0 (or b^0) are the values at atmospheric pressure (p^0). Their change with temperature is sufficiently small that we can safely neglect it. Data for Ω^0, b^0 and K^0 for olivine are given in table 1.

2.2 *Effect of temperature and pressure on the moduli*

To first-order, the moduli increase linearly with pressure and decrease linearly with temperature. We write this in the form

$$\mu = \mu^0 \left\{ 1 - \left[\frac{T_M}{\mu^0} \frac{d\mu}{dT} \right] \frac{(T - 300)}{T_M} \right\} + \left[\frac{d\mu}{dp} \right] (p - p^0); \tag{2.3}$$

$$K = K^0 \left\{ 1 - \left[\frac{T_M}{K^0} \frac{dK}{dT} \right] \frac{(T - 300)}{T_M} \right\} + \left[\frac{dK}{dp} \right] (p - p^0), \tag{2.4}$$

where p^0 is atmospheric pressure, which, for almost all practical purposes, we can ignore.

The coefficients in the square brackets are dimensionless. Table 2 lists means and standard deviations of the temperature coefficients for a number of cubic elements and compounds. Most are metals, though data for some alkali halides and oxides are included; data for olivine are listed separately. The dimensionless coefficients are roughly constant. When no data are available these mean values can be used.

TABLE 2. TEMPERATURE COEFFICIENTS OF THE MODULI

coefficient	number of materials	mean and s.d.	source of data	value for olivine	reference
$\left[\dfrac{T_M}{\mu^0} \dfrac{d\mu}{dT} \right]$ $(\mu = \{\frac{1}{2}C_{44}(C_{11} - C_{12})\}^{\frac{1}{2}})$	11	0.52 ± 0.1	Frost & Ashby (1973)	0.35	Kamazama & Anderson (1969)
$\left[\dfrac{T_M}{K^0} \dfrac{dK}{dT} \right]$ $(K = \frac{1}{3}(C_{11} + 2C_{12}))$	9	0.36 ± 0.2	Huntington (1958)	0.26	Huntington (1958)

TABLE 3. PRESSURE COEFFICIENTS OF THE MODULI

material	$10^{-10} \mu^0/\text{Pa}$	$10^{-10} K^0/\text{Pa}$	$\left[\dfrac{d\mu}{dp}\right]$	$\left[\dfrac{dK}{dp}\right]$	$\dfrac{K^0}{\mu^0}$	$\left[\dfrac{K^0}{\mu^0}\dfrac{d\mu}{dp}\right]$	reference
Al	2.54	7.3	2.2	3.9	2.87	6.3	
Ag	2.64	10	2.3†	6.2‡	3.79	8.8	
Au	—	17.3	1.8†	6.4‡	—	—	
Cu	4.21	13.8	1.4	4.9	3.28	4.6	
Ni	7.89	18.3	1.5	—	2.37	3.6	
Na	0.23	0.83	1.6†	3.6‡	3.61	5.9	
Li	0.35	1.3	1.0†	—	3.71	—	Huntington (1958), Birch (1966)
αFe	6.4	16.8	1.9	4.0	2.63	5.0	
Ge	5.2	7.73	1.3†	4.7‡	1.49	2.0	
Si	6.37	9.88	0.8†	4.2‡	1.55	1.2	
NaCl	1.51	2.35	2.7†	6.0‡	1.56	4.2	
LiF	4.58	7.0	1.4	—	1.53	2.2	
MgO	—	16.7	2.6	3.9	—	—	
SiO_2	—	3.7	2.9†	—	—	—	
Mg_2SiO_4	8.13	12.7	1.8	5.1	1.56	2.8	Graham & Barsch (1969), Kamazama & Anderson (1969)
mean and s.d.			1.8 ± 0.7	4.8 ± 1	2.5 ± 1	4.3 ± 2	

† This is dC_{44}/dp.   ‡ This is $d(\frac{1}{3}(C_{11} + 2C_{12}))/dp$.

Table 3 lists the moduli and their pressure-dependence for a number of materials, including olivine. Again, the dimensionless coefficients are roughly constant; for materials for which no data are available, it is reasonable to use the mean values indicated at the foot of the table.

3. CATACLASTIC FLOW BY CLEAVAGE AND BY ROLLING-PLUS-SLIDING

Almost all crystalline solids are capable of fracture by cleavage if the temperature is sufficiently low; the f.c.c. metals and their alloys appear to be the only exceptions. A confining pressure makes cleavage more difficult but does not necessarily prevent it. It is therefore possible to develop large shearing displacements in a brittle solid, constrained by a pressure, by a process of repeated fracturing, or *cataclasis*.

Once fracturing has started, so that the material has become fragmented, continued shearing may simply break the fragments into ever smaller pieces. But there is an alternative: shearing can continue if the fragments slide and roll over each other.

Simple, and approximate, models for these two alternative processes are considered in this section. From them we learn that rolling is more pressure-sensitive than repeated fracturing, which always becomes the dominant of the two processes when the confining pressure is large.

3.1. *Cleavage fracture*

The tensile stress which will overcome the interatomic forces in a perfect crystal, causing it to separate on a plane normal to the stress axis, defines an upper limiting strength for a crystalline solid. The many calculations of it are in general agreement (Kelly 1966; Macmillan 1972): at an adequate level of accuracy

$$\sigma_{\text{ideal}} = (2E\Gamma_s/\pi b)^{\frac{1}{2}} \approx 0.1\,E, \tag{3.1}$$

where Γ_s is the surface free energy, E is the Young modulus and b is the atomic size. Fracture occurs when the tensile stress exceeds σ_{ideal}.

The ideal strength is rarely realized. Almost always, small cracks pre-exist in brittle and semi-brittle solids, or are created in them by slip as soon as yielding occurs. Such a crack concentrates stress, so that the ideal strength is exceeded at its tip when the applied stress is still much less than σ_{ideal}.

The stress at the tip of an atomically sharp crack of length $2C_0$ in an elastic medium can be calculated: it is $(C_0/b)^{\frac{1}{2}}$ times larger than the applied stress. If the crack propagates when the ideal strength is exceeded at its tip, the fracture stress in simple tension obviously becomes

$$\sigma_f = (2E\Gamma/\pi C_0)^{\frac{1}{2}}, \tag{3.2}$$

a result first developed by Griffith (1924) using an energy argument, and later by Orowan (1934, 1949) using an argument based on stresses (as here). In this equation, 2Γ is the energy absorbed per unit area of crack advance; its minimum value is $2\Gamma_s$. One could, then, describe the cleavage-fracture strength of a material by citing the values of C_0 and Γ which characterize it. But for brittle minerals and rocks this is impractical; it is much easier to treat σ_f, the stress at which cleavage fracture occurs in simple tension, as the material property, and use it to calculate the conditions for fracture under pressure.

3.2. *Effects of pressure*

Griffith (1924) himself extended the fracture criterion to include the effect of biaxial stress states. A material can fracture by cleavage, even when the stress state is a compressive one, because cracks orientated at an angle to the compression axis are loaded in shear, and therefore have regions of tension and compression at their tips, as illustrated in figure 1. Suppose cracks of equal size, in a biaxial stress field, are assumed to have their planes parallel to the unstressed direction, but are otherwise orientated at random. A tensile component in the applied stress will tend to open some of the cracks; but even if the stress field is compressive, any deviatoric component in it will tend to make some cracks slide.

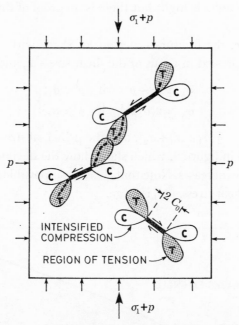

FIGURE 1. Cataclastic flow by cleavage. A deviatoric stress causes shearing displacements across the crack faces, generating tensile stresses in the regions marked T. If these reach a critical level, the cracks extend (broken lines) and ultimately link, even when the stress field is a compressive one.

In either case, a tensile stress appears at the crack tip. Its magnitude can be calculated as a function of the angle of orientation of the crack, assuming the crack faces to be frictionless. If fracture occurs when the maximum value of this stress exceeds σ_{ideal}, then the Griffith fracture criterion becomes (after McClintock & Argon 1966)

$$\left.\begin{aligned} \sigma_1 = \sigma_f \quad \text{if} \quad -3\sigma_f \leqslant \sigma_3 \leqslant \sigma_1; \\ (\sigma_1 - \sigma_3)^2 + 8\sigma_f(\sigma_1 + \sigma_3) = 0 \quad \text{if} \quad \sigma_3 < -3\sigma_f, \end{aligned}\right\} \tag{3.3}$$

where σ_1 is the largest (most tensile) and σ_3 the smallest principal stress.

McClintock & Argon (1966) argue that, in a triaxial field ($\sigma_1 > \sigma_2 > \sigma_3$) the intermediate principal stress σ_2 does not change the criterion. Cracks with normals in the $\sigma_1 - \sigma_3$ plane are exposed both to the largest tensile and largest shear stress; and (in an elastic solid) a normal stress σ_2 in the crack plane is not concentrated by the crack. The criterion above is then unaltered.

The model predicts that the fracture strength in compression should be 8 times larger than that in tension. Tests on rocks and minerals show it to be between 8 and 15 times larger (Jaeger & Cook 1969), so while it is a useful approximation the model is incomplete. There are two deficiencies. First, crack-plane friction is ignored. It can be included (McClintock & Walsh 1962) but the coefficient of friction required to fit experiment is unexpectedly large, perhaps because the crack faces are serrated, and key together when under compression. More important, the crack, when it extends under compression, deviates from the plane originally containing it. Equation (3.3) is an initiation condition, not a criterion for propagation. Final fracture in compression requires an understanding of the interaction between cracks that we do not yet possess. It seems likely that the propagation conditions will depend on pressure in the same way as the initiation condition does, and (as figure 1 suggests) that the two conditions merge when the initial density of cracks is high; but there is no proof of this, and we must treat it as a postulate.

Assuming this to be true, the fracture criterion for shear with a superimposed hydrostatic pressure becomes, when expressed in terms of the shear stress σ_s and the hydrostatic pressure p:

$$\left.\begin{aligned}
\sigma_s &\geqslant \sigma_f + p && \text{if} \quad p < \sigma_f; \\
\sigma_s &\geqslant 2(\sigma_f p)^{\frac{1}{2}} && \text{if} \quad p \geqslant \sigma_f,
\end{aligned}\right\} \tag{3.4}$$

where $\sigma_s = \frac{1}{2}(\sigma_1 - \sigma_2)$, $p = -\frac{1}{3}(\sigma_1 + \sigma_2 + \sigma_2)$, and the principal stresses are $p + \sigma_s$, p, and $p - \sigma_s$. Equations (3.4) are plotted in figure 2, which shows how the deviatoric stress required to cause cleavage rises as the pressure increases. Note that cleavage is possible when the confining pressure is large, provided that the shear stress, too, is large.

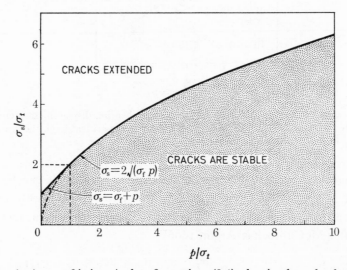

FIGURE 2. Griffith criterion, no friction. A plot of equations (3.4), showing how the deviatoric stress required to cause cleavage (or cataclasis) increases with pressure; σ_f is the fracture strength in tension.

3.3. *Fracture data for olivine*

The material property which appears in equations (3.4) is the fracture strength in tension, σ_f. Values for olivine-containing rocks can be deduced from two sources: the tests of Griggs, Turner & Heard (1960) on dunite, and those of Handin (1966) on peridotite. Both were compressive tests, with a confining pressure. We have used equation (3.3) to calculate the material property, σ_t from their data. The results are tabulated in table 4. In the later calculations of this paper, we have used the value of $\sigma_t = 5 \times 10^{-3}\mu$ (table 1).

3.4. *Rate law for cataclastic flow*

We have modelled cataclastic flow as the result of two independent mechanisms. First we assumed extensive flow is possible by fracturing if equation (3.4) is satisfied; in fracturing, the rock becomes increasingly granulated. Secondly, we assumed that extensive flow is also possible by the rolling and sliding (without further fracture) of the granulated rock. The model for this second process is of the simplest and most approximate kind, and is illustrated by figure 3.

TABLE 4. LOW-TEMPERATURE FRACTURE OF DUNITE AND PERIDOTITE

$T/°C$	T/T_M	σ_1/MPa	σ_3/MPa	σ_t/MPa	$10^3\sigma_t/\mu$	reference
25	0.14	500	2300	650	8.1	Griggs *et al.* (1960)
25	0.14	500	2000	500	6.2	
150	0.2	100	510	400	5.0	Handin (1966)
150	0.2	100	450	350	4.5	

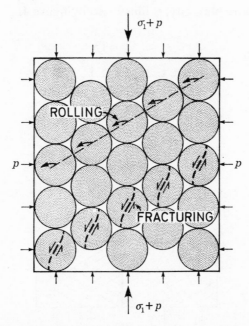

FIGURE 3. Granular or previously fractured materials can deform by the rolling and sliding of the granules or fragments over each other. Such flow is associated with a volume expansion, and because of this, it is more strongly influenced by pressure than is cleavage.

Consider the granules to be cylinders which roll and slide at their points of contact. When the array is sheared, it expands, doing work against the confining pressure and dissipating energy in overcoming friction at points of contact, some of which must slide. This work must be supplied by the applied shear stress. If the cylinders form a close-packed array, the largest stress is that needed to start them shearing; thereafter they will continue to shear. It is easy to show that this requires

$$\sigma_s \geqslant p \left(\tfrac{1}{\sqrt{3}} + \sqrt{3}\,\mu_t\right),\qquad(3.5)$$

where the first term in the parentheses derives from the work done in dilating the material against the confining pressure p, and the second is the work done against friction (coefficient of friction μ_f). The term in parentheses is of order 1, and remains so when the model is broadened to describe the rolling of spheres instead of cylinders. Because of its more rapid dependence on p, rolling is suppressed by even modest pressures, and is replaced by the fracturing mode of flow.

Cataclastic flow is incorporated into the calculations of Section 8 by assuming that:

$$\left.\begin{array}{l} \dot{\gamma} = \infty \text{ if either the rolling or fracturing criteria (equations (3.4) or (3.5)) is satisfied,} \\ \dot{\gamma} = 0 \text{ if neither is satisfied.} \end{array}\right\} \quad (3.6)$$

4. LOW-TEMPERATURE PLASTICITY

At low temperatures, or high rates of strain, plastic flow in crystalline solids is by slip. Dislocations – the 'carriers' of deformation – glide on slip planes; the combination of a plane and a Burgers vector defining a slip system. If a polycrystal is to deform homogeneously without fracturing, five independent slip systems must be available; but if non-homogeneous deformation is allowed, so that although grains deform compatibly, they do not deform uniformly, then four systems may be sufficient (Hutchinson 1976, private communication). This mechanism, which we shall call low-temperature plasticity, is illustrated by figure 4.

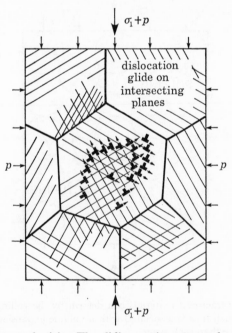

FIGURE 4. Low-temperature plasticity. The gliding motion of several sets of dislocations permits compatible deformation of the grains of a polycrystalline solid.

4.1. *Lattice-resistance and obstacle-controlled plasticity*

Dislocations glide in many pure metals with great ease, and for that reason they are often very soft. But in rocks and minerals this is not so. Many, like the silicates, are covalently bonded, and the covalent bond is hard to break. The result is that the energy of the crystal fluctuates when the dislocation moves (figure 5, top), and a force which is proportional to the slope of the energy–distance curve is required to make it do so.

Obstacles – impurities, precipitates, other dislocations, and so forth – obstruct glide also; but their effect is observable only if the energy hill they place in the path of the moving dislocation is steeper than that due to bond breaking (figure 5, bottom). Sometimes this is so, and the flow strength of the crystal (which reflects an average of the behaviour of dislocations on four or five slip systems) is said to be obstacle controlled. When it is not, the flow stress is lattice-resistance controlled.

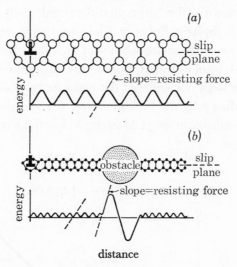

FIGURE 5. The yield strength reflects the force required to move a dislocation. In silicates such as olivine, the structure itself resists the motion, and the strength is said to be 'lattice resistance controlled' (*a*). Strong discrete obstacles – such as other dislocations – can sometimes determine the yield strength, which is then said to be 'obstacle-controlled' (*b*).

Above 0 K, thermal energy is available. It is then possible, by a straightforward application of kinetic theory, to calculate the average drift velocity of dislocations through the crystal, and hence the strain rate. The details of the calculation and of the result depend on the shape of the energy hills of figure 5 (see, for example, Kocks, Argon & Ashby 1976). That which best fits the available experimental data for lattice-resistance controlled glide leads to the rate equation

$$\dot{\gamma} = \dot{\gamma}_p \left(\frac{\sigma_s}{\mu}\right)^2 \exp\left\{-\frac{\Delta F_p}{kT}\left(1 - \left(\frac{\sigma_s}{\hat{\tau}_p}\right)^{\frac{3}{4}}\right)^{\frac{4}{3}}\right\},\tag{4.1}$$

where $\dot{\gamma}_p$ is a pre-exponential rate-constant, ΔF_p is the activation energy for lattice-resistance controlled glide, and $\hat{\tau}_p$ represents the flow stresses at 0 K.

Because of the small activation energy (ΔF_p) the strengthening caused by the lattice resistance drops as the temperature is raised, until its contribution becomes less than that of discrete obstacles (figure 5, lower drawing). When they control the flow stress, flow is better described by the simpler equation

$$\dot{\gamma} = \dot{\gamma}_0 \exp\{(-\Delta F_0/kT)(1 - \sigma_s/\hat{\tau}_0)\},\tag{4.2}$$

where $\dot{\gamma}_0$ is the pre-exponential rate-constant, ΔF_0 the activation energy for cutting or passing the obstacle, and $\hat{\tau}_0$ is the flow stress at 0 K if the obstacles acted alone. In later calculations, we select the slower of these two strain-rates: both lattice resistance and obstacle must be overcome, and it is the more difficult of the two which determines the strength.

Dislocation glide is characterized by a long transient (work hardening), which continues until a saturation flow stress is reached. In tensile tests, fracture occurs before saturation, and, because of this, dislocation glide is not usually thought of as a steady-state mechanism of flow. But experiments in torsion or compression suggest that a saturation flow stress is ultimately reached, though the strains involved are large (> 1).

4.2. *Effect of pressure on low-temperature plasticity*

High strength materials have a yield strength in tension which is lower than that in compression – a phenomenon known as the 'strength differential' (s-d) effect. Recent experimental studies of high strength steels (Spitzig, Sober & Richmond 1975, 1976) have shown that this is simply the effect of pressure on the flow stress. Some of their data are replotted in figure 6, which shows how the shear strength σ_s of a steel at room temperature increases linearly with pressure. The axes of this figure have been normalized: the ordinate by dividing σ_s by σ_s^0 (the value at $p = 0$) and abscissa by dividing p by the bulk modulus, K^0. The slope is a useful dimensionless measure of the pressure-dependence of the yield strength. For this steel, and for two others studied by Spitzig *et al.*, the slope was

$$\mathrm{d}(\sigma_s/\sigma_s^0)/\mathrm{d}(p/K^0) = 6\text{–}10. \tag{4.3}$$

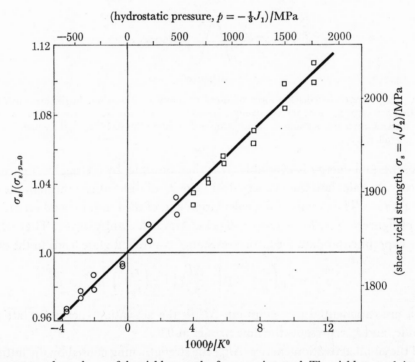

FIGURE 6. The pressure-dependence of the yield strength of a maraging steel. The yield strength increases linearly with pressure. The data are replotted from Spitzig, Sober & Richmond (1976): ○, tension; □, compression; slope = 9.3.

The effect is far too large to be accounted for by the permanent volume expansion associated with plastic flow and must be associated instead with a direct effect of pressure on the motion of dislocations.

This pressure-dependence can be accounted for almost entirely by considering the effect of pressure on the activation energies ΔF_p and ΔF_0 and the strengths $\hat{\tau}_p$ and $\hat{\tau}_0$. It is commonly found (see Kocks *et al.* (1976) for a review) that the activation energy for both obstacle and lattice-resistance controlled glide scales as μb^3 and that the strengths $\hat{\tau}$ scale as μ. For most b.c.c. metals, for example, the activation energy is close to $0.07 \mu b^3$ and the flow stress at absolute zero is close to 0.01μ (Frost & Ashby 1973). As already described, both μ and b depend on pressure, μ increasing more rapidly than b^3 decreases. Pressure, then, has the effect of raising both the activation energy and the strengths, $\hat{\tau}$.

At absolute zero, the shear stress required to cause flow is simply $\hat{\tau}$. Using equation (2.3) for the modulus, and neglecting the pressure p^0, we find by inserting the above proportionalities into equation (4.2),

$$\sigma_s = \hat{\tau}_0(p) = \sigma_s^0 \left(1 + \left[\frac{K^0 d\mu}{\mu^0 dp}\right]\frac{p}{K^0}\right), \tag{4.4}$$

from which

$$d\left(\frac{\sigma_s}{\mu^0}\right) \Big/ d\left(\frac{p}{K^0}\right) = \left[\frac{K^0 d\mu}{\mu^0 dp}\right]. \tag{4.5}$$

Values of this dimensionless quantity are listed in table 3 for a variety of materials. The values range from 2 to 9 compared with the measured coefficient of 6 to 10, but the measurements, of course, were made at room temperature – about $0.2 T_M$ for many of the listed materials.

As the temperature is raised the picture becomes more complicated. The flow stress becomes, on inverting equation (4.2),

$$\sigma_s = \hat{\tau}_0(p) \{1 - [kT/\Delta F_0(p)] \ln(\dot{\gamma}_0/\dot{\gamma})\}, \tag{4.6}$$

in which both $\hat{\tau}_0$ and ΔF_0 increase with pressure, so that at fixed T and $\dot{\gamma}$, the predicted pressure dependence of σ_s *increases* with increasing temperature, and adequately accounts for the observed effects.

There are other contributions to the pressure-dependence of low-temperature plasticity, but they appear to be small. The presence of a dislocation expands a crystal lattice, partly because the core has a small expansion associated with it (about 0.5Ω per atom length) and partly because the non-linearity of the moduli cause any elastic strain-field to produce an expansion (Seeger 1955; Lomer 1957; Friedel 1964). It is this second effect which is, in general, the more important. The volume expansion is roughly

$$V^* = \tfrac{3}{2}\Delta E^{el}/\mu \quad \text{per unit volume,} \tag{4.7}$$

where ΔE^{el} is the elastic energy associated with the strain-field. If the activation energy which enters the rate equations is largely elastic in origin, (as it appears to be) then during activation, there is a small temporary increase in volume, V^*. A pressure further increases ΔF by the amount pV^*. Because they are small, these contributions are neglected in the present treatment.

4.3. *Low-temperature plasticity of olivine*

The slip systems in olivine deformed at temperatures up to 1250 °C have been determined by electron microscopy (Phakey, Dollinger & Christie 1972) and optical examination (Raleigh 1968; Carter & Ave' Lallemant 1970; Young 1969; Raleigh & Kirby 1970). The review by Paterson (1974) gives details of the primary and secondary slip systems, and of how temperature affects their ease of operation and influences the resulting dislocation arrays. At 1000 °C and below, Phakey *et al.* (1972) observed straight dislocations, suggesting a large lattice resistance; but above 1250 °C, they saw subgrains of the kind associated with power-law creep.

The slip system which operates easily at low temperatures is that involving dislocations with $b = 5.98$ [001] Å gliding on (100) planes. If this slip system is suppressed by proper orientation of the crystal, then dislocations with Burgers vector $b = 4.76$ [100] Å gliding on (010) planes appear. At 1000 °C these two systems operate with equal ease. Dislocations with a large Burgers vector (b = 10.21 [010] Å on (100)) appear above 800 °C, particularly in crystals orientated so that the easy slip system is unstressed.

The calculations of § 8 require a value for the Burgers vector, though its value is not critical: it is used largely as a scaling parameter. We have used b = 6 × 10^{-10} m (table 1) since this is broadly typical of the observations. Much more important are numerical values for the low temperature strength, since these are used to set the values of $\hat{\tau}$ and ΔF in equations (4.1) and (4.2). There are three sets of useful observations.

First, the yield stress for a peridotite (60–70 % olivine and 20–30 % enstatite) of grain size 0.5 mm, was measured by Carter & Ave' Lallemant (1970) between 325 and 740 °C. Secondly, Phakey *et al.* (1972) obtained stress–strain curves at 600 and 800 °C for single crystal forsterite. Four compression tests at 800 °C with different orientations of the single crystals produced yield stresses ranging from 570 to 1300 MN/m² (5.7–13 kbar). The highest value was obtained on the specimen orientated to produce no stress on the easy slip system; we have used this because it is the most representative of a polycrystalline material. Finally, some hardness data for olivine exists (Evans, cited by Durham & Goetze 1976), and it is this which gives the most complete picture of the low-temperature strength. It and the yield strengths are plotted on one of the deformation maps of § 8.

The quantities ΔF_{p}, $\hat{\tau}_{\mathrm{p}}$ and $\hat{\tau}_0$ were obtained by fitting equations (4.1) and (4.2) to these data. We have assumed the frequency factor $\dot{\gamma}_0$ for olivine to be comparable with that for other materials with a large lattice resistance. The frequency factor $\dot{\gamma}_0$ and the activation energy ΔF_0 were also set by analogy with metals, but neither are important. The values used in later calculations are listed in table 1.

4.4. *Rate equations for low-temperature plasticity*

The later calculations of § 8 use the equations for lattice-resistance and obstacle-controlled glide as given in equations (4.1) and (4.2). The effects of pressure are included in exactly the way described above, by making ΔF scale as μb^3 and $\hat{\tau}$ as μ, and allowing μ and b to depend on both pressure and temperature as described in § 2. The consequences of doing so appear in the diagrams of § 8 and will be discussed there.

5. DIFFUSIONAL FLOW

The plasticity described in the last section is observed at low temperatures and high stresses. Consider now the opposite extreme: flow at high temperatures and low stresses. Under these conditions metals and ceramics can deform by diffusion alone. Any deviatoric stress applied to them causes ions to flow by lattice diffusion or diffusion in the grain boundaries in such a way that the grains change their shape, permitting the stress to do work: it is this work that drives the diffusive flux. The resulting deformation is called diffusional flow, and is illustrated by figure 7.

5.1. *Diffusional flow*

Creep by diffusion alone can be modelled in detail. A compressive stress raises the chemical potential of ions on the boundaries of the grains. Provided that there is a difference in stress on these boundaries, and that atoms can be detached from them and reattached to them freely, matter will flow through or round the grains at a rate determined by diffusion.

Those who have calculated this rate (Nabarro 1948; Herring 1950; Lifshitz 1963; Raj &

Ashby 1971) are in general agreement: when both lattice diffusion and grain-boundary diffusion are permitted, the shear creep rate is

$$\dot{\gamma} = 42 D_{\text{eff}} \sigma_{\text{s}} \Omega / k T d^2,\tag{5.1}$$

where d is the grain diameter and Ω the atomic volume. D_{eff} is an effective diffusion coefficient, which, for a one-component system, is given by (see Raj & Ashby 1971):

$$D_{\text{eff}} = D_{\text{v}}[1 + (\pi\delta/d)\,(D_{\text{B}}/D_{\text{v}})],\tag{5.2}$$

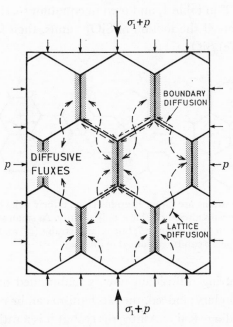

FIGURE 7. Diffusional flow. A deviatoric stress sets up differences in the chemical potential of ions at grain boundaries which then diffuse as shown by the broken lines, depositing in the shaded bands.

where D_{v} is the lattice self diffusivity, and δD_{B} that for boundary diffusion, multiplied by the thickness of the boundary diffusion path, δ. The equation describes a proper superposition of two submechanisms which are sometimes distinguished: when lattice diffusion dominates, it is called Nabarro–Herring creep; when, instead, boundary diffusion dominates, it is termed Coble creep. (The two sub-régimes are separated in the diagrams of § 8 by a vertical broken line.)

The quantities D_{v}, D_{B} and Ω are well defined in one-component solids. In pure copper, for instance, D_{v} is the lattice self-diffusion coefficient, D_{B} is the boundary self-diffusion coefficient, and Ω is the atomic volume. In a two-component system, their definition is more complicated. In ionic solids in which the diffusing species carry a charge, the appropriate diffusion coefficient for transport through the grain is (Lazarus 1971; Ruoff 1965):

$$D_{\text{v}} = \frac{D_{\text{v}}^{\text{A}} D_{\text{v}}^{\text{B}}}{(1 - \chi_{\text{A}})\, D_{\text{v}}^{\text{A}} + \chi_{\text{A}} D_{\text{v}}^{\text{B}}},\tag{5.3}$$

where χ_{A} is the atom-fraction of A in the compound. The important point made by the equation is that it is the slower-moving component which determines the diffusion coefficient, and so limits the creep rate; a similar conclusion holds for systems with more than two components. In most oxides (including olivine) oxygen is the largest ion and the one which diffuses most slowly;

it is oxygen diffusion which will usually control the creep rate. (The silicon–oxygen bond in silicates may present special problems: it is so strong that oxygen might diffuse as an SiO_4^{4-} group, but measurements of creep and diffusion in olivine, reviewed below, do not support this view.)

The value of Ω, too, depends on the nature of the diffusing species. We shall assume that the diffusion of O^{2-} limits the rate of mass transport in olivine. Then, on average, the arrival of one oxygen ion at a grain boundary must be accompanied by $\frac{1}{4}$ of a silicon and $\frac{1}{2}$ of a magnesium or iron ion. Their total volume is $\frac{1}{4}$ of the molecular volume of $(Mg, Fe)_2SiO_4$ ($1.23 \times 10^{-29} \, m^3$); this is the quantity listed as Ω^0 in table 1, and used in equation (5.1). But if (although this seems very unlikely) diffusion involved the motion of SiO_4^{4-} units, then Ω^0 would become the entire molecular volume of $(Mg, Fe)_2SiO_4$.

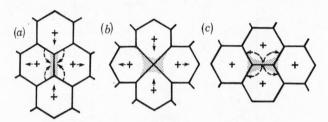

FIGURE 8. Diffusional flow when strains are large. If grains change their neighbours, the sample can undergo a large deformation while the grains alter their shape only slightly. At small strains (a) the diffusive paths, and the creep rate, are identical with those of figure 7, but at large strains ((b) and (c)) the paths differ and the rate is higher. Superplasticity in metals can be explained in this way.

Continuity of material during diffusional flow is maintained only if sliding displacements occur in the plane of the boundary; indeed, the mechanism can be regarded as diffusion-accommodated grain-boundary sliding (Raj & Ashby 1971), but it is a rather special form of this combined mechanism. It is implicit in the derivation of equation (5.1) that grains must suffer the same shape change as the specimen itself, and that they may not change their neighbours. When this constraint is relaxed (Ashby & Verrall 1972), a modified form of diffusional flow becomes possible: the grains slide past each other in the way illustrated by figure 8, changing neighbours and altering their shape (by diffusion) only where it is necessary for continuity to be maintained. This modified mechanism, which appears to be that underlying superplasticity in metals and ceramics, is simply a large-strain adaptation of the more familiar Nabarro–Herring and Coble creep. The rate equation describing it (see Ashby & Verrall 1973) closely resembles that given above. At a satisfactory level of approximation, it has the same form as equation (5.1), but is faster by a constant factor of about 5.

Diffusional flow, then, whether of the small-strain or the large-strain type, is controlled by the rate of diffusion of the slowest-moving species in the crystal. Pressure affects diffusional flow mainly because it slows the rate of diffusion.

5.2. Effect of pressure on diffusional flow

Pressure slows diffusion because it increases the energy required for an atom to jump from one site to another, and because it may cause the vacancy concentration in the solid to decrease. The subject has been extensively reviewed by Lazarus & Nachtrieb (1963), Girifalco (1964) and Peterson (1968); detailed calculations are given by Keyes (1963).

Steady-state diffusional flow requires that all atomic species in the material must move. This appears possible only by a vacancy mechanism, so we shall concentrate on this, and we shall further limit ourselves to pressures of less than $K/10$ – roughly the pressure at the base of the upper mantle.

The application of kinetic theory to self-diffusion by a vacancy mechanism (see, for example, Shewmon 1963) gives, for the diffusion coefficient

$$D = \alpha a^2 n_v \Gamma, \tag{5.4}$$

where α is a geometric constant of the crystal structure (independent of pressure) and a is the lattice parameter (weakly dependent on pressure in the way described by equation (2.2)). The important pressure-dependencies are those of the atom fraction of vacancies, n_v, and the frequency factor, Γ, which we discuss in order.

In a pure, one-component system, a certain atom fraction of vacancies is present in thermal equilibrium because the energy (ΔG_f per vacancy) associated with them is more than offset by the configurational entropy gained by dispersing them in the crystal. But in introducing a vacancy, the volume of the solid increases by V_f, and work $p V_f$ is done against any external pressure, p. A pressure thus increases the energy of forming a vacancy without changing the configurational entropy, and because of this the vacancy concentration in thermal equilibrium decreases. If we take

$$\Delta G_f = \Delta G_f^0 + p V_f, \tag{5.5}$$

where the superscript zero means zero pressure, then

$$n_v = \exp\{-(\Delta G_f^0 + p V_f)/kT\}. \tag{5.6}$$

A linear increase in pressure causes an exponential decrease in vacancy concentration.

It is in the nature of the metallic bond that the metal tends to maintain a fixed volume per free electron. If a vacancy is created by removing an ion from the interior and placing it on the surface, the number of free electrons is unchanged, and the metal contracts. For this reason, the experimentally measured values of V_f for metals are small: about $\frac{1}{2}\Omega^0$ where Ω^0 is the atomic volume. Strongly ionic solids can behave in the opposite way: the removal of an ion exposes the surrounding shell of ions to mutual repulsive forces. The vacancy becomes a centre of dilatation, which is why the observed values of V_f are large: up to $2\Omega_i^0$ where Ω_i^0 is the volume of the ion removed. There are no measurements for oxides or silicates; but, when the bonding is largely covalent, one might expect the close-packed oxygen lattice which characterizes many of them (among them olivine) to behave much like an array of hard spheres. Forming a vacancy then involves a volume expansion of Ω_i^0, the volume associated with an oxygen atom in the structure.

We might, then, expect that V_f should about equal Ω_i^0, the oxygen-ion volume, in a silicate like olivine, but there is a complicating factor. In a multi-component system, vacancies may be stabilized for reasons other than those of entropy. Ionic compounds, for instance, when doped with ions of a different valency, adjust by creating vacancies of one species or interstitials of the other to maintain charge neutrality; pressure will not then change the vacancy concentration. Oxides may not be stoichiometric, even when pure, and the deviation from stoichiometry is often achieved by creating vacancies on one of the sub-lattices. The concentration of these vacancies is influenced by the activity of oxygen in the surrounding atmosphere, so that it is not the net pressure but the partial pressure of oxygen which determines the rates of diffusion. For these reasons it is possible that the quantity V_f in equation (5.6) could lie between 0 and Ω_i.

The jump frequency, too, depends on pressure. In diffusing, an ion, vibrating about a position of equilibrium, jumps to a neighbouring vacant position in which its surroundings are identical. In doing so, it passes through an activated state in which its free energy is increased by the energy of motion, ΔG_m. The frequency of such jumps is then given by

$$\Gamma = \nu \exp\left(-\Delta G_m / kT\right), \tag{5.7}$$

where ν is the vibration frequency of the atom in the ground state (and is unlikely to depend on pressure).

TABLE 5. ACTIVATION VOLUMES FOR DIFFUSION AND POWER-LAW CREEP

material	structure	V^*/Ω_i^0 for diffusion	V^*/Ω_i^0 for creep
Pb	f.c.c.	0.8 ± 0.1	0.76
Al	—	—	1.35
Na	b.c.c.	0.4 ± 0.2	0.41
K	—	—	0.54
In	h.c.p.	—	0.76
Zn		0.55 ± 0.2	0.65
Cd		—	0.63
AgBr	rock salt	1.9 ± 0.5	1.9
Sn	tetragonal	0.3 ± 0.1	0.31
P		0.5 ± 0.1	0.44

Adapted from Lazarus & Nachtrieb (1963), Goldstein *et al.* (1965) and McCormick & Ruoff (1970).

In passing through the activated state, the ion distorts its surroundings, temporarily storing elastic energy; and, if bonding is local, it breaks the bonds with some of its neighbours. In a solid with non-local bonding, like a metal, one might expect that the elastic energy would account for most of the activation energy of motion, so that (by the argument used in §4.2) it would be associated with a maximum volume expansion,

$$V_m \approx 3\Delta G_m / 2\mu, \tag{5.8}$$

per unit volume. Taking the activation energy for motion to be 0.4 of the activation energy of diffusion, we find, typically, $V_m = 0.2 – 0.4 \Omega_i$, where Ω_i is the volume of the diffusing ion. Experimentally, V_m appears to be smaller than this, suggesting that even in metals not all the energy is elastic. And in solids like silicates with localized bonding, most of the energy of motion must be associated with bond-breaking and will not produce a volume expansion.

Assembling these results we find

$$D = D^0(1 - 2p/3K^0) \exp\left(-pV^*/kT\right), \tag{5.9}$$

where $D^0 = \alpha(a^0)^2 \nu \exp\left(-\{\Delta G_f + \Delta G_m\}/kT\right)$ and is the diffusion coefficient under zero pressure, and

$$V^* = V_f + V_m \quad \text{for intrinsic diffusion;}$$

$$V^* = V_m \qquad \text{for extrinsic diffusion.}$$

Because experiments are difficult, there are few measurements of V^*, and these show much scatter. They have been reviewed by Lazarus & Nachtrieb (1963), Keyes (1963), Girifalco (1964) and Goldstein, Hanneman & Ogilvie (1965). The results are summarized in the third column of table 5, in which Ω_i is the volume of the diffusing species: they lie between 0 and $2\Omega_i$.

In summary, both physical reasoning and experiments lead to the conclusion that activation volumes could vary between zero (for a covalent solid in which diffusion is extrinsic) to perhaps

$2\Omega_i$ (for a strongly ionic solid). For silicates, in which oxygen is the slow-diffusing species, we might expect it to lie between zero (if the oxygen sub-lattice has chemically stabilized vacancies on it) and Ω_i, the ionic volume of oxygen (if it does not). These are the two limiting values used in the calculations of § 8.

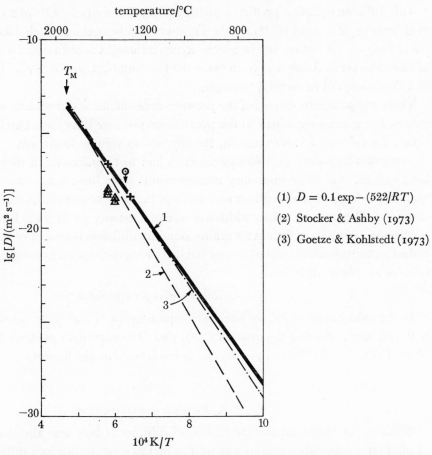

FIGURE 9. Diffusion of oxygen in olivine. The full line is a plot of the diffusion equation (equation (5.10)) used in § 8 to calculate the rates of diffusional flow. Data: +, Goetze & Kohlstedt (1973); ⊙, Borchardt & Schmaltzried (1972); △, Barnard (1975)

5.3. *Diffusion in olivine*

The meagre data for oxygen transport in olivine, and some attempts to fit an equation to them, are shown in figure 9. Borchardt & Schmaltzried (1972) reported that the oxygen-ion diffusivity at 1320 °C was less than 10^{-17} m²/s; Goetze & Kohlstedt (1973) inferred diffusion coefficients from the kinetics of the annealing of prismatic dislocation loops; and Barnard (1975) measured oxygen ion diffusion by a proton activation method.

Studies of power-law creep in olivine, discussed in the next section, are best described by an activation energy of $Q_{cr} = 522$ kJ/mol. If we assume that this creep is diffusion controlled, then Q_{cr} can be identified (after minor corrections for the temperature dependence of the modulus, which we shall ignore) with the activation energy for mass transport in olivine – a process which is almost certain to be limited in its rate by oxygen-ion diffusion. We have, therefore, fitted a line with this slope to the data (full line on the figure) giving

$$D/(\mathrm{m^2/s}) = 0.1 \exp\left(-522\,\mathrm{kJ}/RT\right). \tag{5.10}$$

Also shown are plots of the diffusion equation proposed by Goetze & Kohlstedt (1973) (broken line), and of that by Stocker & Ashby (1973). At high temperature they are all very close, and even at low temperatures the differences are small when compared with scatter in creep and other data.

This diffusion equation predicts a melting point diffusivity for O^{2-} of 2×10^{-14} m²/s, compared to an average of around 10^{-14} m²/s for a number of other oxides which, like olivine, have a close-packed oxygen sublattice, and in which oxygen is thought to diffuse as a single ion, not a complex of ions (Stocker & Ashby 1973). Because the two numbers are similar we believe oxygen diffuses as a single ion, not as an SiO_4^{4-} complex.

There are no measurements of the pressure-dependence of diffusion in olivine, or in any other oxides. For reasons explained in the previous section, we have used two limiting values for the activation volume, V^*: zero and Ω_i, the oxygen-ion volume in olivine.

The grain-boundary diffusion parameters had to be inferred. In those materials for which both volume and grain-boundary measurements have been made, the activation energy for grain-boundary diffusion is about two thirds of that for volume diffusion. This ratio was applied to olivine. The grain boundary width was set equal to twice the average Burgers vector. Although the pre-exponential coefficient for grain-boundary diffusion is generally less than that for volume diffusion, the two were equated, somewhat increasing the importance of grain-boundary diffusion relative to volume diffusion.

5.4. *Rate equation for diffusional flow*

In the calculations of §8, we have used equations (5.1) and (5.2), allowing Ω and Ω_i to vary with pressure according to equation (2.1), and D to vary with pressure according to equation (5.9). The value of V^* used in each calculation is listed on the figures.

6. POWER-LAW CREEP

Between the high-temperature régime of diffusional flow and the low-temperature régime of plasticity, materials – ceramics as well as metals – creep, but in a different way. The strain-rate varies as a power of the stress, the power ranging between 3 and about 10. Microscopy shows a complicated pattern of flow: superimposed on a uniform creeping of the grains is a non-uniform deformation caused by sliding at their boundaries. Electron microscopy shows that dislocations are involved; they contribute to the deformation by gliding, but, at least at high temperatures, they then aggregate to form cells. These cells have a rather small angular misorientation, perhaps 2°, between them; and their size depends on the stress. This mode of deformation is known as power-law creep. It is illustrated by figure 10.

6.1. *Diffusion-controlled power-law creep*

The origin of the power-law behaviour can be understood if the cells are thought of as little grains. Then a kind of diffusional flow is possible using the cell, instead of the grain, boundaries as sources and sinks. Because the cells are smaller, the creep rate is faster than before; and because the cell size itself depends on stress, the creep is no longer Newtonian-viscous. Observations on metals show that the cell size, d_{cell}, is, very roughly, given by

$$d_{cell}/b_i \approx \mu/\sigma_s,$$

where b is the atom size, about $\Omega^{\frac{1}{3}}$. Inserting this into equation (5.1) gives the power-law creep equation:

$$\dot{\gamma} = A(D\mu b/kT)\,(\sigma_s/\mu)^n, \tag{6.1}$$

where $A \approx 21$ and $n = 3$, though we later treat them as constants to be determined by experiment.

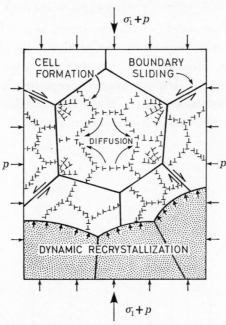

FIGURE 10. Power-law creep. The cells which form are the basis of the model described in the text, but the creep is complicated by grain-boundary sliding and by periodic recrystallization.

There are more sophisticated models for this power-law behaviour: viscous-glide creep (Weertman 1957), dislocation-climb creep (Weertman 1968), and Nabarro creep (Nabarro 1967) are examples; the characteristics of many of the models, including that given above, are discussed by Weertman (1970). All predict a steady-state creep equation of the form of equation (6.1), which varies with temperature roughly as a diffusion coefficient (usually that for lattice self-diffusion), and which depends on a power of the stress.

These models are, at best, a description of one sort of power-law creep. Their weakness lies in their inability to predict with any precision the constants A and (often) n, which must at present be regarded as empirical quantities to be determined experimentally. (The two are related; Stocker & Ashby (1973) find, in a survey of the creep of some 50 materials, that

$$n \approx 3 + 0.3\lg A, \tag{6.2}$$

a relation which may have usefulness in estimating values of n or A when one is known, but which should be used with caution.) There are other difficulties. As the stress is raised above about $10^{-3}\,\mu$, the power-law ceases to be a good description of experiments. Physically, this 'power-law breakdown' is caused by an increasing contribution of dislocation glide to the strain-rate; it is a broad transition from pure power-law behaviour (equation (6.1)) to the exponential stress-dependence of plasticity (equation (4.1)). At high temperatures a second complication arises: the material may recrystallize as it creeps (figure 10), the waves of recrystallization washing out the cells, and giving repeated surges of primary creep.

In spite of these difficulties, equation (6.1) describes well the creep of a wide variety of metals and ceramics, provided that n and A are treated as material properties, to be determined by experiment, and it suggests that pressure should affect creep mainly through its effect on D. Experiments, reviewed in the next section, support this view.

6.2. *Effect of pressure on power-law creep*

There have been a limited number of careful creep tests in which pressure has been used as a variable; they have been reviewed recently by McCormick & Ruoff (1970). Typical of them are the observations of Chevalier, McCormick & Ruoff (1967), who studied the creep of indium under a liquid pressure medium. Their observations are replotted in figure 11. When the pressure was switched between two fixed values, the creep rate changed sharply but reversibly, returning to its earlier value when the additional pressure was removed.

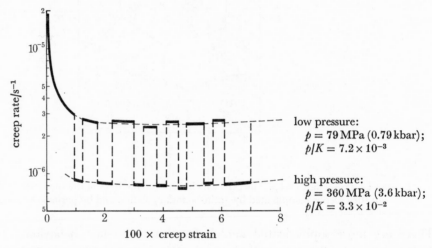

low pressure:
$p = 79\,\mathrm{MPa}\,(0.79\,\mathrm{kbar})$;
$p/K = 7.2 \times 10^{-3}$

high pressure:
$p = 360\,\mathrm{MPa}\,(3.6\,\mathrm{kbar})$;
$p/K = 3.3 \times 10^{-2}$

FIGURE 11. The creep of indium under a superimposed hydrostatic pressure. The creep rate drops when the pressure is applied. Data replotted from Chevalier, McCormick & Ruoff (1967): $T/T_\mathrm{M} = 0.87$; $\sigma/E = 1.6 \times 10^{-5}$.

If the creep is diffusion-controlled, then we might expect that the main influence of pressure would be through its effect on diffusion. This is characterized by the activation volume, so that a comparison of activation volumes derived from diffusion and creep should show them to be about equal, just as the activation energies are found to be. The available data are included in table 5, where it can be seen that, when a comparison is possible, the activation volumes for power-law creep are about the same as those for diffusion.

6.3. *Creep data for olivine*

Published creep data for natural and synthetic olivines span the temperature range from 900 to 1650 °C (0.55–$0.9\,T_\mathrm{M}$) at strain-rates from 10^{-7} to 10^{-4}/s, often under a confining pressure of about 10 kbar (1 GPa). The results are complicated by the fact that talc was sometimes used as a pressure medium: above 800 °C it releases water, and water accelerates the creep of these (and most other) silicates. However, other experiments avoided these problems, by the use of a dry gas to apply pressure when it was required. The data in table 1 are based on these.

All investigators (Carter & Ave' Lallemant 1970; Raleigh & Kirby 1970; Goetze & Brace 1972; Post & Griggs 1973; Kirby & Raleigh 1973; Kohlstedt & Goetze 1974; Durham & Goetze

1976; Durham, Goetze & Blake 1976) observed power-law creep with a power which increased from 3 at low stresses to 5 or more at high. Many of the data were recently re-analysed by Kohlstedt & Goetze (1974), who showed them to be adequately described by the equation

$$\dot{\gamma} = f(\sigma_s) \exp(-Q_{cr}/RT), \tag{6.3}$$

where Q_{cr}, the activation energy for creep, is 522 kJ/mol.

Following their approach, we have examined the function $f(\sigma_s)$ by normalizing all the data for dry olivine to 1400 °C, using an activation energy of 522 kJ/mol, as shown in figure 12. It shows power-law behaviour, with $n \approx 3$ at low stresses, followed by a long transition as stress is raised and glide contributes increasingly to the flow.

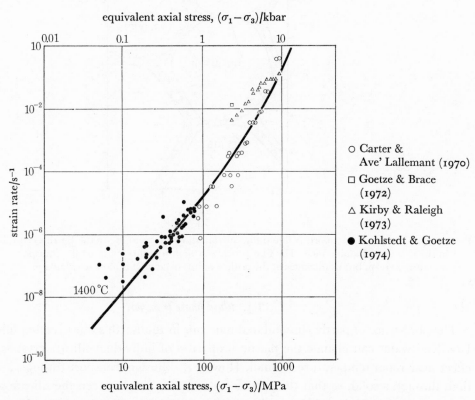

FIGURE 12. The creep of dry olivine. The data, normalized to a single temperature (1400 °C) by using an activation energy of 522 kJ/mol, are replotted from the work of Kohlstedt & Goetze (1974).

6.4. *Rate equation for power-law creep*

Because there is no convincing model for this transition, we have chosen to describe the creep of olivine by the sum of two power-laws:

$$\dot{\gamma} = \frac{\mu b}{kT} \left(A_1 \left(\frac{\sigma_s}{\mu}\right)^3 + A_2 \left(\frac{\sigma_s}{\mu}\right)^5 \right) \exp\left(-\frac{Q_{cr} + pV^*}{kT}\right). \tag{6.4}$$

The equation, evaluated by using the constants listed in table 1, is plotted onto figure 12 as a full line, for zero pressure. In the calculations of § 8, the value of V^* was varied between 0 and Ω_i.

7. OTHER MECHANISMS

The four classes of mechanism described in §§ 3–6 are well enough understood to be modelled in broad outline, but at least three important processes are omitted from the present treatment because we do not yet understand them in sufficient detail to model them properly, and because they are too poorly characterized experimentally to be included in a phenomenological way. They are discussed in this section.

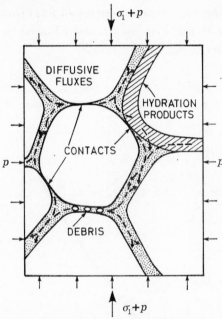

FIGURE 13. Fluid phase transport. A liquid film surrounding the granules: in the Earth's crust, it might be water; in the mantle, a basaltic melt. The film provides a high-mobility path for the transport of ions of the solid, permitting creep, but its thickness at the critical points of contact is hard to calculate.

7.1. *Fluid phase transport*

The pore-space of partly consolidated minerals in the Earth's crust is often filled with water. Dissolved water can change the plastic properties of individual silicate crystals, although the effect near room temperature is small. However, diffusive transport through a liquid is faster than through a solid, so that the presence of a water film between the silicate granules in the aggregate permits measurable creep by solution and re-deposition, even at ambient temperatures. Deep in the mantle, the temperature is such that water has been driven off, but basaltic phases of low melting point can provide a similar fast-diffusing liquid path (figure 13).

There have been a number of attempts to discuss and model the process (Durney 1972, 1976; Elliot 1973; Stocker & Ashby 1973; Rutter 1976). They are all based on the idea, illustrated by figure 13, that a deviatoric stress causes differences in pressure between points of contact between the granules, and that this in turn leads to a gradient in chemical potential of ions of the solid between points where the pressure is high and those where it is lower. If these ions first dissolve in the fluid and then diffuse down the gradient and re-deposit on the grains, the grains will change shape and the material as a whole will deform. The obvious parallel between this and Coble creep (§5) is apparent in all the models: the fluid film simply replaces the grain boundary as a high-diffusivity path.

It is in calculating the width of this path that the models are incomplete. A pressure difference can exist around the grain surface only if the liquid is at some point squeezed out, or at least reduced to a layer of little more than molecular thickness. It is only at these points that the chemical potential is significantly raised, yet it is here that the highly diffusing layer has been almost removed. The thickness of the layer and the solubility of the diffusing ions in it are crucial quantities in calculating the rate of deformation: without them, no meaningful estimate can be made.

The problem may have no easy answer. It appears to us that the most likely explanation for the observed rates of flow in the presence of water films is either that a solid hydration product forms on the grain surfaces which is porous and capable of permitting rapid ionic transport, or that porous debris – clay particles have been suggested by Weyl (1959) – lie in the inter-face and allow free liquid access. Both suggestions are illustrated in figure 13; they resolve the problem of liquid access to an interface under compression, but they do little to make the model quantitative.

7.2. *Dynamic recrystallization*

Above $0.6\,T_M$, minerals, like many metals, recrystallize as they creep. The new grain boundaries sweeping through the material remove the cells and tangles of dislocations, softening the material and allowing a new cycle of primary creep (figure 10). Metals, most of which have 5 easy slip systems, creep faster when they recrystallize, though often the enhancement is slight (it depends on the number of waves of recrystallization per unit creep strain). Typical observations on metals can be found in the publications of Hardwick, Sellars & Tegart (1961), Hardwick & Tegart (1961), Stüwe (1965) and Nicolls & McCormick (1970).

Many non-metals have fewer than five easy slip systems, even at high temperatures. Although we do not understand why, there is some evidence that dynamic recrystallization may have a much more important effect on creep in these materials than it does on metals. Ice, for instance, which has two easy slip systems (those in the basal plane) appears to recrystallize locally through-out creep (see, for example, Steinemann 1958), and may in this way relieve internal stresses which would otherwise lead to cracking.

7.3. *Influence of a texture or fabric*

Mechanisms of flow involving dislocation motion may be accelerated or decelerated by the presence of a texture (which, if sufficiently perfect, tends to make the material behave like a single crystal). The creep rate depends on the degree of perfection of the texture and on the angle between the principal stresses and the preferred directions in it. We shall assume here that the texture is not sufficiently perfect to alter the creep behaviour significantly.

8. Deformation-diagrams and flow in the upper mantle

There are, then, four broad classes of mechanism of flow: cataclasis, plasticity, diffusional flow and power-law creep. All four can be studied in the laboratory, and all must be involved in the deformation of the Earth's crust and mantle, but in a given test, or at a given point in the mantle, one mechanism will be dominant, accounting for most or all of the strain. This dominant mechanism could be identified by evaluating the rate equations and comparing the strain-rate each predicts, but it is far more convenient to use the equations to construct deformation mechanism diagrams which allow the dominant mechanism to be identified, show the overall rate of flow, and allow the depths, pressures or temperatures at which changes of mechanisms occur to be examined.

In laboratory tests the deviatoric stress, the temperature, and the pressure are all independent variables; then the kind of map described in § 8.1 is most useful. It can be regarded as characterizing the material (in this instance, olivine).

The temperature and pressure in the upper mantle, on the other hand, are related, though the nature of the relation beneath a continent may differ from that, say, beneath an ocean. Then more information is conveyed by the maps of the sort described in § 8.2.

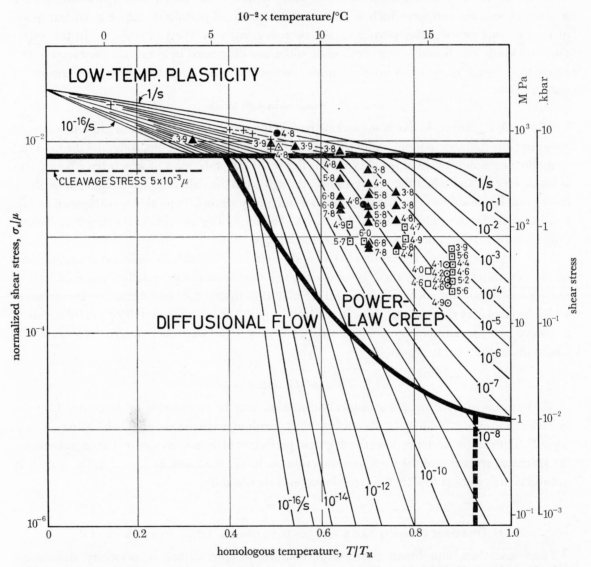

FIGURE 14. A map for olivine with a grain size of 0.1 mm; zero pressure. The symbols identify experimental points, and are labelled with the negative of the logarithm to the base 10 of the observed shear strain-rate. Cleavage was suppressed to show the plasticity field; if a realistic value for σ_t were used, the diagram would be truncated at the level marked 'cleavage stress $5 \times 10^{-3} \mu$'. Data: +, Evans (1976) hardness data; ▲, Carter & Ave' Lallemant (1970) (peridotite, 15 kbar); ●, Durham et al. (1976); △, Phakey et al. (1972) (single crystals, hard direction); ⊙, Kohlstedt & Goetze (1974); ⊡, Durham & Goetze (1976).

8.1. *Maps for olivine*

The first type of map is shown in figure 14. Its axes are the normalized shear stress, σ_s/μ^0 (where μ^0 is the shear modulus at atmospheric pressure and ambient temperature), and homologous temperature T/T_M (where T_M is the melting temperature). The normalization has the advantage of reducing maps for materials of the same crystal class and with similar bonding to a single group (Ashby & Frost 1975).

The construction of the maps involves two steps. We first ask: in what field of stress and temperature is a given mechanism dominant? The boundaries of the fields are obtained by equating pairs of the rate equations and solving for stress as a function of temperature. Figure 14, which describes polycrystalline olivine with a grain size of 0.1 mm, shows these fields; the mechanisms which meet at a field boundary have equal rates there. The figure was constructed from the rate equations listed in §§ 3–6 and the material constants for olivine given in table 1.

The second step is to calculate the net strain-rate at a given point on the diagram: it is the sum of the contributions from each mechanism, provided they operate independently. Cataclasis and diffusional flow are independent: at each point we add their contributions. However, power-law creep and low-temperature plasticity are not: they reflect alternative ways in which dislocations may move. In forming the sum, we include in it only the faster of these two. The superposition law implied by this procedure is obviously over-simplified, but the uncertainty in the rates of individual mechanisms makes more sophisticated superposition pointless. This procedure allows us to plot contours of constant shear strain-rate onto the diagram, from 10^{-16}/s to 1/s.

The stress axis describes either the simple shear stress or, when the stress field is more complicated, the deviatoric stress (equation (1.3)). When this and the temperature are known, the diagram gives the value of the shear strain-rate, or the equivalent shear strain-rate (equation (1.6)), and identifies the mechanism by which the material is deforming.

Figure 14 was constructed for flow under zero pressure, but with cleavage artificially suppressed (by making σ_f very large) to allow low-temperature plasticity to be shown. The map then shows three fields: plasticity, power-law creep and diffusional flow, this last subdivided by a broken line into Coble creep at lower temperatures and Nabarro–Herring creep at higher. If, instead, σ_f is set equal to $5 \times 10^{-3} \mu^0$ (the value arrived at in § 3), the diagram is truncated at the level of the horizontal line labelled 'cleavage stress'. Above this line, the material fractures when sheared under zero confining pressure.

Many of the useful data on the creep and plasticity of olivine are plotted on the diagram. The symbols identify the investigators, and are labelled with numbers which are the negative of the logarithm of the shear strain-rate ($\lg \dot{\gamma}$); these numbers allow the observed strain-rates to be compared directly with those computed from the rate equations. Because there is considerable scatter in the data, we found it best to construct the maps by an iterative procedure, adjusting the constants to give a map which, in our judgement, best fitted the data. Figure 14, and the constants of table 1, are the result.

The effect of a large pressure is shown in figure 15. It was constructed for a pressure of $0.1\,K^0$ (81 kbar), roughly the pressure at the base of the upper mantle. Cleavage was not artificially suppressed, but, because of the effect described in § 3, it appears only at stress levels which are inaccessible because plasticity intervenes. The rate of plasticity has itself been slowed by a factor of 10^6 or more because of the pressure effects described in § 4, but this is equivalent to an increase in flow stress of only a factor of about 1.5. Diffusional flow and power-law creep too, are affected, mainly because pressure slows diffusion on which both depend. Here the important parameter is the activation volume V^*, which for this figure was set equal to the volume occupied by a single oxygen ion, Ω_i. As explained in §5, the likely range for V^*/Ω_i is from zero (represented by figure 14) to 1 (figure 15). A comparison of the two figures shows a reduction of about 10^3 in the rate of creep.

The grain size most affects the rate of diffusional flow (§5): increasing the grain size slows diffusional flow and moves the boundary separating it from power-law creep to lower stresses. The grain size in the upper mantle has been discussed elsewhere (Stocker & Ashby 1973); it is unlikely to be less than 0.1 mm, and because of this the size of the power-law creep field is unlikely to be smaller than that shown on these and later figures.

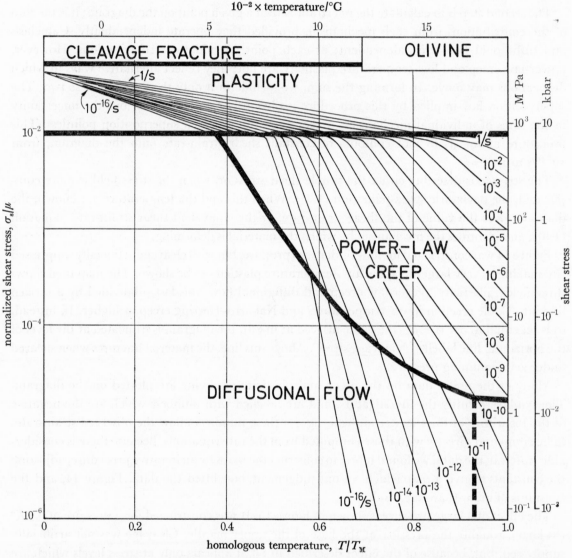

FIGURE 15. A map for olivine, based on the same data as figure 14, but with a pressure of $0.1\,K^0$ (81 kbar) applied. The cleavage stress has been raised by a factor of about 10 to $5 \times 10^{-3}\,\mu$; the stress required for plasticity has increased by a factor of about 1.5 and the creep rates have decreased by a factor of about 10^3.

8.2. *Maps for the upper mantle*

Within the upper mantle the pressure and the temperature increase with depth, the first in a roughly linear way, the second in a way which reflects the steady flow of heat from the interior to the surface. Since p and T are no longer independent, a single diagram can completely describe the mechanical behaviour. We have used depth as the independent variable, relating pressure and temperature to it by

$$p/K^0 = 7.9 \times 10^{-7} + 3.2 \times 10^4 \, \Delta/K^0 \qquad (8.1)$$

and

$$T = 300 + 1579[1 - \exp(-7.6 \times 10^{-6} \Delta)], \qquad (8.2)$$

where Δ is the depth in metres. The first equation identifies the pressure as that due to the atmosphere plus a height Δ of rock of average density $3.25 \times 10^3 \, \mathrm{kg/m^3}$. The second ensures that the temperature is equal to $300\,\mathrm{K}$ at the Earth's surface, and increases with depth with an initial gradient of $12\,\mathrm{K/km}$, reaching the value $1850\,\mathrm{K}$ at a depth of $500\,\mathrm{km}$; it is shown in figure 16. Though reasonable, no claim is made that these equations accurately describe the pressure and temperature of a certain part of the mantle: they are meant simply to illustrate the method and the conclusions which can be drawn from it.

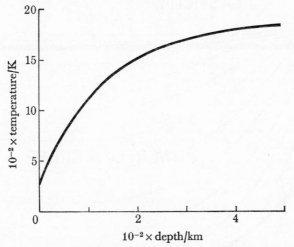

FIGURE 16. The assumed way in which temperature varies with depth in the upper mantle (see equation 8.2).

A map with depth as one axis is shown in figure 17. It describes the flow of olivine with a grain size of $0.1\,\mathrm{mm}$, under conditions defined by equations (8.1) and (8.2). In constructing this figure the activation volume V^* was set equal to zero (the lower bound); it is to be compared with the next figure (figure 18) for which an activation volume equal to Ω_1 (the upper bound) was used.

The two figures illustrate the following points. Near the surface, cleavage is the dominant mode: the mantle, when sheared, will deform by cataclasis rather than plastic flow, to the depth of at least $20\,\mathrm{km}$, even at the slowest strain-rates. Below this, the rising pressure and temperature combine to cause plasticity to replace cataclasis as the dominant flow mechanism. Below $100\,\mathrm{km}$, power-law creep becomes important, and remains so to a depth of $400\,\mathrm{km}$, where olivine transforms to a spinel-structured phase about which little is known. For this rather small grain size, diffusional flow is an important mechanism at low stresses, but if the grain size is increased to $1\,\mathrm{mm}$, it disappears from the diagrams entirely.

The effects of the larger activation volume can be seen by comparing the two figures. The second, with $V^* = \Omega_i$, shows creep rates which are slower by a factor of between a hundred and a thousand than those of the first, and its strain-rate contours loop upwards slightly, because the rising pressure tends to offset the effect of rising temperature; but perhaps the most important point is that for this range of V^* the rate contours are very flat below 200 km, meaning that the viscosity of the upper mantle here is roughly independent of depth. This is in contrast to the results of earlier calculations (Stocker & Ashby 1973) which showed a pronounced viscosity minimum because they used values of V^*/Ω_i of between 3 and 7, appropriate if SiO_4^{4-} were the diffusing unit, but not if O^{2-} is the rate-controlling species (as we now believe).

The depth at which cataclasis stops and plasticity starts is illustrated by the expanded segment of figure 18 which is shown as figure 19. It illustrates first that the depth at which fracturing is

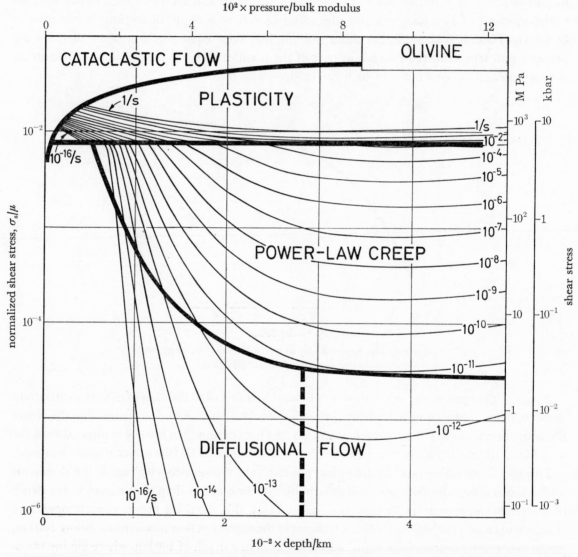

FIGURE 17. Olivine under upper-mantle conditions with $V^* = 0$; cleavage stress = $5 \times 10^{-3}\mu$. In this figure, depth (below the Earth's surface) has replaced temperature as the abscissa; temperature and pressure are both related directly to it. The figure shows that, to a depth of about 30 km, cataclastic flow is the dominant mechanism. Below this, plasticity and creep replace fracturing as the dominant mechanisms.

replaced by plasticity depends on the local strain-rate imposed on the mantle; for our standard conditions (full lines on the figure) it is between 20 and 45 km. If we consider a standard strain-rate – say 10^{-6}/s – then the standard fracture stress of $5 \times 10^{-3}\,\mu^0$ leads to a critical depth of 32 km, and is marked by an arrow on the figure. However, this depends on the pressure and temperature distribution (equations (8.1) and (8.2)) and on the fracture stress σ_f. If the fracture stress is increased to $10^{-2}\,\mu^0$ or reduced to $2 \times 10^{-3}\,\mu^0$, (both are marked on the figure), the depth varies from 20 km to almost 50 km (arrows). It is even more sensitive to the temperature profile (equation (8.2)). The one we have used may be broadly typical of much of the upper mantle, but where, for instance, a plate is subducted and the descending matter cools its surroundings, the temperature profile is depressed and the critical depth can be displaced to below 100 km.

It is tempting to associate the critical depth with the maximum depth at which catastrophic shearing can take place, and it is certainly true that some discontinuity in behaviour must be

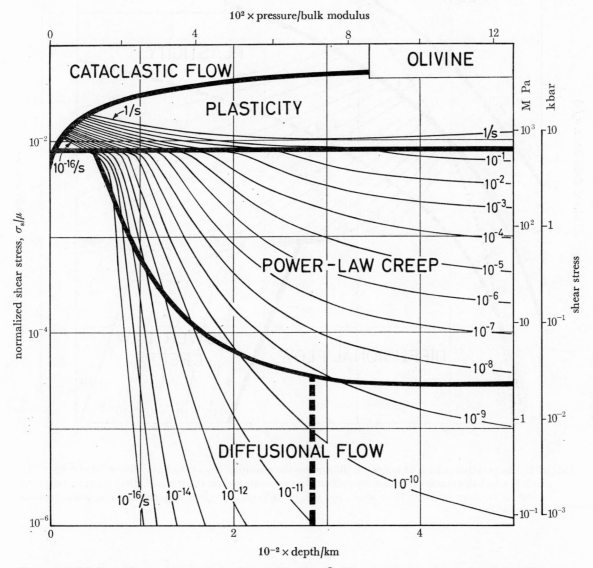

FIGURE 18. Olivine under upper-mantle conditions with $V^* = \Omega_i$. The creep rates are slower by a factor of between 10^2 and 10^3 than those of figure 17. Note that the strain-rate contours are almost flat below 200 km.

expected at this depth. However, as the olivine maps shown in figures 14 and 15 illustrate, a small increase in temperature in the plasticity régime brings with it a very large increase in strain rate: so adiabatic plastic shear, too, could lead to sudden flow.

Figure 19 also illustrates how the rolling criterion of §3 appears on the diagram. The shear stress required for rolling depends more rapidly on pressure than does that for repeated cleavage so that, while rolling is important near the surface, it does not (except for the largest values of σ_f) determine the critical depth.

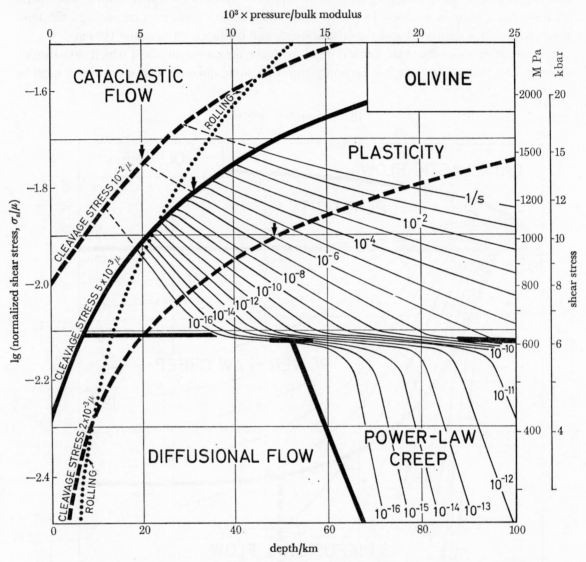

FIGURE 19. An enlarged portion of figure 18, illustrating the transition from cataclastic flow to plastic flow. The depth at which this transition occurs depends on the strain-rate and on the fracture stress, σ_f. It also depends on the temperature profile in the mantle, and will be different in regions where descending plates cool the mantle locally.

9. SUMMARY AND CONCLUSIONS

A number of alternative mechanisms exist which permit flow in minerals and rocks confined by a hydrostatic pressure. Some are well enough understood that they can be modelled approximately: repeated cleavage (cataclasis), low-temperature plasticity, diffusional flow, and power-law creep are examples. Others are less well understood though they are no less important: deformation in the presence of a fluid phase, for instance; and the effects of a texture, or of dynamic recrystallization, on creep. Models for the better-understood of these mechanisms are reviewed.

The mechanical behaviour of olivine is reviewed. Recent data are analysed in the framework of the models, and methods are discussed for inferring data when none exist. A set of material properties for olivine are deduced and listed as table 1.

The models and data are used to construct deformation diagrams of two types, shows as figures 14 and 15, and as figures 17–19. The first can be regarded as characterizing the material, the second as describing its behaviour under the conditions postulated to exist in the mantle.

The study shows that olivine is now a fairly well characterized material (among oxides) and that the maps give a tolerably good description of its behaviour.

When the data and models are used to predict mantle behaviour, a number of conclusions emerge:

(a) Cataclasis should be the dominant mode of flow to a depth of perhaps 35 km, though this depth varies with the assumed temperature profile in the mantle, with the strain-rate, and with material properties, particularly the fracture stress σ_f.

(b) Below this is a region of plasticity. Because of the rapid dependence of strain-rate on temperature in this region, and the large values of stress, adiabatic instabilities are more likely here than at greater depths.

(c) At greater depths, the dominant flow mechanism is power-law creep, with a contribution from diffusional flow which becomes less as the activation volume is increased, and as the grain-size is increased. Diffusional flow cannot be ruled out as the dominant mechanism in mantle convection, though the most plausible values of $V^*/\Omega_i = 1$ and $d > 0.1$ mm make it seem unlikely.

(d) The data used here suggest that, at least below 200 km, the upper mantle viscosity is almost constant with depth, a result which may have implications for the scale of mantle convection.

This work was carried out at the University of Cambridge, and at Harvard University during 1976. One of us (M. F. A.) wishes to thank the Division of Engineering and Applied Physics at Harvard for their hospitality, help and stimulus during this period, and the other (R. A. V.) wishes to thank the Science Research Council for support on contract number B/RG 8058.8. Throughout the course of the work, we had numerous discussions with Professor R. O'Connell of the Hoffmann Laboratory at Harvard, and we wish to thank him, Professor C. Goetze, Dr B. Atkinson and Dr E. Rutter for their helpful and critical comments.

REFERENCES (Ashby & Verrall)

Ashby, M. F. & Frost, H. J. 1975 *Constitutive equations in plasticity* (ed. A. Argon), p. 117. M.I.T. Press.

Ashby, M. F. & Verrall, R. A. 1973 *Acta Met.* **21**, 149.

Barnard, R. S. 1975 Ph.D. thesis, Case-Western Reserve University, Department of Materials Science and Engineering.

Birch, F. 1966 *Handbook of physical constants*. The Geological Society of America Memoir 97, section 7.

Birch, F. 1969 *The Earth's crust and upper mantle*, Geophys. Monogr. Ser. (ed. P. J. Hart), vol. 13, p. 18. Washington D.C. A.G.U.

Borchardt, G. & Schmaltzried, H. 1972 *Ber. dt. keram. Ges.* **49**, 5.

Carter, N. L. & Ave' Lallemant, H. G. 1970 *Bull. geol. Soc. Am.* **81**, 2181.

Chevalier, G. T., McCormick, P. & Ruoff, A. L. 1967 *J. appl. Phys.* **38**, 3697.

Durham, W. B. & Goetze, C. 1976 *J. geophys. Res.* (In the press.)

Durham, W. B., Goetze, C. & Blake, B. 1976 *J. geophys. Res.* (In the press.)

Durney, D. W. 1972 *Nature, Lond.* **235**, 315.

Durney, D. W. 1976 *Phil. Trans. R. Soc. Lond.* A **283**, 229.

Elliott, D. 1973 *Bull. geol. Soc. Am.* **84**, 2645.

Friedel, J. 1964 *Dislocations*, p. 25. Oxford: Pergamon Press.

Frost, H. J. & Ashby, M. F. 1973 *A second report on deformation mechanism maps*. Harvard University Report.

Girifalco, L. A. 1964 In *Metallurgy at high pressures and high temperatures* (eds K. A. Gachneider, M. T. Hepworth & N. A. D. Parlee), p. 260. Gordon and Breach.

Goetze, C. 1971 *J. geophys. Res.* **76**, 1223.

Goetze, C. & Brace, W. F. 1972 *Tectonophysics* **13**, 583.

Goetze, C. & Kohlstedt, D. L. 1973 *J. geophys. Res.* **78**, 5961.

Goldstein, J. I., Hanneman, R. E. & Ogilvie, R. E. 1965 *Trans. A.I.M.E.* **233**, 813.

Gordon, R. B. 1965 *J. geophys. Res.* **70**, 2413.

Gordon, R. B. 1967 *Geophys J.* **14**, 33.

Graham, E. K. & Barsch, G. R. 1969 *J. geophys. Res.* **64**, 5949.

Griffith, A. A. 1924 *Proc. 1st Int. Conf. Appl. Mech., Delft*, p. 55.

Griggs, D. T., Turner, F. J. & Heard, H. 1960 *Geol. Soc. Amer. Mem.* **79**, 39.

Handin, J. 1966 *Handbook of physical constants* (ed. S. P. Clark), revised edn. The Geological Society of America Memoir 97.

Hardwick, D., Sellars, C. M. & Tegart, W. J. McG. 1961 *J. Inst. Met.* **90**, 21.

Hardwick, D. & Tegart, W. J. McG. 1961 *J. Inst. Met.* **90**, 17.

Herring, C. 1950 *J. appl. Phys.* **21**, 437.

Huntington, H. B. 1958 *Solid State Physics* **7**, 213.

Jaeger, J. C. & Cook, N. G. W. 1969 *Fundamentals of rock mechanics*, p. 178. Chapman and Hall.

Kamazama, M. & Anderson, O. L. 1969 *J. geophys. Res.* **74**, 5961.

Kelly, A. 1966 *Strong solids*, ch. 1. Oxford: Clarendon Press.

Keyes, R. W. 1963 *Solids under pressure* (eds W. Paul & D. M. Warschauer), p. 71. New York: McGraw-Hill.

Kirby, S. H. & Raleigh, C. B. 1973 *J. geophys. Res.* **78**, 5961.

Kocks, U. F., Argon, A. S. & Ashby, M. F. 1976 *Prog. Mat. Sci.* **19**.

Kohlstedt, D. L. & Goetze, C. 1974 *J. geophys. Res.* **79**, 2045.

Lazarus, D. 1971 Diffusion in solids. In *Encyclopedia of science and technology*, p. 158. New York: McGraw-Hill.

Lazarus, D. & Nachtrieb, N. H. 1963 In *Solids under pressure* (eds W. Paul and D. M. Warschauer), p. 43. New York: McGraw-Hill.

Lifshitz, I. M. 1963 *Sov. Phys. JETP* **17**, 909.

Lomer, W. W. 1957 *Phil. Mag.* **2**, 1053.

Macmillan, N. H. 1972 *J. Mat. Sci.* **7**, 239.

McClintock, F. A. & Argon, A. S. 1966 *Mechanical behaviour of materials*, p. 490. Addison Wesley.

McClintock, F. A. & Walsh, J. B. 1962 *Proc. fourth U.S. Nat. Conf. on Applied Mechanics*, p. 1015.

McCormick, P. G. & Ruoff, A. L. 1970 *Mechanical behaviour of materials under pressure* (ed. H. Le D. Pugh). New York: Elsevier.

McKenzie, D. P. 1968 In *The history of the Earth's crust* (ed. R. A. Phinney), p. 28. Princeton. N.I.: Princeton University Press.

Nabarro, F. R. N. 1948 In *Strength of solids*, p. 75. London: The Physical Society.

Nabarro, F. R. N. 1967 *Phil. Mag.* **16**, 231.

Nicholls, J. H. & McCormick, P. G. 1970 *Met. Trans.* **1**, 3469.

Orowan, E. 1934 *Z. Kristallographie* **89**, 327.

Orowan, E. 1949 *Rep. Prog. Phys.* **12**, 185.

Paterson, M. S. 1974 *Proc. 3rd Int. Congress Soc. Rock Mech.*, Denver, Sept. 1974, **1**, 521.

Peterson, N. L. 1968 *Solid State Physics* **22**, 409.

Phakey, P., Dollinger, G. & Christie, J. 1972 *Am. geophys. Un., Geophys. Monograph Series* **16**, 117.
Post, R. L. & Griggs, D. T. 1973 *Science, N.Y.* **181**, 1242.
Raj, R. & Ashby, M. F. 1971 *Trans. Met. Soc. A.I.M.E.* **2**, 1113.
Raleigh, C. B. 1968 *J. geophys. Res.* **73**, 5391.
Raleigh, C. B. & Kirby, S. H. 1970 *Mineral. Soc. Am. Spec. Paper* **3**, 113.
Ringwood, A. E. 1970 *Phys. Earth Planet, Interiors* **3**, 109.
Ruoff, A. L. 1965 *J. appl. Phys.* **37**, 2903.
Rutter, E. H. 1976 *Phil. Trans. R. Soc. Lond.* A **283**, 203.
Seeger, A. 1955 *Phil. Mag.* **46**, 1194.
Shewmon, P. G. 1963 *Diffusion in solids*, p. 52. New York: McGraw-Hill.
Spitzig, W. A., Sober, R. J. & Richmond, O. 1975 *Acta Met.* **23**, 885.
Spitzig, W. A., Sober, R. J. & Richmond, O. 1976 *Trans Met. Soc. A.I.M.E.* **7A**, 1703.
Steinemann, von S. 1958 *Experimental study of the plasticity of ice*. Beiträge zur geologischen karte de Schweiz, geotechnische Series Hydrologie, no. 10.
Stocker, R. A. & Ashby, M. F. 1973 *Reviews of Geophysics and Space Physics* **11**, 391.
Stüwe, H. P. 1965 *Acta Met.* **13**, 1337.
Weertman, J. 1957 *J. appl. Phys.* **28**, 1185.
Weertman, J. 1968 *A.S.M. Trans. Quart.* **61**, 681.
Weertman, J. 1970 *Rev. Geophys. Space Phys.* **8**, 145.
Weyl, P. K. 1959 *J. geophys. Res.* **64**, 2001.
Young, C. 1969 *Am. J. Sci.* **267**, 841.

Discussion

H. H. Schloessin (*Department of Geophysics, University of Western Ontario, London, Ontario, Canada*). With regard to the deformation of bodies whose grains are in contact with a liquid phase, it would seem that Gibbs's theory of heterogeneous equilibrium of solids under all states of stress and strain in contact with their solution or melt can provide a suitable basis for its determination (J. W. Gibbs, *Collected Works*, vol. 1, ch. 3, 1928). Similar to the case of Nabarro–Herring creep the deformation will depend on diffusive transport of matter from the stressed faces to the unstressed (free) faces. However, in this particular case the transport will take place through the liquid phase and its rate will be determined by the difference in chemical activity (solubility) of the solid between stressed and free faces. The variation of steady state creep rate with stress should be proportional to the change in activity. This, I suppose, leads to a hyperbolic creep law.

E. H. Rutter (*Geology Department, Imperial College, London SW* 7). During his particularly lucid review of the micromechanisms of flow and fracture, Professor Ashby referred to creep by diffusive mass transfer via an intergranular fluid phase. The microstructures which result from such diffusive transfer in low-grade metamorphic rocks are well known to geologists and are described as being due to pressure solution (see, for example, Heald 1956; Ramsay 1967). Though he mentioned the importance of diffusion path width in determining the kinetics of this process, I felt that special attention should have been given to the question of the diffusivity in the supposed intergranular fluid film. While the effective path width can probably be inferred to within one order of magnitude from microstructural observations, the greatest uncertainty in estimating the rate of rock deformation by this process must be due to uncertainty regarding the magnitude of the diffusivity (Rutter 1976). There are no relevant experimental data to assist here, and it would be wrong to use diffusivities measured for salts dissolved in large volumes o⸱ fluid. If we are reduced to guessing, errors of several orders of magnitude will arise.

Returning to the subject of the diffusion path width, it is relevant to point out that pressure solution in rocks is often characterized by the development of transgranular planar zones (stylolites), extending over many grain diameters and spaced by one or more grain diameters.

Within these zones the less 'mobile' phases, usually the phyllosilicates, become progressively concentrated. The fact that stylolite zones once nucleated are stable, suggests that very fine grained phyllosilicates provide a region of enhanced diffusivity, in part by increasing the effective path width and in part by their ability to adsorb pore water onto their surfaces (see, for example, Heald 1956).

References

Heald, M. T. 1956 Cementation of Simpson and St Peter sandstones in parts of Oklahoma, Arkansas and Missouri. *J. Geol.* **64**, 16–30.
Ramsay, J. G. 1967 *Folding and fracturing of rocks.* New York: McGraw-Hill.
Rutter, E. H. 1976 The kinetics of rock deformation by pressure solution. *Phil. Trans. R. Soc. Lond.* A **283**, 203–219.

K. H. G. ASHBEE (*H. H. Wills Physics Department, University of Bristol, Bristol* 8). Transmission electron microscopy studies of deformed quartz, quartzites and other crystalline silicates demonstrate that creep at moderate temperatures (600–1000 K) and pressures (1–100 atm) is a consequence of dislocation glide mechanisms. If olivine is representative of silicates in general, your deformation map suggests that, in 0.1 mm grain size material, creep under the above conditions should be by diffusion mechanisms. Would increase of grain size (to single crystal material in the limit) extend the plasticity field to the régime for which laboratory data exist?

S. H. WHITE (*Department of Geology, Royal School of Mines, Imperial College, London SW7 2BP*). Professor Ashby has made a significant contribution to olivine deformation studies by considering cataclastic flow. It is noted that this field is positioned mainly in a crustal environment. Leaving aside the question of olivine stability during other than retrograde crustal deformations, the extent of the cataclastic field could perhaps be more accurately determined if the effects of pore fluid pressure were considered. The pore fluid pressure in the crust is normally considered to equal the geostatic pressure (Turner 1968; Price 1975) and it has a pronounced effect on crustal deformation and fracturing processes (Price 1975).

Carbon dioxide may exist in large quantities in the mantle (Roeder 1965; Green 1972) and may be the important fluid phase during the deep crustal and upper mantle deformation of olivine. Bubbles containing CO_2 are present around the grain boundaries of mantle derived polycrystalline olivine rocks and they will influence grain-boundary deformation processes within the mantle and fracture processes within the crust.

References

Green, H. W. 1972 A CO_2-charged asthenosphere. *Nature Phys. Sci.* **238**, 847–852.
Price, N. J. 1975 Fluids in the crust of the earth. *Sci. Prog., Oxf.* **62**, 59–87.
Roeder, E. 1965 Liquid CO_2 inclusions in olivine bearing nodules and phenocrysts from basalts. *Am. Mineral.* **50**, 1746–1782.
Turner, F. J. 1968 *Metamorphic petrology.* New York: McGraw-Hill.

S. A. F. MURRELL (*Department of Geology, University College London, Gower Street, London WC1E 6BT*). Pore pressures and dilatancy during the fracturing process play very important rôles in the deformation of rocks. High fluid pressures may exist down to considerable depths in the Earth, and by the operation of the effective stress principle (Murrell 1964, 1966; Murrell & Digby 1970), may allow fracture and cataclasis to take place at low shear stresses. On the other hand dilatancy caused by the opening of micro-cracks in rock during the fracture process, even under high confining pressures (Brace, Paulding & Scholz 1966), may when pore fluids are present under 'undrained' conditions result in dilatancy hardening, the elimination of macroscopic faulting accompanied by a stress-drop, and the development of more homogeneous processes of cataclastic deformation (Ismail & Murrell 1976). Dilatancy is also likely to play a rôle in the movement of pore fluids and therefore in the dynamics of faulting processes in the crust of the Earth (Nur, Bell & Talwani 1973; Nur & Schulz 1973).

References

Brace, W. F., Paulding, B. W. & Scholz, C. H. 1966 Dilatancy in the fracture of crystalline rocks. *J. geophys. Res.* **71**, 3939–3953.

Ismail, I. A. H. & Murrell, S. A. F. 1976 Dilatancy and the strength of rocks containing pore water under undrained conditions. *Geophys. J. R. astr. Soc.* **44**, 107–134.

Murrell, S. A. F. 1964 The theory of the propagation of elliptical Griffith cracks under various conditions of plane strain or plane stress. *Br. J. appl. Phys.* **15**, 1211–1223.

Murrell, S. A. F. 1966 The effect of triaxial stress systems on the strength of rocks at atmospheric temperatures. *Geophys. J. R. astr. Soc.* **10**, 231–281.

Murrell, S. A. F. & Digby, P. J. 1970 The theory of brittle fracture initiation under triaxial stress conditions. *Geophys. J. R. astr. Soc.* **19**, 499–512.

Nur, A., Bell, M. L. & Talwani, P. 1973 Fluid flow and faulting. 1. *Proc. conf. on tectonic problems of the San Andreas fault system* (eds R. L. Kovach & A. Nur), Geol. Sci. Series, vol. 13, pp. 391–404. California: Stanford University.

Nur, A. & Schulz, P. 1973 Fluid flow and faulting, 2. *Proc. conf. on tectonic problems of the San Andreas fault system* (eds R. L. Kovach & A. Nur), Geol. Sci. Series, vol. 13, pp. 405–416. California: Stanford University.

A. KELLY, F.R.S. (*Vice-Chancellor's Office, University of Surrey, Guildford GU2 5XH*). The question was raised in discussion: what is the limit to cataclasis, i.e. how far does the breakdown of the rock mass by repeated fracturing into smaller pieces proceed? One possible answer is suggested by experience which is described principally if not entirely in the patent literature, e.g. U.K. Patents 137,3214 (1971) and 146,4243 (1973). Mixtures of hard spheres of a wide variety of stony materials, such as chalk, glass beads, aggregate (flint) or china clay, show very high packing densities and great *ease of mixing* provided that populations containing three or four discrete sizes of particle are present in rather specific volume fractions. Such mixtures attain packing fractions of above 80 % and are easily sheared. I interpret this as indicating very small dilatancy hardening of the Reynolds type (see, for example, F. C. Frank (1966), *Rev. Geophys.* **4**, 405) and if cataclasis proceeded until such an array of particles were produced, the array would shear easily, perhaps with rather little further breakdown of particle size occurring. For the relative volume fractions of such arrays showing high packing fractions and easy mixing, the patent literature must be explored. Particles of the following approximate diameter ratios 1:3:7:14 at volume fractions of 40 %, 10 %, 10 %, 40 % show the property I have referred to. J. Ritter (1971) has described the size ratios (*App. Polym. Symp.* **15**, 239, New York: Wiley) as corresponding to those of baseballs, golf balls, acorns and sand.

Phil. Trans. R. Soc. Lond. A. **288**, 97 (1978)   [97]
Printed in Great Britain

Introductory remarks at the afternoon session, 24 February 1977

By Lord Todd, P.R.S.

In taking the Chair at this session of the discussion, I have to do two things in addition to thanking the organizers for inviting me to do so. The first is to apologize for the fact that a variety of other duties in the Society has prevented me from doing more than looking in on some of the papers.

The second is to attempt some kind of defence for being here at all. Actually, when Dr Kelly asked me to take part, it was with some regret that I realized I was unlikely to spend a sufficient amount of time in the meetings really to be a participant. This is because I actually have some remote and rather amateurish interest in some of the things you are discussing.

I am, of course, a chemist, but my subsidiary subjects both in this country and in Germany were geology, mineralogy and metallurgy. Later, I was for a year acting Head of the Metallurgy Department at Cambridge although I doubt whether I made any notable impact on that science during this period. So my first close contact with the parent disciplines of most of you here was between thirty-five and forty years ago. And during the long period since then, I have been aware of some gradual but interesting changes. First metallurgy began to be infiltrated by the expression 'solid state' and then gradually to be almost replaced by the designation 'materials science'. Simultaneously, geology became less descriptive and more physical with the rise of geophysics and the increasing number of its adherents.

Now it has seemed, even to me, that in phenomena like creep, plastic deformation and fatigue the geologist and the metallurgist or materials scientist were dealing with the same problems even if the scenario was different. For this reason I think the present discussion meeting at which both groups can get together is an excellent idea, and I congratulate the organizers who have brought it about and am very pleased that it should be held under the aegis of The Royal Society – a body which throughout its long history has striven to promote 'natural knowledge' without regard to artificial disciplinary barriers.

Introductory remarks at the afternoon session, 21 February 1977

By Dean Cook, F.R.S.

Phil. Trans. R. Soc. Lond. A. **288**, 99–119 (1978)   [99]

Printed in Great Britain

The mechanisms of creep in olivine

By C. Goetze

*Department of Earth and Planetary Sciences, Massachusetts Institute of
Technology, Cambridge, Massachusetts 02139, U.S.A.*

We summarize the progress made in providing experimental verification for the deformation map of polycrystalline olivine published by Stocker & Ashby in 1973 (*Rev. Geophys.* **11**, 391). Porosity-free polycrystalline deformation data, applicable to the mantle, were found to be obtainable only from high-pressure deformation studies. Combination of the results of such studies with hardness measurements and single crystal deformation studies on olivine provides narrow constraints on the flow of olivine resulting from dislocation mechanisms from room temperature to the melting point along a band of experimentally accessible strain rates. A good fit is obtained combining a Dorn law above 2 kbar differential stress,

$$\dot{\epsilon}/s^{-1} = 5.7 \times 10^{11} \exp \left\{ - \frac{128\,\text{kcal/mol}}{RT} \left(1 - \frac{\sigma_1 - \sigma_3}{85\,000} \right)^2 \right\},$$

with a power law below 2 kbar,

$$\dot{\epsilon} = 70(\sigma_1 - \sigma_3) \exp \{ - 122(\text{kcal/mol})/RT \},$$

where stress is measured in bars (1 bar = 10^5 Pa). Indirect data on a mechanism phenomenologically resembling the Coble creep régime are now available from two sources. The observed strain rates are only slightly faster than those predicted by Stocker & Ashby (1973). The 'wet' data, previously believed to show hydrolytic weakening, are found to fall within this Coble field. The asthenosphere is still expected to deform by the dislocation mechanism summarized by the two formulae given above, but higher stress deformation within the lithosphere is almost certainly dominated by this Coble creep régime once dynamic recrystallization sets in.

Introduction

The creep properties of dunite have been the subject of a considerable literature in the geological sciences since the general acceptance of the theory of plate tectonics. Those not familiar with the setting of this creep problem are referred to figure 1, where the principal terms are defined and orders of magnitude of relevant parameters given.

Localized deformation occurs within the lithosphere; however, this region is generally accepted to move as a number of discrete 'plates' over the surface of the Earth at velocities of a few centimetres per year. The corresponding convective motions are continuous and widespread in the asthenosphere to unknown depths. The orthorhombic mineral olivine of the approximate composition $(Mg_{0.91}Fe_{0.09}) SiO_4$ is the dominant phase only to about 375 km depth, below which the spinel phase of the same composition, of unknown creep properties, dominates the creep process. Clearly, a study of the creep of olivine, even if complete, leaves the mechanical properties of the deeper mantle unknown. Nevertheless, a large number of more local problems involving the formation of new plates, the collision of plates, the decoupling of plates from the asthenosphere, and the response of plates to vertical loads, to name a few, can be effectively studied without such knowledge.

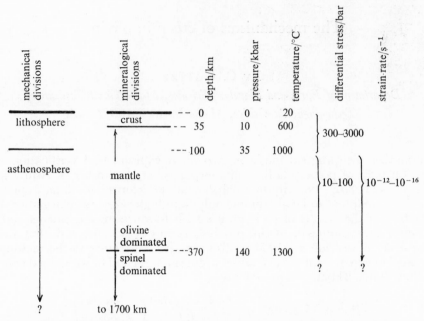

FIGURE 1. Sketch of the principal subdivisions of the Earth in the first 400 km.

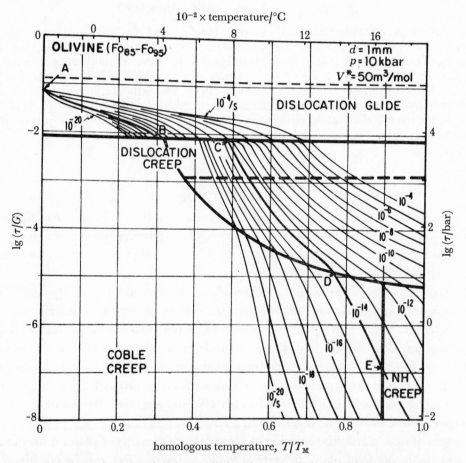

FIGURE 2. Deformation map of olivine predicting the strength developed at each temperature as a function of strain rate. Reprinted from Stocker & Ashby (1973).

Even this upper portion of the mantle is not composed exclusively of olivine but contains 20–50 % pyroxenes, aluminous spinel, garnets, etc. Work has been done on the creep properties of some pyroxene and spinel compositions; however, the creep strengths are similar to those of olivine, and because of their smaller volumetric abundance they are expected to affect the creep properties of the mantle only to a minor extent. The approximation is therefore made that creep of the mantle in the depth range 35–375 km is the creep of olivine.

TABLE 1. APPROXIMATE COMPOSITION, BY MASS,
OF MANTLE-DERIVED OLIVINE

	alpine	xenolith
SiO_2	40.96	40.30
TiO_2	0.01	0.15
Al_2O_3	0.21	0.25
Fe_2O_3	0.00	0.00
FeO	7.86	10.26
MnO	0.13	0.09
MgO	50.45	48.60
CaO	0.15	0.07
Na_2O	0.01	0.04
K_2O	0.00	0.03
H_2O^+	0.29	0.33
	100.35	100.56

The alpine sample, from Dun Mountain, New Zealand, is believed to have reached the Earth's surface from the mantle by tectonic processes. The xenolith sample is obtained from nuggets entrained by a basalt flow originating within the mantle below Ichinomegata, Japan. Data are from Deer, Howie & Zussman (1964).

In 1973 Stocker & Ashby presented a 'deformation map' predicting the flow of polycrystalline olivine as a function of temperature, strain rate, and differential stress. It is reproduced in figure 2. A small number of data were then available near the boundary of the fields labelled 'dislocation glide' and 'dislocation creep'. The remainder of the map was based on systematics (rules of thumb obtained from a broad range of materials). We summarize here the progress made in a number of laboratories around the world to provide experimental verification for this plot. Let me say in advance that, on the whole, only comparatively minor changes have resulted from this work.

The tools employed have included solid medium high pressure deformation apparatus used to 1300 °C, and 15 kbar (Post 1973; Carter & Ave' Lallemant 1970; Raleigh & Kirby 1970; Blacic 1972) gas medium high pressure apparatus used to 1000 °C and 6 kbar (Goetze & Brace 1972), a 1 atm single crystal deformation apparatus used to 1700 °C (Kohlstedt & Goetze 1974; Durham & Goetze 1977 a, b), hot hardness apparatus used to 1500 °C (Evans & Goetze 1977), as well as, of course, optical and electron microscopes for the analysis of both experimentally and naturally deformed specimens. Starting materials have been natural rocks of almost pure olivine composition (dunite) and single crystals of gem quality olivine from Arizona and Egypt. Only recently have artificially grown single crystals of adequate size been available, and these only of the magnesian end member. A comparison creep study showed that these differed only to a very minor extent from the natural material, differences which could be ascribed to their difference in iron content (Durham & Goetze 1977 b). The purity of a typical olivine crystal derived from the mantle can be judged from the compositions given in table 1.

Effect of pressure

The principal purpose for studying the deformation of olivine and olivine-rich aggregates has been to obtain data relevant to the interpretation of motions occurring within the mantle at pressures ranging from 20–130 kbar. Since practically all experiments have been conducted at lower pressures than this, there is a need to extrapolate the results into this range.

It is a common observation that brittle materials such as rocks show a strong pressure sensitivity when deformed in a range of pressure, temperature, stress and strain rate for which some fracturing or cataclasis occurs (Handin 1966). At high temperatures, many ceramic materials are also known to deform through processes in which porosity develops as the specimen strains (Langdon, Cropper & Pask 1971). The development of even a slight porosity introduces a strong pressure sensitivity to the strength which eliminates such mechanism from consideration in the Earth's mantle. Rocks are particularly susceptible to weakening by porosity-forming creep mechanisms, as the primary slip systems of most rock-forming minerals are completely inadequate to meet the Von Mises condition for uniform strain and the formation of porosity permits a considerable relaxation of the Von Mises constraints. This has been shown to be true for MgO, talc, pyrophyllite, graphite, and BN (Paterson & Weaver 1970; Edmond & Paterson 1971; Paterson & Edmond 1972; Edmond & Paterson 1972; Auten & Radcliffe 1976). Early work on rocks at room pressure (Eaton 1968; Murrell & Chakravarty 1973; Misra & Murrell 1965) have all been shown to be porosity-weakened when compared with later high pressure results (Goetze 1971; Goetze & Brace 1972). Unpublished low-pressure creep data on hot-pressed olivine samples of high quality in our laboratory at the lowest practical stresses near the melting point showed clear porosity-weakening. This was evident under the microscope and by comparison with high pressure creep data. The conclusion is therefore that for aggregates of rock-forming minerals, the application of pressures exceeding the differential stresses is required to achieve purely plastic deformation (Edmond & Paterson 1972).

Confining our attention henceforth to pore-free creep experiments, there remains the more modest but not negligible effect of pressure on glide, climb, and the relevant diffusivities. Although creep data have been reported to 30 kbar, they have not so far been of sufficient accuracy to permit any direct observation of these pressure effects. There are at present no less than four groups working on this problem. Under the assumptions that climb limits the creep rate, that oxygen diffusivity limits the climb rate, and that the activation volume for oxygen diffusivity is given by the crystallographically determined volume of the oxygen site in the undisturbed olivine lattice, one predicts a pressure effect on the creep rate of about four orders of magnitude in strain rate at the olivine/spinel boundary (Weertman 1970; Goetze & Brace 1972). The uncertainty in this estimate is believed to be about a factor of 3 in either direction, and seriously limits the usefulness of low pressure creep data below the first 100 km.

As the experimental data are almost exclusively obtained at or below 15 kbar pressure, we restrict the remainder of this discussion to porosity-free but 'low pressure' data. The curves so obtained (figures 3, 4, 6, 10–12) are believed to be applicable to the first 100 km into the Earth without serious correction.

The range of variables of greatest geological interest include strain rates near 10^{-16} to 10^{-12} s^{-1} and stresses in the deep mantle of 10–100 bar and in the upper mantle of up to a few kilobars. As these conditions were predicted by Stocker & Ashby (figure 2) to fall in the 'dislocation creep' field, and because of the greater experimental accessibility of this field, the majority of the data in hand occur within this field along a band of experimentally accessible strain rates in the range 10^{-8} to $10^{-3}\,\mathrm{s}^{-1}$.

A comparison of Stocker & Ashby's (1973) predictions with currently available polycrystalline data is shown in figure 3, where Stocker & Ashby's predictions show as dotted lines. The data sources are indicated in the figure caption. We plot here a subset of the original data points, falling within the range of strain rates from 10^{-6} to $10^{-4}\,\mathrm{s}^{-1}$, corrected to a common strain rate of $10^{-5}\,\mathrm{s}^{-1}$ according to the formula:

$$\frac{1}{T_{\mathrm{corr}}} - \frac{1}{T_{\mathrm{actual}}} = \frac{R\ln(\dot{\epsilon}_{\mathrm{ac}}/10^{-5})}{125\,\mathrm{kcal/mol}} \tag{1}$$

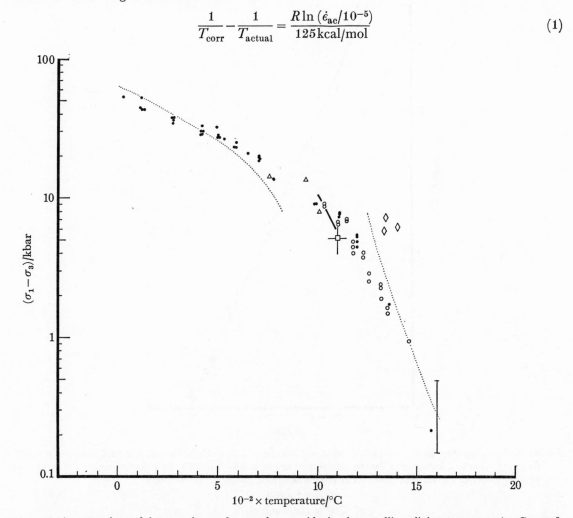

FIGURE 3. A comparison of the experimental creep data on 'dry' polycrystalline olivine aggregates (○, Carter & Ave' Lallemant 1970; solid line, Raleigh & Kirby 1970; △, Blacic 1972; □, Kirby & Raleigh 1973; ◇, Post 1973). Hardness (●, Evans & Goetze 1977) and single crystal bounds (lower right, Durham & Goetze 1977) with Stocker & Ashby's predictions (1973) which are shown as dotted lines. All data have been corrected to a common strain rate of $10^{-5}\,\mathrm{s}^{-1}$ (see text).

where 'actual' refers to the values measured and 'corrected' indicates the corresponding temperature plotted in figure 3. R is the gas constant. As the corrections for one decade in strain rate are comparatively minor and the apparent activation energy (125 kcal/mol) is well established (see figure 10), we regard this step as uncontroversial. A similar comparison would be obtained for other strain rates falling in the experimental range. What is controversial is that we have excluded a second set of data, labelled by most authors as 'wet', which is obtained in a solid medium apparatus, with talc as the pressure medium, under conditions in which the talc breaks down and gives off water. This second set of data differs significantly from the first and will be discussed below under *grain-size sensitive creep* (Blacic 1972; Carter & Ave' Lallemant 1970; Post 1973).

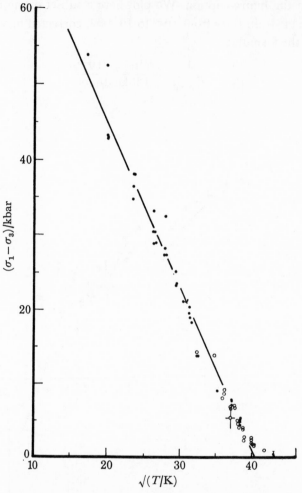

FIGURE 4. The data from figure 3 replotted as $\sigma_1-\sigma_3$ against $\sqrt{T}$. A straight line on this plot indicates a perfect fit to equation (2). Parameters for the line shown are $\sigma_p = 85 - 2.094\,T^{\frac{1}{2}}$. By measurements of the rate sensitivity at 530 °C, the value of Q was set at 128 ± 5 kcal/mol, giving the parameters

$$\dot{\epsilon}_0 = 5.7 \times 10^{11}\,\mathrm{s^{-1}},$$
$$Q = 128\,\mathrm{kcal/mol},$$
$$\sigma_p = 85\,\mathrm{kbar}.$$

A few comments should be made about figure 3.

(*a*) Work-hardening becomes pronounced above *ca.* 10 kbar stress. In this range a nominal strain of 8 % has been used in figure 3 for comparison with the hardness data which are shown as solid circles. At lower stresses a steady-state condition is achieved in less than 5 % strain.

(*b*) Details of the analysis of the hardness data are contained in Evans & Goetze (1977). Hardness data have been corrected to uniaxial yield according to the reduction factors given by Johnson (1970). High-pressure (pore-free) yield data, where available, are in very good agreement with single crystal hardness data reduced in this way (Evans & Goetze 1977).

(c) Stocker & Ashby used the simple Dorn law

$$\dot{\epsilon} = \dot{\epsilon}_0 \exp\left[\frac{-Q\{1 - (\sigma_1 - \sigma_3)/\sigma_\mathrm{p}\}^2}{RT}\right] \qquad (2)$$

for the portion above 8 kbar stress. In fact the Dorn law fits the data quite well down to 1–2 kbar differential stress, as can be seen in figure 4, in which differential stress is plotted against $\sqrt{T}$. A straight line on this plot represents a perfect fit to equation (2). The deviation at low stress corresponds to a transition to power law creep. Clearly, polycrystalline data do not at present constrain this power law region well, as most of them fall within the 'Dorn range' of rapidly varying stress exponent. The experimentally determined parameters for the Dorn range (figure 4 and equation (7)) differ markedly from those assumed by Stocker & Ashby.

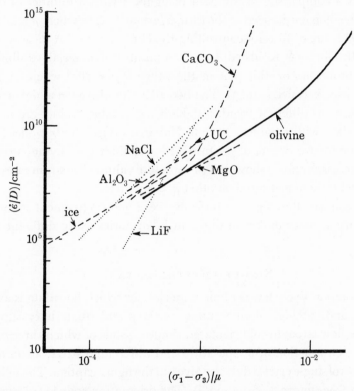

FIGURE 5. Data from figure 3 (solid line) are compared with corresponding data for a few materials showing power law creep (broken lines). Data are from Kirby & Raleigh (1973) except ice (Weertman 1972) and olivine (this study). When normalized in the manner shown (see text), materials exhibiting a cube dependence of strain rate on stress ($n = 3$) show similar absolute values of strain rate at low stresses. Olivine is no exception.

(d) Post's data show an obvious departure, greater than can be accounted for by experimental uncertainties, from the remaining data. The only reported difference between these results and the remaining data is the heat treatment (3 days at 950 °C in vacuum) to which the specimens were subjected before the deformation. Post's interpretation of this discrepancy is that trace quantities of hydrogen within the olivine lattice weaken the structure in a manner analogous to that shown to be true for variously oriented single crystals of quartz (hydrolytic weakening: Griggs 1967; Griggs & Blacic 1965; Heard & Carter 1968; Hobbs, McLaren & Paterson 1972) and that these heated specimens are drier (contain less hydrogen) than the remaining polycrystalline data. No measurements of hydrogen content have, however, been reported for any of the polycrystalline specimens shown in figure 3. The fact that the hardness data were obtained on specimens of very low hydrogen content ($H/Si < 28/10^6$) argues against this interpretation. Post's data is also difficult to reconcile with creep data on single crystals of similarly low hydrogen content. Another possible explanation stems from the fact that the oxygen fugacity in the vacuum furnace was not controlled. It has recently become evident that the stability field of olivine with respect to oxygen fugacity is narrow at all temperatures (Nitzan 1974) and that heterogeneous precipitation of iron oxides onto all dislocations is rapid at 950 °C when the stability field is exceeded (Kohlstedt, Goetze & Durham 1976b; Kohlstedt & Vander Sande 1975). It has also been shown that such pinning of the dislocations increases the creep resistance of olivine (Carter & Ave' Lallemant 1970), and this could explain the unusual hardness of Post's specimens. Yet other explanations may be possible as well.

(e) Figure 5 shows a comparison of the data in figure 3 with a number of other materials displaying dislocation creep normalized to the form σ/μ against $\dot{\epsilon}/D$, where μ is an average elastic shear constant and D is the diffusivity controlling the climb process. A measurement of this D in olivine was made by Goetze & Kohlstedt (1973). A number of materials exhibiting dislocation creep show similar properties on this plot in the range $\sigma/\mu < 10^{-3}$ where climb could be a controlling factor (Kirby & Raleigh 1973). The two alkali halides have a higher stress exponent (n) and follow a somewhat different trend, although the absolute values are not too different from the remaining data with exponent near 3. Nabarro's pure climb model (1967) predicts strain rates about one to two orders of magnitude lower than the olivine curve and of slope $n = 3$. Our purpose here is simply to show that the absolute value of the strain rates for $\sigma/\mu < 10^{-3}$ are quite consistent with those measured on other $n \approx 3$ materials.

In summary, we can say that there is basically very good agreement among 'dry' polycrystalline data reported by a number of different laboratories using different apparatus and starting materials.

SINGLE CRYSTAL DATA

Olivine is one of comparatively few ceramic materials for which flow data is available for the primary, secondary and, at high temperatures, tertiary and quaternary slip systems. The primary data are the flow strengths of orientated single crystals of which three orientations are followed in figure 6 as a function of temperature and differential stress, again at a strain rate of $10^{-5}\,s^{-1}$. The sources of single crystal data are given in the figure caption. The solid line indicates the trend of the data from figure 3 (equation 7). The orientation code $[110]_c$, indicates that σ_1 is orientated 45° from [100] and [010]; $[101]_c$ indicates that σ_1 lies 45° from [100] and [001], etc. The flow curves of single crystals are similar to those of polycrystals in that work hardening becomes prominent at low temperatures, while at the highest temperatures crystals reach a steady-state creep rate in a strain of less than 1%, the transition occurring near 1000 °C (at $\dot{\epsilon} = 10^{-5}\,s^{-1}$).

Slip systems

The slip systems by which these crystal orientations deform is given in the primary references and can be summarized as follows:

At intermediate temperatures a dramatic crossover occurs which is in excellent agreement with earlier reports of changes in the primary slip system with temperature (Raleigh 1965, 1968; Carter & Ave' Lallemant 1970). Near 1000 °C (at $\dot{\epsilon} = 10^{-5}\,s^{-1}$) the slip systems (001)[100] and (010)[100] have very similar critical resolved shear stress, and a complex intermingling of these systems occurs which macroscopically resembles pencil glide, $(0kl)[100]$ (Raleigh 1965, 1967, 1968; Carter & Ave' Lallemant 1970), but may result from extensive cross-slip on an atomic scale (Poirier 1975).

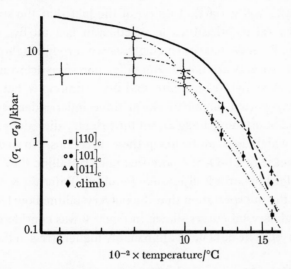

FIGURE 6. A comparison of the creep strength of three single crystal orientations and the trend of the polycrystalline data from figure 3 (solid curve). Temperature scale is linear in T^{-1}. Open circles are from Phakey *et al.* (1972), solid symbols from Durham & Goetze (1977).

TABLE 2

orientation	temperature	slip system(s)	Schmidt factor	reference
$[101]_c$	R.T. to 700 °C	{110} [001]	0.46	Evans & Goetze (1977)
$[101]_c$	800 °C	{110} [001]	0.46	Phakey *et al.* (1972)
$[101]_c$	890 °C	{110} [001]	0.46	Durham (1975)
$[101]_c$	1000 °C	{110} [001]	0.46	
		+ (100) [001]	0.5	Phakey *et al.* (1972)
		+ (?) [100]	?	
$[101]_c$	1600 °C	80% (001) [100]	0.5	Durham (1975)
		+ 20% (100) [001]	0.5	
$[011]_c$	800 °C	{110} [001]	0.21	Phakey *et al.* (1972)
		+ others?	?	
	1600 °C	(010) [001]	0.50	Durham (1975)
$[110]_c$	800–1000 °C	(010) [100]	0.50	Phakey *et al.* (1972)
	1600 °C	(010) [100]	0.50	Durham (1975)

Earlier data on the 'dominant slip systems' in polycrystalline specimens are reported in Raleigh (1965, 1967) and Carter & Ave' Lallemant (1970).

Climb

Six single crystal specimens deformed at 1600 °C (Durham & Goetze 1977a) did so with sufficiently uniform strain that a computer calculation could be used to isolate the proportion strain on the primary, secondary, etc., slip system. Of these the three $[101]_c$ orientation crystals showed that 10–15 % of the strain was accomplished by combined climb of [100] and [001] Burgers vector dislocations, presumably in the manner proposed by Nabarro (1967). The corresponding data are shown as a diamond in figure 6.

The Von Mises criterion

The Von Mises criterion states that five independent slip systems must be activated within polycrystalline grains if homogenous and pore-free deformation is to be achieved by slip alone (Von Mises 1928; Groves & Kelly 1963). In view of the fact that the strain is generally not homogeneous, there is general scepticism as to whether in fact all five systems are needed (Paterson & Weaver 1970). Paterson has convincingly shown through high-pressure deformation tests that two independent systems are not adequate for pore-free deformation, at least in the cases of MgO, NaCl, talc, pyrophyllite, graphite, and BN (Edmond & Paterson 1971, 1972).

The slip systems so far reported for olivine form three independent systems (see Phakey, Dollinger & Christie 1972). Shortening along any of the primary directions ([100], [010], [001]) requires a Burgers vector which does not lie along these directions. No such Burgers vector has so far been found. In the case of MgO, the pore-free polycrystalline flow law follows the flow law of crystals orientated for the hardest slip system required (Paterson & Weaver 1970). At an earlier stage of this study, the interpretation that the polycrystalline flow law approximated the upper envelope of the single crystal curves shown in figure 6 was considered. However, it now seems clear that the polycrystalline data lie at appreciably higher stress in the range 800–1300 °C (at $\dot{\epsilon} = 10^{-5}\,\mathrm{s}^{-1}$).

To our knowledge this is the first material with only three independent slip systems for which pore-free creep data has been reported. We are not aware of any theory for predicting the strength of materials whose individual slip systems do not form a complete set. The complex cross-over in individual slip system strengths shown near 1000 °C in figure 6 should provide an excellent test of such a theory once developed.

At the highest temperatures, climb provides the missing systems and the polycrystalline flow is therefore unlikely to fall outside the range 150–500 bar at 1600 °C ($\dot{\epsilon} = 10^{-5}\,\mathrm{s}^{-1}$).

DISLOCATION MECHANICS

The starting point for most dislocation models is the approximate relation

$$\dot{\epsilon} = \rho b \bar{v}, \tag{3}$$

where ρ is the mobile dislocation density, b the mean Burgers vector, and $\bar{v}$ the mean dislocation velocity. We report here the progress made in understanding ρ and $\bar{v}$ individually. Besides focusing attention on the processes actually controlling the creep mechanism, equation (3), if effective in predicting strain rates, has the potential to circumvent some of the problems inherent in extrapolating to lower strain rates. The dislocation mobilities can be followed to much lower temperatures and stresses than the corresponding strain rates. Actual dislocation velocities in the mantle are in the range 10^{-4} to $10^{2}\,\mathrm{\mu m/d}$ and with some patience an overlap may even be obtained with experimental mobility data. Actual dislocation densities can be inferred from those found in natural rocks. Such data can be combined through equation (3) to give an accurate picture of the flow law in a strain-rate range that cannot be sampled directly. In addition, a study of the dislocation microstructures has the potential for being inverted to permit an interpretation of the unknown deformation history of natural samples (Durham, Goetze & Blake 1977; Goetze 1975; Briegel & Goetze 1977).

Analysis of the dislocation structures has been made by transmission electron microscopy

(Phakey *et al.* 1972; Durham *et al.* 1977; Green & Radcliffe 1972 *a*, *b*, *c*) and, where the dislocation density is low enough to view with the optical microscope ($< 10^8$ cm^{-2}), by a low-temperature decoration technique (Kohlstedt *et al.* 1976 *b*; Durham *et al.* 1977).

Dislocation density

The dislocation density is found to increase rapidly with strain to an equilibrium level, especially at low stress. An example of data from a single crystal sectioned at intervals of strain is shown in figure 7 (Durham *et al.* 1977). Plotting the equilibrium densities so achieved shows a strong correlation with the applied differential stress of the approximate form

$$\sigma_1 - \sigma_3 = \alpha\mu b \rho^{\frac{1}{2}} \quad \text{(Durham } et\ al.\ 1977). \tag{4}$$

In figure 8, equilibrium dislocation density data for olivine are compared with corresponding data for quartz and calcite and a large class of materials displaying power law creep. The broken

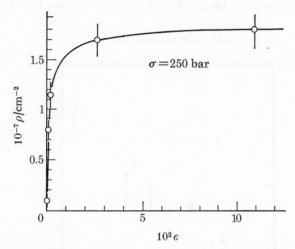

FIGURE 7. Dislocation density recorded in a single crystal of $[101]_c$ orientation from which sections were removed and decorated after strain increments of 0.12, 0.63, 2.9 and 11.1 %.

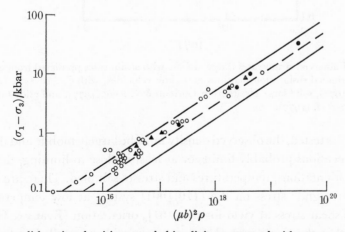

FIGURE 8. Steady-state dislocation densities recorded in olivine compared with quartz, calcite, and a summary of a large class of materials (pure metals, alloys, ionic solids) deforming by power law creep. The broken line is an average of the parameters in equation (4) from 33 such studies; the solid lines represent one standard deviation. In most studies, including olivine, a slope of greater than $\frac{1}{2}$ is indicated. The one shown is 0.62. This figure is from Durham *et al.* (1977): ○, olivine; ●, quartz; ▲, calcite.

line is the mean correlation established from 33 studies on metals, alloys and ionic solids deforming by power law creep (Takeuchi & Argon 1976*b*) and the solid lines represent one standard deviation on either side. Clearly, olivine is no exception to this very general relation.

Glide velocity

One study has been done of the glide mobility on the (001) [100] and (010) [100] slip systems in the temperature range 1100–1600 °C (Durham *et al.* 1977). Unfortunately, the method did not yield data of high accuracy. However, the glide velocities so measured were very sluggish and of low stress sensitivity, as in the case of the diamond structure compounds (Alexander & Haasen 1968). When these measured glide velocities and the dislocation density from figure 8 are entered into equation (3), the predicted strain rates (small circles) fall close to the observed rates (large circles). Figure 9 shows this comparison for crystals of the orientation $[101]_c$ gliding on the (001) [100] slip system. All data are again corrected to $\dot{\epsilon} = 10^{-5}\,\mathrm{s}^{-1}$. A similar result was obtained for the $[110]_c$ orientation (Durham *et al.* 1977). We take this as an indication that,

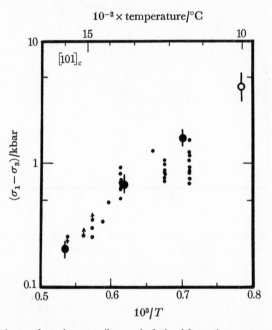

FIGURE 9. Comparison of observed strain rates (large circles) with strain rates predicted from equation (3), using experimentally determined dislocation densities and glide velocities, with $\dot{\epsilon}$ at $10^{-5}\,\mathrm{s}^{-1}$. Large open circle is from Phakey *et al.* (1972), solid large circles from Durham & Goetze (1977), and glide mobility data (small circles) from Durham *et al.* (1977).

for the range of variables tested, the observed density must be largely mobile and the glide velocity of the individual dislocations probably limits, or at least is close to limiting, the strain rate for single crystals of the orientations, temperatures and stresses sampled. There are also indications that the rate-insensitive glide stress on the (110) [001] system at low temperatures coincides with the appropriate shear stress at yield for the $[101]_c$ orientation (Evans & Goetze 1977). It is therefore possible that the single crystal strengths are glide-controlled at low temperatures as well.

Whether the 'Dorn law' portion of the polycrystalline curve is also glide-controlled is a question we do not seek to answer here as we have no direct experimental evidence. The Dorn law formulation was derived to represent the flow of materials exhibiting a high Peierls barrier in glide, and olivine appears to be such a material. However, the slip system of the glide mobility which is limiting the strain rate is not presently known (see discussion under Von Mises criterion). More work is needed to resolve this problem.

Climb velocity

Qualitative indications that climb is important above perhaps 1000 °C include (a) the 'picket fence' tilt boundaries frequently reported, in which the uniformity of the dislocation spacing could only be achieved through climb (Goetze & Kohlstedt 1973; Phakey et al. 1972; Green & Radcliffe 1972 b, c) and (b) the heavily jogged glide loops observed as low stresses near 1600 °C (Durham et al. 1977). Quantitative estimates of climb mobility were obtained by Goetze & Kohlstedt (1973) and in the careful shape change measurements reported in Durham & Goetze (1977) and plotted in figure 6. These latter two cannot be directly compared, but it may be roughly said that glide and climb mobilities are approximately equal near 100 bar but that glide, having a higher stress sensitivity than climb (Durham et al. 1977), dominates at higher stress levels. Computer models of the dislocation mechanics for materials in which glide and climb mobilities are comparable (Takeuchi & Argon 1976 a) show a number of striking similarities with the single crystal results obtained at 1600 °C and few hundred bars stress (Durham et al. 1977).

In summary, while figure 9 shows promise in separating the effect of dislocation density and dislocation mobility on the measured strain rates, there is ambiguity at low stresses whether glide or climb, or more likely some combination, limit the strain rates. The fact that both processes appear to have very similar rate sensitivities (see next section) indicates that a comparable mix will probably occur at much lower strain rates at the same stress level.

Strain rate sensitivity

It is of crucial importance for geological applications to be able to predict how figure 3 will change in going from a strain rate of 10^{-5} to geologically relevant strain rates in the range 10^{-15} to $10^{-10}\,\text{s}^{-1}$. Qualitatively this curve will shift to lower temperatures. The extent to which it does so must at present be determined from strain rate sensitivity measurements within the experimentally accessible range of rates. These are assembled in figure 10.

The simple Dorn law,

$$\dot{\epsilon} = \dot{\epsilon}_0 \exp\{-Q[1-(\sigma_1-\sigma_3)/\sigma_\mathrm{p}]^2 RT\}, \tag{5}$$

has an apparent activation energy

$$Q_\mathrm{app} = -R\Delta \ln \dot{\epsilon}/\Delta(1/T) = Q(1-(\sigma_1-\sigma_3)/\sigma_\mathrm{p})^2 \tag{6}$$

In view of the good fit of the hardness and polycrystalline data to this law (figure 4), the apparent activation energies for these data (1, 2, 3) were corrected by the factor $(1-(\sigma_1-\sigma_3)/\sigma_\mathrm{p})^{-2}$ to yield Q. This correction is comparatively minor below 10 kbar but substantial for the hardness point at 530 °C (1). We take the agreement between points 1, 2 and 3 as further evidence that the Dorn law is an appropriate form to use for the higher stress data. Points 6, 7 and 8 are apparent activation energies obtained from creep tests on single crystals, orientated for both single and multiple slip, deformed in the stress range in which glide and climb mobilities are comparable.

It is therefore uncertain which process controls the activation energy. Point 5 was obtained from the collapse rate of sessile dislocation loops and is presumed to represent the appropriate activation energy for the climb process alone. Point 4 was obtained from the temperature sensitivity of the annealing process in highly deformed olivine, a process which unquestionably involves a complex mix of glide and climb.

The single crystal data of Durham & Goetze (1977) contain data for the parameter

$$\left(\frac{\Delta \ln \sigma}{\Delta (1/T)}\right)_{\dot{\varepsilon}} = -\left(\frac{\Delta \ln \dot{\varepsilon}}{\Delta (1/T)}\right)_{\sigma} \cdot \left(\frac{\Delta \ln \sigma}{\Delta \ln \dot{\varepsilon}}\right)_{T} = \frac{Q_{app}}{nR}$$

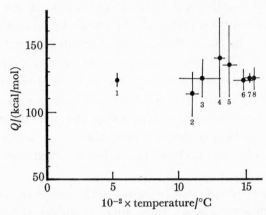

FIGURE 10. Activation energies measured for various processes. Horizontal bars show the temperature range over which the data were determined; vertical bars give standard deviation.

    (1) Hardness data (Evans & Goetze 1977);
    (2) polycrystalline flow (Kirby & Raleigh 1973);
    (3) polycrystalline flow (Carter 1976);
    (4) dislocation recovery (Goetze & Kohlstedt 1973);
    (5) climb, from collapse of sessile dislocation loops (Goetze & Kohlstedt 1973);
    (6) [011]$_c$ orientation single crystals (Durham & Goetze 1977);
    (7) single crystals orientated for multiple slip (Kohlstedt & Goetze 1974);
    (8) single crystals orientated for multiple slip (Durham & Goetze 1977).

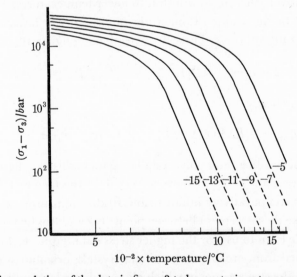

FIGURE 11. Extrapolation of the data in figure 3 to lower strain rates according to equation (7).

(n is the stress exponent in a local power law representation of the creep data), where Q_{app}/n drops at temperatures below about $1300\,°C$ for several orientations. We do not feel that n is well enough determined for these specimens, however, to warrant any conclusion about the activation energy itself.

Apparently the processes likely to control the flow in the low stress region (0.1–1 kbar) is controlled by a similar, possibly identical, activation energy to that applicable to the Dorn region. We will not speculate here why this is so, but the practical consequence is that a simple extrapolation can be made using the Dorn law above 2 kbar and a power law below 2 kbar which will form continuous curves at all strain rates.

$$\sigma > 2\,\mathrm{kbar}: \dot{\epsilon}/\mathrm{s}^{-1} = 5.7 \times 10^{11} \exp\left\{-\frac{128\,\mathrm{kcal/mol}}{RT}\left(1 - \frac{\sigma_1 - \sigma_3}{85\,000}\right)^2\right\};$$

$$\sigma < 2\,\mathrm{kbar}: \dot{\epsilon} = 70(\sigma_1 - \sigma_3)^3 \exp\left\{-122(\mathrm{kcal/mol})/RT\right\}. \tag{7}$$

These curves are plotted in figure 11. A Q of $125 \pm 10\,\mathrm{kcal/mol}$ appears to represent both processes; the values 122 and 128 kcal/mol were arbitrarily chosen to make a perfect match at all strain rates for numerical convenience. Stresses are given in bars in equations (7). In the stress and strain rate range of geologic interest the formal accuracy of equation (7) is better than $\pm 50\,°C$.

GRAIN-SIZE SENSITIVE CREEP

The boundary between dislocation creep and Coble creep is perhaps the least well determined in the deformation map shown in figure 2, as there are very meagre data on which to base the necessary systematics. Bouillier & Gueguen (1975) presented convincing evidence that some form of grain-size sensitive creep has occurred in a number of highly mylonitized natural specimens. Schmid (1976) and Schmid, Boland & Paterson (1977) have found a grain-size sensitive flow field in a recent laboratory deformation study on very fine-grained limestones, but so far there has been no comparable study, to my knowledge, on fine-grained olivine aggregates. Recently two indirect pieces of evidence have come to light that constrain the boundary between dislocation creep and a grain-size sensitive creep field, and in view of the importance of this subject these will now be briefly reviewed.

It has long been known that creep experiments on olivine aggregates in the differential stress range 1–10 kbar occasionally show 'ductile faulting' (Blacic 1972; Post 1973), a process in which the development of a fine-grained shear zone, resembling a fault but apparently devoid of porosity, is accompanied by a marked drop in stress, presumably indicating a grain-size sensitive enhancement of the flow within the shear zone. Twiss (1976) has extracted from Post's thesis four ('dry') specimens for which the relevant variables can be estimated. The four points so obtained are, however, sufficiently close together in stress, temperature, grain size, and strain rate that the dependence of none of these variables can be extracted from the data alone.

A second source comes from hot-pressing studies on samples of olivine powders ranging in grain size from 5–2000 µm (Schwenn & Goetze 1977; Schwenn 1976). Hot-pressing characteristics were found to be controlled by power law creep at large grain sizes and by a very grain-size sensitive flow law at small grain sizes for which strain rate varied as grain size raised to the power -2 to -4.5 and stress raised to the power 1.1–1.9. Using the relations between hot-pressing and creep characteristics (Wilkinson & Ashby 1975, 1976; Coble 1970), the creep law for the corresponding dense material was derived. It predicts the four data points extracted by Twiss within a factor of 3 in strain rate. This flow law is given below, together with the corresponding flow law predicted by Stocker & Ashby on the basis of systematics.

Schwenn & Goetze:

$$\dot{e} = (3\text{–}10) \times 10^4(\sigma_1 - \sigma_3)\, s_g^{-3} \exp\{-85(\text{ckal/mol})/RT\}. \tag{8a}$$

Stocker & Ashby:

$$\dot{e} = 0.5(\sigma_1 - \sigma_3)\, T^{-1} s_g^{-3} \exp\{-103(\text{kcal/mol})/RT\}. \tag{8b}$$

(Parameters are for stress in bars and grain size (s_g) in centimetres.)

As the creep rate in these laws is presumably controlled by the grain boundary diffusivity, it is not yet clear to what extent impurities along the grain boundaries will cause a wide scatter among different determinations of the flow parameters. However, the several groups now working on this problem should provide some answer to this question within a year or two.

TABLE 3. EXPERIMENTALLY DETERMINED CONSTANTS FOR EQUATION (9)

reference	A	b
Post (1973)	40	-0.67
Kohlstedt et al. (1976a)	76	-0.63
Carter & Mercier (1976)	49	-0.71

Constants are for stress in bars and grain size in centimetres

One important question to investigate is whether any of the experimental data now in hand in fact fall within this Coble creep field. Taking the reported starting grain sizes as representative of the run, we find that only the data for creep of Mt Albert peridotite reported by Goetze & Brace (1972) fall in this category. However, many of the data between 1 and 10 kbar are obtained from specimens which partially recrystallized during the deformation experiment.

A number of authors have investigated this dynamic recrystallization process in olivine (Ave' Lallemant & Carter 1970; Kirby & Raleigh 1973; Post 1973; Carter & Mercier 1976) and the results may be briefly summarized as follows. With increasing strain a new generation of grains appears, principally along the grain boundaries of the starting material, and gradually replaces the starting grain structure. The new grain size so formed appears to correlate with stress according to the relation

$$\sigma_1 - \sigma_3 = Ad^b, \tag{9}$$

where table 3 gives the parameters A and b according to three different studies. Post's are the only parameters for which flow data and grain size are reported on the same specimens. The parameters given by Kohlstedt et al. (1976a) are based primarily on naturally deformed specimens. All three make very similar predictions in the grain size range in which the data overlap. Similar results have been reported for α-iron (Glover & Sellars 1973) and Ni–Fe alloys (Luton & Sellars 1969).

Combining Post's parameters in equation (9) with equation (8a) provides a prediction for the strain rate of a fully recrystallized specimen. It is an empirically established Coble creep law:

$$\dot{e} = 3.6 \times 10^{11}(\sigma_1 - \sigma_3)^{5.47} \exp(-85\,000/RT). \tag{10}$$

We use Post's parameters because we will shortly make a comparison with creep data obtained on the same specimens. In figure 12 we compare this flow law with the law appropriate to coarse grain sizes taken from figures 3 and 11, again for a constant strain rate of $10^{-5}\,\text{s}^{-1}$. The clear indication is that specimens which are fully recrystallized will be softer than the coarser-grained starting material over a range of differential stress from approximately 1 to 15 kbar (at $\dot{\epsilon} = 10^{-5}\,\mathrm{s}^{-1}$). It is apparent in retrospect, therefore, that the interpretation of data in this range is potentially treacherous since specimens might show any strength between that predicted by the Coble law and that predicted by the dislocation creep law of figure 11, depending on the degree of recrystallization, a fact which was not appreciated at the time the experiments were done. The reasons for a variety of complexities in the original records, such as ductile faulting,

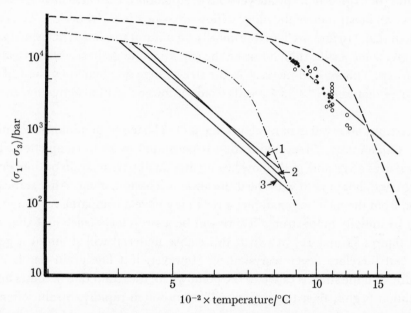

FIGURE 12. On the right: a comparison of the data from figure 3 (broken line) with the strengths predicted from the experimentally determined Coble creep parameter (equation (8a)) and grain sizes (equation (9)) for $\dot{\epsilon} = 10^{-5}\,\mathrm{s}^{-1}$ (solid line, Post 1973). The 'wet' data of Post (1972) (●) and Carter & Ave' Lallemant (1970) (○) appear to coincide with this line. On the left: the corresponding lines extrapolated to a strain rate of $10^{-15}\,\mathrm{s}^{-1}$: 1, Kohlstedt *et al.* (1976a); 2, Murrell & Chakravarty (1973); 3, Post (1973).

are now readily understood. It is also instructive to plot the 'wet' olivine data which are shown as circles in figure 12 (data for $\dot{\epsilon} = 10^{-6}$ to $10^{-4}\,\mathrm{s}^{-1}$ are corrected to 10^{-5} as before). The tendency for these data to fall along the Coble creep line is marked especially in the case of Post's wet data which show low scatter and for which the grain size parameters leading to the Coble line shown were determined. Carter & Ave' Lallemant's (1970) data can be interpreted as showing a wider scatter about the same line. This suggests an interpretation of the 'wet' data as simply more completely recrystallized than the 'dry' data, rather than a hydrolytic weakening of the crystals themselves, as suggested by Post (1973), Carter & Ave' Lallemant (1970), Blacic (1972), and others. This is also consistent with our own inability to find any evidence of hydrolytic weakening in olivine single crystals in spite of repeated attempts.

The effect of moisture is, in this interpretation, only indirectly related to the strength through the degree of recrystallization. It appears unlikely at present that water affects the Coble creep process directly since the hot-pressing data were obtained under both hydrous and non-hydrous atmospheres at 1 bar total pressure and Twiss's data were extracted from ductile faulting which occurred in Post's 'dry' experiments. In fact, Post systematically rejected 'dry' runs which showed a falling stress, believing them not to be 'completely dry'. Moisture therefore does not appear to be a prerequisite for observing the Coble creep process given by equation (10). It is very likely that new data will modify this simple picture, but the recognition that the 'wet' data fall within the Coble creep field could help unravel this longstanding problem.

As a last bit of speculation we might consider what implications the empirically determined Coble creep law has for deformation within the mantle at geological strain rates. In figure 12 we have plotted the same comparison between dislocation and Coble creep strengths extrapolated to $\dot{\epsilon} = 10^{-15}\,\mathrm{s}^{-1}$. All three sets of parameters shown in table 3 have been inserted in equations ($8a$) and (9) to give appropriate versions of equation (10). These have all been plotted in figure 12 to give an illustration of the range of flow stress predicted by table 3. The parameters given by Kohlstedt *et al.* (1976a) are based primarily on natural specimens deformed at geologic strain rates and give some justification for using the experimental grain size/stress relation under geological conditions. The marked increase in the stress range dominated by the Coble process at $\dot{\epsilon} = 10^{-15}\,\mathrm{s}^{-1}$ over that at $\dot{\epsilon} = 10^{-5}\,\mathrm{s}^{-1}$ results from the lower apparent activation energy of the Coble process.

At the highest stresses observed in natural specimens (1–3 kbar) the dominant creep mechanism is almost surely Coble creep, *if* recrystallization is permitted to go to completion. The introduction of grain size as a variable in the flow law creates an apparent strain or time dependence of lithospheric strength not present with the dislocation mechanism alone. After static annealing or crystallization from the melt, for example, a rock may have a comparatively large grain size ($>500\,\mu\mathrm{m}$) and be initially quite strong. There will be a stress level (such as 1 kbar at 600 °C, for example, in figures 11 and 12) at which this coarse material will strain at a geologically negligible rate and therefore never recrystallize. However, if a fine grain size is introduced through a fracture, or if the stress is raised to the point where the strain rate becomes appreciable and recrystallization begins, then this region will strain-soften rapidly. Strain-softening is the principal requirement for concentrating strain into shear zones which could form the downward extension of the brittle faulting found at shallow depths. A fault, once begun, can extend itself downwards by the stress concentrations near its leading edge to result in a band of fine recrystallized material – a permanent zone of weakness through the lithosphere until static annealing returns it to a coarse grain size. Such a material can show a surprisingly complex repertoire of mechanical behaviour, much of which has its analogue in geological field evidence. Understanding these lithospheric deformations remains a promising but undeveloped field.

According to figure 12, the region below the intersection near $\sigma_1 - \sigma_3 = 200$ bar should be dominated by power law creep. This region in stress corresponds to the asthenosphere in the Earth. One may confirm this conclusion by substituting the grain size observed in rocks from the upper mantle (1–10 mm) in equation (8) rather than by using the empirical grain size/stress correlation, equation (9). Twiss (1976) predicts that the Coble process will dominate to stresses as low as 20 bar. This discrepancy between our conclusions does not result from different predictions about the Coble process, however, but from the very 'hard' dislocation creep formula used by Twiss which, although apparently based on Post's 'dry' measurements (see figure 3), is more creep-resistant than any experimental data of which this author is aware. Nevertheless, an improved value for this important boundary between the two mechanisms awaits better data.

Conclusions

(1) There is fairly good agreement and understanding of the dislocation creep field, at least within the range of strain rates which can be explored in the laboratory.

(2) Some experimental data now exist on a grain-size sensitive flow law, phenomenologically resembling the Coble creep mechanism. In the Earth this mechanism is expected to dominate the deformation of olivine at the high stresses (0.3–3 kbar) characteristic of the lithosphere if dynamic recrystallization to small grain sizes occurs. It appears unlikely that it will dominate in the differential stress range 10–100 bar characteristic of the asthenosphere.

(3) The previously designated 'wet' data probably correspond to this grain-size sensitive flow law.

Support from National Science Foundation grant no. DES72-01676 A03 is gratefully acknowledged.

References (Goetze)

Alexander, H. & Haasen, P. 1968 Dislocations and plastic flow in the diamond structure. *Solid State Phys.* **22**, 27–158.

Auten, T. A. & Radcliffe, S. V. 1976 Deformation of polycrystalline MgO at high hydrostatic pressure. *J. Am. ceram. Soc.* **59**, 249–253.

Ave' Lallemant, H. G. & Carter, N. L. 1970 Syntectonic recrystallization of olivine and modes of flow in the upper mantle. *Bull. geol. Soc. Am.* **81**, 2203–2220.

Blacic, J. D. 1972 Effect of water on the experimental deformation of olivine. In *Flow and fracture of rocks* (eds H. C. Heard, I. Y. Borg, N. L. Carter & C. B. Raleigh), pp. 109–115. Am. geophys. Union Monograph no. 16.

Bouillier, A. M. & Gueguen, Y. 1975 Origin of some mylonites by superplastic flow. *Contrib. Mineral. Petrol.* **50**, 93–104.

Briegel, U. & Goetze, C. 1977 Estimate of stress in Lochseiten limestone with disolocation densities. *Tectonophysics* (In the press.)

Carter, N. L. & Ave' Lallemant, H. G. 1970 High temperature flow of dunite and peridotite. *Bull. geol. Soc. Am.* **81**, 2181–2202.

Carter, N. L. 1976 Steady-state flow of rocks. *Rev. Geophys. & Space Phys.* **14**, 301–360.

Carter, N. L. & Mercier, J.-C. C. 1976 Stress dependence of olivine neoblast grain sizes. *Trans. Am. geophys. Un., Eos* **57**, 322.

Coble, R. L. 1970 Diffusional models for hot-pressing with surface energy and pressure effects as driving forces. *J. appl. Phys.* **41**, 4798–4807.

Deer, W. A., Howie, R. A. & Zussman, J. 1964 *Rock-forming minerals*, vol. 1, pp. 12–13. London: Longmans Green.

Durham, W. B. 1975 Plastic flow of single-crystal olivine. Ph.D. thesis, Massachusetts Institute of Technology, Cambridge, Massachusetts.

Durham, W. B. & Goetze, C. 1977*a* Plastic flow of oriented single crystals of olivine. I. Mechanical data. *J. geophys. Res.* (In the press.)

Durham, W. B. & Goetze, C. 1977*b* A comparison of the creep properties of pure forsterite and iron-bearing olivine. *Tectonophysics* (In the press.)

Durham, W. B., Goetze, C. & Blake, B. 1977 Plastic flow of oriented single crystals of olivine. II. Observations and interpretations of the dislocation structures. *J. geophys. Res.* (In the press.)

Eaton, S. F. 1968 The high temperature creep of dunite. Ph.D. thesis, Princeton University.

Edmond, J. M. & Paterson, M. S. 1971 Strength of solid pressure media and implications for high pressure apparatus. *Contrib. Mineral. Petrol.* **30**, 141–160.

Edmond, J. M. & Paterson, M. S. 1972 Volume changes during deformation of rocks at high pressures. *Int. J. rock Mech. & min. Sci.* **9**, 161–182.

Evans, B. & Goetze, C. 1977 Manuscript in preparation.

Glover, G. & Sellars, C. M. 1973 Recovery and recrystallization during high temperature deformation of α-iron. *Metall. Trans.* **4**, 765–775.

Goetze, C. & Brace, W. F. 1972 Laboratory observations of high-temperature rheology of rocks. *Tectonophysics* **12**, 583–600.

Goetze, C. 1971 High temperature rheology of Westerly granite. *J. geophys. Res.* **76**, 1223–1230.

Goetze, C. 1975 Sheared lherzolites: from the point of view of rock mechanics. *Geology* **3**, 172–173.

Goetze, C. & Kohlstedt, D. L. 1973 Laboratory study of dislocation climb and diffusion in olivine. *J. geophys. Res.* **78**, 5961–5971.

Green, H. W. & Radcliffe, S. V. 1972*a* The nature of deformation lamellae in silicates. *Bull. geol. Soc. Am.* **83**, 847–852.

Green, H. W. & Radcliffe, S. V. 1972*b* Dislocation mechanisms in olivine and flow in the upper mantle. *Earth & planet. Sci. Lett.* **15**, 239–247.

Green, H. W. & Radcliffe, S. V. 1972*c* Deformation processes in the upper mantle. In *Flow and fracture of rocks* (eds H. C. Heard, I. Y. Borg, N. L. Carter & C. B. Raleigh), pp. 139–157. Am. geophys. Un. Monograph, no. 16.

Griggs, D. T. & Blacic, J. D. 1965 Quartz: anomalous weakness of synthetic crystals. *Science N.Y.* **147**, 292–295.

Griggs, D. T. 1967 Hydrolytic weakening of quartz and other silicates. *Geophys. J. R. astron. Soc.* **14**, 19–31.

Groves, G. W. & Kelly, A. 1963 Independent slip systems in crystals. *Phil. Mag.* **8**, 877–887.

Handin, J. 1966 Strength and ductility, in *Handbook of physical constants* (ed. S. P. Clark, Jr). Geol. Soc. Am. Memoir, no. 97.

Heard, H. C. & Carter, N. L. 1968 Experimentally induced 'natural' intragranular flow in quartz and quartzite. *Am. J. Sci.* **266**, 1–41.

Hobbs, B. E., McLaren, A. C. & Paterson, M. S. 1972 Plasticity of single crystals of synthetic quartz. In *Flow and fracture of rocks* (eds H. C. Heard, I. Y. Borg, N. L. Carter & C. B. Raleigh), pp. 29–53. Am. geophys. Un. Monograph no. 16.

Johnson, K. L. 1970 The correlation of indentation experiments. *J. Mech. Phys. Solids* **18**, 115–126.

Kirby, S. H. & Raleigh, C. B. 1973 Mechanisms of high-temperature, solid state flow in minerals and ceramics and their bearing on creep behaviour of the mantle. *Tectonophysics* **19**, 165–194.

Kohlstedt, D. L. & Goetze, C. 1974 Low-stress high-temperature creep in olivine single crystals. *J. geophys. Res.* **79**, 2045–2051.

Kohlstedt, D. L., Goetze, C. & Durham, W. B. 1976*a* Experimental deformation of single crystal olivine with application to flow in the mantle. In *The physics and chemistry of minerals and rocks* (ed. S. K. Runcorn), pp. 35–49. London: John Wiley.

Kohlstedt, D. L., Goetze, C., Durham, W. B. & Vander Sande, J. B. 1976*b* A new technique for decorating dislocations in olivine. *Science, N.Y.* **191**, 1045–1046.

Kohlstedt, D. L. & Vander Sande, J. B. 1975 Heterogeneous precipitation on dislocations in olivine. *Contrib. Mineral. Petrol.* **53**, 13–25.

Langdon, T. G., Cropper, D. R. & Pask, J. A. 1971 Creep mechanisms in ceramic materials at elevated temperatures. In *Ceramics in severe environments*, vol. 5, pp. 297–315. Materials Science Research, New York: Plenum Press.

Luton, M. J. & Sellars, C. M. 1969 Dynamic recrystallization in Ni and Ni–Fe alloys during high temperature deformation. *Acta metall.* **17**, 1033–1043.

Misra, A. K. & Murrell, S. A. F. 1965 An experimental study of the effect of temperature and stress on the creep of rocks. *Geophys. J. R. astron. Soc.* **9**, 509–535.

Murrell, S. A. F. & Chakravarty, S. 1973 Some new rheological experiments on igneous rocks at temperatures up to 1120 °C. *Geophys. J. R. astron. Soc.* **34**, 211–250.

Nabarro, F. R. N. 1967 Steady-state diffusional creep. *Phil. Mag.* **16**, 231–239.

Nitzan, U. 1974 Oxidation and reduction of olivine. *J. geophys. Res.* **79**, 706–711.

Paterson, M. S. & Edmond, J. M. 1972 Deformation of graphite at high pressures. *Carbon* **10**, 29–34.

Paterson, M. S. & Weaver, C. W. 1970 Deformation of polycrystalline MgO under pressure. *J. Am. ceram. Soc.* **53**, 463–471.

Phakey, P., Dollinger, G. & Christie, J. 1972 Transmission electron microscopy of experimentally deformed olivine crystals. In *Flow and fracture of rocks* (eds H. C. Heard, I. Y. Borg, N. L. Carter & C. B. Raleigh), pp. 117–138. Am. geophys. Un. Monograph, no. 16.

Poirier, J. P. 1975 On the slip systems of olivine. *J. geophys. Res.* **80**, 4059–4061.

Post, R. L., Jr 1973 High temperature creep of Mt Burnet dunite. Ph.D. thesis (geophysics), University of California at Los Angeles.

Raleigh, C. B. 1965 Glide mechanisms in experimentally deformed minerals. *Science, N.Y.* **150**, 739–741.

Raleigh, C. B. 1967 Plastic deformation of upper mantle silicate minerals. *Geophys. J. R. astron. Soc.* **14**, 45–49.

Raleigh, C. B. 1968 Mechanisms of plastic deformation of olivine. *J. geophys. Res.* **73**, 5391–5407.

Raleigh, C. B. & Kirby, S. H. 1970 Creep in the upper mantle. *Mineral. Soc. Am. Spec. Paper,* **3**, 113–121.

Schmid, S. M. 1976 Rheological evidence for changes in the deformation mechanism of Solenhofen limestone towards low stresses. *Tectonophysics* **31**, T21–T28.

Schmid, S. M., Boland, J. N. & Paterson, M. S. 1977 Superplastic flow in fine-grained limestone. *Tectonophysics* (In the press.)

Stocker, R. L., & Ashby, M. F. 1973 On the rheology of the upper mantle. *Rev. Geophys.* **11**, 391–497.

Schwenn, M. B. 1976 The creep of olivine during hot-pressing. S.M. thesis in geophysics, Massachusetts Institute of Technology, Cambridge, Massachusetts.

Schwenn, M. B. & Goetze, C. 1977 Creep of olivine during hot-pressing. *Tectonophysics* (In the press.)

Takeuchi, S. & Argon, A. S. 1976a Steady-state creep of alloys due to viscous motion of dislocations. *Acta metall.* **24**, 883–890.

Takeuchi, S. & Argon, A. S. 1976b Steady-state andrade creep of single-phase crystals at high temperature. *J. mater. Sci.* **11**, 1542–1566.

Twiss, R. S. 1976 Structural superplastic creep and linear viscosity in the earth's mantle. *Earth & planet. Sci. Lett.* **33**, 79–86.

Von Mises, W. 1928 Mechanik der plastischen Formänderung von Kristallen. *Z. angew. Math. Mech.* **8**, 161.

Weertman, J. 1972 Creep of ice. In *The physics and chemistry of ice* (eds E. Whalley, S. J. Jones & L. W. Gold), pp. 320–337. Ottawa: Royal Society of Canada.

Weertman, J. 1970 The creep strength of the earth's mantle. *Rev. Geophys.* **8**, 145–168.

Wilkinson, D. S. & Ashby, M. F. 1975 Pressure sintering by power law creep. *Acta metall.* **23**, 1277–1285.

Wilkinson, D. S. & Ashby, M. F. 1976 The development of pressure sintering maps. *Proc. 4th Int. Cong. on Sintering and Related Phenomena, May* 26–28, 1975. (In the press.)

Discussion

J. P. POIRIER (*Saclay, France*). Following Professor Goetze's interesting considerations on the glide and climb of edge dislocations, I should like to make a few remarks on the mobility of screw dislocations in olivine:

(1) Numerous long and straight [100] screw dislocations are very often observed in olivine crystals experimentally or naturally deformed at high temperature. These dislocations obviously move with great difficulty.

(2) Similar straight screw dislocations are known to be present in body-centred cubic metals at low temperatures. This has been explained by the fact that dislocations are split on several planes. We have recently shown that such a multiple splitting is also possible for [100] screw dislocations in olivine on (010) and (001) planes.

(3) Thus, the mobility of [100] screws in glide and/or cross-slip is probably controlled by the pinching of one of the stacking fault ribbons.

(4) If the mobility of slow moving screws contributes to the control of high temperature creep of olivine and if hydrostatic pressure favours the pinching of the stacking fault ribbons as is the case with sodium chloride, then the effective activation volume could be much smaller than the activation volume for diffusion controlled climb.

Phil. Trans. R. Soc. Lond. A. **288**, 121–146 (1978)   [121]

Printed in Great Britain

High-temperature deformation of two phase structures

By J. H. Gittus

United Kingdom Atomic Energy Authority, The Reactor Group,
Springfields, Salwick, Preston, Lancashire, U.K.

Some of the ways in which two phase crystalline substances deform are reviewed. By the addition of a small amount of a finely dispersed second phase the resistance of a pure material to deformation can be spectacularly increased. The effect is quite disproportionate to the amount of second phase added. It stems from the fact that crystal dislocations have to increase in length to circumvent the particles and the energy needed to produce this effect must be supplied by the external machine causing deformation.

If equal amounts of equally deformable phases are present (as in certain alloys of eutectic or eutectoid composition) then the material has a low resistance to deformation but a very high ductility: the phenomenon of superplasticity. Deformation now occurs by sliding at the interphase boundaries (i.p.bs). Superdislocations whose glide causes this sliding are effectively trapped in the boundary – a condition of thermodynamic equilibrium which (it is shown) gives rise to the observed deformation properties.

At higher deformation rates dislocations appear in the grains (particles) and may form into cells. They are not, however, trapped in the cell boundaries and are observed to produce strain by traversing the cells. The aggregate now exhibits the same deformation as the cells of which it is composed. The cell size is that which minimizes the free energy for a given rate of deformation and the ductility is high, but not so high as that produced in the superplastic state.

A new mechanism of deformation appears if transformation continues during straining. Large, self-cancelling internal stresses and strains occur and the external strain rate polarizes the latter, producing external strain. The external strain rate bears the same proportionality to the external stress as does the internal strain rate to the internal stress. An analogy is provided by irradiation creep where crystallization of self-interstitials created by bombardment occurs continuously from the supersaturated solid solution. Here the rate of transformation and the internal stress have actually been measured and the predicted creep characteristics correlate well with in-pile creep data.

Introduction

Much of the existing fabric of our knowledge, both theoretical and experimental, about the deformation of crystalline materials derives from pure substances. By contrast, many commercial alloys and naturally occurring minerals comprise two or more phases. In this paper I review some of the modes of deformation for two phase materials paying special attention to areas where recent advances in understanding have occurred.

The simplest case might be thought to be that in which a very small amount of one of the phases is present and so we shall start with this. In the event, we shall see that despite its superficial simplicity this case has many interesting and unexpected features, most of them stemming from the properties of the crystal lattice.

Dispersion hardening

The simplest hypothesis about the effect of a second phase upon deformability is that Einstein's (1911) equation would apply:

$$\eta_c = \eta(1 + 2.5\,C_p), \tag{1}$$

where C_p is the volume fraction of dispersed second-phase particles and η is the viscosity of the medium in which the particles are dispersed. This equation predicts that the viscosity, η_c, of a dilute suspension is a linear function of the amount of second phase which is dispersed. As C_p approaches zero, the effect of the dispersed phase on viscosity becomes vanishingly small.

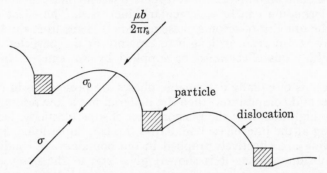

FIGURE 1. Dislocation held up by dispersed particles and subject to the external stress (σ), a friction-stress due to the particles (σ_0) and a stress due to neighbouring dislocations ($\mu b/2\pi r_s$).

Dilute suspensions of particles in liquids obey equation (1) quite well. However, dispersion-strengthened crystalline materials do not seem to obey it at all. For example, consider the case of silver containing a dispersion of fine alumina particles (Oguchi, Oikawa & Karashima 1974): here the addition of only 0.05 % of the dispersed particles (i.e. $C_p = 5 \times 10^{-4}$) was enough to introduce a threshold stress estimated by Evans & Knowles (1976) to be 29 MPa. Below this stress deformation occurred very slowly. By contrast equation (1) would predict a negligible effect of such a small concentration of particles upon deformability. The reason for this difference is that, in a fluid, flow will occur around any small cluster of atoms whose deformation is impeded by a particle, since the flowing atoms are not on a permanent lattice. By contrast, in a crystal, the atoms are constrained to remain on their lattice sites by large interatomic forces. A particle which hinders deformation in one part of the crystal therefore prevents movements of contiguous regions too.

From an analysis of the interaction between particles and dislocations in crystalline materials the theoretical equation for the creep of dispersion strengthened crystalline materials has been derived (see Gittus 1975 a).

Thus a dislocation is subjected to the stress ($ = \mu b/2\pi r_s$) due to its neighbours and to stress σ due to an externally applied force (figure 1). If the specimen contains a dispersion of impenetrable obstacles then the condition for the dislocation to glide is that the difference between the external stress and the internal stress due to the other dislocations must equal or exceed a friction stress σ_0:

$$\sigma - \mu b/2\pi r_s \geqslant (\mu b/xL = \sigma_0).$$

Here L is the distance between adjacent particles, μ is the shear modulus, r_s is the spacing between adjacent network dislocations (the dislocation density ρ then equals $1/r_s^2$), b is the Burgers vector and x is a factor of order unity whose exact magnitude depends on the details of the mechanism whereby the dislocation penetrates the array of dispersed particles (Brown & Ham 1971). Following the lines of the analysis detailed in previous papers (Gittus 1974 *a*, *b*) then the steady-state creep rate ($\dot{\epsilon}$) should, theoretically, be

$$\dot{\epsilon} = (8\pi^3 c_j) \frac{D_v \mu b}{kT} \left(\frac{\sigma}{\mu} - \frac{b}{xL} \right)^3 ; \tag{2}$$

c_j is the jog concentration, D_v is the volume self-diffusion coefficient at temperature T (K) and k is Boltzmann's constant.

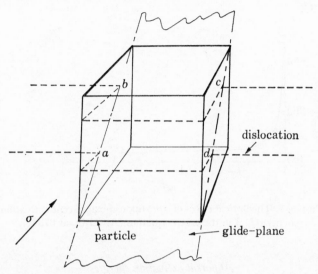

FIGURE 2. Brown & Ham's (1971) model of a dislocation surmounting a cubic particle. The glide plane is *abcd*. As the dislocation moves from *ad* to *bc*, additional line length must be created.

If the material contains no dispersed particles, then equation (2) reduces to that for a pure material developed in the earlier work (Gittus 1974*a*), as it should. The condition for creep to occur is that the external stress should exceed $\mu b/xL$. The dispersion-strengthened material creep-characteristics are in fact simply those of a pure material subjected to a smaller stress:

$$\dot{\epsilon}(\mathrm{disp}, \sigma) = \dot{\epsilon}(\mathrm{mat}, \sigma - \sigma_0). \tag{3}$$

In equation (3), $\dot{\epsilon}(\mathrm{mat}, \sigma - \sigma_0)$ is the creep rate of the matrix material in the absence of a strengthening dispersion and under an external stress of magnitude $\sigma - \sigma_0$; and $\dot{\epsilon}(\mathrm{disp}, \sigma)$ is the creep rate of the disperson-strengthened material at a stress σ. Equation (3) offers us an alternative method of estimating the creep resistance of a dispersion strengthened material if we have creep data for the pure matrix: in the absence of such data we use equation (2).

As an example of the absolute magnitude predicted by equations (2) and (3) we consider the dispersion-strengthened material of Hansen & Clauer (1973) at 400 °C ($0.72 T_m$).

The following equation is taken for the jog concentration (Gittus 1974*a*):

$$c_j = \exp \left(-\mu b^3 / 8\pi kT \right). \tag{4}$$

Figure 2 shows how a dislocation can surmount a particle by climb-plus-glide. Climb occurs most readily at high temperatures and facilitates particle bypassing. Following Brown & Ham (1971) we can show that the following approximations are, theoretically, valid:

$$\begin{aligned} \tfrac{1}{2}T_m \leqslant T < T_m &: x = 2T/T_m; \\ \tfrac{1}{2}T_m > T &: x = 1. \end{aligned} \tag{5}$$

Then using the values of the various parameters of equations (2), (3) and (4) collected together in Table 1 of Ashby's (1972a) paper on deformation maps we find that for $\sigma = 28.76$ MPa and $T = 673$ K with $b/xL = 1.36 \times 10^{-3}$, equation (2) gives a predicted steady-state creep rate of 5.8×10^{-6} s^{-1} while equation (3) gives the almost identical value of 4.1×10^{-6} s^{-1}. We note in passing that at this temperature and stress the theory predicts the creep rate of the pure aluminium matrix with high accuracy. Comparison with Hansen & Clauer's data then shows that the 'theoretical material' has at 400 °C (673 K) the creep rate which the real material would exhibit at about 440 °C.

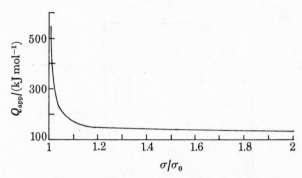

FIGURE 3. Theoretical effect of σ/σ_0 upon the apparent activation energy (Q_{app}) in the creep-equation (Al–Al$_2$O$_3$ at 673 K).

Apparent activation energies

An empirical equation attributed to Nutting and Scott-Blair which is often used (Gittus 1964) to fit dislocation creep data has the following form:

$$\dot{\epsilon} = A\sigma^n \exp[-Q_{app}/RT]. \tag{6}$$

In estimating the value of Q_{app} in equation (6) by the temperature-cycling method it is usual to assume that Q_{app} and n are both independent of temperature and stress so that we can calculate Q_{app} by differentiating equation (6) with both Q_{app} and n assumed constant:

$$Q_{app} = -R\,\mathrm{d}\,(\ln \dot{\epsilon})/\mathrm{d}(T^{-1}). \tag{7}$$

Now, it is in fact incorrect, according to the theoretical equation for the creep of a dispersion strengthened alloy (equation (2)), to assume that Q_{app} and n are constants. If we differentiate equation (2) we can predict the value of Q_{app} which will be produced if the results of a temperature-cycling experiment are used to solve equation (6):

$$-Q_{app} = -(Q_v + Q_j) + RT + 2RK_2 T^2 - 3RT\frac{1 - K_2 T}{\sigma/\sigma_0 - 1}. \tag{8}$$

Here Q_v is the activation energy for volume self-diffusion and Q_j the activation energy for thermal jog-formation. Clearly the apparent activation energy, Q_{app}, depends on both the applied stress and on the mean temperature. Here

$$Q_j = \mu R b^3/k8\pi \tag{9}$$

and

$$\mathrm{d}\mu/\mathrm{d}T = K_2\mu. \tag{10}$$

Using these equations together with data for aluminium taken from Ashby's (1972a) Table 1 for $T = 673$ K we find that $(Q_v + Q_j) = 130$ kJ/mol while at this temperature $K_2 = -5.4 \times 10^{-4}$.

From equation (5) we have $x = 1.44$ so we can calculate a value of Q_{app} for the dispersion-strengthened material of Hansen & Clauer (1973), using the parameter values of the previous section. Substituting these values into equation (8) we find that the apparent activation energy at 400 °C and a stress σ of 28.76 MPa should, theoretically, be 585 kJ/mol: this is the value which Hansen & Clauer calculated from their data for a stress of 27.68 MPa and temperatures in the range 400–500 °C. Figure 3 shows the general effect of stress on Q_{app}.

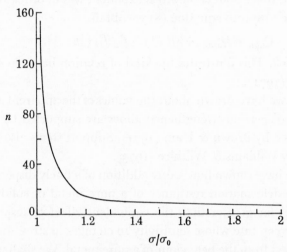

FIGURE 4. Theoretical effect of the ratio of applied stress to threshold-stress (σ/σ_0) upon the exponent of stress (n) in the creep-equation.

According to equation (8), the apparent activation energy should fall as the value of the externally applied stress is increased. This is what McLauchlin (1971) found in his studies of a stainless steel strengthened by a dispersion of NbC. McLauchlin's results also exhibit the high levels of apparent activation energy which the theory predicts. A high activation energy is found also in the results of Gulden & Shyne (1963) while Russell, Ham, Silcock & Willoughby (1968) found that in their experiments the apparent activation energy has the kind of stress dependence which our theory would have led us to expect. Similarly, Vickers & Greenfield (1968) report that in the case of Mg–MgO the apparent activation energy for steady-state creep is about 420 kJ/mol while that for self diffusion is only 135 kJ/mol. They review work on Al–Al$_2$O$_3$ which like the studies of Hansen & Clauer, yields activation energies (in the range 630–1260 kJ/mol) much higher than the self diffusion value.

Stress-sensitivities

Differentiating equation (6) with respect to stress we obtain

$$n = \mathrm{d}/(\ln \dot{\epsilon})/\mathrm{d}(\ln \sigma), \tag{11}$$

which with equation (2) gives $\qquad n = [3/(\sigma - \mu b/xL)]\,\sigma, \tag{12}$

i.e. $\qquad\qquad\qquad\qquad n = 3/(1 - \sigma_0/\sigma), \tag{13}$

and so $\qquad\qquad n \to \infty \quad \text{as} \quad \sigma \to \sigma_0; \quad n \to 3 \quad \text{as} \quad \sigma \to \infty. \tag{14}$

Evidently the limiting behaviour summarized in relations (14) explains why, with increasing external stress, the value of n falls from the near-infinite values at low stresses towards an asymptotic value (see figure 4) ($n = 3$). A numerical example can be obtained by considering again the alumina dispersion strengthened aluminium at 400 °C and a stress of 28.76 MPa (the conditions which led, in the previous section, to a predicted activation energy of 585 kJ/mol). Then equation (12) gives $n = 78$. For comparison, Hansen & Clauer (1973) found that their data conformed to values of n ranging from 50 at 600 °C to over 100 at low stresses and 400 °C.

McLauchlin (1971) found that for his NbC-hardened stainless steel, the value of n was high and exhibited the kind of temperature dependence which equation (12) would predict. Crossland & Jones (1972) found that above 350 °C the stress exponent was of order 10 for Mg–MgO.

By substituting equation (13) into equation (8) we obtain

$$Q_{app} = Q_{app,0} + 3RT(1 - K_2 T) (\tfrac{1}{3}n - 1), \tag{15}$$

where $Q_{app} \to Q_{app,0}$ as $n \to 3$. This illustrates the kind of relation between n and Q_{app} which was observed by McLauchlin (1971).

The conclusions which we have drawn about the values of the apparent activation energy and the exponent of stress for dispersion-strengthened alloys are supported by most of the relevant experimental data reviewed by Brown & Ham (1971). Support stems also from later work on a Nimonic alloy reported by Williams & Wilshire (1973).

In this section, then, we have shown how a tiny addition of a finely dispersed second phase can dramatically increase the deformation resistance of a pure metal or solid solution at elevated temperature. The effect is far greater than the equivalent effect of suspending particles in a liquid and it produces a creep rate whose sensitivity to changes in stress and temperature is not at all what one would expect from the behaviour of a pure metal. We shall now go on to examine what happens if the second phase, instead of being present in small amounts, is almost equal in volume to the first phase.

SUPERPLASTICITY IN TWO PHASE STRUCTURES

The materials which exhibit what Johnson (1970) terms 'fine-grain superplasticity' may be single phase but those exhibiting the highest stability during deformation coupled with the largest extensions are more commonly multiphased. In the latter case it is frequently observed that the two phases are present in approximately equal proportions and that they appear to have approximately equal ductilities. Many of the superplastic alloys are therefore eutectics or eutectoids.

In these superplastic materials, individual grains or phase particles translate large distances relative to one another without exhibiting any large permanent changes in shape. This is illustrated in figure 5. In their studies of superplastic deformation in the Pb–Sn eutectic, Geckinli & Barrett (1976) used a special tensile stage in a scanning electron microscope. They photographed consecutive positions of the grains continuously during straining and it is from measurements made during these experiments that figure 5 was prepared. The grains rotated up to ± 30° from the tensile axis during the tests, often reversing the direction of their rotation during the course of straining. The grains remained equiaxed during straining although no voids were observed to form between them: evidently sufficient localized deformation was occurring to maintain coherency between adjacent grains as they slid past one another.

These observations caused Geckinli & Barrett to conclude that the dominant mode of deformation during superplastic flow is grain boundary sliding with localized fluctuating vibrations in grain shape. This view is consistent with the results of the majority of experiments in which superplastic flow has been produced (Gittus 1975 b).

There are several theories, but two have attracted the most interest. One of these, due to Ashby & Verrall (1973) proposes that a process akin to diffusion creep occurs. It allows the grains to make the fluctuating changes in shape that permit them to remain in contact during sliding. In effect all that need happen is for the material from some grain corners to diffuse to adjacent grain-faces, and vice versa. The interior of the grain does not alter in shape at all and may be regarded as a sphere which slides relative to the other spheres (grains). At low stresses and small grain sizes, Ashby & Verrall noted that the ability of the boundary to act as a source and sink for point defects might become limiting. The creep rate may then depend on the density of dislocations in the boundary. Now the Ashby & Verrall model, for the case in which the interface reaction is *not* rate controlling, leads to the prediction that the creep rate will depend on the first power of the applied stress, whereas experimentally it usually depends on the square or the cube. Interface-reaction control coupled with a sink-density which increases with stress, leads to a dependence of the strain-rate on the square of the stress.

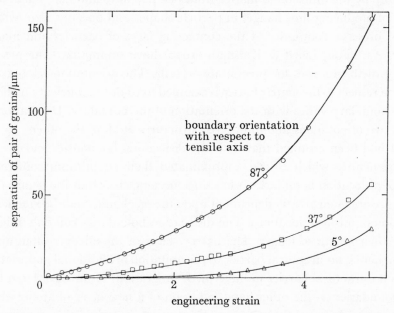

FIGURE 5. Separation of pairs of grains as a function of engineering strain. Pb–Sn eutectic. Reproduced by permission of Dr C. R. Barrett. Initial strain rate $5 \times 10^{-4} \mathrm{s}^{-1}$; grain size 3.5 μm.

This kind of stress dependence was in fact a characteristic of an earlier model of superplastic flow put forward by Ball & Hutchison (1969). Their model, as they remark, is virtually the same as one suggested earlier by Friedel (1964) to explain the fact that the activation energy for the creep of fine grained zinc is that for grain-boundary diffusion. They suggested that at high temperatures dislocations from the head of a pileup would climb into the boundary against which they were piled up. Sources on the slip plane, by operating to replace the dislocations so 'lost', produced creep. They were led to a creep equation having the observed type of dependence of strain-rate upon both stress and grain size. Mohamed, Shei & Langdon (1975) have examined the Ball & Hutchison model, particularly in relation to superplasticity in the Zn–Al eutectic and have pointed to two drawbacks: first, the model assumes that adjacent grains slide in groups and this is not supported by direct observations in the scanning electron microscope; secondly, the model predicts the presence of many intergranular dislocations although transmission electron microscopy and internal marker experiments have been interpreted to indicate that there is very little within-grain dislocation activity during superplastic flow. Another problem is that superplastic flow is sometimes observed to cease at a low stress and neither theory can account for the magnitude of this 'threshold' stress.

Neither of the two theories then is entirely satisfactory, even for the case of superplastic flow in a single phase alloy. In the case of a two phase alloy there are additional difficulties due to the nature of the sliding and climb processes at interphase boundaries. We consider this matter in the next section.

Sliding, growth and dissolution at interphase boundaries

It is, of course, established that particles of a second phase can precipitate and dissolve. In addition processes such as 'ripening' are observed in which material from one particle of a second phase diffuses to another through the intervening matrix. Again, the shape of a particle can change, presumably by the diffusion of matter from one region to another. These mechanisms are capable then of producing the changes in particle shape (and possibly, size) which have to occur in order to preserve continuity at the contacting faces of particles and matrix during superplastic flow. Aaronson, Laird & Kinsman (1970) have summarized the mechanisms of these diffusional growth processes for precipitate crystals. The orientation relation of the precipitate lattice with respect to the matrix lattice is assumed fixed during nucleation. The structure of the interphase boundary varies with the orientation of the boundary. Dislocation structures appear at one or two orientations and disordered structures at all of the others. In Aaronson's original theory it had been assumed that both the dislocation boundaries and the disordered boundaries had structures which closely approximated their equilibrium configurations. At least in the case of dislocation boundaries it has since become clear that the details of the misfit dislocation structure are often greatly dependent upon the mechanics and the kinetics of generation or introduction of misfit dislocations. The dislocation boundaries can only migrate by the movement along their length of ledges. The narrow faces of these ledges often appear to be disordered and so there is no structural barrier to precipitation of additional material onto them. Similarly there is no structural barrier to the growth (migration) of the disordered boundaries. In the ordered boundaries on the other hand, addition of a new layer of atoms will generally alter the packing relation between the two contacting lattices, introducing a measure of disorder. The energy needed to accomplish this is usually high. It is for this reason that precipitates are often in the form of long plates or rods: the long dimension is a dislocation boundary which can only thicken by the slow process of ledge-growth whereas elongation of the precipitate occurs preferentially by the addition of atoms to regions of the boundary that are already so disordered that the new atoms cannot raise their energy significantly.

Sliding at dislocation interphase boundaries is impeded by the same type of energy barrier. Thus the passage of a glissile dislocation leaves in its wake a disordered structure of high energy. The energy increase could be so high as to prohibit the glide and with it any sliding at the interphase boundary. That is to say the threshold stress for i.p.b. sliding could be very high (Ashby 1972b). Dislocation boundaries in single phase materials can slide more easily. They do this, in most cases, by a combination of translation and migration. Atoms are transferred across the interface during sliding to restore the boundary structure, which would otherwise be altered by the sliding process. In this way one of the grains grows at the expense of the other. If atoms are transferred across the interface between two different phases, then the grain which grows does so by acquiring the wrong kind of atoms – those which constituted the other phase. Equilibrium can be restored by the diffusional redistribution of the components but this introduces a large viscosity term since the diffusion path is of the order of the grain size.

It is now suggested, however, that there *is* a mechanism of sliding available on dislocation interphase boundaries. It involves the glide of a superdislocation in the boundary plane. Figure 6 shows what happens. The first of a pair of boundary dislocations creates disorder but the second dislocation, when it passes along the same path, restores order once more. Nakagawa & Weatherly (1972) have actually observed pairs of dislocations in the interface between two

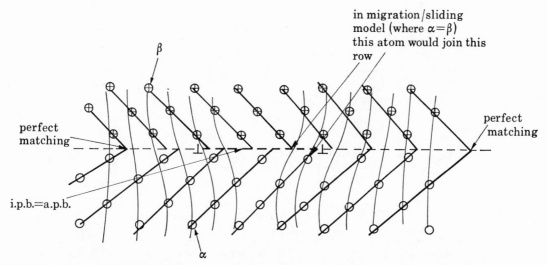

FIGURE 6. Interphase boundary (i.p.b.: – – –) sliding by motion of an i.p.b.–superdislocation contrasted with the Ashby migration/sliding process. The latter involves crystallization of α-atoms on the β-lattice or their diffusion along paths comparable in length to the grain-size: this produces a high i.p.b. viscosity. The viscosity is characteristically much lower if an i.p.b. superdislocation glides in the boundary. The two partial dislocations are then separated by a region of i.p.b. having the structure of an antiphase domain boundary (a.p.b.: – – –).

ordered lattices. The lattices were those of Ni_3Al and Ni_3Nb in a directionally solidified eutectic alloy. The dislocation pairs had separations of 30–40 nm. They were epitaxial dislocations with $b = a/2[1\bar{1}0]_{Ni_3Al}$ and the orientation relation between the phases at their common boundary was $(1\bar{1}0)_{Ni_3Al} \| (100)_{Ni_3Nb}$, $(113)_{Ni_3Al} \| (031)_{Ni_3Nb}$. What we are now suggesting is that the formation of superdislocations during deformation may be a quite general phenomenon in interphase boundaries, even where the contacting phases do not themselves exhibit long-range order. An interphase boundary between two pure substances (α and β) is in fact an ordered solid solution of one phase in the other and the disordered region between the leading and the trailing components of the superdislocation is an antiphase boundary (a.p.b.). The Burgers vector can make any angle to the i.p.b. In figure 6, *b* lies in the i.p.b. If *b* is perpendicular to the i.p.b. then the a.p.b. is equivalent in structure to a faulted dislocation-loop. The superdislocation now moves by climb, vacancies diffusing through the fault to the trailing edge from the leading edge. More generally a mixture of climb and glide will occur, the latter leading to i.p.b.-sliding.

The equilibrium width (w) of the disordered region can be calculated from a knowledge of the elastic constants and a fault energy, Γ (Gittus 1977a). Thus for the case where superdislocations occur with separation l, we can define a long-range order parameter, S, as

$$S = 1 - w/l,$$

(16)

so that, when the dislocations are so close together that the partials touch, $S = 0$ corresponding to the complete absence of long range order (a disordered boundary). The width of the super-dislocation is then given, to the first approximation, by an equation essentially similar to that (Haasen 1965) for the width of the superdislocations in an ordered solid solution:

$$w = \frac{\mu b^2}{2\pi \Gamma S^2}.$$

(17)

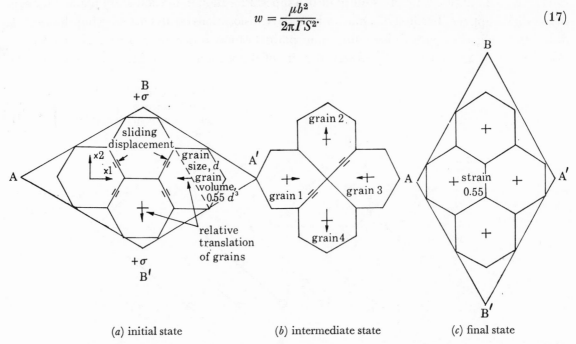

(a) initial state   (b) intermediate state   (c) final state

FIGURE 7. Unit step in the deformation process observed by Naziri *et al.* (1973) and modelled by Ashby & Verrall (1973). The lozenge $ABA'B'$ undergoes a shear of magnitude 0.55 by the moving together of A and A' and the simultaneous moving apart of B and B'.

As an example consider the case where the energy of the disordered boundary in the fault, Γ, is $100\,\mathrm{erg\,cm^{-2}}$ and let $b = 2 \times 10^{-8}\,\mathrm{cm}$, $\mu = 5 \times 10^{11}\,\mathrm{dyn\,cm^{-2}}$, and $S = 0.75$. Then

$$w = 5\,\mathrm{nm}.$$

The disordered region is identical in structure with the narrow face of a growth ledge and we may in fact regard such ledges as superdislocations in which the fault and glide-planes are no longer coparallel. Such 'staggered' superdislocations are certainly observed in grain-interiors.

A new theory for superplastic flow in two phase materials

In the previous section we have argued that particles of differing phases can slide relative to one another by the movement of interfacial superdislocations. We have also remarked that a particle can change its shape or size by the diffusion of matter to and from disordered regions of its boundary. These principles have now been incorporated (Gittus 1977a) in a model of the superplastic flow process for two phase materials. That model is as follows:

The unit process is taken to be the grain-switching event observed during the superplastic flow of the Zn–Al eutectoid by Naziri, Pearce, Henderson-Brown & Hale (1973). It is the unit process which Ashby & Verrall analysed (1973) and is illustrated in figures 7–9. Figure 7 shows what Naziri *et al.* observed in the high-voltage microscope. Grains of the zinc-rich phase (labelled 1 and 3 in figure 7b) push into the boundary between two grains of the Al-rich phase, separating the latter. The aggregate of four grains undergoes a shear of magnitude 0.55 in this process but the final shape of each grain is the same as its starting shape. In fact (figures 7 and 8) each face of a grain becomes a corner and each corner a face. Figure 8 also shows the length of the diffusive paths, λ, along which matter flows to accomplish the shape-changes. It is on average $0.3\,d$. By contrast, in Herring–Nabarro diffusion creep the mean diffusion path is approximately equal to

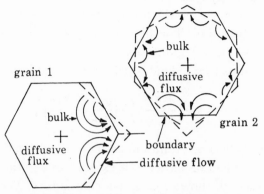

FIGURE 8. In order to maintain contact at the interphase boundaries during the grain switching process of figure 7, material must diffuse between the points and facets of the grains. This flow occurs over short distances and only enough material need be transported to prevent cavitation. After Ashby & Verrall (1973).

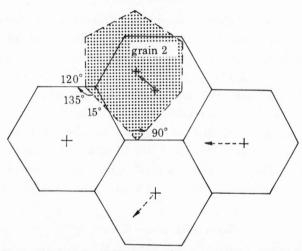

FIGURE 9. Here, superimposed on the initial state (figure 7a), is shown the intermediate position of grain 2. Matter has diffused to produce a new apex (point). In addition the centre of the grain has translated and a shear displacement equal to the amount of this translation has occurred at the interphase boundary. After Ashby & Verrall (1973).

half the grain diameter. Again only about one sixth of the volume of the grain has to diffuse during the unit process in order to produce unit of strain. In the Herring–Nabarro process on the other hand the whole volume of a grain must suffer diffusion transport to produce unit of strain. Figure 9 shows that the grains slide relative to one another during the unit process. It is these shear displacements in the boundary that permit the grains to translate past each other. Their magnitude can be seen by examining the motion of grains relative to a fixed origin, say the centre of mass of grain 2 in figure 9.

Suppose that i.p.b. superdislocations can be generated by (for example) the expansion of intrinsic dislocation loops under the applied stress. In the boundary the dislocations are effectively trapped and can only emerge into the matrix if they reassociate to form matrix dislocations of appropriate Burgers vector. The dislocation boundary is therefore potentially a very effective pileup of length d. The movement of dislocations in this dynamic pileup causes i.p.b. sliding and consequently creep. It proceeds at a rate determined by the climb of the dislocations away from the head of the pileup. They cannot generally climb out of the boundary; therefore they must climb in it. The most likely situation is one in which the dislocations pile up at a triple point where a dislocation boundary on which they are gliding meets two disordered boundaries into which the dislocations then climb. For this climb to occur matter must diffuse to or from the dislocations (diffusion coefficient = D_B, diffusion distance λ).

Analysis of this model (Gittus 1977a) finally yields the *constitutive equation for superplastic-flow*:

$$\dot{\epsilon} = 4.8 \left(\frac{d}{\lambda}\right)^2 \frac{D_B \mu b}{kT} \left(\frac{\sigma - \sigma'_0}{\mu}\right)^2 \left(\frac{b}{d}\right)^2, \tag{18}$$

where d is the particle size, D_B the i.p.b. diffusion-coefficient and σ'_0 a threshold stress. In equation (18) the strain rate is independent of particle size except through the effect of d on the diffusion distance λ: on most models of the flow process, $\lambda = $ constant $\times d$. The constant is of magnitude 0.3 in the Ashby–Verrall grain-switching model.

Structural effects

The dislocation-spacing on dislocation-boundaries theoretically decreases with increasing stress. Hence (equation (16)) the long-range order parameter, S, will fall as the stress increases. Now a completely disordered boundary ($S = 0$) has no preferred relation to the lattices of the contacting phases and tends to adopt a spherical shape to minimize its area (and thus to minimize the surface energy). At the other extreme a dislocation-boundary ($S \rightarrow 1$) follows preferred crystallographic directions and is essentially planar. When a two phase material is stressed, then, we expect the gradual build-up of interphase boundary superdislocations towards the equilibrium density (and the associated fall in S) to make more and more of the interphase boundary adopt a curved shape. This is observed (Gittus 1975b) and in the earlier literature led to the idea that the formation of 'bulbous' structures was a prerequisite of superplasticity.

Source of the threshold stress, σ'_0

It is suggested (Gittus 1977a) that the threshold stress, σ'_0, is due to an interaction between i.p.b. superdislocations and boundary-ledges. To traverse a ledged boundary the dislocations will have to bend. Moreover there will be a change in fault energy of magnitude Γwh per intersection, where h is the height of a ledge, when the narrow face of a ledge is a disordered i.p.b. The applied stress will have to do work in order to overcome this energy in order to detach a dislocation from a ledge. Garmong & Rhodes (1975) saw evidence of these interactions between i.p.b. dislocations and ledges in their transmission electron microscope studies of Ni_3Al–Ni_3Nb composites. The interaction was manifested both as a local alteration of the line vector of the dislocation in certain circumstances and as a change in the reponse of the dislocation image in image-contrast experiments.

A suggested model of the ledge-dislocation network is a two dimensional dislocation network. In order to detach itself from two ledges a boundary dislocation may be forced to adopt a curvature C as high as $2/L$, where L is the width of the broad face of a ledge (i.e. the distance between adjacent narrow faces). If the boundary defect is regarded as a flexible line whose energy per unit length is E, a potential

$$\Delta\mu = E\Omega C/b \approx 2E\Omega/bl \qquad (19)$$

is required to do this (Bardeen & Herring 1952; Thompson & Balluffi 1962). The threshold stress, σ_0', can supply this potential when $\sigma_0' \Omega \geqslant \Delta\mu$. Here Ω is the atomic volume. This, together with equation (19), then gives as an estimate of the threshold stress due to the interaction of glissile i.p.b. superdislocations and ledges:

$$\sigma_0 \leqslant 2E/bL. \qquad (20)$$

Precipitates lying in the boundary are another source of a threshold stress. An equation similar to (20) then applies but with L equal to the spacing between the precipitate particles. Solute-drag may also tend to restrict the movement of the boundary super-dislocations. However, this will reduce the rate of glide at all stresses, rather than introduce a threshold stress (Gittus 1974b).

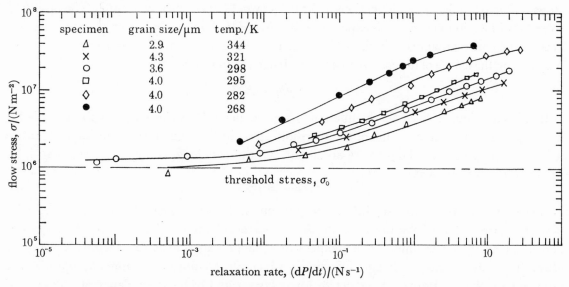

FIGURE 10. Illustrating the 'sigma-shape' of the curve relating log strain rate to log stress. Superplastic Pb–Sn eutectic. The relaxation rate is the rate at which the load fell in a stress relaxation test and must be multiplied by a factor to arrive at a strain rate. In other plots no effect of grain size upon the value of the threshold stress was found. Data of Geckinli & Barrett (1976).

Comparison with the results of experiment

Detailed comparison has been made (Gittus 1977a) between equation (18), the theoretical relation for superplastic flow, and the published results of experiments. Good agreement is generally found and the value of λ $(0.3\,d)$ demanded by the grain-switching model is often closely approximated. For example Misro & Mukherjee (1974) found that equation (18) fitted their data for the Al–Zn eutectoid with a constant of proportionality which in the present terminology implies that $\lambda = 0.19\,d$: this is close to the value of $0.33\,d$ which best fits Ball & Hutchison's data (1969) for the same material. Again Alden has reported on the superplastic flow of a Sn–5 % Bi alloy (1967) and in this case the best value is $\lambda = 0.45\,d$.

Magnitude of the threshold stress

A very clear illustration of the threshold effect is contained in figure 10 (Geckinli & Barrett 1976). In this work the specimens were first deformed in tension until the maximum flow stress was attained. The cross head motion was then stopped and unloading characteristics recorded. The threshold stress is about 1.3 MPa at room temperature. One possibility, which Geckinli & Barrett consider, is that the threshold is determined by the work which has to be done to produce the changes in boundary area which accompany the fluctuating changes in the shape of the *particles*. They show that this would yield a threshold stress about an order of magnitude smaller than the one they measured and Ashby & Verrall (1973) who proposed this source of a threshold stress confirm that this is a typical discrepancy. If we are correct in ascribing the threshold stress to a ledge-dislocation interaction, then (equation (20)) in Geckinli & Barrett's work the value of L is calculated to have been of the order of 400 nm. This is of the order of magnitude observed on dislocation-boundaries. That the ledges do impede the movement of boundary dislocations is established by the electron microscopy reported by Garmong & Rhodes (1975). These workers calculated that in their $Ni_3Al–Ni_3Nb$ eutectic composites the pinning stress acting to pin the i.p.b. dislocations to the ledges would, theoretically, be $\mu/70$. The threshold stress for dislocation movement is expected to be about h/L times this value, for the applied stress acts on the entire unpinned length of dislocation (minimum length L) while the pinning stress only acts on the length of dislocation near to the ledge (height h). So

$$\sigma_0' \approx \mu h/70L.$$

If this equation, derived for a different material, nevertheless holds good for the Pb–Sn eutectic then the observed value of the threshold stress in that material would imply that the ledges on the dislocation boundaries are about 100 times as long as they are high. This is in reasonable accord with experiment.

Activation energies

A change in temperature can alter the proportions of the phases in a superplastic alloy. As the new structure will not have the same strain rate–stress relation as the old one, the apparent activation energy determined by a temperature-change experiment can include a spurious effect due to the structural change. Of work in which an attempt has been made to avoid this latter pitfall, that of Herriot, Baudelet & Jonas (1977) and of Suery & Baudelet (1975) is of special interest. These workers found in their studies of the two phase Cu–P alloys that the activation energy for superplastic flow, calculated with no endeavour to correct for change in structure, was close to that for grain boundary diffusion in copper. This is the kind of result that appears to fit in quite well with the present theory. When, however, allowance was made for changes in *structure* the activation energy for flow rose to a value quite close to that for volume self diffusion in copper. On the face of it this would seem then to run contrary to our theory. However, Perinet (1975) and Bondy, Regnier & Levy (1971) have found that a high value of the activation energy is characteristic of impurity diffusion at interphase (as opposed to grain-) boundaries. Thus Perinet found that the activation energy for diffusion in the boundary between copper and silver was 65 kcal mol^{-1} and this is more than twice the grain boundary activation energy for either of the pure elements by itself. Again a similar finding was reported for the diffusion of Fe and Ag in the interface between these two metals. Experiments to measure self-diffusion in i.p.bs have been planned by Chadwick & Gittus (1977) and are now in progress.

Initial results show that self-diffusion at an i.p.b. has much in common with impurity-diffusion, being a spectacularly rapid process with an activation energy nearer to that for volume diffusion than to that for grain boundary diffusion. These findings clearly accord, on the present theory, with the high value of activation energy for superplastic flow at constant structure found by Baudelet and his collaborators. Indeed this may be a general feature of superplastic flow in two phase materials, having passed unnoticed in earlier investigations because care has not always been taken to compensate for the effect of changing structure when calculating the activation energy for superplastic flow. When Baudelet omitted this precaution he obtained a spuriously low activation energy which is, coincidentally it now appears, similar to that for grain boundary diffusion (as opposed to phase boundary diffusion) in one of the pure components of which his Cu–P alloy was composed.

THE TRANSITION TO NON-SUPERPLASTIC DEFORMATION

If the strain rate is increased sufficiently, dislocations start to appear in considerable numbers inside the grains or particles of a superplastic material. It loses its high ductility and the strain rate varies now with a higher power of the externally applied stress. 'Conventional' dislocation creep has begun. In recent work a method of deducing the critical stress at which this change in deformation character will occur has been developed (Gittus 1977b). Details are given below.

Thermodynamic considerations

Consider a specimen undergoing dislocation creep at a constant strain-rate, $\dot{\epsilon}$, above that at which superplasticity is observed. Characteristically it will, at the outset of the creep process, contain a random three dimensional dislocation network. With the passage of time the dislocations form themselves into cells. Typically the cell diameter and dislocation density both move towards an equilibrium value during this initial period (McElroy & Szkopiak 1972). The large extensions which can occur during this steady state strongly suggest that it is a condition of thermodynamic equilibrium, characterized by a minimum free energy.

At this thermodynamic equilibrium:

$$(\delta F)_{T,V} \geqslant 0. \tag{21}$$

Here, after Shewmon (1965), the lower case delta (δ) is used to indicate that the free energy is unchanged or increased for any small (finite) displacement of the system.

Components of the free energy of the specimen

The dislocations which are contained in the creep specimen make an important contribution to its free energy. That contribution depends on the dislocation density and upon the arrangement of the dislocations characterized by a quantity K. If $K = 1$ then the dislocations are arranged in a uniform three dimensional network. If $K > 1$ then the dislocation arrangement consists of cells having most of the dislocations in their walls or boundaries. The larger is K, the bigger is the cell diameter (for a given dislocation density).

Another component of the free energy of the specimen is the elastic plus viscoelastic strain-energy produced by the externally appplied stress. If this were the only component, then the lower the applied stress the lower the free energy. So at a fixed strain rate the dislocations would tend to move into that arrangement which minimizes the stress that the external machine has to exert. The condition of thermodynamic equilibrium could then be the condition of minimum creep-strength. We can show that there is a certain value of K (of order $c_j^{-\frac{1}{3}}$) which produces this minimum creep strength. If we now consider both components of the free energy (that due to dislocations and that due to elastic-plus-viscoelastic strain) then we can calculate a more exact estimate of the value of K which minimizes the free energy. It transpires (Gittus 1977b) that

$$\frac{dF}{dK} = \frac{-1}{\mu}\left(\frac{\dot{\epsilon}}{B}\right)^{\frac{2}{3}}K^{-3} + \frac{1}{\mu}\left(\frac{\dot{\epsilon}}{B}\right)^{\frac{2}{3}}(-2K^{-3} + \tfrac{1}{5}c_j^2 K^3 + \tfrac{3}{10}C'c_j^2 K^5)(K^{-2} + \tfrac{1}{20}c_j^2 K^4). \tag{22}$$

Here
$$B = c_j D_v b/(\mu^2 kT), \tag{23}$$

$$C' = ((\mu/\mu_R) - 1)/K^2, \tag{24}$$

and μ_R is the relaxed modulus.

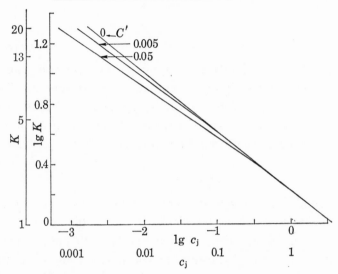

FIGURE 11. Theoretical value of K as a function of c_j for various values of C'. Here $C'K^2 = (\mu - \mu_R)/\mu_R$ where μ and μ_R are the shear modulus and the relaxed modulus respectively. Theoretically $0 \leqslant C' \leqslant 0.05$. In practice its value is usually less than 0.005.

The condition set by equation (21) is met if $dF/dK = 0$ and $d^2F/dK^2 = $ positive. Equation (22) with $dF/dK = 0$ gives

$$c_j^2 = \{-b + (b^2 - 4ac)^{\frac{1}{2}}\}/2a, \tag{25}$$

where
$$a = 10^{-2}K^7 + \tfrac{3}{200}C'K^9; \atop b = \tfrac{3}{10}C'K^3 + \tfrac{1}{10}K; \atop c = -K^{-3} - 2K^{-5}. \Big\} \tag{26}$$

Differentiating equation (22) a second time we find that the condition $d^2F/dK^2 = $ positive is met. So equations (25) and (26) correspond to a state of thermodynamic equilibrium. From them the theoretical value of K may be calculated.

Theoretical relation between K, C' and c_j

In figure 11 equation (25) has been used to provide the theoretical relation between K and c_j for $C' = 0$, 0.005 and 0.05. $C' = 0.05$ is the maximum theoretical value of this parameter, corresponding to the case in which all of the dislocations in the cell walls bow out reversibly under stress, contributing to the viscoelastic strain (Friedel 1964). In practice the measured ratios of relaxed to unrelaxed modulus correspond to C'-values of order 0.005 (i.e. only about one tenth of the dislocations bow in this manner).

For materials in use at the temperatures where dislocation-creep is the dominant deformation mechanism we generally have $0.1 > c_j > 0.001$ and so theoretically $20 > K > 5$: the values of K that have been measured experimentally do in fact lie generally in this range (Gittus 1975 *b*).

TABLE 1. THEORETICAL RELATION BETWEEN THE VALUE OF K AND
THOSE OF c_j AND K' FOR VARIOUS VALUES OF C'

C'	K	c_j	c_j (approx.)	K'
0	1	3.61	3.16	1.65
	5	5.50×10^{-2}	5.62×10^{-2}	3.36
	13	5.11×10^{-3}	5.13×10^{-3}	7.30
	20	1.75×10^{-3}	1.74×10^{-3}	10.80
0.001	10	9.42×10^{-3}	9.04×10^{-3}	5.44
0.005	1	3.60	9.12	1.65
	5	5.20×10^{-2}	7.24×10^{-2}	3.11
	13	3.80×10^{-3}	4.17×10^{-3}	4.48
	20	1.25×10^{-3}	1.15×10^{-3}	6.60
0.05	1	3.49	5.13	1.61
	5	3.87×10^{-2}	4.07×10^{-2}	2.17
	13	2.37×10^{-3}	2.34×10^{-3}	2.36
	20	6.87×10^{-4}	6.46×10^{-4}	2.51

Note: if $C' = 0$ then $K \approx 1.58\,c_j^{-0.4}$; $K' = 1 + K/2$.

Approximations to K

For finite values of C' and for $K > 5$, equation (25) reduces to the following approximation:

$$K \approx [0.03\,C']^{-\frac{1}{12}}c_j^{-\frac{1}{3}}. \tag{27}$$

For $C' = 0$ and for $K \geqslant 1$ the following alternative approximation to equation (25) is valid:

$$K \approx 1.58\,c_j^{-0.4}. \tag{28}$$

To illustrate the range of validity of these approximations, equations (27) and (28) have been used to produce estimates of c_j (c_j (approx.)) for selected values of K and C'. The results are compared in table 1 with the exact values of c_j derived from the full solution of equation (25). Only in the case of $K = 1$ for $C' = 0.005$ and 0.05 are the approximations poor: for most purposes, the approximations are acceptable and lead to some simplification of the creep equation.

Theoretical relation between cell diameter and stress

It has been shown (Gittus 1976) that, according to the present theory, the following equation should relate the stress causing creep to the diameter of the cells, L':

$$\sigma = (\mu b/L')\,(1 + \tfrac{1}{20}c_j^2\,K^6). \tag{29}$$

Experiments indicate (Staker & Holt 1972) that

$$\sigma \approx K'\mu b/L', \tag{30}$$

with

$$K' \approx 10. \tag{31}$$

A theoretical value of K' can be obtained by solving equation (29) for K' using the values of K and c_j from table 1. This has been done and the theoretical values of K' so obtained are there tabulated. The theoretical K' values are seen to be of the expected magnitude.

Deformation maps for the transition

We can now begin to see what happens as the stress is reduced in a two phase material undergoing dislocation creep. In each of the particles of the two phases there is a network of dislocations arranged in cells. As the stress is reduced, the equilibrium cell diameter, L, increases (equation (29) and table 1 with equation (30)). We can see that at a low enough value of the stress the cell diameter will become equal to the particle size and at that point the whole of the strain must be accounted for by superplastic flow. At some higher stress the strain due to superplastic deformation mechanisms will be equal to that due to dislocation creep. This will occur when the *cell* boundary area is about equal to the *interphase* area, or more precisely when

$$L' \approx d/3. \tag{32}$$

So from equations (30) and (32): $d/b = 3K' (\sigma/\mu)^{-1}.$ (33)

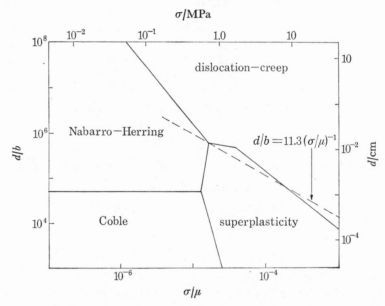

FIGURE 12. Deformation mechanism map prepared by Mohamed & Langdon (1976) for Zn–22% Al. Theoretical line for the boundary between superplasticity and dislocation-creep.

The validity of equation (33) can be assessed by means of deformation maps prepared by Mohamed & Langdon (1976) for Zn–22 % Al and Pb–62 % Sn. These are reproduced here as figures 12 and 13 and on them have been drawn the theoretical lines corresponding to equation (33). The appropriate values of K' were calculated using table 1 and were as follows:

$$\text{Zn–22 \% Al,} \quad K' = 3.78; \tag{34}$$

$$\text{Pb–62 \% Sn,} \quad K' = 2.31. \tag{35}$$

The theoretical lines according to equation (33) with the values given in equations (34) and (35) have been plotted in figures 12 and 13 and are seen to give a good indication of the boundary where superplastic deformation produces the same contribution to the flow rate as does dislocation creep.

Source of the high ductility in superplastic flow

It has long been difficult to explain just why the ductility in the superplastic régime should be so high. Certainly if $m = 1$ then one would expect quasi-infinite ductility (Newtonian viscous flow) since it is easy to show that in such a case a 'neck' will not propagate. A neck is a local region of reduced diameter in a tensile specimen. It forms as a result of some local inhomogeneity of structure or temperature. If thinning occurs faster at the neck than elsewhere, the specimen will eventually neck-down to a point and fracture. This is what happens in ordinary creep. It does not happen (and is not expected) if $m = 1$. The anomaly has been that necks do not propagate in superplastic materials where m can be as low as 0.3 and is theoretically (we have shown) equal to 0.5.

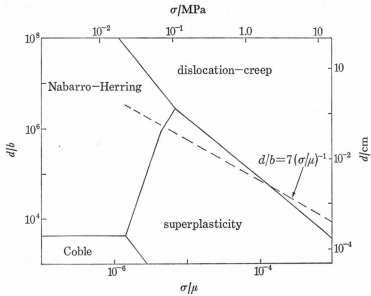

FIGURE 13. Deformation mechanism map prepared by Mohamed & Langdon (1976) for Pb–62% Sn. Theoretical line for the boundary between superplasticity and dislocation-creep.

An explanation can now be offered. It is simply that when the cell-diameter becomes equal to the particle (grain) size *there is a further reduction in the free energy of the specimen*. This is partly due to the low Burgers vector values of i.p.b. dislocations and partly because the dislocation-pileups on the i.p.bs, by concentrating the stress causing climb, reduce the stress needed to cause a given strain rate. We have shown that cells adopt a size which minimizes free energy during ordinary dislocation creep. When that size is equal to the particle (grain) size there is a further step-reduction in free energy. If a neck does try to propagate in the superplastic state then the associated local increase in stress, by reducing the equilibrium cell size below the particle size, produces a large increase in the free energy of the system. So neck propagation is strongly resisted.

DISLOCATION CREEP

Equation (29) is an approximation, taken from a more exact analysis of dislocation creep (Gittus 1976) which has led to the following equation:

$$\sigma = \sigma_{\perp} + F',$$

(36)

where $\qquad\qquad\qquad\qquad\qquad \sigma_\perp = (\dot{\varepsilon}/B)^{\frac{1}{3}}/K^2;$ $\qquad\qquad\qquad\qquad\qquad$ (37)

$$F' = (kT/b^2 l')\,\mathrm{arsinh}\,(\sigma_\perp\,b^3 c_{\mathrm{j}}\,K^6/20\,kT).$$ $\qquad\qquad$ (38)

Here l' is the jog-separation. The equation agrees quantitatively with creep data.

The sensitivity of the creep rate to a change in the externally applied stress, predicted by these equations, falls as the stress moves towards zero (figure 14). So (from table 2) $n = \mathrm{d}\lg\dot{\varepsilon}/\mathrm{d}\lg\sigma$ for gold (at a temperature T equal to half the absolute temperature of melting) is predicted to fall from a value of about 5 toward its low-stress asymptote of 3 as the applied stress falls from 10^{-3} to $10^{-4}\mu$. As K (the ratio of the cell size to the dislocation-spacing) becomes large the equations reduce to one found (by Garofalo) to give a good fit to creep data over an extended range of temperatures and stresses. Most of the temperature dependence arises from that of the volume self-diffusion coefficient, D_{v}; again this is in general accord with the results of experiment.

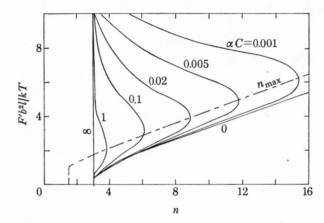

FIGURE 14. Relation between $F'b^2 l/kT$ (proportional to stress, σ, at high stress) and the value of n in the approximation $\dot{\varepsilon} \propto \sigma^n$.

TABLE 2. COMPARISON OF THE ACTUAL AND THEORETICAL VALUES OF STRESS WHICH PRODUCE A GIVEN CREEP-RATE IN GOLD AT $T_{\mathrm{m}}/2$.

Predicted and actual values of n; theoretical value used for K.

actual, experimentally determined stress, σ	theoretical stress	$\mathrm{d}\ln\dot{\varepsilon}/\mathrm{d}\ln\sigma = n$	
		theoretical	experimental
$10^{-3}\mu$	$0.96 \times 10^{-3}\mu$	5.0	5.5
$10^{-4}\mu$	$0.37 \times 10^{-4}\mu$	3.0	5.5

It seems that during dislocation creep in a two phase material the particles of the two phases deform like the specimen itself and so a simple law of mixtures, based on the deformability of the phases in isolation, often yields a good account of the creep rate of the composite.

DEFORMATION DURING TRANSFORMATION FROM ONE PHASE TO ANOTHER

An additional mechanism of deformation can operate in a two phase material if it is subjected to stress while in the course of actually transforming from one phase to another. Imagine that the transformation from beta to alpha occurs by the formation, in a matrix of beta, of a small region of alpha and let there be an increase in volume due to the phase change. Then transformation will cause an internal stress (I) and it will also deform the material (at rate $\dot{\epsilon}_i$ say). We can use these values to form an approximate estimate of the constant λ_L in the Levy–Mises stress–strain increment relations:

$$\dot{\epsilon}_{ij} = \lambda_L \sigma_{ij}, \tag{39}$$

i.e.
$$\lambda_L \approx \dot{\epsilon}_i/I. \tag{40}$$

Equation (39) with λ_L defined by equation (40) then becomes the creep equation for a material which is transforming. The condition for the applicability of equation (40) is seen to be that the second invariant of the strain tensor is not significantly altered by the external stress system. Or more simply:

$$\sigma \ll I. \tag{41}$$

Roberts & Cottrell (1958) were the first to suggest this type of creep and a more exact derivation of the equation was given by Anderson & Bishop (1962).

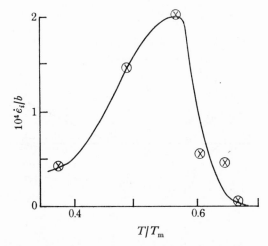

FIGURE 15. Dislocation-flux ($\dot{\epsilon}_i/b$) measurement for copper against homologous temperature (T/T_m) for a displacement-rate ($\dot{\phi}$) of 1.5×10^{-2} atom per lattice-site per second.

There is good support for the equation (Gittus 1975b). For example, it represents quite well creep in ferrous alloys during transformation and the creep of uranium during transformation. Indirect evidence indicates that the effect is also present during the transformation of dilute alloys of zirconium. In order to produce significant strains it is necessary to cycle the material through the phase change. If this is not done then creep by this mechanism ceases as soon as the transformation is complete. An exception is provided by the analogous case of irradiation creep in cubic materials such as stainless steels and nickel based alloys. Here atoms which have been displaced from their lattice sites by bombardment tend to drift towards edge dislocations, arriving there in somewhat greater numbers than do the vacant lattice sites from which the atoms were displaced. The surplus vacancies then form internal voids and as a result the material *swells*.

The process is akin to transformation in the sense that atoms are continually crystallizing onto dislocations. Unlike a normal transformation, however, it continues as long as irradiation continues and so the associated irradiation creep should continue also. Its magnitude should be given by equations (39) and (40) and as an estimate of the value of $\dot{\epsilon}_i$ we can use the experimentally measured value of the swelling rate, $\dot{S}$. Alternatively we can measure the rate of climb of the dislocations by time lapse photography in the million volt electron microscope (Gittus 1975 *b*), a technique which is particularly attractive since we can arrange matters so that some of the electrons displace the atoms whilst others form the magnified image. Both approaches have been used and they provide good confirmation of equations (39) and (40). Thus figure 15 shows the value of $\dot{\epsilon}_i/b$ measured for copper in the electron microscope and figure 16 gives values for the internal stresses deduced from measurement of dislocation-bow in similar experiments. When these results are used in equations (39) and (40), they lead to predicted creep rates which are in the order of magnitude observed in-pile creep experiments. Again from the results of such creep experiments and the rate of swelling (which occurred simultaneously) it is possible, via the theoretical equations, to calculate the internal stress. This has been done for two sets of creep data and (Gittus 1975 *b*) the estimates so produced are close to the irradiated yield point of the material: arguably a reasonable estimate of the internal stress which operated in-pile.

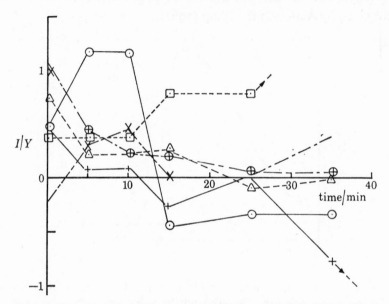

FIGURE 16. Internal stress (I) divided by the Young modulus (Y) deduced from measurements of dislocation bow at various instants.

DISCUSSION

In this paper we have drawn together the threads of contemporary understanding about the deformation properties of two phase materials. We see that these differ from single phase materials in several, on the face of it unexpected, ways.

First, there are the dispersion strengthened materials whose characteristic is that a tiny proportion of dispersed particles can produce an enormous reduction in deformability. The effect is explained by the way in which dislocations have to behave in order to circumvent the particles. Generally, to produce a significant strain rate, the dislocations must increase in length in order to get past the particles. The work which is needed to produce this increase in length is supplied by the external machine that produces the deformation. The smaller the strain (movement of the externally applied load) produced as a result of each bypassing event, the larger the external stress must be if it is to supply the work needed to increase the line length of the dislocation and thus permit it to bypass the particle.

The next class of materials which we have considered contain almost equal amounts of the two phases. They are the eutectics and eutectoids and they exhibit superplasticity under suitable conditions. We show that this is a process in which the particles of the two phases slide over one another like grains of sand in an hour-glass. Sliding is provisionally attributed to the glide of interphase boundary superdislocations and an equation which fits the available data and which is based on an analysis of this sliding process is presented. It contains a threshold stress which, it is shown, can reasonably be identified with the stress needed to overcome the friction between i.p.b. superdislocations and ledges in the interphase boundary. The high value of the activation energy for superplastic flow in two phase materials is seen to be in accord with recent measurements of i.p.b. self-diffusion coefficients.

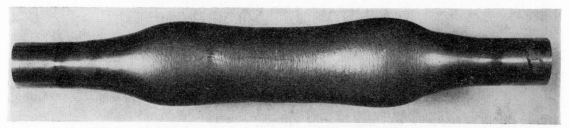

FIGURE 17. Specimen of the zirconium–tin alloy Zircaloy-2, illustrating the weakness and ductility which characterize superplasticity. The tubular specimen was heated by passage of a current of electricity: the ends were maintained colder than the centre by clamping them in water-cooled copper grips. Gas pressure was used to inflate the tube. The end regions were in the alpha-plus-beta phase field and swelled superplastically. The centre was in the beta phase-field and swelled by dislocation-creep: although much hotter than the ends it was evidently stronger (a consequence of the different deformation mechanisms).

If the strain rate is increased, then eventually within the grains (particles) there appears a network of dislocations arranged in cell–boundaries. The equation for creep in such a structure is developed. It invokes the movement of dislocations *across* the cells and so the cells shear during this type of creep ('dislocation creep') and the strain of the specimen (an aggregate of cells) equals that of an individual cell. In contrast, the dislocations which cause superplastic deformation were effectively trapped in the particle boundaries. So their glide gives rise, not to shear of the particle but to sliding on its boundary. It is this difference which accounts for the change from heterogeneous to homogeneous flow as we go from superplastic flow (at low strain rates) to dislocation creep (at higher strain rates).

The size of the cells increases, theoretically, as the applied stress is diminished. Experiments have confirmed this and agree numerically with the theory. The cell size is that which minimizes the free energy of the creeping specimen, by reducing both the external stress (needed to maintain a given creep rate) and the elastic energy of the dislocations themselves. The specimen is in a state of thermodynamic equilibrium and is resistant to perturbations such as those which lead to neck-propagation and fracture.

An even deeper minimum in the free energy occurs when the cell size becomes equal to the particle (grain) size and it is this which accounts for superplasticity – the property of extending by a factor of 2–20 without fracture (figure 17).

Finally we consider what happens if, during transformation from one phase to another, a material is subject to stress. This, it is shown, gives rise to an additional source of deformation. What happens is that the external stress perturbs slightly the conditions (high internal stress and internal deformation) set up by the transformation, leading to a component of external strain. The ratio of external stress to external strain rate is theoretically equal to the ratio of internal stress to internal strain rate (for small enough external stresses). Again it seems that experiments support the theory. In particular, when transformation is simulated by using irradiation to generate a constant super-saturation of self interstitials (which continually crystallize out on dislocations) then irradiation creep occurs at the theoretical rate and can be tied to measurements of transformation rate and internal stress made during irradiation in the high voltage microscope.

I am indebted to the United Kingdom Atomic Energy Authority for permission to publish this paper.

REFERENCES (Gittus)

Aaronson, H. I., Laird, C. & Kinsman, K. R. 1970 In *Phase transformations*, chapter 8. Metals Park, Ohio: Am. Soc. Metals.

Alden, T. 1967 *Acta metall.* **15**, 469.

Anderson, R. G. & Bishop, J. F. W. 1962 *Inst. Metals. Symp. on Uranium and Graphite*, paper 3. London: Institute of Metals.

Ashby, M. F. 1972a *Acta metall.* **20**, 887.

Ashby, M. F. 1972b *Surface Science* **31**, 498–542.

Ashby, M. F. & Verrall, R. A. 1973 *Acta metall.* **21**, 149–163.

Ball, A. & Hutchison, M. M. 1969 *Met. Sci. J.* **3**, 1–7.

Bardeen, J. & Herring, C. 1952 In *Imperfections in nearly perfect crystal* (ed. W. Shockley), p. 279. New York: Wiley.

Bondy, A., Regnier, P. & Levy, V. 1971 *Scr. metall.* **5**, 345–350.

Brown, L. M. & Ham, R. K. 1971 In *Strengthening methods in crystals* (eds A. Kelly & R. B. Nicholson). London: Applied Science Publishers.

Chadwick, G. A. & Gittus, J. H. 1977 *Interface Diffusion.* (U.K.A.E.A. document).

Crossland, I. G. & Jones, R. B. 1972 *Met. Sci. J.* **6**, 162.

Einstein, A. 1911 *Annln Phys.* **34**, 59.

Evans, H. E. & Knowles, G. 1976 *Central Electricity Generating Board report*, RD/B/N3590.

Friedel, J. 1964 *Dislocations.* Oxford: Pergamon.

Garmong, G. & Rhodes, C. G. 1975 *Metall. Trans.* A **6** A, 2209–2216.

Geckinli, A. E. & Barrett, C. R. 1976 *J. Mater. Sci.* **11**, 510–521.

Gittus, J. H. 1964 *Phil. Mag.* **9**, 749.

Gittus, J. H. 1974a *Acta metall.* **22**, 789.

Gittus, J. H. 1974b *Acta metall.* **22**, 1179.

Gittus, J. H. 1975a *Proc. R. Soc. Lond.* A **342**, 279–287.

Gittus, J. H. 1975b *Creep, viscoelasticity and creep-fracture in solids.* London: Elsevier Applied Science Publishers.

Gittus, J. H. 1976 *Phil. Mag.* **34**, 401.

Gittus, J. H. 1977a *A.S.M.E. J. Engng Mater. & Technol.* **99**, 244.

Gittus, J. H. 1977b *Phil. Mag.* **35**, 293.

Gulden, T. D. & Shyne, J. C. 1963 *Trans. metall. Soc. A.I.M.E.* **227**, 1088.

Haasen, P. 1965 In *Physical metallurgy* (ed. R. W. Cahn), see page 867, eqn. (16). Amsterdam: North Holland.

Hansen, N. & Clauer, A. H. 1973 *Proc. 3rd Int. Conf. on the Strength of Metals and Alloys*, Cambridge, England, 20–25 August 1973, vol. 1, paper 65.

Herriot, G., Baudelet, B. & Jonas, J. J. 1977 (submitted for publication).

Johnson, R. H. 1970 *Superplasticity*, Met. Rev. No. 146.

McElroy, R. J. & Szkopiak, Z. C. 1972 *Int. Metall. Revs* **17**, 175.

McLauchlin, I. R. 1971 *Central Electricity Generating Board (England) Report* RD/B/N2151.

Misro, S. C. & Mukherjee, A. K. 1974 In *Dorn memorial symposium on rate processes in plastic deformation.* Am. Soc. Metals.

Mohamed, F. A., Shei, S.-A. & Langdon, T. G. 1975 *Acta metall.* **23**, 1443–1450.

Mohamed, F. A. & Langdon, T. G. 1976 *Scripta metall.* **10**, 759–762.

Nakagawa, Y. G. & Weatherly, G. C. 1972 *Mater. Sci. & Engng* **10**, 223–228.

Naziri, H., Pearce, R., Henderson-Brown, N. & Hale, K. F. 1973 *J. Microsc.* **97**, 229–238.

Oguchi, K., Oikawa, H. & Karashima, S. 1974 *Technology Report, Tohuku University* **39**, 375.

Perinet, F. 1975 CEA-R-4657, Saclay, Gif sur Yvette, France.

Roberts, A. C. & Cottrell, A. H. 1958 *Phil. Mag.* **1**, 711–717.

Russell, B., Ham, R. K., Silcock, J. M. & Willoughby, G. 1968 *Met. Sci. J.* **2**, 201,

Shewmon, P. G. 1965 In *Physical metallurgy* (ed. R. W. Cahn), esp. p. 268. Amsterdam: North Holland.
Staker, M. R. & Holt, D. L. 1972 *Acta metall.* **20**, 569.
Suery, M. & Baudelet, B. 1975 *J. mater. Sci.* **10**, 1022–1028.
Thompson, R. M. & Balluffi, R. W. 1962 *J. appl. Phys.* **33**, 803.
Vickers, W. & Greenfield, P. 1968 *J. nucl. Mater.* **27**, 73.
Williams, K. R. & Wilshire, B. 1973 *Met. Sci. J.* **7**, 176.

Discussion

ALAN H. COOK, F.R.S. (*Department of Physics, Cavendish Laboratory, Madingley Road, Cambridge 0B3 0HE*). It is very likely that the materials of the mantle of the Earth form not just two-phase but multiphase systems and so the ideas of strain hardening and superplasticity may be very important especially as the existence of a threshold strength could have significant consequences for our ideas of the topology of convection and of the conditions for its onset. Can Dr Gittus say how the phenomena scale in time and length? Will there still be a threshold strength when deformations extend over many millions of years and hundreds of kilometres?

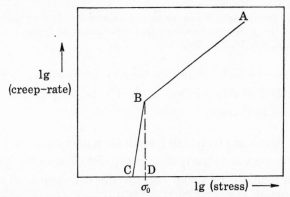

FIGURE 18. Plot of creep rate against stress illustrating the presence of a pseudothreshold stress σ_0.

J. GITTUS. The situation is best understood by reference to figure 18 which is a plot of strain rate against stress on logarithmic axes. Line *ABD* corresponds to the case in which a threshold stress, σ_0, exists. Line *ABC* is the real case to which *ABD* is a good approximation for 'laboratory' timescales. However, when deformations extend over many millions of years and hundreds of kilometres the approximation is no longer adequate and portion *BC* shows that below the 'threshold stress' σ_0 there will still be slow but finite strain rates leading to finite displacements on geological timescales.

J. P. POIRIER (*Saclay, France*). The equation Dr Gittus gives for superplastic deformation of two phase materials is adapted from the Ashby–Verrall equation for single phase materials, mainly by introducing a diffusion coefficient for heterophase boundary diffusion. Now, in the interface between A and B, we may define two diffusion coefficients for the diffusion of species A or B respectively. Which of these, or which combination of these, should be used?

J. GITTUS. The derivation is an adaptation of Ball & Hutchison's (1969) model. It uses the Raschinger grain-switching process which was also used by Ashby & Verrall in their (somewhat different) model.

In cases (there is at the moment only one such case) where both diffusion coefficients are known, their average is used in the equation.

S. A. F. Murrell (*Department of Geology, University College London, Gower Street, London, WC1E 6BT*). Professor Weertman suggested that isostatic adjustment (due for example to the melting of an ice cap) takes place by transient creep processes, so that the mantle viscosity estimated from such data is a lower limit to the true viscosity. I have discussed the rôle of transient creep in lithosphere deformation in a recent paper (Murrell 1976). However, one of the problems in this field is that there is no recent general theory which relates transient creep to steady state creep, or defines the important parameters in transient creep. In my work I have used an early theory due to Mott (1953), but have also developed a new theory of transient creep based on Friedel's (1964) dislocation network growth theory (Murrell & Chakravarty 1973). Would Professor Weertman, Professor Ashby or Dr Gittus care to comment on the theory of transient creep?

References

Friedel, J. 1964 *Dislocations*. Oxford: Pergamon Press.
Mott, N. F. 1953 A theory of work-hardening of metals. II. Flow without sliplines, recovery and creep. *Phil. Mag.* **44**, 742–765.
Murrell, S. A. F. 1976 Rheology of the lithosphere – experimental indications. *Tectonophysics* **36**, 5–24.
Murrell, S. A. F. & Chakravarty, S. 1973 Some new rheological experiments on igneous rocks at temperatures up to 1120 °C. *Geophys. J. R. astr. Soc.* **34**, 211–250.

J. Gittus. A recent general theory which relates transient creep to steady state creep has been developed by me and is described in detail in my book (Gittus 1975 *b*). See particularly § 3.3 on page 65 and equation 3.40 of that book.

J. Weertman. Obviously the transient creep field has not been developed, both experimentally and theoretically, to the extent that the steady-state creep field has been. Transient creep is much more complicated than steady-state creep because the deformational flow history of a sample can be very important. That is, if a sample is suddenly stressed its transient creep response will be different if the sample is well annealed or if the sample previously had undergone large plastic flow. Laboratory and theoretical investigations on transient creep phenomena that have been made generally are for the case of stresses suddenly applied to well annealed material. However, in glacial rebound, stress is 'suddenly' applied to mantle rock that presumably has been deformed to large plastic strains.

Phil. Trans. R. Soc. Lond. A. **288**, 147–158 (1978) [147]
Printed in Great Britain

Recrystallization of metals during hot deformation

By C. M. Sellars

Department of Metallurgy, University of Sheffield, St George's Square, Sheffield, S1 3JD, U.K.

[Plate 1]

Recovery processes tend to counteract the effects of work hardening during plastic deformation at high temperatures and at strain rates ranging from those of slow creep to those of rapid hot working operations. However, in metals in which recovery is relatively slow, sufficient stored energy can be accumulated to cause the occurrence of dynamic recrystallization during deformation once a critical strain is exceeded. This process then occurs repeatedly with continued straining. If any metal that has been deformed at high temperatures by a dislocation mechanism is held at temperature after deformation, static recrystallization tends to occur with time.

The effects of dynamic and static recrystallization on microstructure and on the flow stress or creep rate of the metal are considered in this paper and particular attention is given to the range of deformation conditions under which these recrystallization processes are expected to occur.

When metals deform plastically by crystal slip at elevated temperatures, the work hardening produced by deformation tends to be counteracted by recovery processes. These recovery processes cause rearrangement and annihilation of dislocations so that, as strain increases, the dislocations tend to form into two dimensional subgrain walls. In some metals and alloys the recovery entirely balances work hardening, and steady state is achieved and can be maintained to large strains before fracture occurs. In other metals in which recovery is less rapid, certain conditions of stress and temperature of deformation can result in the accumulation of sufficiently high local differences in dislocation density to nucleate recrystallization during deformation. This recrystallization is referred to as *dynamic* recrystallization to distinguish it from the *static* recrystallization that can occur in all metals when deformation is discontinued but the elevated temperature is maintained, or when deformation is carried out at low temperature and the metal is subsequently annealed.

In this paper, both types of recrystallization and their effects on deformation behaviour and microstructure will be outlined and the range of deformation conditions under which they are likely to occur will be considered.

Dynamic recrystallization

The occurrence of dynamic recrystallization was first observed and studied systemmatically during the creep of lead (Greenwood & Worner 1939; Andrade 1948; Gifkins 1958–9). Surveys of work carried out on other metals under creep conditions of constant stress of load (Hardwick, Sellars & Tegart 1961–2) and under constant strain rate deformation conditions (Jonas, Sellars & Tegart 1969) have shown that dynamic recrystallization of initially annealed metals also occurs in nickel, copper, gold, γ-iron, austenitic steels and high purity α-iron, but does not appear to occur in aluminium, zinc, magnesium, tin, low purity α-iron or ferritic steels.

Dynamic recrystallization has also been observed in ice (Steinemann 1958) and has recently been considered as a contributory mechanism in the deformation of quartzites (White 1976).

Effect on deformation behaviour

During creep deformation, the characteristic effect of dynamic recrystallization is to produce one or more transient periods of accelerated creep (Hardwick *et al.* 1961–2). In constant strain rate deformation, analogous fluctuations in flow stress are observed at relatively low strain rates (Rossard & Blain 1958; Jonas *et al.* 1969). The effect of dynamic recrystallization will be illustrated by results from the latter type of tests as these have frequently been carried out in torsion, which enables much higher strains than in tensile creep tests to be attained before fracture processes intervene.

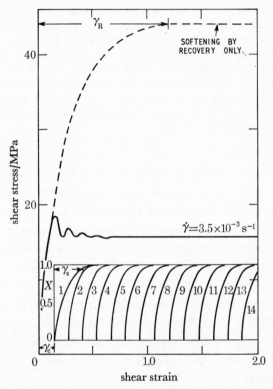

FIGURE 1. Shear stress–shear strain curve for 99.9% nickel deformed in torsion at a surface shear strain rate of $3.5 \times 10^{-3}\ \mathrm{s^{-1}}$ at 934 °C. The inset shows the cycles of dynamic recrystallization during deformation.

A typical stress–strain curve for nickel at 0.7 absolute melting temperature (Luton & Sellars 1969) is shown in figure 1. The curve rises initially as a result of work hardening and recovery processes, but at a critical strain, γ_c, dynamic recrystallization is nucleated and the curve passes through a maximum and then oscillates several times before settling to a steady state value. The broken curve indicates the stress–strain behaviour expected if recovery were the only operative dynamic softening process. The much higher steady state level is that deduced for the recovery creep mechanism by Ashby (1972, 1973) and is in accord with observations by Weertman & Shahinian (1956) on lower purity nickel in which dynamic recrystallization did not occur at this temperature. The important additional softening effect of dynamic recrystallization is thus clear, when the critical strain for recrystallization, γ_c, is much less than the strain expected for the onset of steady state by recovery softening only, γ_R.

The inset on this figure shows the recrystallization curves from which the observed stress–strain behaviour can be computed (Sah 1971). When the shear strain reaches γ_c, new strain-free grains are nucleated and with increasing strain (time) recrystallization proceeds along curve 1 leading to a decrease in flow stress. However, the concurrent deformation causes work hardening of these grains so that as the rate of recrystallization falls, the flow stress passes through a minimum and starts to rise again. When γ_c is again reached in the grains that recrystallized earliest in the first cycle, these grains recrystallize again along curve 2, leading to a second maximum in flow stress. This process is repeated each time γ_c is reached in the recrystallized grains, leading to further oscillations in flow stress until the process becomes sufficiently out of phase in different local regions of the material to make recrystallization effectively 'continuous'. This results in an overall steady state flow stress.

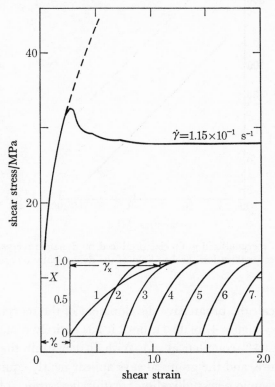

FIGURE 2. Shear stress–shear strain curve for 99.9% nickel deformed in torsion at a surface shear strain rate of $1.15 \times 10^{-1}\,\mathrm{s}^{-1}$ at 934 °C. The inset shows the cycles of dynamic recrystallization during deformation.

At higher strain rates, only one maximum in flow stress is observed (figure 2), but as shown by the inset, this again arises from the repeated occurrence of recrystallization. The difference in this case is that the strain interval over which the first cycle of recrystallization occurs, γ_x, is now considerably larger than γ_c so that several cycles of recrystallization overlap, giving effectively 'continuous' recrystallization and a smoother fall in flow stress to steady state. It was originally suggested (Luton & Sellars 1969) that the changeover from initially 'periodic' to initially 'continuous' recrystallization, with the accompanying change in form of stress–strain curve, would occur when $\gamma_x \approx \gamma_c$. Developments of the early model (Sah 1971) to allow for differences in recrystallization kinetics in the first and subsequent recrystallization cycles, which are expected to arise from differences in grain size, show that the changeover probably occurs when $\gamma_x \approx 2\gamma_c$.

Microstructural changes

During deformation up to γ_c, work hardening and recovery lead to the development of a dislocation subgrain structure, but typically dislocations in the subgrain boundaries remain tangled (figure 3, plate 1), rather than forming the clean two dimensional networks observed in metals in which recovery is more rapid (Jonas *et al.* 1969). The existance of these higher energy tangled subgrain boundaries has been shown (Sandström & Lagneborg 1975) to be essential to obtain sufficient stored energy differences in local regions to nucleate dynamic recrystallization.

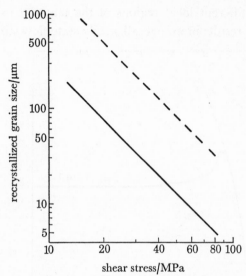

FIGURE 5. Stress dependence of recrystallized grain size produced by dynamic recrystallization (solid line) and by static (metadynamic) recrystallization (broken line) of the dynamically recrystallized structure in nickel (after Sah *et al.* 1974).

Optical microscopy reveals the progressive development of the recrystallized structure with increasing strain beyond γ_c (figure 4, plate 1), until it has entirely replaced the original grain structure (recrystallization 95 % complete at γ_x). With further strain there is no further apparent change in grain structure, and the grains always appear nearly equiaxed as a result of the repeated occurrence of dynamic recrystallization creating new grains of a size determined by the steady state deformation conditions. This contrasts with the situation in metals which do not recrystallize and in which the original grains become progressively more elongated with increasing strain.

Measurements of dynamically recrystallized grain size have been carried out on nickel (Luton & Sellars 1969; Sah, Richardson & Sellars 1974), copper (McQueen & Bergerson 1972; Bromley & Sellars 1973) and α-iron (Glover & Sellars 1973). In all cases it was found that the grain size is uniquely determined by the stress, independent of the temperature of deformation. As illustrated in figure 5, the relation between recrystallized grain size (d) and stress (τ) may be represented by an equation of the form

$$\tau \propto d^{-n}, \tag{1}$$

where n is a constant of value between $\frac{1}{2}$ and 1. The exact value is difficult to ascertain experimentally and it may vary with purity of the metal. Certainly, the level of the curve depends on

FIGURE 3. Electron micrograph of the subgrain structure in pure nickel deformed in creep at 20.7 MPa and 800 °C to a strain less than the critical value for dynamic recrystallization (Richardson *et al.* 1965).

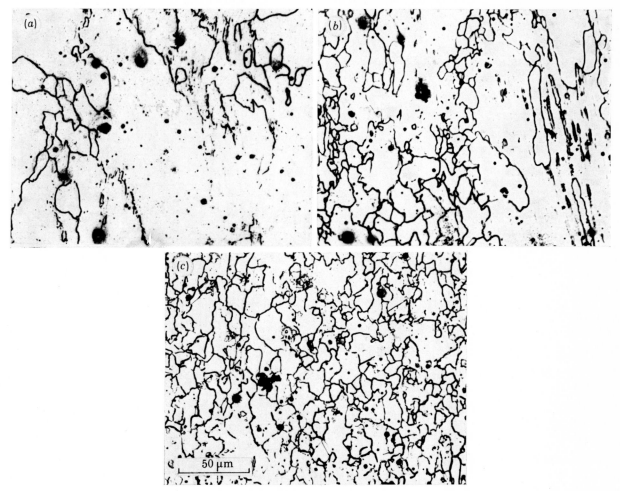

FIGURE 4. Optical micrographs of dynamically recrystallized grains developed in nickel deformed in torsion at a surface shear strain rate of 6.6×10^{-2} s^{-1} at 880 °C to surface shear strains of (*a*) 1.3, (*b*) 4.7 and (*c*) 8.8. Critical shear strain for dynamic recrystallization 1.0 (Sah 1971).

impurity or alloy content (Luton & Sellars 1969; Bromley & Sellars 1973; Glover & Sellars 1973), increasing levels giving a smaller grain size at a given flow stress. The dynamically recrystallized grain size does, however, appear to be independent of the original grain size of the material, although this influences the kinetics of the first cycle of recrystallization (Sah *et al.* 1974). Dynamic recrystallization, particularly under deformation conditions leading to high flow stresses, is therefore a very potent mechanism of grain refinement.

Electron microscope observations of dynamically recrystallized grains reveal that they contain tangled dislocation substructures analogous to those observed at strains less than γ_c (Luton & Sellars 1969; McQueen & Bergerson 1972). This distinguishes them from statically recrystallized grains and is to be expected as the grains are deformed as they develop.

It appears probable that, at least at fairly high strain rates, the concurrent deformation destroys the driving force for growth after an initial burst of growth from the nucleus (Sah *et al.* 1974). Recrystallization then proceeds by the continual formation of new nuclei and their restricted growth. This mechanism differs from that for classical (static) recrystallization, in which nucleation takes place initially and recrystallization proceeds by continued growth from these nuclei until impingement of the growing grains occurs.

The repeated nature of dynamic recrystallization means that during steady state the degree of development of the dislocation structure is heterogeneous and in different local regions there will be grains which have just recrystallized, grains which have been deformed by a further increment of γ_c and are just about to recrystallize again, and a spectrum of grains between these two limiting conditions.

This heterogeneity of substructure again contrasts with the uniform subgrain structure that is present during steady state when recovery is the only dynamic softening process. Under high-temperature deformation conditions, dynamic recovery involves subgrain boundary migration or 'repolygonization' which maintains an equiaxed subgrain structure with misorientations that reach a steady state value of only a few degrees (Jonas *et al.* 1969; Warrington 1976). However, in lower temperature deformation to high strains, there is evidence that subgrain misorientations may continue to increase with strain until they effectively form grain boundaries (Cairns, Clough, Dewey & Nutting 1971; Nutting 1974). The structure then has the appearance of being recrystallized, although only recovery has taken place and no separate nucleation and growth events are involved. This mechanism, which may be considered as 'recrystallization *in situ*' will not be discussed further in this paper.

Deformation conditions for dynamic recrystallization

From the combined observations on nickel deformed in compression creep and in torsion (Richardson, Sellars & Tegart 1965; Luton & Sellars 1969) the critical strain for the onset of recrystallization, γ_c, is found to pass through a minimum as a function of stress (figure 6). The reasons for this are not yet fully understood, but the rising curve at low stresses appears to be associated with the decrease in dislocation density with decreasing stress. Higher strains are then necessary to give sufficient misorientation, and hence dislocation density, in the subgrain boundaries to provide the stored energy for nucleation. The rising curve at high stresses appears to result from the need for increasing stored energy with increasing strain rate to ensure that boundary migration is sufficiently rapid for growth of the nuclei to occur at all before the dislocation density behind the moving boundary has been increased sufficiently by concurrent deformation to destroy the initial driving force.

Over the same range of stress, the strain that takes place during recrystallization, γ_x, increases continuously. This leads to the changeover from periodic to 'continuous' recrystallization with increasing stresses.

The expected strain at which steady state would be obtained by recovery softening only, γ_R, is also shown by the broken curve in this figure. Observations on α-iron (Glover & Sellars 1973) have shown there to be an upper stress limit for the occurrence of dynamic recrystallization and have suggested that this occurs when $\gamma_c \approx \gamma_R$. This would arise because, when a steady state structure is obtained by dynamic recovery, the stored energy no longer rises with

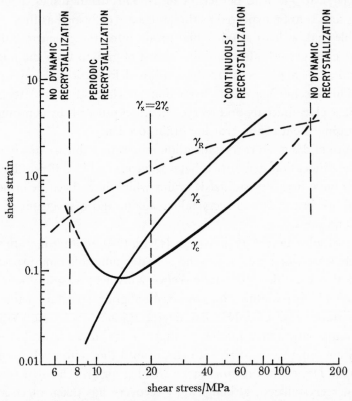

FIGURE 6. Relation between the critical strain γ_c for the onset of dynamic recrystallization, the strain interval γ_x for a large fraction of dynamic recrystallization to take place and the strain γ_R expected for the onset of steady state if softening were by recovery only, in 99.9 % nickel of initial grain size 200 μm at *ca.* 930 °C.

strain so that if conditions for nucleation are not met at strains less than γ_R, they are unlikely to be met at higher strains. A cut-off for the occurrence of dynamic recrystallization at low stresses has also been observed during creep of impure nickel at 1100 °C and a shear stress of about 7 MPa (Weertman & Shahinian 1956). It is suggested in figure 6 that it also arises when $\gamma_c \approx \gamma_R$.

From diagrams of the form shown in figure 6 for different temperatures, the limits of dynamic recrystallization in pure nickel of initial grain size 200 μm have been deduced and are super-imposed on an Ashby (1972, 1973) deformation map in figure 7. The lower temperature limit is uncertain, but has been made to be consistent with that observed on lower purity nickel of fine grain size (Jenkins, Digges & Johnson 1954).

From figure 7 it can be seen that the stress limits fall well within the area where recovery creep deformation is expected. These limits are expected to be sensitive to both alloy content and initial grain size as both influence γ_c (Gifkins 1958–9; Luton & Sellars 1969; Bromley & Sellars 1973; Glover & Sellars 1973; Sah *et al.* 1974). The lowest possible stress limit even with high purity materials of fine initial grain size will, however, be given by the lower bound of the area for recovery creep, which is itself sensitive to grain size (Ashby 1972, 1973), as diffusional flow does not involve dislocation movement and therefore cannot generate a driving force for recrystallization.

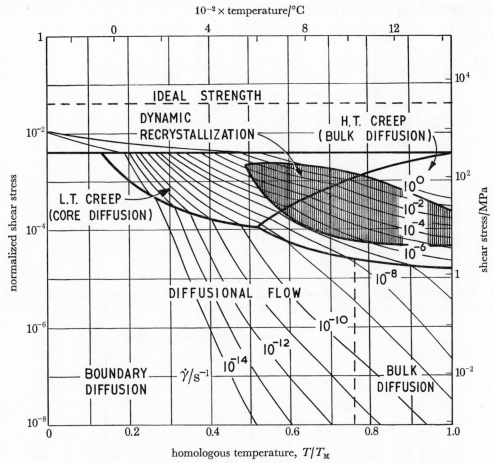

FIGURE 7. Ashby deformation map for 99.9 % nickel of initial grain size 200 μm showing the limiting conditions for the occurrence of dynamic recrystallization.

Within the area where dynamic recrystallization does occur, deformation tends to be heterogeneous (Cottingham 1966; Sah *et al.* 1974). The softer recrystallized grains formed first cause concentration of strain within them which leads to favourable conditions for nucleation in adjacent regions where a strain rate gradient exists. This gives conditions for instability and localization of strain as recrystallization proceeds.

STATIC RECRYSTALLIZATION

If deformation is discontinued but the material is retained at high temperature, the structure undergoes further change as a function of time by the occurrence of static recovery, recrystallization and grain growth.

These structural changes are reflected by softening of the material as a function of time after deformation. This is shown clearly by a change in form of the stress–strain curve obtained if deformation is continued after different rest periods, as shown in figure 8, which is taken from current work on austenitic stainless steel deformed in plane strain compression. Softening proceeds until recrystallization is complete, when the stress–strain curve is similar in form to that observed during the initial deformation. Minor differences remain when the recrystallized grain size differs from the initial grain size.

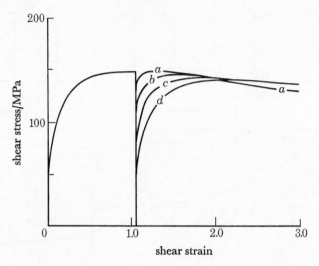

FIGURE 8. Effect of increasing rest period on the stress–strain curve for 18:8 austenitic stainless steel deformed in plane strain compression at an equivalent shear strain rate of 34 s^{-1} at 916 °C: a, 1 s; b, 20 s; c, 50 s; d, 300 s.

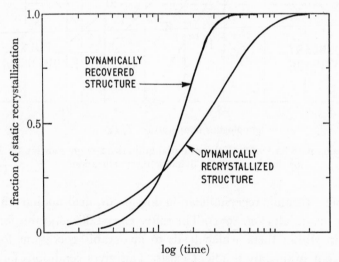

FIGURE 9. Time dependence of static recrystallization of dynamically recovered structures and of dynamically recrystallized structures produced during high-temperature prestrain.

Direct metallographic observations (Glover & Sellars 1972) and the technique of interrupting the deformation (Djaic & Jonas 1972, 1973; Petkovic, Luton & Jonas 1974) have established that the kinetics of static recrystallization differ fundamentally when the deformed structure has developed by work hardening and recovery only and when dynamic recrystallization has occurred (figure 9). The reasons for this are discussed later. Static recrystallization following dynamic recrystallization has been referred to as 'metadynamic' recrystallization (Djaic & Jonas 1972) to distinguish the process from the more 'classical' recrystallization of dynamically recovered structures which is closely related to the processes that occur on annealing metals after cold deformation.

Interrelation of static softening processes

The way in which static softening occurs in metals which undergo dynamic recrystallization during deformation is shown schematically as a function of prestrain in figure 10.

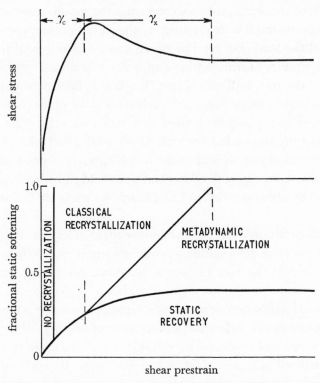

FIGURE 10. Schematic representation of the interrelation between the static softening mechanisms as a function of prestrain in a material that dynamically recrystallizes (after Djaic & Jonas 1973).

Static recovery

After all prestrains, static recovery takes place immediately the deformation is halted and proceeds at a decreasing rate with time. The recovery process involves the annihilation of dislocations in individual events and accounts for up to 40–50 % of the total softening at high strains (Evans & Dunston 1971; Petkovic *et al.* 1974). At low strains the stored energy is insufficient to cause static recrystallization and limited softening takes place by static recovery alone. The critical strain for static recrystallization after deformation at relatively high strain rates appears to be about 0.05–0.1, depending on deformation conditions (Djaic & Jonas 1973). This value is well below the critical strain for dynamic recrystallization, γ_c. The difference is consistent with the earlier conclusion that at relatively high strain rates (high stresses), excess stored energy, which requires higher strains, is necessary for dynamic recrystallization as growth of nuclei must occur more rapidly than some critical rate.

Equivalent measurements do not appear to have been carried out after deformation at low strain rates, but it would be expected that under these conditions the critical strain for static recrystallization will more nearly approach γ_c. Similar considerations to those discussed for γ_c will determine the absolute lower limit of strain rate (or stress) during deformation for static recrystallization to occur at all after any amount of prestrain.

Classical recrystallization

For the deformation conditions appropriate to figure 10, prestrains greater than the critical value for static recrystallization but less than γ_c result in classical recrystallization after an incubation period in which recovery processes create the recrystallization nuclei. The rate of recrystallization in a given material is determined by the stored energy, the density of favourable nucleation sites and the temperature. The stored energy increases with increasing strain rate and decreasing temperature of deformation and is strongly dependent on strain. All these factors therefore increase the recrystallization rate (English & Backofen 1964; Morrison 1972; Glover & Sellars 1972; Djaic & Jonas 1973). As nucleation takes place preferentially at grain boundaries, the density of favourable nucleation sites increases with decrease in grain size, leading to more rapid recrystallization in finer grained materials (Barraclough 1974). Recrystallization is a thermally activated process and so the rate is strongly dependent on the temperature of holding after deformation, e.g. with other conditions held constant the rate of recrystallization of α-iron decreases by about one order of magnitude for each 50 °C drop in temperature (Glover & Sellars 1972).

It is important to distinguish between the two independent rôles that temperature plays in influencing recrystallization after deformation at constant strain rate. Whereas decreasing the holding temperature decreases the rate of recrystallization, decreasing the temperature of deformation increases the stored energy at a given strain and therefore tends to increase the rate. This means that recrystallization occurs more rapidly when the temperature at which deformation has taken place is well below the temperature of holding.

Apart from the temperature of holding, the variables which increase recrystallization rate also decrease the recrystallized grain size (Glover & Sellars 1972; Barraclough 1974). The lack of effect of temperature is, however, often masked by the occurrence of grain growth after recrystallization is complete as this is a strongly temperature dependent process.

Metadynamic recrystallization

When the prestrain exceeds γ_c, an increasing fraction of the softening occurs by metadynamic recrystallization until at prestrains in the steady state range this becomes the only static recrystallization process.

During dynamic recrystallization at relatively high strain rates there are always nuclei present in the material and some grain boundaries are migrating. When the deformation is halted, these boundaries continue to migrate and nuclei continue to grow without the need for an incubation period. This form of recrystallization therefore proceeds very rapidly after deformation (figure 9), but the rate of recrystallization falls with time as the grains grow progressively into less dislocated material in the heterogeneous structure produced by the repeated cycles of dynamic recrystallization.

The overall rate of metadynamic recrystallization is influenced by the same factors as classical recrystallization, but after steady state deformation the stored energy is no longer dependent on strain and the dynamically recrystallized grain size is independent of the original grain size.

The important variables are therefore the strain rate and temperature of deformation and the temperature of holding (Glover & Sellars 1972; Djaic & Jonas 1972; Barraclough 1974).

The grain size produced by metadynamic recrystallization is uniquely determined by the steady state flow stress during the prestrain. As shown in figure 5, the relation follows the form given in equation (1), although the stress exponent is not necessarily identical to that for dynamic recrystallization (Glover & Sellars 1972, 1973). Because repeated nucleation does not occur during metadynamic recrystallization, the grain size is larger than that produced by dynamic recrystallization (Glover & Sellars 1972, 1973; McQueen & Bergerson 1972).

CONCLUSION

This survey of recrystallization during and after high-temperature deformation of metals has shown that three types of recrystallization may occur. Their effects on microstructure and strength, or creep rate, have been considered, and the fact that the deformation must always take place by dislocation movement rather than by diffusion mechanisms has been emphasized.

The range of deformation conditions over which dynamic recrystallization is observed is generally more restricted than that for recovery creep and, even under favourable conditions, a critical strain must be exceeded before either dynamic recrystallization or static recrystallization after deformation takes place. Within these limitations, equivalent recrystallization processes would be expected in other crystalline solids in which dynamic recovery is relatively slow. They will be favoured when the material is also of high purity and fine grain size.

REFERENCES (Sellars)

Andrade, E. N. da C. 1948 *Nature, Lond.* **162**, 410.

Ashby, M. F. 1972 *Acta metall.* **20**, 887–897.

Ashby, M. F. 1973 *Third Int. Conf. on Strength of Metals and Alloys*, vol. 2, pp. 8–42. London: Institute of Metals; Iron and Steel Institute.

Barraclough, D. R. 1974 Ph.D. Thesis, University of Sheffield.

Bromley, R. & Sellars, C. M. 1973 *Third Int. Conf. on Strength of Metals and Alloys*, vol. 1, pp. 380–385. London: Institute of Metals; Iron and Steel Institute.

Cairns, J. H., Clough, J., Dewey, M. A. P. & Nutting, J. 1971 *J. Inst. Metals* **99**, 93–97.

Cottingham, D. M. 1966 *Deformation under hot working conditions*, pp. 145–156. Iron and Steel Institute Publication, no. 108. London.

Djaic, R. A. P. & Jonas, J. J. 1972 *J. Iron Steel Inst.* **210**, 256–261.

Djaic, R. A. P. & Jonas, J. J. 1973 *Metall. Trans.* **4**, 621–624.

English A. T. & Backofen, W. A. 1964 *Trans. T.M.S.–A.I.M.E.* **230**, 396–407.

Evans, R. W. & Dunston, G. R. 1971 *J. Inst. Metals* **99**, 4–14.

Gifkins, R. C. 1958–9 *J. Inst. Metals* **87**, 255–261.

Glover, G. & Sellars, C. M. 1972 *Metall. Trans.* **3**, 2271–2280.

Glover, G. & Sellars, C. M. 1973 *Metall. Trans.* **4**, 765–775.

Greenwood, J. N. & Worner, H. K. 1939 *J. Inst. Metals* **64**, 135–158.

Hardwick, D., Sellars, C. M. & Tegart, W. J. McG. 1961–2 *J. Inst. Metals* **90**, 21–22.

Jenkins, W. D., Digges, T. G. & Johnson, C. R. 1954 *J. Res. natn. Bur. Stand.* **53**, 329–352.

Jonas, J. J., Sellars, C. M. & Tegart, W. J. McG. 1969 *Metall. Rev.* **14**, 1–23.

Luton, M. J. & Sellars, C. M. 1969 *Acta metall.* **17**, 1033–1043.

McQueen, H. J. & Bergerson, S. 1972 *Metal Sci. J.* **6**, 25–29.

Morrison, W. B. 1972 *J. Iron Steel Inst.* **210**, 618–623.

Nutting, J. 1974 *Eighth International Conference on Electron Microscopy*, vol. 1, pp. 580–581.

Petkovic, R. A., Luton, M. J. & Jonas, J. J. 1974 *Int. Symposium on the hot formability of steels*, Strbske Pleso, Czechoslovakia.

Richardson, G. J., Sellars, C. M. & Tegart, W. J. McG. 1965 *Acta metall.* **14**, 1225–1236.

Rossard, C. & Blain, P. 1958 *Rev. Métall.* **55**, 573–594.

Sah, J. P. 1971 Ph.D. Thesis, University of Sheffield.
Sah, J. P., Richardson, G. J. & Sellars, C. M. 1974 *Metal Sci.* **8**, 253–331.
Sandström, R. & Lagneborg, R. 1975 *Acta metall.* **23**, 387–398.
Steinemann, S. 1958 *Beitr. Geol. Schweiz (Hydrologie)* **10**, 1–22.
Warrington, D. H. 1976 Computer simulation for materials science applications. In *Nuclear metallurgy*, vol. 20, pt 2, pp. 672–683. National Bureau of Standards.
Weertman, J. & Shahinian, P. 1956 *Trans. A.I.M.E.* **206**, 1223–1226.
White, S. 1976 *Phil. Trans. R. Soc. Lond.* A **283**, 69–86.

Phil. Trans. R. Soc. Lond. A. **288**, 159–176 (1978) [159]

Printed in Great Britain

Deformation and recrystallization textures in metals and quartz

By R. W. Cahn

School of Engineering and Applied Sciences, University of Sussex, Brighton

The nature and graphical representation of preferred orientations, or textures, in polycrystalline assemblies is outlined. The genesis of textures during plastic deformation of metals is examined in terms of single crystal behaviour, with special reference to the formation of deformation bands, and the influence of pre-existing texture on subsequent plastic deformation (including creep) is exemplified. Next, the nature and origin of textures formed by annealing after plastic deformation are analysed and the relative rôles of oriented nucleation and oriented growth assessed.

A different kind of texture in stressed polycrystalline quartz is formed as a consequence of Dauphiné twinning. This form of twinning is explained in some detail and the origin of the textures explained in terms of the driving force that brings about Dauphiné twinning; these driving forces are compared with those that determine recrystallization in metals.

1. Introduction

A *texture* (metallurgy) or *fabric* (geology) represents a non-random distribution of the orientations of the constituent crystal grains in a polycrystalline aggregate, which may be a drawn wire, rolled sheet, cast ingot, electroplated layer, a compressed or sheared mass of rock or a rock cylinder deformed in the laboratory under hydrostatic constraint. The texture is defined relative to an identifiable geometrical feature, such as the wire axis or the sheet normal and rolling direction. A *wire texture* has rotational symmetry about the wire axis and can be referred to that axis alone, whereas a *rolling texture*, which does not have rotational degeneracy, needs the *two* cited vectors to define it completely. Ingot and electroplated textures resemble wire textures.

Textures have hitherto been displayed in *pole figures*. These are stereographic projections of a specified crystal form, generally {100} or {111} in the case of metals of cubic symmetry, shown in terms of contours of population density. Figure 1, which refers to rolled copper, is an example. The metallurgical grade of copper has to be specified because the texture is sensitive to minor impurities. Figure 1 also contains the positions of several ideal textures, each of which in effect represents the orientation of a single crystal in simple crystallographic relation to the defining vectors. Thus (112) [$\overline{1}\overline{1}1$] denotes (112) parallel to the rolling plane and [$\overline{1}\overline{1}1$] parallel to the rolling direction. Metallurgical textures are commonly denoted in terms of ideal textures, though actual textures are scattered as shown in figure 1, commonly clustering about more than one ideal texture (Kalland & Davies 1972) but sometimes, as in the 'cube texture' of copper, (100) [001], about one ideal texture only.

Metallurgical textures can in practice only be determined by X-ray diffraction (unlike the geologist's recourse to polarized light and the universal stage). This task was formerly achieved photographically, but now it is always done with the aid of a specially constructed X-ray counter diffractometer, and the process can be largely automated. A pole figure such as figure 1 determined in this way has a series of numbered contours representing multiples of the count corresponding to a random orientation distribution. An alternative form is the *inverse pole figure*, in which a unit triangle of the crystal is displayed in standard orientation and the distribution of a vector such as the rolling direction is shown by means of contours. Figure 10, below, is an instance of this.

Neither form of pole figure shows complete information about a real texture. In principle, if there are 10000 grains, a {100} pole figure could display 30000 poles, serially numbered, so that each grain can be individually identified. This could be done – if anyone had the patience – by a geologist using the universal stage and polarized light with a transparent,

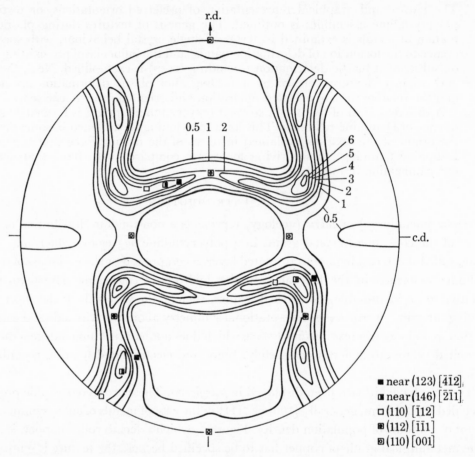

FIGURE 1. {100} pole figure of electrolytic copper rolled 96.6 % at 25 °C. Density contour of unity corresponds to randomness. r.d. and c.d. are rolling direction and cross direction, respectively, in the rolling plane. Five ideal textures are indicated. (Hu & Goodman 1963.)

optically anisotropic mineral, or by a metallurgist using the Kossel X-ray technique or selected area electron diffraction on a thin foil. A half-way approach to this is to construct an *orientation distribution* (Bunge 1965, 1969; Roe 1965). Each grain requires 3 angles to define its position precisely, whereas an ordinary pole figure embodies 2 angles, i.e. those defining a particular crystal vector. Bunge and Roe invented a three-dimensional plot which can be displayed in serial sections, to define the full orientation distribution of a population of grains. Such a plot can be constructed, without the labour of orientating thousands of individual grains, by first constructing several pole figures for distinct {*hkl*} and using those to compute the orientation function by an iterative method. The more such pole figures that are employed as input data, the better is the angular resolution of the resultant function. This is analogous to the improvement of the spatial resolution of a crystal structure determination when more structure factors are included in the Fourier summations. The detailed information contained in an orientation function greatly eases the rigorous testing of theories of texture formation. The details are too involved for this review; good illustrations of the use of these functions can be found in papers by Bunge (1969) and by Dillamore & Katoh (1974).

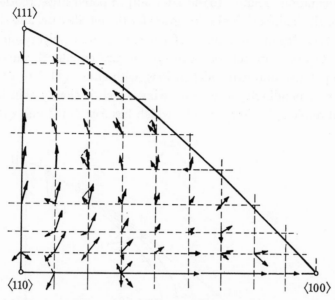

FIGURE 2. Lattice rotation for axisymmetric tension determined by Taylor (1934) for {110}⟨1̄10⟩ slip in a face-centred cubic metal, for various initial orientations. 2.37 % strain.

The enormous volume of labour which has been expended on textures of metals and alloys (Grewen & Wassermann 1962; Davies, Dillamore, Hudd & Kalland 1975) is directly due to the influence of textures on mechanical and magnetic behaviour. For many industrial purposes it is important to prevent the formation of textures altogether (thus, uranium fuel rods for certain types of nuclear reactor must be randomly orientated if catastrophic distortion in service is to be avoided), but for some major applications, such as deep drawing and the production of transformer laminations, a pronounced texture is a necessity.

2. DEFORMATION TEXTURES

In a book which has had lasting influence and is still frequently cited, Schmid & Boas (1935) showed how the orientation of a metal rod consisting of a single grain alters when it is plastically stretched. Typically only one slip† system is active. The operative slip *direction* rotates into the rod axis. Correspondingly, a monocrystalline cylinder on compression tends to align the operative slip *plane* normal to the acting stress. As between crystallographically equivalent slip systems, that which is most highly stressed in shear is found to operate.

On the naïve view that an assembly of separate crystal grains behaves as though the grains were independent, any consistent form of plastic deformation should evidently generate a texture. In fact, the interconnected nature of the grains requires slip on multiple systems. Taylor, in a classic study (1934), showed that any one grain requires the uniform operation of at

† The mineralogist's term is 'translation glide'.

least 5 independent systems to achieve an arbitrary imposed strain; 'independent' here is a rather subtle concept which places restrictions on shared slip directions or planes. Taylor introduced a criterion of minimum work: that combination of 5 systems will function which minimizes $M = \Sigma\gamma_j/\epsilon$. Here M is the *Taylor factor*, γ_j is the shear in the jth slip system and ϵ is the macroscopic tensile or compressive strain. On this basis, Taylor was able to show how grains of different initial orientations, relative to the externally imposed stress, reorientated in different ways. Figure 2 is a version of his original stereogram for the tensile case, and shows his computed rotations for different starting orientations of a face-centred cubic polycrystal. Where more than one arrow appears, two or more sets of 5 slip systems are equally favoured. It is clear that orientations tend to polarize into two end-positions, with either $\langle 111 \rangle$ or $\langle 100 \rangle$ parallel to the tensile stress axis. Correspondingly, in wire-drawing which is closely akin to tension, a cylindrically symmetrical duplex *fibre texture* results which has *both* $\langle 111 \rangle$ and $\langle 100 \rangle$ parallel to the wire axis.

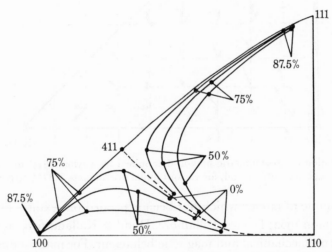

FIGURE 3. Slip rotation paths for axisymmetric compressive deformation of a body-centred cubic metal, deforming by pencil glide. Three different initial orientations are indicated, with rotations corresponding to various compressive strains for each initial orientation. (Dillamore & Katoh 1974.)

Taylor's model has generated a host of variants, of which the most mathematically satisfactory is that due to Bishop & Hill (1951). They were able to prove that Taylor's theory fufils Schmid's empirical yield criterion, derived from single crystal experiments: the resolved shear stress reaches the critical level for the 5 systems chosen by Taylor without reaching it for any of the other potential systems (except where pairs of systems happen to be equally favoured). As is pointed out by Chin (1969) in his particularly clear account of the Taylor theory and its descendants, most of the derivative versions do not satisfy the Schmid yield criterion. Taylor's (and Bishop & Hill's) theory presumes that each grain deforms as does the assemblage, i.e. strain continuity is preserved, but this then renders *stress* continuity impossible. Accordingly, it is to be expected that the strain in each grain is non-uniform to allow stress continuity across grain boundaries. This requires more than 5 slip systems to operate, each non-uniformly across the grain, which in turn implies variable orientations in different parts of the deformed grain. This, indeed, is observed experimentally (see example in figure 4, below). In spite of these ineradicable defects, the Taylor/Bishop & Hill theory gives a good approximation to the textures observed in a variety of deformed aggregates.

Figure 3 shows a theoretical diagram constructed by Dillamore & Katoh (1974) which predicts, on the basis of the Taylor theory, how various starting orientations will rotate in a polycrystalline assemblage of iron crystals subject to compression. Iron deforms by so-called *pencil glide* slip along a rigidly defined direction, ⟨111⟩, but on a number of alternative planes. The same predicted polarization into two alternative end-orientations is evident. In fact, individual grains are apt to split into irregularly shaped regions, called *deformation bands*, which strive towards different orientational destinations. Barrett & Levenson (1940), in a now classic study, were the first to reveal such bands. (The terminology of these bands is today in a confused state, with several rival usages.) The use of the Kossel X-ray diffraction technique (Ferran,

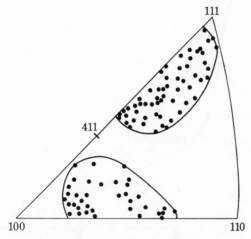

FIGURE 4. Orientations of various positions in a single grain of polycrystalline iron deformed 40 % in compression. Orientations were determined by individual X-ray Kossel diffraction photographs. The two groups of orientations correspond to the matrix and to a series of deformation bands. (Inokuti & Doherty 1977 *b*.)

Doherty & Cahn 1971) allows the orientation to be rapidly mapped at many points of a single deformed grain, and when this technique is applied to grains in a moderately deformed iron polycrystal, results such as those in figure 4 are found (Inokuti & Doherty 1977 *a*, *b*). The lattice rotation is very variable indeed; the maximum rotation is much greater than would be expected for 40 % strain according to the theoretical stereogram, figure 3, and the grain splits into deformation bands. (The high orientational spread of deformed grains is, as explained above, an expected departure from Taylor's simple model.) The trend of figure 3 is, however, qualitatively correct, and this applies generally to similar predictions based on a combination of the Taylor theory and slip systems of known crystallographic nature. It is easier to interpet wiredrawing (roughly equivalent to simple tension) and simple compression than it is to deal with rolling of sheets, since this is a more complex deformation mode. Similar interpretations have been attempted for much more complex processes. The most involved is Calnan & Clews's (1952) interpretation of the rolling texture of uranium, which has orthorhombic symmetry; this metal undergoes copious twin-gliding according to several distinct twinning laws, and this plays a major part in determining the rolling texture. Agreement between theory and experiment is only moderately good.

The detailed interpretation even of the supposedly simple wire textures is, in fact, a good deal more subtle than at first appears. Figure 5 (English & Chin 1965) illustrates this point. It shows that the relative proportions of ⟨111⟩ and ⟨100⟩ fibre textures in drawn wires in a number of different face-centred cubic metals and solid solutions varies as a function of the normalized energy of a stacking-fault (s.f.e.) (where the stacking of successive close-packed planes is locally anomalous). The value of the s.f.e. determines the fine structure of the individual dislocations in these materials, and this in turn governs the details of the slip morphology (e.g. 'cross-slip' on rogue {111} planes) and also determines how rapidly the latent (potential) slip systems, which do not actually operate, themselves work-harden. Finally, and most important, the s.f.e.

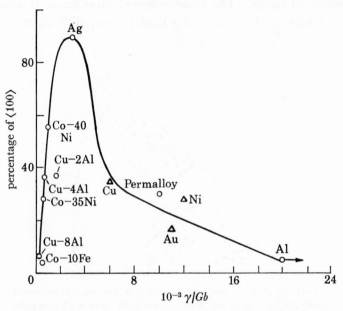

FIGURE 5. The variation of the proportions of ⟨111⟩ and ⟨100⟩ textures in duplex wire textures of face-centred cubic metals as a function of the specific stacking-fault energy parameter γ/Gb, where γ is the stacking fault energy, G is the shear modulus and b the magnitude of the Burgers vector. (English & Chin 1965.)

determines the ease of twin-gliding (deformation-twinning in metallurgical parlance). Chin (1969) explains in detail how these various factors can be used to interpret figure 5. In particular, the almost pure ⟨100⟩ texture of drawn silver wire can be firmly attributed to the intermediate ease of twinning in this metal when intensely deformed (Ahlborn & Wassermann 1963). At still smaller stacking fault energies, twinning becomes so easy relative to slip that the pattern of lattice rotations changes again. For further particulars, Chin's paper must be consulted. Generally, the rôle of twinning in contributing to reorientation of a single crystal is quite involved, because the orientation dependence of the stress required to initiate twinning is itself a function of stacking-fault energy (Narita & Takamura 1974).

The details of rolling textures of the various families of metals are far too involved even to summarize here, as are the rival theories advanced to interpret them. An excellent outline will be found in Barrett & Massalski's book (1966), and a fuller account in the survey by Hu, Cline & Goodman (1966). The most thoroughly studied feature is the transition, in face-centred cubic metals and alloys, from a (123) [$\bar{4}$12] ideal texture to a (110) [$\bar{1}$12] ideal texture (both shown in figure 1) as a result of extensive alloying; lowering the rolling temperature also plays a part. This transition is again linked to the specific stacking-fault energy, which is plainly a major determinant of textures of all kinds in face-centred cubic metals and alloys.

3. COLD AND HOT DEFORMATION OF TEXTURED SHEETS

The control of texture in sheets has become a major industrial necessity, primarily because of the widespread use of the deep drawing process. This is illustrated in figure 6. The extent of deformation possible in drawing is intimately related to the mechanical anisotropy of the sheet, and this in turn is linked to the nature and sharpness of the texture. The preferred measure of

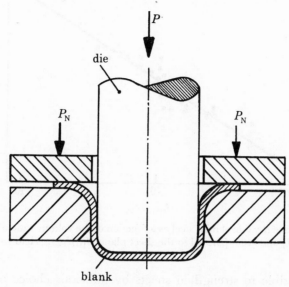

FIGURE 6. Arrangement for deep drawing of a circular sheet-blank.

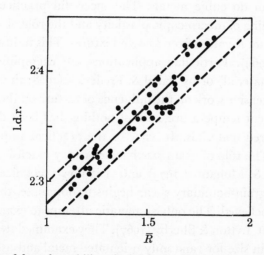

FIGURE 7. The dependence of deep drawability of mild steel, represented by the limiting drawing ratio (l.d.r. = ratio of the diameter of the largest blank which can be completely drawn without fracture to the diameter of the punch) as a function of the average R-value for the material. (Atkinson & McLean 1965.)

mechanical anisotropy is the *R-value* defined as ϵ_w/ϵ_t; ϵ_w, ϵ_t are width and thickness strains, respectively, in a sheet plastically deformed in uniaxial tension. For isotropic (texture-free) material, $R = 1$, while the extent of anisotropy in a textured sheet is best measured by $\bar{R}$, which is the mean of R-values averaged for all possible tensile directions in the plane of the sheet. Figure 7 shows that deep-drawability is enhanced by a high $\bar{R}$-value, and figure 8 in turn shows how the $\bar{R}$-value of a steel sheet is determined by texture. A high $\bar{R}$-value implies a good resistance to sheet-thinning, and it can be shown from macroscopic plasticity theory (Panknin 1969) that under these circumstances a larger drawing strain is feasible before the fracture stress is reached in the most thinned part of the drawn disk.

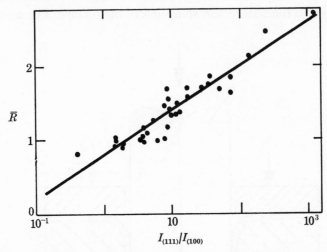

FIGURE 8. Average R-value of mild steel as a function of the volume ratio of grains having (111) and (100) parallel to the sheet plane (I_{111}/I_{100}). (Held 1967.)

Conversely, it is possible to strengthen sheets by judicious choice of texture: this applies especially to metals of hexagonal symmetry such as titanium, which have a smaller number of alternative slip systems than do cubic metals. The successful practice of *texture-strengthening* requires a subtle understanding of macroscopic plasticity and the rôle of anisotropy, and ideally requires an ability to calculate R-values from known textures. This industrially important topic is outlined by Hosford (1969). Particular applications are exemplified by Chin, Hart & Wonsiewicz (1969; spring material) or by Babel & Frederick (1968; titanium sheet).

In contrast to the very extensive work on the influence of texture on the ductility and strength of sheets deformed at ambient temperature, almost nothing has been done to investigate the resistance to creep of textured materials. In view of the practical importance of creep, this is a remarkable omission. The rôle of grain *size* has been very extensively examined over the years (review by Langdon & Mohamed 1975) and it is now recognized that, except for the smallest grain sizes, where grain-boundary shear begins to dominate, this factor plays a much smaller rôle than formerly believed. The only systematic attempt to examine the rôle of texture on creep was made by Barrett, Lytton & Sherby (1967). They examined steady-state creep in pure copper as a function of grain size for randomly orientated metal and as a function of texture – (100)[001] or random – at constant grain size (0.03 mm). The grain-size effect was appreciable only for grain diameters smaller than 0.2 mm, and was attributed primarily to a sensitivity of the incidence of grain-boundary shear to grain size at small grain sizes only. The replacement of random sheet by intensely textured sheet reduced the creep rate at all stresses by a factor of only two. The reduction was attributed to the exclusion of grain-boundary shear in the textured material.

There is need for further investigation, in a range of metals, especially non-cubic, of the rôle of texture in conferring resistance to creep.

4. Annealing textures

If a drawn wire or rolled sheet has been deformed substantially enough to generate a pronounced deformation texture, then on subsequent annealing it will always be found to have a pronounced annealing texture. Recrystallization does not generate random orientations, and indeed texture-free annealed sheets are difficult to make. The texture is often quite distinct, in a crystallographic sense, from the original deformation texture, and its nature can be affected by annealing temperature and time and even by the nature of the atmosphere. The annealing textures are not determined by the deformation texture alone: for instance, the addition of a modest amount of solute may sharply alter the annealing texture without creating any detectable changes in the prior deformation texture. Nevertheless, since all the 'genetic' information leading to the annealing texture must be present in the deformation texture, a change in one implies some change in the other, even if undetectable. Thus two sheets giving the same pole figure may have a different spatial distribution of the texture constituents. The theoretical problem is to understand how the 'genes' do their work or – to change the metaphor – how 'memory' is transmitted from deformed to recrystallized structure.

There is no need to go into details of particular textures; a very full account was published by Grewen & Wassermann (1962) and a more concise outline by Barrett & Massalski (1966). Of the many surveys dealing with the mechanisms of annealing texture formation, one which is particularly comprehensive and clear is that by Hutchinson (1974). The underlying mechanisms of recrystallization have been surveyed by Cahn (1970).

For many years, the theory of texture formation was characterized by a conflict between two extreme views – the *orientated nucleation* and *orientated growth* hypotheses. The former view is based on the central principle that new recrystallized grains generated by annealing a deformed structure have orientations which were present in the latter – that is, the nuclei are present all along, and annealing merely activates their growth. (It is reminiscent of the aesthetic dogma that a sculpture is latent in the raw block of marble.) On this hypothesis, textures are determined by what orientations are available. On the orientated growth hypothesis in its pure form, when a deformed metal is annealed all orientations are generated with equal probability at the nucleation stage. This evidently presupposes a nucleation mechanism different from that outlined above. Some nuclei then *grow* faster than others because of the restraining effect of the deformation texture on some of the growing grains. If a new grain is favourably orientated with respect to all or most components of the deformation texture, it will grow apace.

The orientated growth theory is buttressed by a body of experimental information which establishes that the growth rate of a strain-free grain into a dislocated grain is a function of mutual orientation, especially in the presence of small amounts of solutes which are believed to segregate to the grain boundary. These experiments have the limitation that the deformed grains used all had very low dislocation densities. The experiments have been surveyed by Aust (1969) and the corresponding theory of impurity-limited grain boundary migration by Gordon & Vandermeer (1966).

The orientated nucleation theory is buttressed by a growing body of microstructural investigation of the early stages of recrystallization which has established that nuclei do indeed grow from pre-existing orientations. This has been established particularly by the technique of Kossel X-ray diffraction, which allows the orientations of small new grains to be related to the local orientation of the deformed parent grain nearby. (It is essential to know the *local*

orientation of a deformed grain, because overall such a grain has a surprisingly wide orientation range.) This technique has been applied to aluminium (Ferran *et al.* 1971; Doherty & Cahn
1972; Bellier & Doherty 1977) and iron (Inokuti & Doherty 1977 *a*, *b*). In both metals, deformed
in the range 20–40 % reduction, it is unambiguously clear that new grains grow from pre-
existing orientations at locations where there is a steep orientation gradient: this implies
either intergranular boundaries or the narrow *transition bands* between adjacent deformation
bands within individual grains. (These bands might be described as 'artificial grain boundaries', with very large misorientations.) At smaller strains, nucleation was preferentially at true
grain boundaries; at larger strains, 'artificial grain boundaries' were preferred. The physical
process involved in the nucleation is *strain-induced boundary migration*; a deformed grain or
deformation band bulges into its neighbour, achieving a low dislocation density as it does so.
The early stages of this well-documented process have been analysed by Doherty & Cahn
(1972) and the energy balances which govern the process were originally examined by Bailey
(1960) and more comprehensively (in connection with the special problem of nucleation at
transition bands) by Dillamore, Morris, Smith & Hutchinson (1972).

In the past, attempts to interpret annealing textures were based on analysis of pole figures
'before and after', with attempts either to verify that the two major texture components of the
annealing texture were present in the deformation texture (to support orientated nucleation) or
to show that appropriate components of the two textures were related in such a way as to
favour rapid growth (to support orientated growth). This is a blunt approach which can be
sharpened by recourse to the more sophisticated orientation distributions (see above) to replace
ordinary pole figures. This has been done recently with regard to textures in copper and its
alloys (Schmidt, Lücke & Pospiech 1974). The conclusion is that in general there is an element
of growth selection operating on a non-random supply of available nuclei, themselves the
product of orientated nucleation. There are particular textures the origin of which is still
debated, and chief among these is the peculiar 'cube texture', (100) [001], in copper and its
alloys. Small composition changes which do not detectably change the deformation texture
(R. K. Kay, quoted by Hutchinson 1974) completely alter the annealing texture. The orientated
growth and nucleation theories have alternately held the field and, as Hutchinson shows, this
debate continues apace.

Orientated nucleation – with its concomitant element of growth selection – is only part of
the story. It is becoming plain that different components of a deformation texture may have
differing proclivities to nucleate new grains on annealing. Dillamore, Smith & Watson (1967)
were the first to establish, by electron microscopy, that in iron the stored energy of various
grains in a rolled sheet increases in the sequence (100)[011], (211)[0$\bar{1}$1], (111)[uvw], (011)[0$\bar{1}$1].
(The last is a very minor component.) The stored energy increases with the average strain
needed to bring randomly oriented initial grains into the particular orientation under
consideration. The most highly deformed texture components will recrystallize first, and any
processing variable which delays nucleation gives the first-comers more time to grow. This will
sharpen the annealing texture and (in the case of iron) enhance the (111) component which is in
fact desired for its high $\bar{R}$ value and good drawability. Precise control of precipitates such as
AlN, NbC and Cu has this desirable effect. The complex balance of variables involved has been
ably reviewed by Hutchinson (1974) and, in somewhat greater detail, by Hatherly & Dillamore
(1975).

Annealing textures may alter even after recrystallization is complete, during the subsequent stage of grain growth. There are two entirely distinct forms of grain growth: *normal* grain growth leads to a gradual *uniform* increase in size of surviving grains, while *abnormal* grain growth (secondary recrystallization, coarsening) involves extreme growth of a few grains and stasis of all others. Surprisingly, almost no research has been done on textural changes during *normal* grain growth. Dunn & Walter (1966) explain why only normal grain growth would be expected to remove grains furthest from the average texture, and thus sharpen the texture. Rauch, Thornberg & Foster (1977) have very recently established that precisely this happens in low-alloyed iron: the (110)[001] texture is sharpened during normal grain growth.

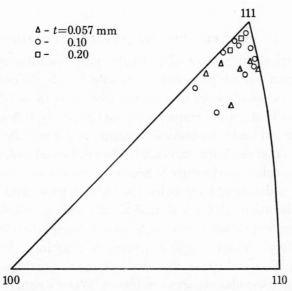

FIGURE 9. Poles of planes parallel to the free surface of individual secondary grains in sheets of high-purity platinum of three different sheet thicknesses, after annealing at 1500 °C. (McLean & Mykura 1965.) The factors governing the sharpness of the texture are discussed by Dunn & Walter (1966).

Secondary recrystallization, however, has been copiously examined, and excellently reviewed by Dunn & Walter (1966) and Walter (1969). The selective growth of a few grains 'takes off' when these grains exceed the average size by a sufficient factor, between 2 and 3. This arises from a size-dependent inbalance of tensions at grain-boundary triple points. A variety of structural factors can bring a population of recrystallized grains to this 'take-off' condition': a sharp annealing texture with a few deviant grains, a critical dispersion of precipitates which will restrain most grains but not all, or anisotropy of surface energy. In all cases, the large grains which win the race have a pronounced texture which is generally quite different from the original annealing texture. The production of the celebrated 'Goss texture' in iron–silicon alloys, (110)[001], which confers superior magnetic properties on transformer laminations and is of major industrial importance, is generated by secondary recrystallization controlled through subtle manipulation of dispersed precipitates.

A particularly interesting form of texture generation during secondary recrystallization is associated with surface energies. Thus, thin sheets of platinum (McLean & Mykura 1965) will develop large secondary grains all with (111) closely parallel to the surface (figure 9).

This happens because this orientation has a particularly low surface energy. An even more remarkable variant of this behaviour is the change of texture which results from changing the composition of the atmosphere in which an iron–silicon alloy is annealed: (110) parallel to the surface is favoured *in vacuo* and (100) in oxygen-bearing argon. Hydrogen containing H_2S favours (100). The very complex factors determining relative surface energies of different orientation as a function of atmosphere are fully analysed by Dunn & Walter (1966). They also show that for such surface energy factors to play a major rôle, the sheet has to be thin and the grain size large, because only in this way is the driving force associated with surface energy large enough to outweigh other driving forces associated with grain boundary energy and drag by precipitate particles. The matter of driving forces is one to which we shall return at the end of this review.

5. Textures in quartz and the rôle of Dauphiné twinning

In view of the geological importance of quartzite (polycrystalline quartz) and its possible rôle as an indicator of past tectonic processes, it is worthwhile, in the context of this Discussion, to compare the textural characteristics of quartzite with those of metals.

While quartzite is brittle at normal temperatures and pressures, it flows by dislocation movement readily when hot and under hydrostatic pressure, and especially if weakened by water. Hobbs (1968) has shown that single quartz crystals when deformed under such conditions in the laboratory and then annealed, recrystallize in much the same way as aluminium or iron, from nuclei forming at deformation band boundaries. (Hobbs calls them 'kink bands', adding further to the terminological luxuriance.) Optical analysis of *c* axis orientations shows that strain-induced boundary migration occurs here also. Similar behaviour has been reported in ortho-pyroxene (Etheridge 1975). White (1976) has recently examined, by electron and optical microscopy, the deformation and recrystallization of quartzite and has found that nucleation at grain boundaries (mantles) also occurs as in metals. White's paper shows a comprehensive familiarity with the metallurgical literature and the resemblances between quartzite and polycrystalline metals are firmly established. The textures (fabrics) formed in quartzite appear to be generated similarly to those in metals, though the experimental information is much sparser. In particular, polarized-light techniques as normally used in the examination of quartzite fabrics reveal the orientation of the *c* axis only and do not permit the construction of a complete pole figure, let alone orientation distributions. This would require X-ray diffraction which does not seem to be much used in this field of study.

An exception to this rule is found in the particularly important textural studies of Tullis (1970) and Tullis & Tullis (1972). They hot-deformed quartzite by uniaxial compression under superimposed large hydrostatic pressure to render the rock plastic. They used standard X-ray methods to obtain inverse pole figures such as figure 10*a*, and present persuasive evidence that the texture revealed in such pole figures is primarily due *not* to plastic deformation by dislocation mechanisms or indeed to normal recrystallization mechanisms but to the unfamiliar process of *Dauphiné twinning*.

Low or α-quartz, crystal class 32, is subject to several distinct forms of twin, of which one, the Dauphiné twin, is generated by a 180° rotation of the lattice about the threefold axis. The resultant rearrangement of the crystal structure is shown in figure 11. The Dauphiné twin is the prototype of what Klassen-Neklyudova, in her review of mechanically induced twinning (1964), calls 'twinning without change of form'. It is *not* a shear-twin, formed by twin gliding like the familiar lamellar twins in iron, zinc or uranium, but instead involves only atom shifts small compared with interatomic separations. (In M. J. Buerger's well-known terminology, Dauphiné twinning is *displacive* as distinct from the *reconstructive* process involved in twin-gliding.) In natural quartz crystals, Dauphiné twins occur as complex intergrowths. Tsinzerling was the first to observe that Dauphiné twin configurations could be generated by sustained application of concentrated loads on various faces of twin-free crystals, most easily at high temperatures; details may be found in Klassen-Neklyudova's book. Figure 12 summarizes the morphology of these pressure-induced twins, which are generated the more easily, the higher the temperature. Subsequent experimenters (Wooster, Wooster, Rycroft & Thomas 1947; Thomas & Wooster 1951; Aizu 1973; Newnham, Miller, Cross & Cline 1975) have established that the twin interface (probably controlled by a distribution of ultrafine precipitate particles) will move under an effective stress of *ca.* 50 kgf/mm² at room temperature, reducing to 0.1–1.0 kgf/mm² at 400–500 °C. (At 573 °C, α-quartz transforms into hexagonal β-quartz, and this is not subject to Dauphiné twinning.) The twin interface tends to be plane (Aizu 1970; Newnham & Cross 1974) but easily breaks away to adopt an arbitrary configuration like an ordinary grain boundary. (In fact, it is helpful to regard complex Dauphiné twin configurations as though they were grain configurations involving two fixed orientations only.)

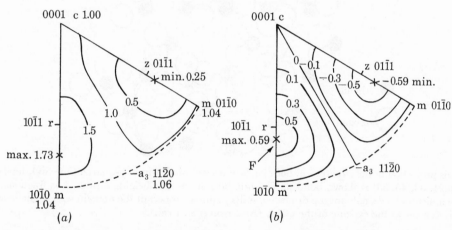

FIGURE 10. (*a*) *Inverse* pole figure for a flint (fine-grained form of quartzite) loaded at 500 °C and a confining pressure of 4 kbar to a differential compressive stress of 13 kbar for 3 h. The pole figure shows the distribution of the stress direction, and contours are in multiples of a uniform (random) distribution. (*b*) *Inverse* pole figure showing the variation with crystal direction in α-quartz of the *difference* in S'_{11} between original and twinned orientations, in units of 10^{-12} cm²/dyn. Equal-area projections are used as distinct from conventional stereographic projections. (Tullis 1970.)

Since Dauphiné twins do not form by shear, how does it come about that an applied stress can create and cause to grow a twin-orientated region? Thomas & Wooster (1951), in a classic study, answered this question when they discovered that if a particular applied stress-pattern was able to move a twin boundary, then the same stress pattern with signs of all stresses reversed would move the twin boundary in the *same* direction as before. This implies that the important factor is the square of the stresses, and this led Thomas & Wooster to the recognition that the crucial factor was the stored elastic energy, which involves the squares of the applied stresses. The twinned domains grow or contract in such a way that, in the end result, *the whole crystal yields elastically as much as possible under a given stress.* This is simply a corollary of the Le Chatelier principle. Many subsequent investigators have examined this criterion both experimentally and theoretically (see Paterson 1973 for a critical survey) and it is now well established for the case of constant applied stress. (For the special case of constant applied *strain*, it does not apply; see Tullis & Tullis 1972.) Because domains grow under applied stress, Thomas & Wooster coined the name *piezocrescence* for the process, but this expressive term does not seem to have been widely adopted.

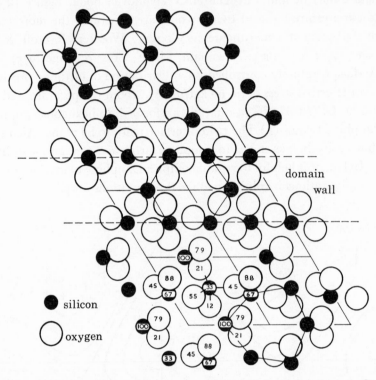

FIGURE 11. *c*-axis projection of the structure in both parts of a Dauphiné twin in α-quartz. One SiO_4 tetrahedron (atoms 12, 21, 33, 45, 55) is shown bonded. The numbers are atomic heights expressed as fractions of 100. The figure indicates how small are the displacive shifts required to permit the domain wall to move and one orientation to grow at the expense of the other. (Anderson *et al.* 1976.)

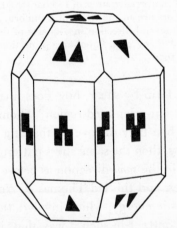

FIGURE 12. Surface traces of Dauphiné twins on the principal faces of a α-quartz crystal, generated by concentrated normal loading for prolonged periods. All Dauphiné twins have the same orientation. (Tsinzerling, as reproduced by Klassen-Neklyudova 1964.)

Dauphiné twinning in quartz is a special case of what has recently been recognized (Aizu 1970, 1973) as *ferroic* behaviour in a variety of crystals which can be transformed from one bistable state to the other by applied stress, magnetic or electric field, or combinations of these. The very involved physics and crystallography of the family of ferroic crystals is very clearly outlined by Newnham (1975).

For the simple case of uniaxial compression under superimposed hydrostatic pressure, the driving force ΔW causing a Dauphiné twin boundary to move is given by $\Delta W = \frac{1}{2}\Delta S'_{11}\, \sigma^2$, where σ is the (differential) compressive stress, S'_{11} is the reciprocal of Young's modulus (i.e. the elastic compliance) in the direction of the applied stress and $\Delta S'_{11}$ is the difference in the value of S'_{11} in the two constituent orientations of the twin. ΔW measures the difference in stored elastic energy for the two extremes of a crystal consisting entirely of one twin orientation or entirely of the other. ΔW has the dimensions of energy per unit volume or force per unit area (Tullis & Tullis 1972). If the stress system is more complicated, then a complex tensor criterion takes the place of the simple Tullis criterion (Newnham 1975; Anderson, Newnham, Cross & Laughner 1976). The considerable interest by physicists in this topic stems from the importance of quartz slices as oscillators in electrical circuits. Formerly (Wooster *et al.* 1947) the overriding aim was to remove all twins, but now it has been recognized that controlled twin configurations, generated by local laser-heating, can remove unwanted harmonics and improve oscillator performance (Newnham *et al.* 1975; Anderson *et al.* 1976).

The general concept of a driving force – or, more exactly, the force on an interface – has been examined by Eshelby (1970).

Tullis & Tullis uniaxially compressed randomly orientated fine-grained quartzites or flints at elevated temperatures under a confining pressure to make plastic deformation possible. Figure 10*a*, an *inverse* pole figure, shows the distribution of the stressing axis relative to the α-quartz unit cell for a flint loaded at 500 °C and a confining pressure of 4 kbar† to a differential compressive strain of 13 kbar (*ca.* 130 kgf/mm²). In figure 10*b*, values of $\Delta S'_{11}$ are shown for the same range of compression axes. It can be seen that the two pole figures are quite similar, which is a clear indication that the texture was formed by stress-induced Dauphiné twinning, i.e. that the more compliant directions tend, statistically, to be aligned with the stress. Whether this should be called a deformation texture or an annealing texture is a moot question: perhaps 'dynamic annealing texture' would best meet the circumstances of its genesis.

Tullis (1970) and Tullis & Tullis (1972) systematically varied the temperature and differential compressive stress and also used quartzites and flints with different impurity contents, and examined the resultant textures. (They established a simple statistical test to determine the contribution of factors other than Dauphiné twinning to texture formation. In all instances where a texture formed, it was found to be primarily due to twinning; a conventional deformation texture would generate a concentration of stress-axes near the centre of the pole figure (a '*c*-axis texture'), and thus a mixture of Dauphiné texture and conventional deformation texture causes a shift of the 'maximum' position from *F* nearer to the *c*-axis.) The experimental findings can be summarized as follows:

(1) A sample such as that of figure 10*a*, showing only a very small permanent strain (no exact figures were quoted), had a texture entirely due to Dauphiné twinning. A substantial plastic strain (30–50 %) led to contributions from a *c*-axis texture, but even then the Dauphiné contribution remained strong.

† 1 kbar = 10^8 Pa.

(2) For small strains, increasing the temperature at constant compressive stress led to stronger Dauphiné texture, up to a limiting temperature beyond which no further changes were found. This limiting temperature was much lower for purer quartzites, which confirmed that twin boundaries are constrained by impurities: the effect of temperature is essentially to provide thermal activation to overcome the twinning action of impurities. Finer-grained rocks (flints) formed stronger textures than coarse-grained quartzites, other things being equal.

(3) For small strains and constant temperature and purity, increasing stress leads to stronger Dauphiné texture. A large increase of stress is needed to achieve a modest increase of texture, i.e. texture intensity is not proportional to ΔW.

(4) Some samples which were deformed to substantial plastic strains had undergone dynamic recrystallization, others had not. In both cases there was a strong Dauphiné texture component, i.e. the Dauphiné twinning pattern can be repeatedly created after successive waves of dynamic recrystallization.

It follows from these observations that Dauphiné twinning, controlled by the elastic strain energy criterion, to a substantial extent controls texture in stressed quartzites and flints, *even in the absence of permanent strain*. In this last respect, these textures are different from anything known in metals. Tullis & Tullis (1972) conclude tentatively that Dauphiné textures offer scope for the assessment of the deformation history of naturally occurring quartz-rich rocks. It would be interesting to know whether very intense plastic deformation, as in mylonite zones in quartzites, proves to cause c-axis textures strong enough to swamp the Dauphiné component.

Driving forces

To a metallurgist, it is surprising that even in the presence of large plastic strains, the elastic strain criterion can nevertheless control a major component of the resultant texture. To conclude this section, therefore, we shall address this question by estimating the driving force for Dauphiné twin-texturing and comparing it with typical driving forces for recrystallization into a plastically deformed material. We will undertake the former calculation for a typical condition used by Tullis & Tullis, namely 15 kbar (differential) uniaxial stress. This is just enough to produce substantial permanent strain over a period of some hours at *ca.* 500 °C.

$\Delta S'_{11}$ (in the expression $\Delta W = \frac{1}{2}\Delta S'_{11}\sigma^2$) will be taken as the value for the most favourably orientated grain (position F in figure 10b). On this basis (compare figure 10b),

$$\Delta S'_{11} = 0.5 \times 10^{-12} \text{ cm}^2/\text{dyn}, \quad \sigma = 15 \text{ kbar} \approx 150 \text{ kgf/mm}^2 \approx 15 \times 10^9 \text{ dyn/cm}^2,$$

whence
$$\Delta W \approx 5 \times 10^7 \text{ dyn/cm}^2 \quad \text{or} \quad 5 \text{ J/cm}^3.$$

In terms of a gram-molecular mass, the driving energy ≈ 130 J/mol. This surprisingly high driving force is directly attributable to the very high differential stress needed to deform quartzite plastically even at a temperature as high as 500 °C. No measurements are available of the stored energy of plastic deformation in quartzite, so we must take metallurgical data for comparison (see Martin & Doherty 1976). Thus copper, deformed to a logarithmic strain of 1.2, has a stored energy (= driving force for primary recrystallization) of *ca.* 50 J/mol. The maximum strain imposed in Tullis & Tullis's experiments was much less than this, so that one can conclude with confidence, even allowing for the unknown differences between copper and quartz, that the driving force for Dauphiné-twinning is somewhat greater than that for conventional recrystallization of plastically deformed rock. We can now understand why a strong Dauphiné texture forms even in permanently deformed and recrystallized quartzites.

For comparison, the following are representative driving forces for other microstructural processes: a phase change, such as the precipitation of $CuAl_2$ from an Al–Cu solid solution at 600 K; 1200 J/mol. The driving force for normal grain growth, in a metal of grain diameter 0.03 mm and boundary energy of 0.5 J/m², is only about 0.5 J/mol. This is the energy liberated by complete elimination of the boundaries. For secondary recrystallization controlled by differential surface free energy (for an iron sheet 0.1 mm thick, an initial grain size of 1 mm, a surface energy ratio of $\sigma_{100}/\sigma_{110} = 0.9$ and $\sigma_{100} \approx 2.1$ J/m²), the driving force $\approx 3 \times 10^{-2}$ J/mol. It is thus not surprising that this effect is found only after primary recrystallization is complete: if any appreciable plastic deformation were to remain, the stored energy associated with the dislocations would utterly swamp the frail driving force favouring (100)-orientated grains over (110)-orientated grains! The driving force associated with a magnetic field applied to a ferromagnetic polycrystalline sheet leading in principle to the selection of grains with a particularly high induction, is smaller still, and indeed there is no convincing evidence that a magnetic field can affect the annealing texture of a steel. The directed migration of a grain boundary in a (diamagnetic) bismuth bicrystal under the influence of a very large magnetic field has been recorded (Mullins 1956) but there is no record of any influence on a texture.

It might have been supposed that elastic anisotropy might be able to generate an annealing texture in a highly anisotropic metal during grain growth following primary recrystallization, if a small stress (too small to cause creep) were to be applied during the anneal. Thus for zinc, the largest value of $\Delta S'_{11}$ would be $S_{33} - S_{11}$, about 2×10^{-12} cm²/dyn (larger than ΔS for quartz). A typical flow stress at ambient temperature for polycrystalline zinc is 25 kg/mm²; if a compressive stress of one quarter of this value were applied at say 250 °C, then $\Delta W = \frac{1}{2} \Delta S'_{11} \sigma^2$ would be about 1.5 J/mol, and grain growth should proceed under this directing influence. This value of ΔW is a much smaller value than we have found for quartz (because the stress is necessarily so small), but it exceeds the driving force for normal grain growth and in the absence of any plastic deformation to swamp the driving force, it should be quite sufficient to generate a texture by selection of favourably orientated grain during grain growth. This kind of texture in metals remains to be established.

I am grateful to Dr R. D. Doherty and Dr Y. Inokuti for allowing me to reproduce a stereogram from their work in advance of publication, and to Prof. R. E. Newnham for information about his research in advance of publication.

References (Cahn)

Ahlborn, H. & Wassermann, G. 1963 Z. Metallk. **54**, 1.

Aizu, K. 1970 Phys. Rev. B **2**, 754.

Aizu, K. 1973 J. phys. Soc. Japan **34**, 121.

Anderson, T. L., Newnham, R. E., Cross, L. E. & Laughner, J. W. 1976 Phys. Stat. Sol. (a) **37**, 235.

Atkinson, M. & McLean, I. M. 1965 Sheet Metal Industries **42**, 290.

Aust, K. T. 1969 In Textures in research and practice (eds J. Grewen & G. Wassermann), p. 24. Berlin: Springer-Verlag.

Babel, H. W. & Frederick, S. F. 1968 J. of Metals (A.I.M.E.), October, p. 32.

Bailey, J. E. 1960 Phil. Mag. **5**, 833.

Barrett, C. R., Lytton, J. L. & Sherby, O. D. 1967 Trans. met. Soc. Am. Inst. min. metall. Engrs **239**, 170.

Barrett, C. S. & Levenson, L. H. 1940 Trans. Am. Inst. min. metall. Engrs **137**, 112.

Barrett, C. S. & Massalski, T. B. 1966 Structure of metals, 3rd edn, chs 20, 21. New York: McGraw-Hill.

Bellier, S. P. & Doherty, R. D. 1977 Acta metall. **25**, 521.

Bishop, J. F. W. & Hill, R. 1951 Phil. Mag. **42**, 414, 1298.

Bunge, H. J. 1965 *Z. Metallk.* **56**, 872.

Bunge, H. J. 1969 In *Textures in research and practice* (eds J. Grewen & G. Wassermann), p. 24. Berlin: Springer-Verlag.

Cahn, R. W. 1970 *Physical Metallurgy*, 2nd edn, ch. 19. Amsterdam: North-Holland.

Calnan, E. A. & Clews, C. J. B. 1952 *Phil. Mag.* **43**, 93.

Chin, G. Y. 1969 In *Textures in research and practice* (eds J. Grewen & G. Wassermann), p. 51. Berlin: Springer-Verlag.

Chin, G. Y., Hart, R. R. & Wonsiewicz, B. C. 1969 *Trans. met. Soc. Am. Inst. min. metall. Engrs* **245**, 1669.

Davies, G. J., Dillamore, I. L., Hudd, R. C. & Kallend, J. S. (eds) 1975 *Texture and the properties of materials*. London: The Metals Society.

Dillamore, I. L. & Katoh, H. 1974 *Met. Sci.* **8**, 21, 73.

Dillamore, I. L., Morris, P. L., Smith, C. J. E. & Hutchinson, W. B. 1972 *Proc. R. Soc. Lond.* A **329**, 405.

Dillamore, I. L., Smith, C. J. E. & Watson, T. W. 1967 *Met. Sci. J.* **1**, 49.

Doherty, R. D. & Cahn, R. W. 1972 *J. less-common Metals* **28**, 279.

Dunn, C. G. & Walter, J. L. 1966 In *Recrystallization, grain growth and textures* (ed. H. Margolin), p. 295. Metals Park: American Society for Metals.

English, A. T. & Chin, G. Y. 1965 *Acta metall.* **13**, 1013.

Eshelby, J. D. 1970 In *Inelastic behaviour of solids* (eds M. F. Kanninen *et al.*), p. 77. New York: McGraw-Hill.

Etheridge, M. A. 1975 *Tectonophysics* **25**, 87.

Ferran, G., Doherty, R. D. & Cahn, R. W. 1971 *Acta metall.* **19**, 1019.

Gordon, P. & Vandermeer, R. A. 1966 In *Recrystallization, grain growth and textures* (ed. H. Margolin), p. 205. Metals Park: American Society for Metals.

Grewen, J. & Wassermann, G. 1962 *Texturen metallischer Werkstoffe*. Berlin: Springer-Verlag.

Hatherly, M. & Dillamore, I. L. 1975 *J. Austral. Inst. Metals* **20**, 71.

Held, J. F. 1967 *Trans. met. Soc. Am. Inst. min. metall. Engrs* **239**, 573.

Hobbs, B. E. 1968 *Tectonophysics* **6**, 353.

Hosford, W. F. 1969 In *Textures in research and practice* (eds J. Grewen & G. Wasserman), p. 464. Berlin: Springer-Verlag.

Hu, Hsun & Goodman, S. R. 1963 *Trans. Am. Inst. min. metall. Engrs* **227**, 627.

Hu, Hsun, Cline, R. S. & Goodman, S. R. 1966 In *Recrystallization, grain growth and textures* (ed. H. Margolin), p. 295. Metals Park: American Society for Metals.

Hutchinson, W. B. 1974 *Met. Sci.* **8**, 185.

Inokuti, Y. & Doherty, R. D. 1977*a Textures crystall. Solids* **2**, 143.

Inokuti, Y. & Doherty, R. D. 1977*b Acta metall.* **25**. (In the press.)

Kalland, J. S. & Davies, G. J. 1972 *Phil. Mag.* **25**, 471.

Klassen-Neklyudova, M. V. 1964 *Mechanical twinning of crystals*, pp. 87, 150. New York: Consultants Bureau.

Langdon, T. G. & Mohamed, F. A. 1975 In *Grain boundaries in engineering materials* (*Proc.* 4th Bolton Landing Conf.) (eds J. C. Walter & J. H. Westbrook), p. 339. Baton Rouge: Claitor's Publishing Division.

McLean, M. & Mykura, H. 1965 *Acta metall.* **5**, 628.

Martin, J. W. & Doherty, R. D. 1976 *Stability of microstructure in metallic systems*, pp. 5, 82. Cambridge University Press.

Mullins, W. W. 1956 *Acta metall.* **4**, 421.

Narita, N. & Takamura, J. 1974 *Phil. Mag.* **29**, 1001.

Newnham, R. E. 1975 *Structure–property relations*, ch. 4. Berlin: Springer-Verlag.

Newnham, R. E. & Cross, L. E. 1974 *Mater. Res. Bull.* **9**, 1021.

Newnham, R. E., Miller, C. S., Cross, L. E. & Cline, T. W. 1975 *Phys. Stat. Sol.* (a) **32**, 69.

Panknin, W. 1969 In *Textures in research and practice* (eds J. Grewen & G. Wasserman), p. 464. Berlin: Springer-Verlag.

Paterson, M. S. 1973 *Rev. Geophys. & Space Phys.* **11**, 355.

Rauch, G. C., Thornburg, D. R. & Foster, K. 1977 *Metall. Trans.* **8 A**, 210.

Roe, R. J. 1965 *J. appl. Phys.* **36**, 2024.

Schmid, E. & Boas, W. 1935 *Kristallplastizität*. Berlin: Springer-Verlag.

Schmidt, W., Lücke, K. & Pospiech, J. 1974 In *Texture and the properties of materials* (eds G. J. Davies *et al.*), p. 147. London: The Metals Society.

Taylor, G. I. 1934 *J. Inst. Metals* **62**, 307.

Thomas, L. A. & Wooster, W. A. 1951 *Proc. R. Soc. Lond.* A **208**, 43.

Tullis, J. 1970 *Science, N.Y.* **168**, 1342.

Tullis, J. & Tullis, T. 1972 In *Flow and fracture of rocks*, Geophysical Monograph Series, no. 16 (eds H. C. Heard *et al.*), p. 67. Washington: American Geophysical Union.

Walter, J. L. 1969 In *Textures in research and practice* (eds J. Grewen & G. Wassermann), p. 227. Berlin: Springer-Verlag.

White, S. 1976 *Phil. Trans. R. Soc. Lond.* A **283**, 69.

Wooster, W. A., Wooster, N., Rycroft, J. L. & Thomas, L. A. 1947 *J. Instn Elect. Engrs* **94** (part IIIA), 927.

Phil. Trans. R. Soc. Lond. A. **288**, 177–196 (1978) [177]
Printed in Great Britain

Stresses and deformation at grain boundaries

By W. Beeré

*Central Electricity Generating Board, Berkeley Nuclear Laboratories,
Berkeley, Gloucestershire, U.K.*

[Plates 1 and 2]

Grain boundary sliding is frequently observed during the creep of polycrystals and this can alter both the internal stresses and the creep rate. Sliding arises because the shear and normal forces acting on the grain boundaries can be relaxed by separate mechanisms at different rates. If sliding is easy the shear forces can in the limit tend to zero. The normal forces are then relaxed more slowly by plastic deformation inside the grain or by diffusion creep. While the former distorts the interior of the grain the latter does not. Several two dimensional models of the diffusion mechanism have appeared in which rigid slabs slide past each other. Diffusion plates out material on the boundaries and controls grain movement normal to the boundary. It is also possible to solve the 'rigid grain' situation in three dimensions when rapid diffusion at boundaries relaxes the normal forces. The shear process then controls the grain motion and it is necessary that the grains roll over each other.

1. Introduction

Grain boundary sliding is frequently observed in metals and ceramics deformed at elevated temperatures. Localized strains within the material can be highly inhomogeneous with extensive shear at the boundary planes. Sliding never occurs on its own in polycrystals, but has to be accompanied by other deformation mechanisms. The sliding and accommodating processes can conveniently be classified into three categories.

First, the grains can deform by diffusion creep, figure 1 (*a*). Material removed from one boundary is deposited at another thus moving adjacent grain centres normal to their common boundary.

Examination of the grain geometry shows that sliding must occur simultaneously to prevent gaps appearing at grain boundaries (Lifshitz 1963; Gibbs 1965; Stephens 1971; Gates 1975). The grain interior does not deform and all displacements of adjacent grain centres parallel to a common boundary have to take place by boundary sliding. Diffusion creep and grain boundary sliding are mutually accommodating. Usually the sliding process has a short relaxation time compared with diffusion across the grain. Thus the shear stresses on the boundaries necessary to cause sliding at a rate compatible with the diffusion process are vanishingly low (Raj & Ashby 1971).

A similar type of deformation can be envisaged again in which the interior of the grain does not deform, but in which boundary shear forces are large, figure 1 (*b*). An important feature of this type of deformation is grain rotation. Frictional forces developed on the boundary rotate the atomic lattice (Beeré 1976). Possible applications are in superplastic creep and diffusion creep near a threshold stress.

Lastly, at higher stresses dislocation creep in the grain interior becomes important, figure 1 (*c*).

Shear between grain centres takes place both by shear inside the grain and by boundary sliding (Bell & Langdon 1969; Crossman & Ashby 1975; Speight 1976).

Examples of all three categories are available. Figure 2, plate 1, is a scanning electron micrograph showing sliding offsets during the diffusion creep of hyperstoichiometric UO_2 (Reynolds, Burton & Speight 1975). The scratch lines produced during polishing of the originally flat surface show clear offsets at grain boundaries. The scratch lines remain straight in the grain interior indicating a negligible contribution from dislocation creep. Some grains have moved normal to the surface and a few grain boundaries have separated. The UO_2 was crept under

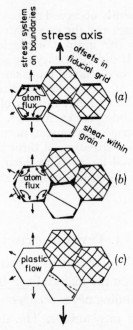

FIGURE 1. Three categories of grain boundary sliding: (a) diffusion creep (zero boundary shear stress); (b) rigid grain creep (finite boundary shear stress); and (c) dislocation creep (with grain boundary sliding). Types (a) and (b) have rigid grain interiors while the grain interior deforms plastically in (c). Grain boundary shear forces are absent in (a) and can be fully relaxed in (c) but are present in (b).

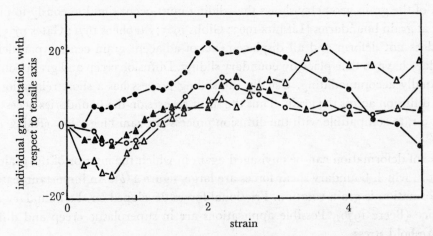

FIGURE 3. Grain rotations observed during creep of a Pb/Sn eutectic
(Geckinli & Barrett 1976).

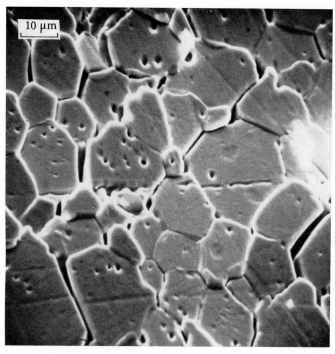

FIGURE 2. A scanning electron micrograph of UO_2. The initially flat polished surface shows ledges, sliding offsets of surface scratches and cracks resulting from grain boundary sliding during creep. (Reynolds *et al.* 1975.)

FIGURE 4. Deformation by grain boundary sliding and dislocation creep of an initially rectangular 8 µm grid on aluminium. (Pond *et al.* 1976.)

Phil. Trans. R. Soc. Lond. A, volume 288

Beeré, plate 2

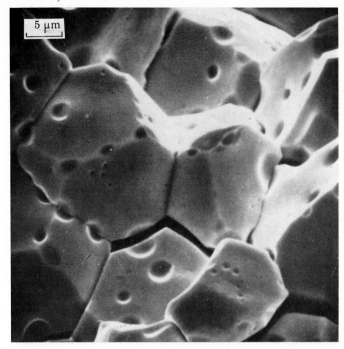

FIGURE 8. Cracks in a UO_2 compression specimen. The wedging action of adjacent grains coupled with grain boundary sliding produces tensile forces and cracking. (Reynolds *et al.* 1975.)

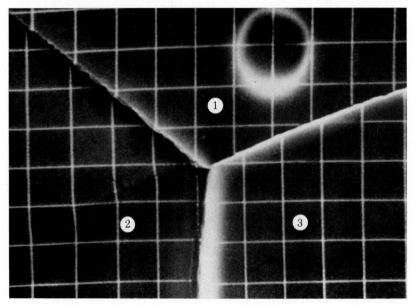

FIGURE 13. Grain boundary sliding in aluminium revealed by displacements in an initially rectangular 8μm grid. Sliding between grains 1 and 3 has created a slip band in grain 2. (Pond *et al.* 1976.)

compression, but relaxation of the shear stresses results in tensile forces developing across certain boundary orientations.

Significant grain rotation occurs during stage II superplastic creep. Figure 3 shows the angular variation of four grains in a lead tin eutectic (Geckinli & Barrett 1976). Quite large rates of rotation are observed although the angular variation seldom exceeds 30°.

Dislocation creep and grain boundary sliding are common to a very large number of systems. Figure 4, plate 1, shows a scanning electron micrograph of pure aluminium crept at high temperature *in situ* in the scanning electron microscope (Pond, Smith & Southerden 1976). The grid bars were originally a rectangular array spaced 8 μm apart before creep. After deformation the grid shows internal grain deformation and sliding along the three boundaries present.

Analysis of combined sliding and accommodating deformation is facilitated by knowledge of the boundary stresses and the possible sliding displacements. The two types of sliding with rigid grain interiors, figure 1 (a, b), are now examined in detail.

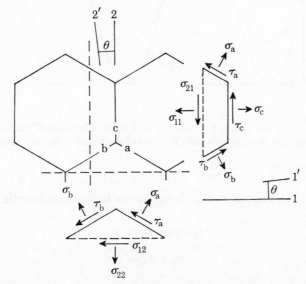

FIGURE 5. The stresses acting on idealized hexagonal grains and elements of grains.

2. STRESSES ACTING ACROSS GRAIN BOUNDARIES

The normal and shear stresses on a boundary in a homogeneous body behave as tensors and their value can be calculated simply from the boundary orientation. When the boundaries slide the body becomes inhomogeneous, the boundary stresses change and have to be recalculated taking the grain geometry into account. The complex morphology of real grains has not been treated, but typically a much simpler grain shape is substituted to ease calculations. Hexagonal grains are often chosen because they are the simplest shape in which three grain boundaries meet in a position of stable equilibrium. The stress system on the hexagon boundaries is calculated for an aggregate of many regular grains randomly orientated with respect to an applied uniaxial tensile stress. The same procedure could be followed for any superimposed system of stresses.

Figure 5 shows three grains of the aggregate. Each grain is assumed to be indistinguishable from its neighbours. Hence there are three types of boundary since for instance a vertical boundary (figure 5) behaves identically to any other vertical boundary. The three boundary types are labelled a, b and c.

The stresses are calculated by removing elements of the grains and balancing forces. The stress distribution inside the grains is not known, but the internal stresses along the dashed lines (figure 5) repeat cyclically in each grain. Hence the average stress on the dashed face of one of the elements is the same as the average stress on the complete dashed line in the aggregate. Since the line passes completely through the aggregate the average stress must support the applied load and so its value is the same as the stress on a similarly orientated plane in a homogeneous

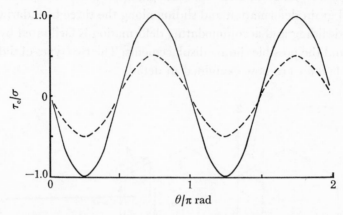

FIGURE 6. The shear force acting on a grain boundary of an idealized hexagonal grain for a homogeneous solid, broken line, and a grain with fully relaxed normal boundary forces.

body. Putting $\sigma_{11} = \sigma \cos^2 \theta$, etc., where σ is the uniaxial stress and θ the angle between the stress axis and the 1 axis, the forces on the elements can be balanced vertically and horizontally giving

$$\tau_a + \tau_b + \tau_c = 0, \tag{1}$$

$$\sigma_a + \sigma_b + \sigma_c = \tfrac{3}{2}\sigma, \tag{2}$$

where τ_a and σ_a are the average shear and normal stresses on an 'a' boundary respectively and the subscripts b and c refer to the other boundaries.

The calculation can proceed further only if assumptions are made about the material properties. It is simplest to deal with complete relaxation of either the normal or the shear stresses on the boundary.

If for instance the normal stresses are relaxed they are everywhere equal because only an infinitesimal deviatoric stress is necessary. From equation (2) their value must be

$$\sigma_a = \sigma_b = \sigma_c = \tfrac{1}{2}\sigma.$$

The shear stress on a 'c' boundary is then (Beeré 1976)

$$\tau_c = 2\sigma \sin \theta \cos \theta, \tag{3}$$

which is exactly twice that for a homogeneous solid (figure 6). The stresses on the 'a' and 'b' boundaries can be found by adding multiples of $\tfrac{2}{3}\pi$ to θ. Conversely, if the shear stresses are relaxed, $\tau_a = \tau_b = \tau_c = 0$, then the normal stress on a 'c' boundary is

$$\sigma_c = \sigma(\tfrac{3}{2} - 2 \sin^2 \theta). \tag{4}$$

The normal stress is illustrated in figure 7 along with the stress in a homogeneous solid. Relaxing the shear stresses doubles the normal stresses about their mean value. During tensile creep compressive forces appear across certain boundaries. Likewise, if the external load is compressive tensile forces appear which can be large enough to cause fracture on the boundary. This is illustrated in figure 8, plate 2 (Reynolds *et al.* 1975), which shows a $UO_{2.2}$ specimen crept at 80 MPa and 1350 °C.

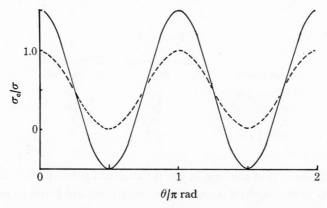

FIGURE 7. The normal force acting on a grain boundary of an idealized hexagonal grain for a homogeneous solid, broken line, and a grain with fully relaxed boundary shear forces.

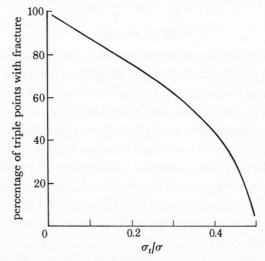

FIGURE 9. The percentage of triple points associated with a cracked boundary plotted against the ratio of average tensile fracture stress to applied compressive stress.

Significant fracture is expected to take place when the average tensile fracture stress is less than half the applied compressive stress. If it is assumed that fracture takes place when the average tensile stress exceeds a critical value σ_t, then assuming a random distribution of boundary orientations the percentage of boundaries fractured can be calculated from equation (4). The result is illustrated in figure 9. Thus if σ_t/σ is 0.45, the initial application of the external load causes about 30 % of triple points to be associated with a fractured boundary. An examination of the stress system near a fractured boundary reveals that the locality can no longer support an applied uniaxial stress. The grain can rapidly relax the local stress system by grain boundary sliding redistributing part of its load onto unfractured grains. This causes more grains to fracture. If the fracture stress is sufficiently high this redistribution mechanism eventually stabilizes. Below a critical value of σ_f/σ the mechanism becomes unstable and the polycrystal suffers catastrophic fracture. This is best observed in doped ceramics in which the dopant segregates preferentially to the grain boundary reducing the fracture stress without significantly altering the creep properties of the grain. These materials are susceptible to catastrophic intergranular fracture during the application of large compressive stresses (Reynolds 1977).

Table 1

invariant	homogeneous solid	fully relaxed shear stresses
$\sigma_a + \sigma_b + \sigma_c$	$\frac{3}{2}\sigma$	$\frac{3}{2}\sigma$
$\sigma_a^2 + \sigma_b^2 + \sigma_c^2$	$\frac{9}{8}\sigma^2$	$\frac{9}{4}\sigma^2$
$\tau_a + \tau_b + \tau_c$	0	0
$\tau_a^2 + \tau_b^2 + \tau_c^2$	$\frac{3}{8}\sigma^2$	0

Equations (1) and (2) show that the sum of the normal and shear stresses are both invariant with respect to rotation of the applied stress. Other invariants are listed in table 1 with values when the boundary shear stresses are zero and for a homogeneous material. The 'invariants' can alter in value during boundary relaxation.

3. Grain motion

The previous section considered the stresses acting on grain boundaries. Next it is pertinent to ask how the boundary stresses plastically deform the grain. First, the geometry of the possible motions is considered followed by the rate of motion resulting from the material properties.

When the dominant deformation mechanism is diffusion creep the interiors of the grains do not deform. Material is transferred preferentially between boundaries usually by a vacancy mechanism. If again we treat a two dimensional array of hexagons, the hexagons behave as rigid slabs with some boundaries gaining material while others lose material. If the increase in distance between grain centres across an 'a' boundary is N_a and the sliding displacement is S_a then there are six separate movements (figure 10). These are not all independent. If gaps do not appear on the boundary during creep then analysis shows that the sum of the normal and sliding displacements are both zero, i.e.

$$N_a + N_b + N_c = 0, \tag{5}$$

$$S_a + S_b + S_c = 0, \tag{6}$$

where the subscripts refer to the boundary type. Equation (5) is simply a statement that the volume of material remains constant.

If the motion of one grain is indistinguishable from its neighbours then the motion between adjacent grain centres is identical to the bulk motion of the aggregate. When the aggregate deforms uniaxially by a strain ϵ the strain in the coordinate system of the grain centres, ϵ_{ij}, is given by the equation:

$$\epsilon = \epsilon_{11}\cos^2\theta + \epsilon_{22}\sin^2\theta + (\epsilon_{12} + \epsilon_{21})\sin\theta\cos\theta. \tag{7}$$

The strain between grain centres, e_{ij}, can also be written in terms of the boundary displacements. Substitution into equation (7) gives the desired relation between the aggregate strain and the normal and sliding boundary displacements.

$$ed = -(N_a + N_b)(2\cos^2\theta - 1) + \frac{1}{\sqrt{3}}(N_a - N_b)2\sin\theta\cos\theta \tag{8}$$

$$= \frac{1}{\sqrt{3}}(S_a - S_b)(2\cos^2\theta - 1) + \frac{1}{\sqrt{3}}(S_a + S_b - 2S_c)2\sin\theta\cos\theta, \tag{9}$$

where d is the hexagon diameter.

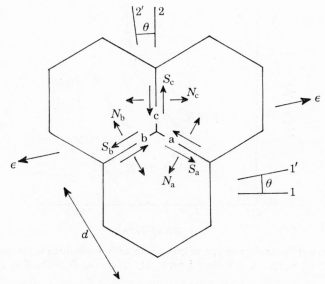

FIGURE 10. The three sliding and three normal displacements of adjacent grains at a triple point.

The bulk specimen strain can be defined completely in terms of the normal displacements or the sliding displacements. Thus, if diffusion changes the grain centre distances resulting in creep, sliding must also take place and both types of displacement are mutually accommodating.

4. DIFFUSION CREEP

The equilibrium concentration of vacancies adjacent to a grain boundary depends on the normal stress acting on the boundary. A difference in vacancy concentration between boundaries sets up a vacancy flux. If we consider an 'a' boundary the rate of separation $\dot{N}_a$ is given by

$$\dot{N}_a = \text{const } D\Omega(\sigma_a - \tfrac{1}{2}\sigma_b - \tfrac{1}{2}\sigma_c)/kTd, \tag{10}$$

where D is the volume self-diffusion coefficient, Ω the atomic volume, k Boltzmann's constant and T absolute temperature. The bracketed term takes account of the flux between both 'a' and 'b' and 'a' and 'c' boundaries. A detailed calculation gives the constant a value $3^{\frac{3}{2}}\pi^3/16$ (Beeré 1976).

When diffusion creep is the rate controlling mechanism the boundary shear forces are considered to be vanishingly small and the normal stresses will be given from the previous calculation. From equations (4), (8) and (10) the creep rate is

$$\dot{e} = 15D\sigma\Omega/kTd^2, \tag{11}$$

which is seen to be independent of orientation of the applied stress.

This equation has been derived previously by a number of workers (Gibbs 1965; Raj & Ashby 1971; Herring 1950). In the models, like the present one, which consider total relaxation of boundary shear stresses the numerical constant is about 50 % higher than in those calculations which assume a stress distribution typical of a homogeneous solid. The equation can be generalized further by inclusion of the Coble or grain boundary creep (Coble 1963) component by replacing the diffusion coefficient D with $(D + \pi D_g \delta / d)$, where D_g is the grain boundary diffusion coefficient and δ the boundary width.

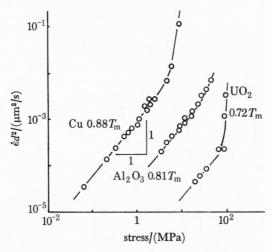

FIGURE 11. The observed linear dependence of creep rate with applied stress for three materials deforming in diffusion creep (Cu; Burton & Greenwood 1970; Al_2O_3: Davies & Sinha Ray 1972; UO_2: Poteat & Yust 1968).

The linear relation between creep rate and stress has been recently reviewed as well as the $1/d^2$ dependence for Nabarro–Herring creep and $1/d^3$ dependence for Coble creep (Burton 1977). Figure 11 shows creep data for Cu (Burton & Greenwood 1970), UO_2 (Poteat & Yust 1968) and Al_2O_3 (Davies & Sinha Ray 1972). The creep rate is compensated for grain size to allow for simultaneous grain growth during the test. The data show the linear relation between creep rate and uniaxial stress at low stresses. Increasing the stress rapidly increases the rate of dislocation creep, which is an independent process, and this eventually becomes the dominant deformation mechanism.

The value of the numerical constant in equation (11) is usually difficult to assess accurately from experimental data because of uncertainty in the diffusion coefficient. There are exceptions, and five independent observations of the volume self-diffusion coefficient in copper agree within a factor of 1.5 (Butrymowicz, Manning & Read 1974) giving an observed value from figure 11 of 18 with an upper and lower limit of 27 and 12 respectively. The agreement with theory is good considering the two dimensional nature and regularity of hexagonal grains.

The stress on the boundary calculated earlier was an average value for a particular boundary. When the grains deform by diffusion creep the vacancy flux leaving unit area of boundary is constant over the entire length of a particular boundary. The flux is proportional to the difference in normal tensile forces acting locally on source and sink and inversely proportional to the diffusion distance. Near a triple point the diffusion distance is small so the boundary stress is also small and tends to zero at the triple point. This also satisfies the requirement that there is no instantaneous change in stress at the triple point. At the centre of the boundary the stress is largest reaching a value of 1.44 times the average value. This is illustrated in figure 12.

The distribution of stress is sensitive to the deformation mode of the grains. If for instance the dominant mechanism is dislocation creep, grain boundary sliding concentrates stress at the triple points, in complete contrast to diffusion creep. Figure 13, plate 2, illustrates this point in aluminium containing precipitates (Pond *et al.* 1976). The perpendicular grid lines super-imposed on the surface before creep show signs of grain boundary shear between grains 1 and 3. The shear on this boundary has concentrated the stress at the triple point creating a slip band in the adjacent grain. High localized stresses can lead to triple point cracking. Alternatively if diffusion creep were rapid the stress concentrations would be absent, suppressing cracking.

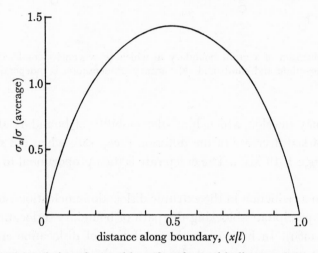

FIGURE 12. The variation of normal boundary force with distance along the boundary for a material deforming by Nabarro–Herring creep.

5. GRAIN BOUNDARIES AS SOURCES AND SINKS FOR VACANCIES

A second type of creep can occur in which the grain interior does not deform and in which shear forces play an important part in determining the grain trajectories. The application to physical situations is best understood by considering the operation of grain boundary vacancy sinks and sources. Several possible models are in existence and the mechanism of vacancy absorption on grain boundary dislocations is not completely clear, but situations where boundary shear stresses develop can often be envisaged.

Two types of boundary will be considered here, one in which vacancies are absorbed by dislocation (or a similar defect) climb in the boundary (McLean 1971; Gates 1973; Das & Marcinkowski 1972; Ashby 1969) and the second in which dislocation motion is unnecessary. The first is illustrated in figure 14 (*a*) in which just two sets of parallel dislocations with arbitrary Burgers vectors inhabit the boundary. If one set of dislocations is mobile and moves by simul-taneous glide and climb the material on both sides of the boundary is displaced normally and sheared relative to the boundary plane. A vacancy flux is required which may have to come from another part of the crystal. Motion of the second set of dislocations can provide the vacancies enabling sliding to take place at a rate limited only by vacancy diffusion between dislocations. If the material is deforming by diffusion creep the supply of vacancies from external sources is limited by diffusion across the grain. The much shorter diffusion path between dislocations enables sliding to rapidly release the boundary shear forces. In three dimensions three arbitrary sets of dislocations are required to enable independent boundary sliding in any direction. In general, three sets of intrinsic dislocations are necessary to provide coincidence when the lattices on either side of the boundary are at some random orientation.

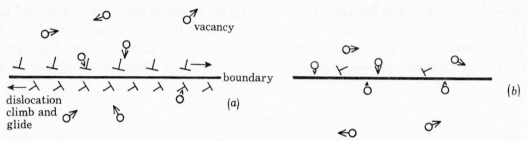

FIGURE 14. A schematic diagram of a grain boundary in which (a) vacancies condense on dislocation cores producing simultaneous glide and climb and (b) vacancy condensation is independent of the dislocation structure.

Several situations may develop which limit the mobility of boundary dislocations. Many materials exhibit a marked decrease of the diffusion creep rate when the stress is reduced to a value often in the range 1–10 MPa. The creep rate is then proportional to $(\sigma - \sigma_0)$, where σ_0 is the threshold stress.

This may result from a reduction in the extrinsic dislocation nucleation rate in the boundary (Burton 1973) or from precipitates blocking the path of boundary dislocations (Harris, Jones, Greenwood & Ward 1969). In both cases the inhibition of dislocation motion inhibits the ability of boundaries to emit vacancies and also prevents the dislocations from rapidly relaxing the shear forces. Diffusion creep rates are invariably slow near the threshold stress which limits the ultimate specimen strain. As a consequence morphological changes in grain structure are small and unlikely to be observed.

In direct contrast the strains realized in stage II superplastic creep are extremely large. Deformation takes place predominantly at the grain boundary and the grain interior often remains underformed. Low stress stage II creep has the following attributes: (a) no evidence of internal slip lines; (b) scratch offsets are sharp; (c) an almost equiaxed structure; (d) grains relatively dislocation free; (e) grains rotate; (f) grains switch neighbours (see Edington, Melton & Cutler 1976). Several deformation mechanisms are based on grain boundary dislocations with motion limited by diffusion barriers in the grain boundary structure or grain boundary dislocation pileups near the triple point. In each case the mechanism limits the relaxation of both shear forces and normal boundary forces.

If boundary dislocation motion is unnecessary for vacancy emission or absorption the sliding mechanism will be completely independent of diffusion creep (figure 14b). If in this case the sliding process has a different activation energy or stress dependence from diffusion creep, situations can occur where sliding is rate controlling. Shear forces will develop on the boundaries but the normal forces will everywhere relax to some constant value dependent on the externally applied stress system.

6. Grain boundary sliding control of creep

Previously it was shown that shear forces are likely to develop on grain boundaries during diffusion creep at low stresses near a threshold stress and during stage II superplastic creep. The creep rate will be controlled either by the shear process with rapid relaxation of normal forces or by simultaneous slow relaxation of both normal and shear forces. In both cases the important feature from the point of grain motion is the effect of the shear forces. The friction which develops between grains leads to rotation of the grain. This is not to be confused with apparent rotation resulting from motion of the grain boundaries. The process to be discussed rotates the lattice whereas boundary migration leaves the lattice orientation unchanged. The principle of grain rotation is illustrated in figure 15 for square grains. The dotted lines (figure 15a) join the grain centres. When deformed uniaxially (figure 15b) material is redistributed between boundaries and the boundaries slide. When the boundary shear forces are not zero the externally applied forces do work sliding the boundaries.

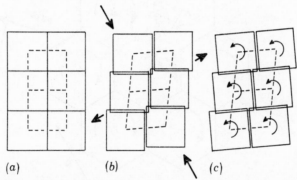

(a)   (b)   (c)

FIGURE 15. Six square grains (a) are deformed uniaxially, (b) producing grain boundary sliding on all faces. Grain rotation (c) allows reduction of the sliding displacement of the vertical faces without change in specimen strain.

In the example figure 15(b) the vertical and horizontal boundaries slide by equal amounts. The boundary viscosity, in general, will not be exactly uniform but can show variation due to a number of effects such as mis-orientation and precipitate density. If the vertical boundaries are more viscous than the horizontal boundaries, sliding can be reduced on the vertical boundaries by allowing the grains to rotate. In figure 15(c) sliding has been reduced to zero on the vertical boundaries while maintaining the same bulk or grain centre strain. In this way a given strain can be achieved with a smaller expenditure of energy.

The optimum rotation can be calculated by minimizing the rate of doing work during deformation. This is now done for regular hexagons. The same problem has been treated in three dimensions for a cubic array of grains (Beeré 1977). The results are very similar with the exception that when the boundaries are all equally resistant to sliding the cubic grains must still rotate by a small amount (see appendix).

The hexagonal array is illustrated in figure 16. The deformation of a grain is indistinguishable from its neighbours and again the three types of boundary are labelled a, b and c. (This analysis differs from the one previously where the grain deformation was allowed to vary between grains (Beeré 1976).)

The sliding displacement strain rate $\dot{x}_a$ on the 'a' boundaries is given by

$$\dot{x}_a = \dot{S}_a - \dot{\omega}d, \tag{12}$$

where $\dot{S}_a$ is the shear displacement rate at the grain centre. The sum of the grain centre shears is zero (equation (6)), hence the sum of the grain boundary shears is

$$\dot{x}_a + \dot{x}_b + \dot{x}_c = -3\dot{\omega}d. \tag{13}$$

From equations (12) and (13) the grain centre shear is

$$3\dot{S}_a = 2\dot{x}_a - \dot{x}_b - \dot{x}_c. \tag{14}$$

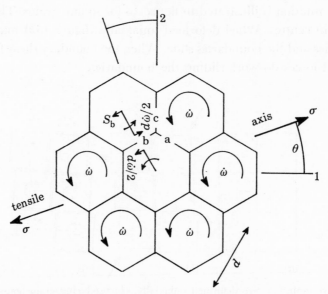

FIGURE 16. The system of grain rotations for hexagons.

Previously the bulk strain ϵ was found in terms of the grain centre shears (equation (9)). Substituting the grain boundary shears,

$$2\sqrt{3}d\dot{\epsilon} = (\dot{x}_a - \dot{x}_b)\,(4\cos^2\theta - 2) + (\dot{x}_a + \dot{x}_b - 2\dot{x}_c)\,(4\sin\theta\cos\theta/\sqrt{3}). \tag{15}$$

The creep rate may now be calculated if it is known how the shear rate varies with stress. When the normal and shear displacements on the boundary are interdependent processes the boundary shear rate will depend on both normal and shear stress. When normal and shear displacements are independent processes the normal stresses can be relaxed independently of the shear forces. In the following calculation it is assumed that the normal stresses assume their fully relaxed values (i.e. $\sigma_a = \sigma_b = \sigma_c$). This case is developed because the calculation is simplified, but the results are not likely to be greatly different for the case of a general system of normal stresses.

The shear rate will be given by an equation of the type

$$\dot{x}_a = \lambda_a |\tau_a^{n-1}| \tau_a, \tag{16}$$

where the modulus is to ensure the correct shear direction when n is even and λ depends on the material and creep mechanism

When the boundary shear rate varies linearly with stress ($n = 1$) and all boundaries are equally resistant to shear, the creep rate is given by

$$\dot{\epsilon} = \lambda\sigma/d, \tag{17}$$

independent of the orientation of the grains to the applied uniaxial stress, σ. When $n = 2$ the creep rate is given by

$$\dot{\epsilon} = \tfrac{1}{2}\sqrt{3}\,\lambda\sigma^2/d. \tag{18}$$

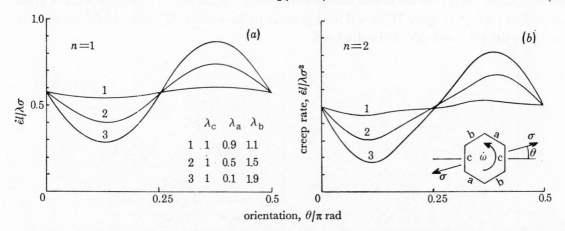

FIGURE 17. The variation of creep rate with orientation of the applied stress for grain boundary sliding control of creep when the sliding mechanism depends linearly on stress (a) and on stress squared (b). The numbers 1, 2 and 3 refer to variation in the anisotropy of sliding resistance.

If boundary viscosity varies with boundary orientation the creep rates have to be calculated numerically. This is done for stress exponents of $n = 1$ and $n = 2$ (figures 17a and b respectively). The average value of sliding resistance is kept constant, but a variation of up to a factor of ten is allowed between boundaries. The creep rate is dependent on orientation although the average creep rate for a random distribution of orientations is unchanged.

The optimum rate of grain rotation is found by minimizing the work done in achieving a fixed strain. This is readily found provided that the normal forces are relaxed, that is if the work done by the externally applied stress is primarily spent in grain boundary sliding. The rate of working on unit area of boundary is then the product of shear velocity and shear stress

$$W \propto \dot{x}_a\tau_a + \dot{x}_b\tau_b + \dot{x}_c\tau_c;$$

written in terms of the sliding shears, this becomes

$$W \propto (1/\lambda_\alpha)^{1/n}\,|\dot{x}_\alpha|^{(1+n)/n} \quad (\alpha = a, b, c),$$

and in terms of grain centre motion and rotations

$$W \propto (1/\lambda_\alpha)^{1/n}\,|\dot{S}_\alpha - \dot{\omega}d|^{(1+n)/n} \quad (\alpha = a, b, c).$$

The optimum rate of rotation is found by minimizing the rate of work done, W, at constant specimen strain rates, $\dot{S}_a$, $\dot{S}_b$ and $\dot{S}_c$.

Putting $\partial W/\partial\dot{\omega} = 0$ and solving numerically gives the results illustrated in figure 18 (a, b) for $n = 1, 2$, respectively. The resistance to sliding, λ, is varied by up to a factor of 10 to 1 with orientation. As expected, increasing the anisotropy of the grains increases the rotation.

The calculations can be compared with observed rotations in the scanning electron microscope (Geckinli & Barrett 1976). Differentiating the variation of angle with strain (figure 3) gives the strain rate compensated rate of rotation, $\dot{\omega}/\dot{\epsilon}$, which is shown in figure 19. The maximum observed rate of rotation is about $0.6\dot{\epsilon}$. This rate would be achieved in the present model when the boundaries vary by an order of magnitude in viscosity.

Although quite high rates of rotation are momentarily achieved the maximum angular rotation is seldom more than 30° (figure 3). The grains appear to oscillate about a mean orientation. This behaviour is predicted by the present model. If the angular rotation, $\dot{\omega}$, is positive (figure 18b), then the rate of change of the angle θ is negative. This implies that a grain situated on curve 3 in figure 18 (b) will tend to move to the position of stable equilibrium where $\dot{\omega} = 0$ on the left hand side of the diagram.

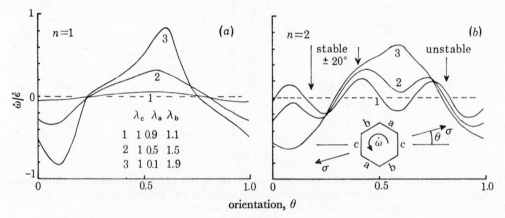

FIGURE 18. The rates of grain rotation for hexagons with anisotropic sliding resistance.
The stable orientation is shown in (b).

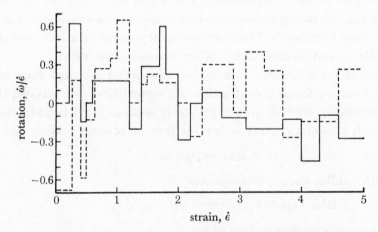

FIGURE 19. The observed rate of grain rotation calculated from figure 3.

In this example the 'a' boundary is the most viscous by an order of magnitude. The position of stability, $\dot{\omega} = 0$, is achieved when the 'a' boundary is perpendicular to the applied stress. Little sliding then takes place on this boundary and intuitively we have the correct result.

The stable configuration is disturbed by grain rearrangement during creep. Grain boundaries are often mobile and migrate during deformation resulting in simultaneous grain growth. Relative movement of the grain centres during deformation also requires grain boundary migration to maintain equilibrium angles at grain edges. This is illustrated in figure 20 in which hexagons undergo a shear strain of 0.25. The boundaries have been constructed such that they always meet at angles of $\frac{2}{3}\pi$ at the triple point and pass midway between grain centres. The latter maintains a constant number of indistinguishable grains. In practice, limits on boundary mobility restrict the approach to ideal configurations. The important feature, though, is the rotation of the boundary.

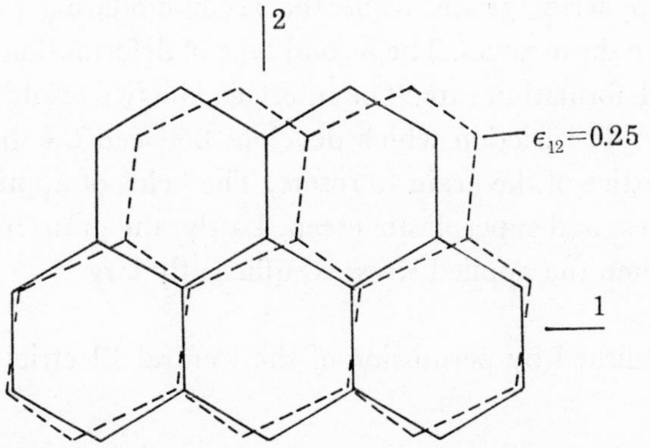

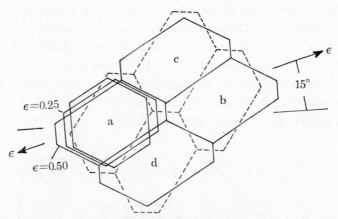

FIGURE 20. Regular hexagons deformed in shear. The position of the boundaries has been calculated assuming a constant number of identical grains maintaining equal angles at the triple points.

FIGURE 21. As figure 20, but deformed uniaxially. The broken lines show neighbour exchange maintaining a more equiaxed grain shape.

In contrast, figure 21 shows the same hexagonal configuration subjected to uniaxial strains of 0.25 and 0.50. Here the ideal configuration shows no boundary rotation. At 50 % strain the grains can reduce their boundary surface area (length) by neighbour exchange (Rachinger 1952; Ashby & Verrall 1973). This now results in boundary rotation.

In three dimensions boundary migration is complicated by the appearance of new grains at a surface. Also boundary sliding viscosity will vary with change in boundary angle even when adjacent lattices are kept at the same misorientation. Allowing for this grain rotation will still orientate normally to the stress axis those boundaries most resistant to sliding. Grain rearrangement will disturb these configurations, but equilibrium will be restored by rapid grain rotation.

7. Summary

Non-uniform deformation resulting from grain boundary sliding can be divided into three categories, namely (a) rigid grain interior with zero shear forces on boundary; (b) rigid grain interior with non-zero shear forces on boundary; (c) plastic deformation within grain. The first type, situation (a), is the classical picture of diffusion creep. Deformation is limited by the rate of vacancy diffusion across grains while the accommodating grain boundary sliding independently relaxes the shear forces. The second type of deformation (b) occurs when an interface reaction limits the deformation rate. The interface kinetics result from the nature of defects in the grain boundary. The friction which develops between the grains during grain boundary sliding causes the lattice of the grain to rotate. The fields of application are in diffusion creep near a threshold stress and superplastic creep. Lastly, the grain interiors deform plastically by dislocation creep when the applied stress is sufficiently large.

This paper is published by permission of the Central Electricity Generating Board.

Appendix

Figure A 1 shows one cube of an array located in a tensile specimen at some random orientation to the tensile axis. The interior of the grain is considered to be rigid with all deformation taking place at or near the grain boundary. Shear displacement between adjacent grain centres takes place by shear on the boundary. If the rate of grain boundary sliding is independent of the normal stress across the boundary (sliding an independent process) the shear rate is related to the boundary shear stress by a relation of the type $\dot{\epsilon} = \lambda \sigma^n / d$, where λ depends on the boundary viscosity, d is the cube edge length and n is the stress exponent.

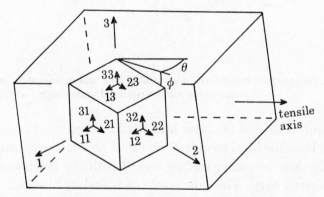

FIGURE A1. A cube randomly orientated in a tensile specimen.

If σ_{31} and σ_{21} are the shear stresses on face 1 (figure A 2b), then the shear in the 2 direction is given by

$$\dot{\epsilon}_{21}^s = (\lambda/d) \, (\sigma_{31}^2 + \sigma_{21}^2)^{(n-1)/2} \, \sigma_{21}, \tag{A 1}$$

where the superscript s denotes sliding and $(d\dot{\epsilon}_{21}^s)$ is the actual displacement on the face. In the absence of grain rotation $\dot{\epsilon}_{21}^s$ is equal to $\dot{\epsilon}_{21}$, the shear strain between grain centres. An equiaxed polycrystal does not undergo rotation when stretched uniaxially and so $\dot{\epsilon}_{21} = \dot{\epsilon}_{12}$. From equation (A 1) and the similar expression for $\dot{\epsilon}^s_{12}$ it follows that

$$(\sigma^2_{31} + \sigma^2_{21})^{(n-1)/2} \, \sigma_{21} = (\sigma^2_{12} + \sigma^2_{32})^{(n-1)/2} \, \sigma_{12}, \tag{A 2}$$

but $\sigma_{12} = \sigma_{21}$ and so from equation (A 2)

$$\sigma_{31} = \sigma_{32} \quad (n \neq 1). \tag{A 3}$$

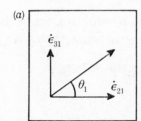 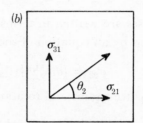

FIGURE A 2. The shear displacement and shear stress on face 1 of the cube.

Equating the other shear strains in the same manner it is found that all the shear stresses must be equal. This absurdity is removed when the grains are allowed to rotate at rates $\dot{\omega}_1$, $\dot{\omega}_2$ and $\dot{\omega}_3$ about the 1, 2 and 3 axis respectively (figure A 3). The rate of sliding on a face then depends on the grain centre shear and the rate of rotation. For instance the sliding displacement rate in the 1 direction on face 2 is given by

$$\dot{\epsilon}^s_{12} = \dot{\epsilon}_{12} - \dot{\omega}_3. \tag{A 4}$$

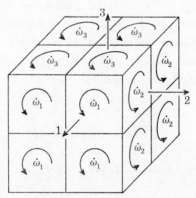

FIGURE A 3. The system of cube rotations.

When the array deforms by a given strain the cubes rotate by an amount which minimizes the energy dissipation required to achieve that strain.

If W_1 is the rate of working on face 1 then

$$W_1 \propto [(\dot{\epsilon}^s_{31})^2 + (\dot{\epsilon}^s_{21})^2]^{\frac{1}{2}} [\sigma^2_{31} + \sigma^2_{21}]^{\frac{1}{2}}, \tag{A 5}$$

or in terms of strains and rotations

$$\left.\begin{aligned}
W_1 &\propto [(\dot{\epsilon}_{31} + \dot{\omega}_2)^2 + (\dot{\epsilon}_{21} - \dot{\omega}_3)^2]^{(1+n)/2n}; \\
W_2 &\propto [(\dot{\epsilon}_{12} + \dot{\omega}_3)^2 + (\dot{\epsilon}_{32} - \dot{\omega}_1)^2]^{(1+n)/2n}; \\
W_3 &\propto [(\dot{\epsilon}_{13} - \dot{\omega}_2)^2 + (\dot{\epsilon}_{23} + \dot{\omega}_1)^2]^{(1+n)/2n},
\end{aligned}\right\} \tag{A 6}$$

where W_2 and W_3 are the rates of working or faces 2 and 3 respectively. The total rate of energy dissipation is given by $W = W_1 + W_2 + W_3$ and the most favourable rotations are found by partially differentiating with respect to the rotations at constant cube centre strain rate, i.e. $\partial W/\partial \dot{\omega}_1 = \partial W/\partial \dot{\omega}_2 = \partial W/\partial \dot{\omega}_3 = 0$. The derivative with respect to $\dot{\omega}_1$ is

$$\partial W/\partial \dot{\omega}_1 \propto [(\dot{e}_{13} - \dot{\omega}_2)^2 + (\dot{e}_{23} + \dot{\omega}_1)^2]^{(1-n)/2n} (\dot{e}_{23} + \dot{\omega}_1)$$
$$- [(\dot{e}_{12} + \dot{\omega}_3)^2 + (\dot{e}_{32} - \dot{\omega}_1)^2]^{(1-n)/2n}(\dot{e}_{32} - \dot{\omega}_1). \tag{A 7}$$

Next the shear stresses are written in terms of the strain rates. Since the shear stresses are related to sliding strains by an equation of the type $\sigma = (\dot{e}^s d/\lambda)^{1/n}$, σ_{23} is given by

$$\sigma_{23} = (d/\lambda)^{1/n} [(\dot{e}_{23}^s)^2 + (\dot{e}_{13}^s)^2]^{(1-n)/2n} \dot{e}_{23}^s, \tag{A 8}$$

or, in terms of grain centre strains and rotations,

$$\sigma_{23} = (d/\lambda)^{1/n} [(\dot{e}_{13} - \dot{\omega}_2)^2 + (\dot{e}_{23} + \dot{\omega}_1)^2]^{(1-n)/2n} (e_{23} + \dot{\omega}_1), \tag{A 9}$$

and similarly for σ_{32},

$$\sigma_{32} = (d/\lambda)^{1/n} [(\dot{e}_{12} + \dot{\omega}_3)^2 + (\dot{e}_{32} - \dot{\omega}_1)^2]^{(1-n)/2n} (\dot{e}_{32} - \dot{\omega}_1). \tag{A 10}$$

Reference to equations (A 7), (A 9) and (A 10) shows that putting $\sigma_{23} = \sigma_{32}$ implies that $\partial W/\partial \dot{\omega}_1 = 0$. Identically when $\sigma_{12} = \sigma_{21}$, $\partial W/\partial \dot{\omega}_3 = 0$ and when $\sigma_{13} = \sigma_{31}$, $\partial W/\partial \dot{\omega}_2 = 0$. Thus, when the cubes deform they follow the path of least energy expenditure which simultaneously satisfies the balance of shear stresses on the cube faces.

The cubes were considered to be equally resistant to sliding on all faces. If, however, sliding is easier say on face 1 than face 2 the material parameter λ takes on different values λ_1, λ_2 and λ_3 for faces 1, 2 and 3 respectively. If the above arguments are repeated with the new values of sliding resistance the same conclusions are reached.

The rotations and strain rates calculated for the cube model are in substantial agreement with the rates calculated for the hexagon model.

REFERENCES (Beeré)

Ashby, M. F. 1969 Scr. met. 3, 837.
Ashby, M. F. & Verrall, R. A. 1973 Acta metall. 21, 149.
Beeré, W. 1976 Met. Sci. 10, 133.
Beeré, W. 1977 (To be published.)
Bell, R. L. & Langdon, T. G. 1969 Interfaces Conference (ed. R. C. Gifkins), p. 115. Sydney: Butterworths.
Burton, B. 1973 Met. Sci. J. 11, 337.
Burton, B. 1977 Diffusion creep of polycrystalline materials. Switzerland: Trans. Tech. Pub.
Burton, B. & Greenwood, G. W. 1970 Acta metall. 18, 1237.
Butrymowicz, D. B., Manning, J. R. & Read, M. E. 1974 J. phys. Chem., Ref. Data 2, 643.
Coble, R. L. 1963 J. appl. Phys. 34, 1679.
Crossman, F. W. & Ashby, M. F. 1975 Acta metall. 23, 425.
Das, E. S. P. & Marcinkowski, M. J. 1972 Acta metall. 20, 199.
Davies, C. K. L. & Sinha Ray, S. K. 1972 Spec. Ceram. 5, 193.
Edington, J. W., Melton, K. N. & Cutler, C. P. 1976 Prog. mater. Sci. 21, 61.
Gates, R. S. 1973 Acta metall. 21, 855.
Gates, R. S. 1975 Phil. Mag. 31, 367.
Geckinli, A. E. & Barrett, C. R. 1976 J. mater. Sci. 11, 510.
Gibbs, G. B. 1965 Mém. Sci. Rev. Mét. 62, 781.
Harris, J. E., Jones, R. B., Greenwood, G. W. & Ward, M. J. 1969 J. Aust. Inst. Met. 14, 154.
Herring, C. 1950 J. appl. Phys. 20, 437.

Lifshitz, L. M. 1963 *Sov. Phys. J.E.T.P.* **17**, 909.

McLean, D. 1971 *Phil. Mag.* **23**, 182.

Pond, R. C., Smith, D. A. & Southerden, P. W. J. 1976 *4th Int. Conf. Strength Metals and Alloys.* Nancy, p. 378.

Poteat, L. E. & Yust, C. S. 1968 *Ceramic microstructures* (eds R. M. Zulraith & J. A. Pask). New York: Wiley.

Rachinger, W. A. 1952–3 *J. Inst. Met.* **81**, 33.

Raj, R. & Ashby, M. F. 1971 *Metall. Trans.* **2**, 1113.

Reynolds, G. L. 1977 Private communication.

Reynolds, G. L., Burton, B. & Speight, M. V. 1975 *Acta metall.* **23**, 573.

Speight, M. V. 1976 *Acta metall.* **24**, 725.

Stephens, R. N. 1971 *Phil. Mag.* **23**, 265.

Discussion

E. H. RUTTER (*Geology Department, Imperial College, London, SW7*). Dr Beeré has shown that oscillatory grain rotations occur during diffusion creep of polycrystalline aggregates loaded along an irrotational finite strain path. In natural rock deformation a situation of particular interest is that of simple shear within a narrow zone (shear zone). In mylonite belts associated with major overthrusts, rocks may have suffered shear displacements in excess of *ca.* 1 km over shear zone widths often less than 100 m. A common microstructure involves early, severely flattened grains either partly or totally replaced by an aggregate of small (*ca.* 30 μm) recrystallized grains (White 1976), recrystallization being concentrated at old grain boundaries. One is tempted to wonder if this microstructure is stable over large ranges in strain because the new, small grains 'roll' over one another, the potential for dilation being counteracted by diffusion. A simple shear deformation involves a vorticity, so one might expect a bias in the sense of grain rotations which accompany grain boundary sliding. Does Dr Beeré know of observational data on any materials which indicate that such a rolling process occurs in high temperature creep during a vortical strain history?

Reference

White, S. H. 1976 The effects of strain on the microstructures, fabrics and deformation mechanisms in quartzites. *Phil. Trans. R. Soc. Lond.* A **283**, 69–86.

W. BEERÉ. The treatment of grain boundary sliding was confined to uniaxial tension but the principles can easily be applied to a shear deformation. Considering a two dimensional square array of grains by viewing the cubic grains, figures A 1, A 2 and A 3, along the 1 axis, the shear strain rates between cube centres $\dot{\epsilon}_{23}$ and $\dot{\epsilon}_{32}$ are given by:

$$\dot{\epsilon}_{23} = \dot{\epsilon}_{23}^s - \dot{\omega}_1;$$

$$\dot{\epsilon}_{32} = \dot{\epsilon}_{32}^s + \dot{\omega}_1,$$

where $\dot{\epsilon}_{23}^s$ is the surface sliding velocity on a unit square and $\dot{\omega}_1$ is the angular velocity. Subtraction gives the latter:

$$\dot{\omega}_1 = \tfrac{1}{2}(\dot{\epsilon}_{32} - \dot{\epsilon}_{23}) + \tfrac{1}{2}(\dot{\epsilon}_{23}^s - \dot{\epsilon}_{32}^s).$$

The sliding rates $\dot{\epsilon}_{23}^s$ and $\dot{\epsilon}_{32}^s$ depend on the boundary viscosity. If for instance $\dot{\epsilon}_{23}^s = \lambda_3 \sigma_{23}^n$ where λ_3 depends on the boundary properties and n is the stress exponent, then since $\sigma_{23} = \sigma_{32}$, the angular velocity is given by

$$\dot{\omega}_1 = \tfrac{1}{2}(\dot{\epsilon}_{32} - \dot{\epsilon}_{23}) + \tfrac{1}{2}\sigma_{23}^n(\lambda_3 - \lambda_2).$$

Clearly if the boundaries are all equally resistant to sliding, $(\lambda_3 - \lambda_2)$ is zero and the angular velocity $\dot{\omega}_1$ is identical to the bulk macroscopic rotation.

Rotation of a grain relative to the macroscopic axes requires differences in sliding resistance on different faces of a grain. Two phase superplastic materials consisting of α and β grains contain α–α, α–β and β–β boundaries. An α grain will have α–α and α–β boundaries which can account for the differences in sliding resistance. Superplastic alloys are usually deformed in tension. The mechanism proposed for grain rolling operates equally in both tension and torsion.

Phil. Trans. R. Soc. Lond. A. **288**, 197–212 (1978)   [197]

Printed in Great Britain

Geological faults: fracture, creep and strain

By G. C. P. King†

Department of Geophysics, Faculty of Science, Ferdowsi University, Mashad, Iran

To extend our understanding of faulting in the Earth's crust it will be necessary to describe the various physical processes of faulting in terms of boundary and initial value problems.

This is not easy to do. Field evidence indicates that faults form geometrically complex systems and time histories depend on the highly nonlinear processes of fracture and friction.

The phenomenon of faulting is reviewed starting with a description of the work of E. M. Anderson who demonstrated that a partial knowledge of the boundary conditions under which faulting could occur allows fault types to be classified.

However, many commonly observed features of fault behaviour are unexplained by Anderson's ideas. These features are described and the various attempts to explain them or reproduce them by modelling are discussed.

Seismic studies are briefly covered and it is noted that seismically determined stress drops can also be interpreted to show that earthquake faults have displacement to length ratios close to 10^{-4}. A similar value has also been found from field observation of intersections of earthquake faults with the ground surface. It is also pointed out that faults observed in the field are always significantly more complex than the simple geometrical models of earthquake sources used in seismology and that this deserves greater study.

1. Introduction

To understand the mechanical behaviour of the Earth's crust we must understand the role of faulting in the processes of crustal deformation. Until the advent of plate tectonics in the 1960s there was a tendency for faulting to be regarded as a secondary phenomenon with ductile deformation being the principal deformation mechanism (Anderson 1951, p. 1). The discovery of major fault systems in the oceans with displacements of more than 1000 km (Mason & Raff 1961; Raff & Mason 1961) showed that at least some faults play a major rôle in the deformation of the crust. A further source of interest in faulting has followed the full realization that earthquakes are due to fault motion and that the only real hope for effective earthquake prediction lies in gaining a proper understanding of fault mechanics. This realization has gained ground steadily since Reid (1910) published his observations of ground deformations associated with the 1906 San Francisco earthquake.

The theory of faulting, such as it is, does not come from global tectonics but from the study of industrial materials. Even seismology, as yet, does no more than provide field measurement of parameters that owe their origin to laboratory sample experiments and it is questionable whether such use is entirely justifiable. Seismology allows active faults to be mapped to depth and is an important adjunct to geological field observation of fault systems. However, the greatest body of literature on faulting consists of geological description and it seems appropriate to review these observations first. This task is hampered by the fact that only a small part of a fault is actually seen in the field and this can lead geologists to interpret their data with the help

† Present address: Department of Geodesy and Geophysics, Cambridge University, Madingley Rise, Madingley Road, Cambridge CB3 0EZ.

of inadequate or incorrect theories. The importance of such prejudices is emphasized by an account given by King–Hubbert (1972, p. 9) of a geologist losing his job for mapping faults (correctly) in a place where theory (incorrectly) was thought to demonstrate they could not occur. I shall, therefore, start with an account of the Anderson fault classification, since, as far as it goes, it is the only correct theory available and it is fairly easy to see the nature of its limitations. It is widely used in geological interpretation.

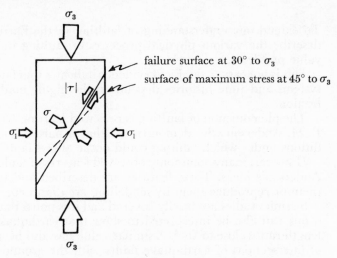

FIGURE 1. The Coulomb failure criterion: $|\tau| = \tau_0 + \mu\sigma$, where $|\tau|$ is critical shear stress, τ_0 internal cohesion, μ internal friction, and σ normal stress.

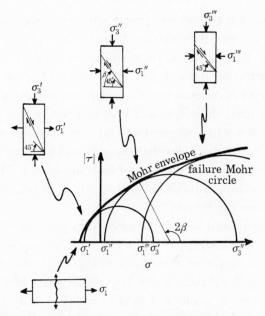

FIGURE 2. The Mohr construction. The cartoons of rock samples show the way in which the angle of failure varies for different parts of the Mohr enveloping curve (envelope).

2. THE ANDERSON CRITERIA

The Anderson criteria are produced by combining empirical results for the failure of rock (derived from the study of laboratory samples) with constraints on stress systems inside the Earth that arise from the existence of the Earth's stress free surface.

Figure 1 shows the Coulomb failure criterion (see, for example, Jaeger & Cook 1969) for a sample under triaxial test conditions. The largest stress is σ_3 and the smallest σ_1. Provided that σ_2 remains such that $\sigma_3 > \sigma_2 > \sigma_1$, then σ_2, the intermediate stress, appears to play no significant part in the failure process. Failure does not occur on the plane of maximum shear but at an angle to it. Coulomb explained this by the introduction of a coefficient of internal friction.

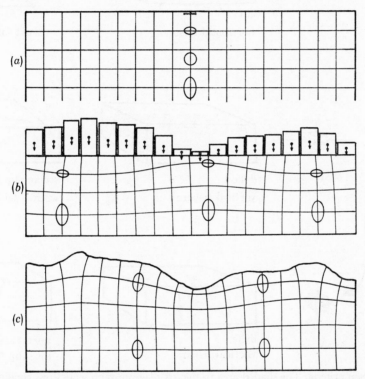

FIGURE 3. Trajectories of normal stress near the Earth's surface. The conditions near a flat surface are shown in (a). These are modified by the affects of the weight of surface topography (b), and internal density variations (not shown). External stresses are modified by topography (c) and internal rigidity variations (not shown).

The relation between this coefficient in unfractured material and the coefficient between two discrete surfaces is purely notional and it is important to realize that this 'friction' operates even when there are tensile forces acting across the future failure plane. The internal friction does not obey a linear law very well, particularly at low confining pressure, an effect that is conveniently demonstrated with the aid of Mohr's construction (figure 2) (Jaeger & Cook 1969). Mohr's construction is so widely employed to describe failure that the term Mohr–Coulomb failure is often used. Figure 2 shows the angle of failure for idealized rock samples for different stress conditions. For crustal rocks a rough average of 30° to the direction of maximum principle stress was assumed by Anderson.

The only boundary close to a fault about which certain information is known is the Earth's stress free surface. At a stress free surface there are no normal or tangential forces. Thus the surface stress tensor reduces to a flat ellipse in the plane of the surface. One principal axis is vertical and the other two are horizontal (figure 3*a*). Since gravity acts vertically this configuration of principal axes does not change with depth although the magnitudes of the principal stresses change. The Earth's surface is not, in general, flat and this causes the principal stresses to be deflected from the vertical and horizontal. There are two separate topographic effects shown in figures 3(*b*) and (*c*). The first is a deviation of directions due to the spatial

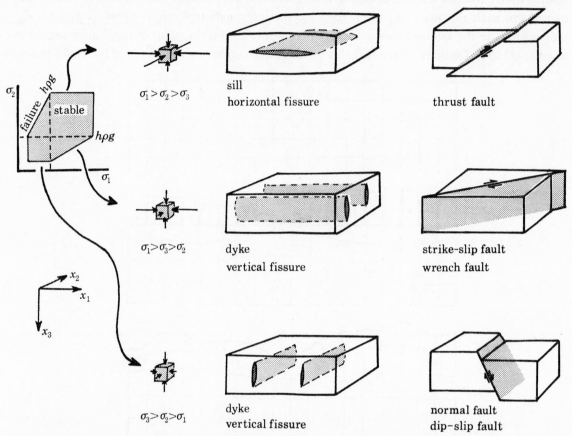

FIGURE 4. The Anderson criteria. The shaded area in the left diagram represents stress conditions under which failure will not occur. The broken lines marked $h\rho g$ indicate the pressure of rock overburden and the diagram is therefore only correct for one depth. The central set of cartoons indicate the tensional structures that can develop if a fluid is present with a pressure intermediate between the greatest and least principal rock stress. At depth, the only fluid with sufficient pressure to produce large fissures is molten rock. Micro-fissures filled with water extend to depths of several kilometres. The cartoons on the right show the faulting that occurs under the same stress conditions in the absence of fluid of sufficient pressure. Conjugate faults are not shown; neither are the fractures that might occur (but are not observed in the field) when two principal stresses have identical values.

variation of the weight of the topography and the second is the effect of the irregular surface on stress systems that act on the region from a distance. This latter effect has been discussed in the tidal literature (for example, Harrison 1976). It is easy to see that the two effects are calculable for any given conditions and to concur with Anderson that, except under conditions of 'Alpine' topography, one principal stress will be nearly vertical and the other two nearly horizontal.

In figure 4 the consequences of combining the stress free surface conditions and the failure criteria are summarized. Three stress conditions are identified, vertical stress being the least, the intermediate, or greatest of the three principal stresses in turn. These give the three classes of faulting shown in the block diagrams on the right. The central diagrams show the fissures that can form under conditions where a fluid is available with a pressure intermediate between the greatest and least principal stresses. Thus horizontal fissures or sills form under the same stress conditions as thrust faults and vertical fissures or dykes under the same conditions as strike-slip or dip-slip faults. The graph on the left shows, for one depth, the stress conditions under which different types of faulting or fissuring occur. It is interesting to notice that the stresses involved in thrust formation are very much greater than those associated with normal faulting and that strike slip faulting can be initiated over a wide range of intermediate stress conditions.

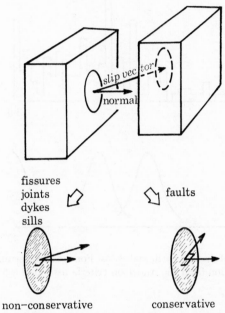

FIGURE 5. Conservative and non-conservative dislocations. Both conservative and non-conservative rock dislocations occur. Non-conservative dislocations that involve a volume increase are common but those that involve
a volume decrease are not common because there are no very effective mechanisms of a macroscopic scale for
melting, dissolving (or diffusing) material away from cracks.

The combination of the surface boundary conditions and rock failure criteria places substantial constraints on the type of rock dislocations that should occur. Both conservative and
non-conservative dislocations (Nabarro 1967) can occur (cf. figure 5) but the Anderson criteria
place severe constriants on the permissable orientations of the fault planes and directions of the
slip vector.

3. DEVIATION OF FIELD DATA FROM THE ANDERSON CRITERIA

Field observations of faults and dykes broadly fit Anderson's classifications. Dykes are generally
vertical or nearly vertical and sills are close to horizontal. Dip-slip faults generally dip at close to
60°, thrust faults at 30° and strike-slip faults have nearly vertical fault planes. The direction of
slip appears to be more variable, dip-slip faults may have to up 30 % strike-slip motion and
strike-slip faults sometimes have a similar proportion of dip-slip motion. However, field observations of faults define the direction of slip by the direction of scratching (slickensides) on exposed

small fault surfaces and a local direction may not be representative of a broader average. Figure 6 shows a histogram of the normals to the nodal planes of fault plane solutions of shallow earthquakes in the Mediterranean (McKenzie 1972) together with a predicted distribution for comparison. There is some similarity between the two distributions but the fit is far from perfect. From a fault plane solution alone it is impossible to distinguish the slip vector from the normal to the fault plane and it is therefore impossible to determine whether the deviation from the Anderson conditions is predominantly due to deviation in the orientation of the fault alone or to the angle of slip.

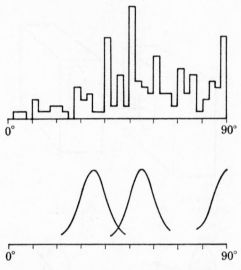

FIGURE 6. The dip of nodal planes of shallow earthquakes. (a) For the Mediterranean (from McKenzie 1972) and (b) a crude prediction from the Anderson criteria assuming a 5° normal error.

Explanations for deviations from the Anderson criteria fall into three categories:

(a) *Inhomogeneity of material properties*. Faults often appear to follow old lines of weakness which either arise from the depositional characters of the rocks or due to pre-existing faults. There is evidence that this is one significant reason for fault planes and slip directions deviating from those predicted (Bott 1958; McKenzie 1969). Very shallow angle thrust faults are common in major mountain belts and are attributed to weakening of sediments due to high water pressure (Hubbert & Rubey 1957; Raleigh & Griggs 1963).

(b) *Inhomogeneity of stress pattern*. The stress systems assumed by Anderson are homogeneous and semi-infinite. Topographic effects can cause stress concentrations or these can arise from material properties or as a result of previous fracture history. A stress inhomogeneity which is small in dimensions compared to its depth from the surface is not constrained by the surface boundary conditions and a fracture in it can occur in any direction.

(c) *Inhomogeneity of fault motion and Poisson ratio effects*. The Anderson theory assumption of homogeneous stress implies faulting of infinite extent and homogeneous (and strictly speaking infinite amplitude), fault displacement. That faults are not infinite in extent is presumably due to variations of material properties or localization of stresses. A consequence of spatially varying slip amplitudes on faults is Poisson ratio motions perpendicular to the predominant slip direction. Although there are no vertical stresses, vertical strains and hence motion can result from horizontal stresses. The effect is frequently observed on the surface breaks of strike-slip earthquake faults and is known as 'scissoring'. Although the predominant motion of the fault is strike-slip, alternate sides of the fault are up-thrown or down-thrown (Richter 1958).

4. OTHER FIELD OBSERVATIONS

Faulting appears to occur on all scales. Faults range from those with displacements of thousands of kilometres to those with negligible displacements. Perhaps the simplest argument suggesting that faulting occurs on a wide range of scales is the Gutenberg & Richter frequency–magnitude relation which is obeyed both for large and small shocks (Scholtz 1968):

$$\lg N = a + bM, \tag{1}$$

where N is the number of earthquakes per unit time, a and b are constants and M is earthquake magnitude

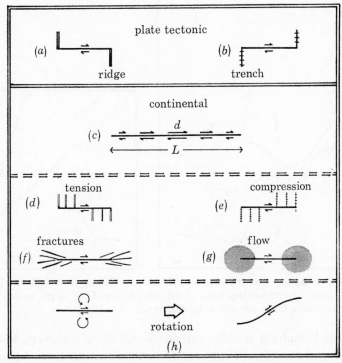

FIGURE 7. Ways of ending faults; in (c), d = displacement and L = length.

Faults vary in length between many thousands of kilometres to crystalline dimensions. Many faults observed geologically, however, have a surprisingly constant displacement to length ratio in the region of $1:10$. This may be compared with the displacement to length ratio, of about $1:10^4$, associated with earthquake motion as observed on the surface breaks of large earthquakes or inferred seismically (Kasahara 1975, in Japanese; see Ohnaka 1976). In both cases the amplitude of the motion decreases progressively towards the ends of the fault in the manner shown in figure 7 (c). This does not apply to the great oceanic transform faults, nor perhaps, to the great thrust and strike-slip faults on continents (Freund 1974; Ranalli 1977). The large-scale geometric behaviour of the transform faults of plate tectonics is well understood; they have constant displacement along their length and terminate at clearly defined features which take up the motion, such as ocean trenches or ridges (Cox 1973). This is shown in figure 7 (*a*, *b*). No such clearly defined features are observed in association with most continental faults. Significant features are not necessary to explain the ending of earthquake faults since the length to displacement ratio is sufficiently small for all the motion to be accommodated elastically. This is not the case for geological faults on continents and figure 7 (*d–g*) shows some of the possible ways that displacement may be accommodated. There is no evidence for any one of these mechanisms predominating in general.

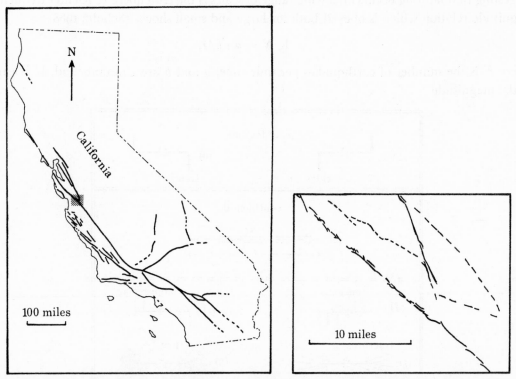

FIGURE 8. The San Andreas fault system (adapted from Moody & Hill 1956, and R. O. Burford, private communication.) The small map is an enlargement of the shaded part of the larger map. The maps are certainly incomplete and many more faults are still to be discovered.

The depth to which faulting usually extends is not clearly known, but is thought to be about 20 km. There are two lines of evidence. The clearest comes from the depth of epicentres. Except in subduction zones (where deep faults do occur), earthquakes are rarely deeper than 15–20 km (Brace & Byerlee 1966). However, since ductile fault processes could take over below this depth, the limit of seismicity does not necessarily delineate the maximum depth of faulting. Geodetic methods have been used to estimate the depth of earthquake faulting but lack sensitivity unless it is assumed that the fault has an abrupt lower boundary (Chinnery 1966).

Another line of evidence comes from the examination of ancient fault zones that have been exposed at the surface by erosion. It is, however, rather difficult to assess the depth at which these zones were active. Depth estimates are based on the pressure and temperature stability of minerals associated with the zone and assumptions about temperature and pressure conditions that existed in the crust at the time of fault formation (Sibson 1977). These cannot be determined accurately. The Anderson theory suggests that faults are simple structures; this is never in practice true. Figure 8 shows the San Andreas fault system on two scales and it can be seen that at both scales it appears as a complex system of fractures. Figure 9 illustrates the general features discernable near a major branch of a *predominantly* strike-slip fault system. Disturbance of the rock is encountered some distance before the fault zone is entered. The rock disturbance may take the form of crushing or shattering of the rock or more gentle warping known as 'drag folding'. The term 'drag folding' arises because it was originally believed that such structures resulted from drag on the fault plane. This cannot be true and these structures must result either from pre-existing weakness near the fault zone or from vertical or horizontal inhomogeneity of fault motion. Inhomogeneous motion causes transient high stress concentrations to occur near the fault zone. This problem has been partly discussed by Garfunkel (1966).

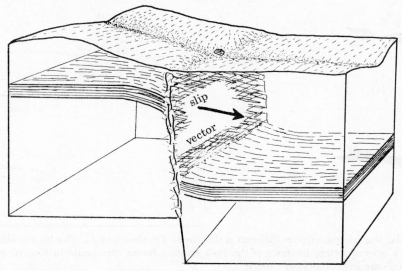

FIGURE 9. Some of the features of a major fault traversed by a river valley.

As the fault zone is approached more closely, the disturbance becomes more intense until a zone is reached where the rocks have suffered so much cataclastic damage that they are no longer obviously similar to the rocks surrounding the fault. This change results partly from mechanical crushing and partly from the chemical changes that the crushing facilitates (Sibson 1977). Within the fault zone are one or more planes across which it is clear that even more intense deformation has been concentrated. On a detailed examination of a fault these are often identified as the 'true' fault plane. However, they do not appear to extend far and are probably not continuous with similar features that can be identified on a traverse across the same fault zone a few hundred metres distant. At the surface, very little of a fault is actually exposed (perhaps a small exposure in a river valley), and faults are generally traced from the slight topographic expression that results from mechanical processes or from selective erosion of different rock types.

5. THE PROBLEM OF SCALING

A striking feature of faults is the observation that large fractures appear to be composed of smaller fractures which are, in turn, made up of smaller fractures still. This had been studied by Tchalenko (1970), who demonstrates substantial similarities between fracture patterns on different scales. An example is shown in figure 10. His observation emphasizes what could be

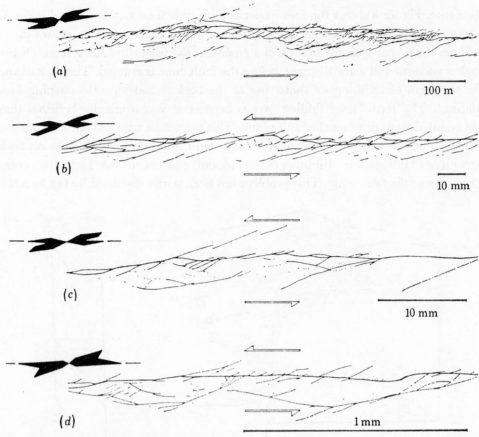

(a)

100 m

(b)

10 mm

(c)

10 mm

(d)

1 mm

FIGURE 10. Similar fracture patterns on different scales (from Tchalenko 1970). The large-scale diagram (a) is taken from a map of surface fractures of the 1968 Dasht-e Bayaz earthquake in Eastern Iran. The other diagrams (b–d) are from laboratory models.

finite brittle strain

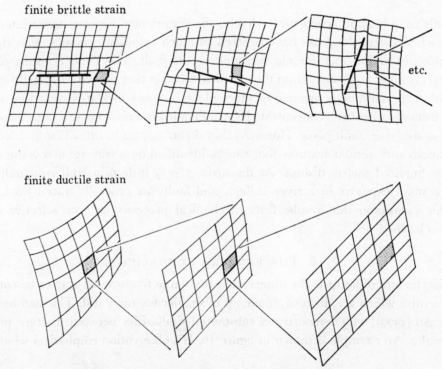

etc.

finite ductile strain

FIGURE 11. Finite brittle and finite ductile strain.

described as a 'Russian doll' effect in cataclastic deformation, a behaviour which suggests that the modelling of fault zones cannot be simple. The nature of this difficulty is illustrated by figure 11 which compares the difference between finite strain in a ductile medium (Ramsey 1967) and finite strain in a brittle medium. In the former, a region that is subject to inhomogeneous strain boundary conditions can be divided into smaller regions in which the strain is effectively homogeneous, while in the latter, even large scale homogeneous boundary conditions will result in inhomogeneous strain fields on smaller scales. This emphasizes the fact that Anderson's assumption of stress boundary conditions is not correct except for the instant of initiation of a fracture. It is also clear that strain boundary conditions are not an alternative to stress, and that faulting must take place in an environment intermediate between the two. This has been discussed by anaology to rock press stiffness by Walsh (1971) and Ohnaka (1973). It is also clear that the boundary conditions cannot normally be expected to be homogeneous. This interaction of scales makes fault modelling, either mathematically, or with model materials, rather difficult. I shall briefly outline some of the approaches that have been taken.

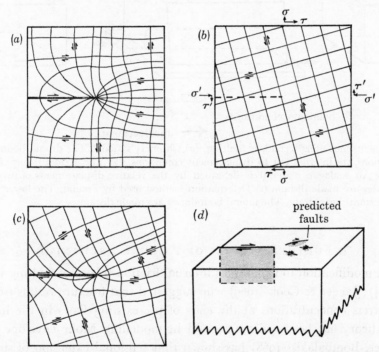

FIGURE 12. Secondary faulting (after Chinnery). Shear stress trajectories of an external field (b), summed with the stresses due to a dislocation surface (a), give the stress conditions (c). The model is shown in (d) on approximately the same scale. Some of the predicted secondary faulting is also shown.

6. FURTHER THEORIES OF FAULTING

Chinnery (1966) adapted an idea initiated by Anderson to explain the origin of complex fault structures. The argument depends on the assumption that a fault has for 'some reason' ended leaving a high stress concentration and for 'some reason' the next fracture does not simply extend the main fault. The conditions are shown in figure 12. Chinnery superposes an external field on the stress field produced by an internal cut representing the fault that has already moved. Assuming that the original plane does not extend, he determines, using Coulomb–

Mohr criteria, where new fractures might initiate. Many of these are not in direct line with the original cut but form at angles to it. His principle justification for his model, in particular for his use of a simple cut with uniform displacement across its faces and for the non-propagation of the original fault, is that his predictions accord with field data. This may be true in some field cases. However, it is questionable whether his theory explains field data better than that of Moody & Hill (1956) whose ideas he correctly shows to be wrong.

Figure 13 (b, c) shows some experiments carried out with analogue materials designed to explain the same features that Chinnery seeks to model. Arguably they all fit the field observations and there appears to be, as yet, no reason to favour any one model on this basis. The most interesting feature of all the models, surprising in the case of figure 13 (b), is that fracture initiates at the surface. (In the case of (b), surface fractures appear before the fracture, which starts at the base, reaches the surface.)

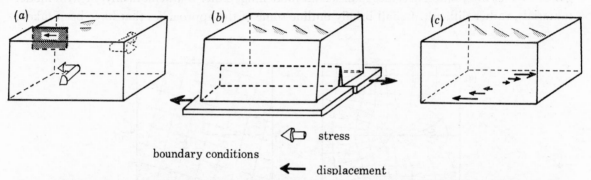

FIGURE 13. Experiments modelling secondary faulting. (a) Chinnery's model. The distant boundaries are subject to stress conditions, the internal cut to displacement conditions. (b) The Reidel deformation method used by Tchalenko. An analogue material is deformed by the relative displacements of two basal slabs. The other boundaries are made distant. (c) Deformation method used by Freund. The lower boundary is constrained to constant shear strain. The lateral boundaries are made distant.

7. FURTHER STUDIES OF FAILURE MECHANISMS

An important modification to Coulomb's internal friction theory of faulting was introduced by Griffith (1924); Jaeger & Cook (1969) who suggested that fracture results from extensional failure due to stress concentrations at the ends of pre-existing flaws in the material (figure 14 (a, c)). The theory has been quite successful in modelling Mohr envelope shapes at low stresses. However, Bombalakis (1968) has shown that tensional extension of single fissures or simple arrays of fissures (figure 14 b) do not lead to failure but to another stable condition. This renders these theories, while empirically satisfactory, physically unsatisfactory. McClintock & Walsh (1962) have modified Griffith's theory by considering the effect of friction across crack surfaces. At high stresses this produces results identical to those of Coulomb but they are no more satisfactory physically.

Since it was appreciated that the mechanical stiffness of the deformation apparatus used to experiment with rock samples has a significant affect on the results, many tests have been carried out over a range of stiffnesses. The two end conditions are constant stress and constant displacement. Under stress conditions any negative gradients in the slope of the stress–strain curve leads to large amounts of energy being 'dumped' into the sample with a consequent catastrophic failure. The same sample, however, subject to displacement conditions can fail progressively, retaining considerable strength after the onset of failure. This led to the discovery that highly fractured rock could support shear stresses approaching that of virgin material (Hobbs 1966; Byerlee 1967; Brace 1968).

Recent interest has centred on the effect of machine stiffness on stick-slip sliding of precut surfaces (Byerlee & Brace 1968; Jaeger & Cook 1971) and how this relates to the conditions experienced by rocks at faults (see §5). Another approach to the study of failure was initiated by Mogi (1962) and taken up by Scholtz (1968). They attached transducers to rock samples under stress to examine acoustic (mainly ultrasonic) emission. Small events begin at stresses considerably less than failure stress and these events concentrate progressively on the plane where failure will occur. Scholtz examined frequency magnitude relations and sequences of small events following more substantial fracture and demonstrated a striking similarity between this microfracturing behaviour and the behaviour of earthquake aftershock sequences.

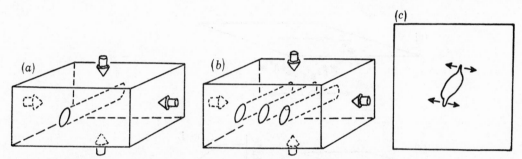

FIGURE 14. Griffith failure. Fracture initiates by extensional failure at crack tips (c). These are assumed to be elliptical initially, to facilitate the mathematics. Arrays of cracks (b) do not develop fissures that join unless they are so close that their interaction is strong and complex.

8. THE SEISMOLOGICAL VIEW OF FAULTING

At wavelengths large compared to the fault dimension, the spectral displacement amplitude Ω of a seismic wave is given by a function of the type

$$\Omega = Mo R_{\theta\phi}/4\pi\rho R v^3, \tag{2}$$

where $R_{\theta\phi}$ is the radiation pattern of P or S waves, ρ is the near source density, R is a distance function, and v is the P or S velocity. (Keilis-Borok 1959). Burridge & Knopoff (1964) have shown that exactly similar functions result if the source is considered to be a pair of counteracting couples (figure 15 b) or a dislocation surface (figure 15 a). In the former case the interpretation of Mo, the seismic moment, is clear; it is the moment of the couples, and has the dimensions of force times distance. In the case of the dislocation interpretation it is physically clearer to use a geometric moment $M = Mo/\mu$. This has the dimensions of length cubed and equation (2) can be rewritten

$$\Omega = \frac{1}{4\pi R}\left(\frac{v_s}{v}\right)^2 \frac{M}{v} R_{\theta\phi}, \tag{3}$$

where v_s is S velocity. The relation between these two moments is similar to that between moment of inertia and moment of cross section in beam theory. The geometric moment is simply $M = S\bar{d}$, where S is the area of the fault plane and $\bar{d}$ the average displacement. From equation (3) it can be seen that seismic amplitudes provide geometric information about the dislocation motion. It is necessary to know the seismic velocity in the source region but it is not necessary to know the shear modulus, or density, separately.

Some of the most useful seismic information in recent years has come from examining the radiation function $R_{\theta\phi}$. This provides the method of fault plane solutions (Honda 1962) which was of such importance in establishing plate tectonics (Cox 1973). The method depends only on establishing the sense of motion of seismic arrivals and is not very sensitive to path effects. The determination of seismic moment, on the other hand, is much more path dependent and the determination and application of appropriate corrections is critical.

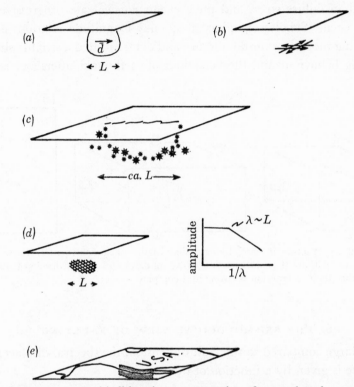

FIGURE 15. Seismic fault parameters: (a) dislocation representation of an earthquake source; (b) double couple source; (c) determination of the fault dimensions from aftershock area or surface fractures; (d) determination of fault dimensions from the spectral content of seismograms; (e) summing seismic moment along a plate boundary to determine average slip rate.

Seismic moment alone is not a very useful parameter. It is necessary to establish additionally either displacement or fault area. Figure 15 (c, d) illustrates some of the ways in which this can be done. For large near-surface earthquakes, the length of the surface break is a guide to the fault dimensions. Fault dimensions can also be estimated from the extent of the aftershock sequence although this generally extends over a larger region than the original fault. A method developed by Brune (1970) estimates the dimensions of the fault from the wavelength (and hence frequency) at which equation (2) breaks down because of the effect of finite source size. The spectral character of a seismogram changes at this frequency giving a guide to source dimensions. Either of these methods allows average slip to be determined provided that assumptions about the fault geometry are made. A common assumption is that the fault plane is circular giving $S = \frac{1}{4}\pi L^2$, where L is the fault length. A quantity frequently discussed is stress drop, defined as

$$\Delta\sigma = \mu\bar{d}/L = Mo/L^3.$$

This is no more than the ratio of the average displacement divided by the fault length multiplied by a local elastic modulus (and, if the fault geometry is considered carefully, a shape factor close to unity). Average stress drops of 30 bar are not accepted as the mean but stress drops between 10 and 100 bar are common (Kanamori & Anderson 1975). Alternatively, stress drop can be viewed in terms of the displacement: length ratio

$$\bar{d}/L = M/L^3,$$

which gives ratios of around $1:10^{-4}$ (Ohnaka 1973). The moduli assumed near different faults do not vary much so that stress drop and displacement to length ratio are, in practice, a measure of the same thing and perhaps it is easier to visualize the significance of the latter rather than the former. Although it does not seem surprising that stress drop should vary by an order of magnitude, it does seem surprising that displacement to length ratio should be as constant as that.

Seismic moment has an additional use on plate boundaries (figure 15e). Over a period of time, all of the seismicity moment along a boundary can be divided by the area of that boundary and the time period. This gives a slip rate that can be compared with the slip rates determined by other methods (see, for example, Davies & Brune 1971; North 1974).

As yet seismology has contributed relatively little to an understanding of failure processes. However, recent studies of the waveform and high frequency spectra of seismic radiation (see, for example, Kanamori & Anderson 1975; Ohanaka 1976) together with greatly increased computing power may produce important results.

9. CONCLUSION

Aspects of the behaviour of faults and theories of faulting have been reviewed. The Anderson theory of faulting has been shown to have been a great success and an important aid to field geologists, although it can be shown to have limitations both in theory and practice.

Seismologists have demonstrated that earthquakes result from dislocation motion that occurs when a fault is abruptly initiated or an existing fault plane abruptly increases its displacement. At present, a stick-slip process on a uniform plane is assumed. However, there is substantial field evidence that fault geometries are more complex. This may be of importance in developing seismic source models or it may be useful to use seismology to study the apparent complexity of geological faults.

The author would like to thank the staff of the Faculty of Science, Ferdowsi University, for their assistance during the preparation of this paper, to the Natural Environment Research Council of the U.K. and the United States Geological Survey for some support, to Raphael Freund for very valuable discussions and to Paul Davis for critically reading the manuscript.

REFERENCES (King)

Anderson, E. M. 1951 *The dynamics of faulting.* New York: Hafner. Reprinted 1972.
Bombalakis, E. G. 1968 *Tectonophysics* **6**, 461–473.
Bott, M. P. H. 1958 The mechanics of oblique slip-faulting. *Geol. Mag.* **96**, no. 2.
Brace, W. F. 1968 *Tectonophysics* **6**, 75–87.
Brace, W. F. & Byerlee, J. D. 1966 *Science, N.Y.* **153**, 990–992.
Brace, W. F. & Byerlee, J. D. 1967 *J. geophys. Res.* **72**, 3639–2648.

Brace, W. F. & Byerlee, J. D. 1968 *J. geophys. Res.* **77**, 3690–3697.

Brune, J. N. 1970 *J. geophys. Res.* **75**, 4997–5009.

Burridge, R. & Knopoff, L. 1964 *Bull. seism. Soc. Am.* **54**, 1874–1888.

Byerlee, J. D. 1967 *J. geophys. Res.* **72**, 3639–3648.

Byerlee, J. D. & Brace, W. F. 1968 *J. geophys. Res.* **77**, 3690–3697.

Chinnery, M. A. 1966 *Can. J. Earth Sci.* **3**, 163–190.

Cox, A. 1973 *Plate tectonics and geomagnetic reversals.* San Francisco: Freeman.

Davies, G. & Brune, J. N. 1971 *Nature, phys. Sci.* **229**, 101.

Freund, R. 1974 *Tectonophysics* **21**, 93–134.

Garfunkel, Z. 1966 *Tectonophysics* **3**, 457–473.

Griffith, A. A. 1924 Theory of rupture. *Proc. Inst. Int Congress appl. Mech.*, Delft, 55, 63.

Harrison, J. C. 1976 *J. geophys. Res.* **81**, 319–328.

Hobbs, D. W. 1966 A study of the behaviour of a broken rock under triaxial compression. *Int. J. Rock Mech. & Mining Sci.* **3**, 11–43.

Honda, H. 1962 Earthquake mechanism and seismic waves. *Geophys. Notes, Tokyo* **15**, Supplement.

Hubbert, M. 1972 *Structural geology.* New York: Hafner.

Hubbert, M. K. & Rubey, W. W. 1957 *Trans. Am. Inst. min. Engrs* **210**, 153–158.

Jaeger, J. C. & Cook, N. G. W. 1969 *The fundamentals of rock mechanics.* London: Methuen.

Jaeger, J. C. & Cook, N. G. W. 1971 *Friction in granular materials, Int. Conf. Struct. Solid Mech. Eng. Des. Civ. Eng. Mater.* (Southampton Univ., 1969), p. 22.

Kanamori, H. & Anderson, D. L. 1975 *Bull. seism. Soc. Am.* **65**, 1073–1095.

Keilis-Borok, V. I. 1959 *Annali Geofis.* **12**, 205–214.

Mason, R. G. & Raff, A. D. 1961 *Bull. geol. Soc. Am.* **72**, 1259.

McClintock, F. A. & Walsh, J. 1962 *U.S. National Congress Appl. Mech.*, Berkeley 1962.

McKenzie, D. P. 1969 *Bull. seism. Soc. Am.* **59**, 591–601.

McKenzie, D. P. 1972 *Geol. J. R. astr. Soc.* **30**, 109–185.

Mogi, K. 1962 *Bull. Earthq. Res. Inst. Tokyo Univ.* **40**, 125–173.

Moody, J. D. & Hill, M. J. 1956 *Bull. geol. Soc. Am.* **67**, 1207–1246.

Nabarro, F. R. N. 1967 *Theory of crystal dislocations.* Oxford: Clarendon Press.

North, R. G. 1974 *Nature, Lond.* **252**, 560.

Ohnaka, M. 1973 *J. Phys. Earth* **21**, 285–303.

Ohnaka, M. 1976 *Bull. seism. Soc. Am.* **66**, 433–451.

Raff, A. D. & Mason, R. G. 1961 *Bull. geol. Soc. Am.* **72**, 1267.

Raleigh, C. B. & Griggs, D. I. 1963 *Bull. geol. Soc. Am.* **74**, 819–830.

Ramsey, J. G. 1967 *Folding and fracturing of rocks.* New York: McGraw-Hill.

Ranalli, G. 1977 *Tectonophysics* **37**, 71.

Reid, H. F. 1910 *The mechanics of the earthquake: the Californian earthquake of April* 18 1906. Report of the State Investigation Commission, Carnegie Institute of Washington, **2**.

Richter, C. F. 1958 *Elementary seismology.* San Francisco: H. Freeman.

Scholtz, C. H. 1968 *Bull. Earthq. Res. Inst. Tokyo Univ.* **40**, 125–173.

Sibson, R. H. 1977 *J. geol. Soc. Lond.* **133**(3) (in the Press.)

Tchalenko, J. 1970 *Bull. geol. Soc. Am.* **81**, 1625–1640.

Walsh, J. B. 1971 *J. geophys. Res.* **76**, 8597–8598.

Phil. Trans. R. Soc. Lond. A. **288**, 213–227 (1978) [213]

Printed in Great Britain

Fracture during creep

By G. W. Greenwood

Department of Metallurgy, University of Sheffield, St George's Square, Sheffield S1 3JD

Although materials generally become more ductile with increase in temperature and decrease in strain rate, many metals have a local minimum in ductility at specific temperatures and strain rates. There is thus a region where a decrease in stress level results in a decrease of elongation before fracture. Failure is observed to occur at interfaces and particularly at grain boundaries nearly perpendicular to the maximum principal tensile stress. The nucleation of cavities and cracks, however, appears to arise initially from shear processes at these interfaces at temperatures and stresses where grain boundary sliding is significant but transverse boundary movement is negligible. Thus, tensile and shear stresses play different rôles. Since cavities have a finite volume and link together to cause failure, the hydrostatic component of stress is important and so, in evaluating criteria for fracture, a precise definition of the stress system is necessary. Further, chemical composition, segregation and microstructural features play a vital rôle and materials that are only slightly different may show quite dissimilar behaviour.

Introduction

Brittleness is often associated with rapid impact and high stresses and with sharp cracks in hard materials. The effects of temperature rises to levels, say, of half the absolute melting temperature, are normally regarded as softening materials significantly to permit stresses to be relieved, cracks to heal and voids to sinter. The overall situation, however, is now widely recognized to be much more complex (Garofalo 1965). At relatively low stresses, at temperatures when materials are soft and where they are capable of slow deformation, it is not uncommon to find that they fracture suddenly, particularly when tensile forces are applied.

Despite the apparent suddenness of fracture at elevated temperatures, it is now clear (Leckie 1978, this volume) that it originates from the slow deformation, or creep, progressively causing 'damage' in the material that can be identified by metallographic techniques. The most general observations have indicated that the separation that ultimately causes fracture occurs at interfaces and particularly at grain boundaries that are nearly normal to the principal tensile stress. Because these interfaces are involved, it appears that the crystallography of the material is relatively unimportant, and even metals of face centred cubic structure and close packed hexagonal structure have been observed to fail by this mode (Greenwood 1973). Nevertheless, such fracture is not observed universally and aluminium, titanium and lead have only fractured in this manner when specific alloying elements or impurities were present.

The time scale of laboratory experiments is such that fracture during creep takes place in a temperature range roughly in the range from 0.4 to 0.7 of the absolute melting temperature (Taplin 1973). Where the time scale is enlarged, for example, by a factor of about 10^5, it is expected that this temperature range would be greatly exceeded, and that creep fracture could occur at quite low fractions of the melting temperature.

It is often observed that the ductility is a minimum at a given temperature and at a certain rate of creep. This minimum, however, is usually a shallow one and the strain before fracture does not vary greatly over a rather wide range of temperature and creep rate (Perry 1974).

The strain before failure, nevertheless, is strongly dependent on the stress system. Under compressive forces there is often a very large ductility and some increase in the strain to fracture is also observed if a tensile component of stress is partly offset by hydrostatic compression. It seems necessary to distinguish between the rôles of the stresses that change shape and those that affect the volume.

The broadest interpretation of the influence of strain rate and of temperature is that, for fracture to occur, there should be significant amount of sliding on interfaces but these interfaces should not move transversely. It is for this latter reason that ductility almost invariably appears to be restored at low strain rates when the temperature rises above about 0.7 of the absolute melting temperature.

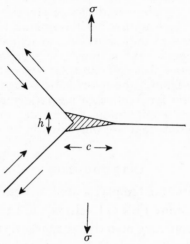

FIGURE 1. The sliding at grain boundaries to open up a crack of length c and maximum thickness h perpendicular to an applied stress σ.

In this paper the creation of cracks by sliding at grain boundary junctions will first be explored and the condition evaluated whereby these may lead to a complete separation.

It is often observed, however, that fracture does not occur purely by progressive crack growth at an interface but by the formation of cavities at specific points along a boundary. Proposals are next considered that attempt to account for cavity nucleation. The rôle of particles at the boundaries is often found to be of particular significance in this respect. The segregation of impurities can also have a large influence. Interfacial separation can be made easier when gases are precipitated as bubbles along a grain boundary, but their rôle as nucleants for fracture is less potent than has sometimes been concluded. Geometrical factors make it possible to envisage ways in which deformation processes can directly lead to fracture but these cannot provide a complete approach.

It is now widely recognized that much of the deformation in creep resistant materials over long periods of time, and possibly many materials in the Earth's crust, depends more on the flow of vacancies than on the effects of dislocation mobility. It is pointed out in this paper that the vacancy flow effects can be even more important in assisting the growth of cracks and cavities because the diffusion paths for flow of vacancies along grain boundaries are somewhat easier than through the lattice; these paths are relatively short and the vacancies can be produced within the boundary itself. Such vacancy flows cannot, however, produce the initial step of interfacial separation. The mode of nucleation thus requires another mechanism to operate until a critical size of cavity nucleus has been reached. The continued growth by vacancy accumulation can be analysed by well established procedures. This also permits some account to be taken of the type of stress system as well as of its magnitude.

FAILURE BY SLIDING AT INTERFACES

It is easy to visualize geometrically how sliding at interfaces can cause cracks to open and to propagate from points where three interfaces meet. This situation has been explored in depth by considering the effects of the coalescence of dislocations slipping on different planes to form a crack on a third plane. This approach has led to interesting developments of the theories of fracture arising from the concepts of Zener (1948), Stroh (1954, 1957) and Cottrell (1958) in particular.

It is more recently that a similar approach has been adopted to interpret one of the possible ways in which cracks can grow during creep.

Following the analysis of Williams (1967 a, b), where grain boundaries intersect as shown in figure 1, a wedge-type crack can be formed of length c and of maximum thickness h under a tensile stress σ with the equilibrium condition given by

$$c = \frac{4\mu\gamma}{\pi(1-\nu)\sigma^2}\left\{\left(1 - \frac{\sigma h}{4\gamma}\right) - \left(1 - \frac{\sigma h}{2\gamma}\right)^{\frac{1}{2}}\right\},$$

where μ is the shear modulus, ν is the Poisson ratio and γ is the energy per unit area to create the surfaces. Now continued sliding at the grain boundaries increases the value of h and the corresponding rate of increase in crack length c is given by

$$\frac{\mathrm{d}c}{\mathrm{d}h} = \frac{\mu}{\pi(1-\nu)\sigma}\left\{\left(1 - \frac{\sigma h}{2\gamma}\right)^{-\frac{1}{2}} - 1\right\}.$$

It is immediately clear from this that the crack becomes unstable and can propagate rapidly when $\sigma = 2\gamma/h$. Now $\mathrm{d}h/\mathrm{d}t$ is related to the rate of grain boundary sliding $\dot{e}_s$ and to the grain size x by the expression $\mathrm{d}h/\mathrm{d}t = x\dot{e}_s$.

Where t_f is the time to fracture, the maximum value of h is then given by

$$h = x\dot{e}_s t_f.$$

Now h cannot exceed $2\gamma/\sigma$ and so it follows that

$$2\gamma/\sigma \approx x\dot{e}_s t_f.$$

This provides a simple evaluation of the life-time before failure occurs under a tensile stress σ, with

$$t_f \approx 2\gamma/x\dot{e}_s\sigma.$$

The above situation can also be developed to evaluate the case where the crack may traverse a complete grain boundary before becoming unstable. It may now be assumed, purely on a statistical basis, that when grain boundaries are completely cracked this is a criterion of a complete failure of the material. This situation is not always true, for in superplastic materials it is well known that complete interfaces can be fractured whilst the material continues to elongate up to 20-fold (Ritchie 1970). This case, however, only appears true for small grains of only a few micrometres in diameter.

To evaluate the time necessary for a crack to propagate to a length equal to a grain diameter we may proceed as follows: the rate of crack growth $\mathrm{d}c/\mathrm{d}t = (\mathrm{d}c/\mathrm{d}h)\,(\mathrm{d}h/\mathrm{d}t) = x\dot{\epsilon}_s\,\mathrm{d}c/\mathrm{d}h$. Hence

$$\frac{\mathrm{d}c}{\mathrm{d}t} = x\dot{\epsilon}_s\frac{\mathrm{d}c}{\mathrm{d}h} = \frac{x\dot{\epsilon}_s\mu}{\pi(1-\nu)\,\sigma}\left\{\left(1-\frac{x\dot{\epsilon}_s t\sigma}{2\gamma}\right)^{-\frac{1}{2}} - 1\right\}.$$

Since in this instance $x\epsilon_s t\sigma$ is substantially less than 2γ,

$$\mathrm{d}c/\mathrm{d}t \approx x^2\dot{\epsilon}_s^2\mu t/4\pi(1-\nu)\,\gamma.$$

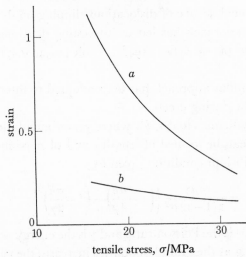

a, with superimposed hydrostatic pressure numerically equal to σ

b, without hydrostatic pressure

tensile stress, σ/MPa

FIGURE 2. The effect of superimposed hydrostatic pressure on strain at fracture at 260 °C as observed by Williams (1967c) for an Al (20 mass %)–Zn alloy heat-treated to give precipitate free zones about 4 μm wide adjacent to the grain boundaries. The applied tensile stress was numerically equal to the independently superimposed hydrostatic pressure. Failure at the higher tensile stress levels was mainly by triple point cracking.

It follows that $4\pi(1-\nu)\,\gamma\int_0^x\mathrm{d}c = \int_0^{t_f}x^2\dot{\epsilon}_s^2\mu t\,\mathrm{d}t$ and so the time to fracture is given on integration by

$$t_f^2 \approx 8\pi(1-\nu)\,\gamma/\mu x\dot{\epsilon}_s^2;$$

thus we may write

$$t_f \approx (4/\dot{\epsilon}_s)\,(\gamma/\mu x)^{\frac{1}{2}}.$$

The above approaches are clearly very much simplified and leave out many factors that can influence the failure process. Nevertheless, they do indicate some important features that have been observed experimentally in studies of the failure of metals and alloys at elevated temperatures.

First, it is noted that $\dot{\epsilon}_s t_f \approx \epsilon_f$ and this is a measure of the strain to fracture although it does not take into account the primary and tertiary creep stages (Feltham & Meakin 1959). It was mentioned earlier that although ductility minima occur, these minima are shallow. It is only when grain boundary sliding is absent or when grain boundaries may move transversely that substantial ductility can be restored and the latter situation is not included in the analyses.

A further factor is that for a given boundary sliding rate both the time and strain at fracture are decreased by a decrease in the interfacial energy. There is a good deal of evidence that indicates that the segregation of impurities at interfaces can reduce γ and consequently reduce the fracture strain. Further, both the above expressions suggest that brittleness is enhanced by an increase in grain size and this again is observed experimentally (Taplin 1973).

It would be wrong to regard these equations as a means of evaluating accurately the conditions under which complete interfacial separation occurs, even under the simplest case of a tensile stress. The principal reason is that they ignore other modes of deformation that occur concurrently and that may have a substantial influence on the strain energy distribution created by the crack and the interfacial sliding. Phenomenologically, it may be adequate to group these effects together in the term relating to the energy required to separate the interfaces but this would then give a somewhat artificial meaning to the term γ. Comparison of experimental results with the equations, however, does suggest that the value of γ is somewhat higher than the true surface energy and this may be related to the reasons indicated.

A further feature in the above analysis is that the applied tensile stress does work which includes a term due to the increase in crack volume. It is then clear that with superimposed hydrostatic pressure the energy required for the crack opening is consequently increased. This is indicated in figure 2.

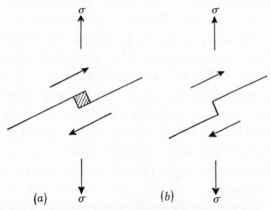

FIGURE 3. If ledges are present or can be created in grain boundaries, then they may have a different reaction to a shear stress depending on whether they are of type (a) where sliding may possibly cause cavity formation or type (b) where the ledge makes boundary movement more difficult.

THE CREATION OF CAVITIES BY INTERFACIAL SLIDING

Many observations relating to fracture during creep show that the growth of a crack from triple point junction, as previously indicated, is only one of the possible failure mechanisms (Gifkins 1959). Instead, discrete cavities may be formed on grain boundaries (Perry 1974), usually nearly perpendicular to the applied stress, and these cavities often preserve their geometrical characteristics as they grow. It is the nucleation of such cavities that has proved difficult to elucidate. This is partly because the critical nucleus size is believed, theoretically, to lie generally between 0.1 and 1 µm which is a particularly difficult region to explore experimentally, for nuclei are too small for light microscopes and too large to be contained in thin films that can be studied easily by transmission electron microscopes operating at 100 kV. Transmission electron microscopes operating at 1 MV hold more promise in this area (Fleck, Taplin & Beevers 1975) and a much used technique in cavity observation is to study the surface by scanning electron microscopy on fracture surfaces or on prepared sections that have in some instances been bombarded by ions to preserve the cavity shape as far as possible.

The character of these cavities has indicated that they have some features similar to those of grain boundary precipitates (Greenwood, Miller & Suiter 1954) and it was natural to assume that their formation and growth arose from precipitation processes; in this case by the agglommeration of vacancies. From considerations of generally accepted nucleation theory, however, it was soon clear that the vacancy supersaturation would not be sufficiently high to result in such precipitation unless nuclei were present (Cottrell 1961). Thus the problem of identifying such nuclei was recognized.

Geometrically, it is clear from figure 3 that ledges in grain boundaries could play a rôle in causing cavity nucleation. The problem then becomes one of deciding whether such ledges can exist or whether they can be caused by the deformation processes. It is conceivable that deformation within the grains may cause the impingement of slip bands at grain boundaries while sliding is taking place and the resultant step formation could result in cavity formation (Gifkins 1956). It is to be noted, however, that ledges are essentially of two types (Perry 1974). As illustrated in figure 3 one type of ledge is capable of opening, and the other simply locks the boundary.

An analysis was proposed (McLean 1963) whereby quite small ledges in grain boundaries that could exist in thermal equilibrium would be of sufficient size to cause cavity nucleation. On this basis, the rate of grain boundary sliding was calculated that would cause the cavity to continue to open and to resist sintering forces. Unfortunately, in the numerical calculations, the product was taken of the vacancy concentration and self-diffusion coefficient in the grain boundary and not simply the self-diffusion coefficient of atoms within the grain boundary and so the sintering rates were underestimated by many orders of magnitude (Harris 1965). Thus, the analysis served to show that there is considerable difficulty in developing a satisfactory quantitative theory to describe the circumstances whereby cavity nucleation may be caused by grain boundary ledges although much discussion remains in this field.

It is much easier to envisage the formation of cavities at particles within the grain boundaries (Cottrell 1961) although metallographic observations have not always been successful in positively identifying the presence of such nuclei where cavities have been observed.

Cavity nucleation at particles must clearly depend on the strength of the interfacial bond between the particles and matrix and the criterion for interfacial separation at particles has been approached in different ways.

The criterion to nucleate a crack is that the shear stress σ_s should exceed a value given by $\sigma_s = \{2\gamma\mu/\pi(1-\nu)x\}^{\frac{1}{2}}$. This expression can be written alternatively by considering the number of dislocations n of Burgers vector b piled up within the grain boundary such that fracture occurs when $\sigma_s = 2\gamma/nb$ (Smith & Barnby 1967). It is worth noting that this value is only one sixth of the value often quoted from earlier work.

An essential feature of an analysis of this type is that the high stresses generated by the sliding interface are supported by an infinite amount of material. It has been pointed out that this assumption may not be valid (Eborall 1961). It follows that nucleation may be possible at much lower stresses than those predicted by the above relation. A detailed calculation (Smith & Barnby 1967) indicates that the stress for nucleation decreases with the width x_b of the barrier to sliding provided that the width of such discontinuities is less than a critical value. The formula above then becomes modified to the form $\sigma_s = \{\pi\mu\gamma x_b/(1-\nu)\}^{\frac{1}{2}}/x$. It has been pointed out that this results in the possibility that the stress for nucleation may be only one twenty-fifth of that previously expected.

Another way of approaching the nucleation problem is simply to consider the influence of the interfacial sliding rate. In such analyses it is necessary to calculate the rate at which a potential nucleus would disappear by sintering forces and to compare this with the rate at which the interfacial sliding would cause opening to occur. Where the latter process was most rapid a nucleus would be formed. As previously pointed out, arguments such as these (Harris 1965) have been used to cast some doubt on ledge mechanisms of cavity formation and they also show that, even for particles, a rather fast rate of sliding is required. The velocity v of grain boundary sliding for cavity opening must be greater than a value given by

$$v > \{D_g w/r^2 \ln (b/r)\} \exp \{(2\gamma\Omega/kTr) - 1\}$$

in order for the cavity of radius r to open, where D_g is the grain boundary self-diffusion coefficient, w the grain boundary width, Ω the atomic volume, k Boltzmann's constant and T absolute temperature. For many metals and alloys this rate of grain boundary sliding is calculated to be somewhat faster than that observed in practice and this is true in situations where cavities have been observed. To overcome this problem, the inhomogeneous nature of sliding processes has been invoked (Chen & Machlin 1959) whereby sliding occurs discontinuously and so, at certain times, the sliding velocity can very considerably exceed that given by the above equation. Analyses of this kind clearly depend on the value of the interfacial energy but it is not true that where this is very low a minute sliding rate would be sufficient to cause permanent separation. Otherwise, any segregation of elements and particularly the formation of gas bubbles would be seen immediately to be an effective nucleus but this is known not to be the case.

Since no interfacial separation is required in the case of gas bubbles it may be considered that these would form ready nuclei for unlimited cavity growth. This, however, is not the situation, because, as the cavity enlarges, its surface area and consequently the surface energy must be increased. The gas pressure soon becomes of little assistance to cavity growth for as a bubble enlarges, the pressure, at constant temperature, falls inversely with the increase in bubble volume (Sykes & Greenwood 1965). Thus the cavity becomes subject to sintering forces unless it exceeds a given size. The evaluation of this size may be calculated by following the treatment of Hyam & Sumner (1962).

If an applied tensile stress σ acts perpendicularly to a grain boundary on which a bubble of radius r is situated, then the equilibrium condition can be written $\sigma = 2\gamma/r - p$ where p is the gas pressure within the bubble. If $r = r_0$ and $p = p_0$ when $\sigma = 0$, then $pr^3 = p_0 r_0^3$ and $p_0 = 2\gamma/r_0$. Hence

$$\sigma = 2\gamma/r - 2\gamma r_0^2/r^3.$$

By differentiation, $d\sigma/dr = 2\gamma(-1/r^2 + 3r_0^2/r^4)$. When $d\sigma/dr = 0$, the bubble will continue to expand, and this condition is met when $r = \sqrt{3}\, r_0$. Thus the tensile stress to cause the gas bubble to grow must be $\geqslant 4\gamma/3\sqrt{3}\, r_0$.

THE GROWTH OF CAVITIES

Where cavities are of an effective radius r greater than the critical size then convincing thermodynamic arguments can be presented to show that vacancy fluxes should permit continued growth principally of the cavities on those grain boundaries that are approximately perpendicular to the applied stress (Balluffi & Seigle 1957; Hopkin 1957). It may be anticipated that such a mechanism would be entirely adequate to account for cavity growth in a majority of cases. The true situation, however, does not seem to be so simple. There is a possibility that deformation mechanisms may enhance or may detract from cavity growth (Dyson 1976) and there are the further possiblities that grain boundaries may not be able to act as perfect sources of vacancies (Ashby 1969). It now seems that these factors are partly responsible for the variety of results and interpretations that have been reported. The implications of these will be considered later but, first, it is appropriate to consider the basis of the theory of cavity growth by vacancy condensation and the extent of the experimental support that it has so far received.

It is some 25 years since it was pointed out that cavities often have the character of precipitates and this led to the proposal that vacancy condensation was required for their growth (Greenwood et al. 1954). Support for such a model also arose independently from studies of inter-diffusion processes between two metals (Barnes & Mazey 1958), where the growth of voids could be correlated with the rate of condensation of vacancies as they flowed to compensate for the unequal diffusion of different atoms across the interface.

Some further implications of the vacancy growth model were subsequently considered and led to the prediction that a hydrostatic pressure numerically equivalent to the applied tensile stress would effectively prevent cavity growth by this process (Hull & Rimmer 1959). The basis of this approach was that near a grain boundary perpendicular to an applied stress σ the average vacancy concentration is increased by a factor $\exp(\sigma\Omega/kT)$ and near a cavity of radius r the vacancy concentration is increased by a factor $\exp(2\gamma\Omega/kTr)$. These equations form the basis of the criterion previously mentioned for cavity growth, namely that $\sigma > 2\gamma/r$. When a hydrostatic pressure P is applied, however, then the vacancy concentration at a grain boundary perpendicular to the principal tensile stress is modified to the form

$$\exp\{(\sigma - P)\,\Omega/kT\}.$$

This implies that if $P = \sigma$ there can be no cavity growth, at least by a vacancy condensation mechanism. There is considerable experimental support (Hull & Rimmer 1959; Ratcliffe & Greenwood 1965).

Some confusion has occasionally arisen between the rôle of vacancy flow in cavity growth and that in Nabarro–Herring or in Coble creep (Coble 1963), where vacancy fluxes are responsible for the deformation. The situation here is to evaluate the effective concentrations of vacancies at the respective grain boundaries in relation to the stress system and also the vacancy concentration at voids, and from these the relative importance of the various fluxes may be determined (Greenwood 1976). It is not difficult to show that most of the vacancy flux can take place simply by vacancy creation in grain boundaries in which the voids are situated. Conversely, this may be visualized as a process whereby atoms leave the cavities to cause their enlargement and subsequently plate on those grain boundaries that are nearly perpendicular to the tensile stress (Harris, Tucker & Greenwood 1974). This process can in itself cause a component of creep strain that can be significant. The entire situation, however, may be difficult to analyse in complete detail because of the overall adjustments that may be required to accommodate the increase in cavity volume (Dyson 1976).

When cavities are growing in this prescribed manner, then the stress distribution is modified. The stress is relaxed particularly at places on the grain boundary close to voids and is a maximum at points midway between the voids on grain boundaries that are nearly perpendicular to the principal tensile stress. It is this stress concentration gradient that gives rise to the vacancy concentration gradient and so to the vacancy flux. This aspect may also be related to the fact that cavities often remain quite distinct until a time near to the final fracture.

Although theories have generally been based on spherical cavities, the theory is not modified if cavities have crystallographic form or where the ends of the cavities are pointed to preserve an equilibrium configuration. In such an instance, effectively, it can be considered that surfaces of the cavity form sections of spheres and it is the radius of these spheres that corresponds to the effective radii of cavities in growth models. It may be highly significant, however, that such cavities need fewer vacancies per unit area of grain boundary that they occupy than do spherical cavities and so they may cause a more severe form of creep damage that leads to a reduced creep life. Segregation of those elements that cause a decrease in surface energy may be especially relevant here.

Several assessments have been made of the rate of cavity growth and the theory is now widely regarded as well established, although detailed treatments lead to slightly different results (Hull & Rimmer 1959; Weertman 1973; Raj & Ashby 1975; Speight & Beere 1975). Where x_c is half the cavity spacing, the equation which now appears to be most widely accepted is

$$\frac{dv}{dt} = \frac{2\pi D_g w\{\sigma - P - (2\gamma/r_0)\}\,\Omega}{kT} \Big/ [\ln\{x_c/r_0\} - \tfrac{1}{4}\{1 - (r_0/x_c)^2\}\{3 - (r_0/x_c)^2\}];$$

when cavities are relatively widely spaced and $v \gg \tfrac{4}{3}\pi r_0^3$, this formula reduces to the expression

$$dv/dt \approx 10 D_g w\sigma\Omega/kT \quad \text{when} \quad P = 0.$$

Although this equation clearly cannot hold throughout creep life it has the important characteristic that it predicts that the volume of an individual cavity would be expected to grow approximately proportionally with time over the greater part of creep life.

If it is further assumed that all cavities are nucleated at the start of creep and that creep failure occurs when a certain fraction of the grain boundary is covered by cavities, then it follows from this approach that the time to fracture is given approximately by $t_f \propto 1/\sigma$.

In practice, this relation has generally been found not to hold and the time to facture has been noted to be a much higher power of the applied tensile stress (Feltham & Meakin 1959). A further discrepancy between this theory and experiment has been that quite often (Woodford 1969; Needham, Wheatley & Greenwood 1975) the volume increase due to cavitation in creep is given by a formula of the type $\Delta V/V \propto \epsilon t\sigma^n$. This contrasts with the equation $\Delta V/V \propto \sigma t$ that would be expected to hold on the theory so far stated.

The simplest explanation of such discrepancies is that the cavities are not all nucleated at the start of the test (Greenwood 1963) and that a finite amount of deformation is necessary for their creation as well as a given magnitude of the applied stress.

When cavity nucleation is taken to be some function of strain then it is possible to obtain a much closer fit between theory and experiment. Most exhaustive studies in the field of high-temperature fracture have been carried out on copper. In some respects this may be considered an ideal material because it is well established that grain boundaries readily act as a source of vacancies which emerge with a negligible threshold stress (Greenwood 1976). There is good evidence for the operation of Nabarro–Herring and Coble creep processes in the régimes anticipated (Burton & Greenwood 1970) and results on α-particle injection into copper (Barnes 1960) and sintering experiments all lead to the view that vacancies may be readily emitted or absorbed at grain boundaries under conditions previously mentioned.

Although a number of aspects of the cavity nucleation mechanism remain obscure, an increasing amount of evidence (Kelly 1975; Morris 1977) is pointing to an approximate relation for a number of cavities given by $N \propto \sigma^2 \epsilon$. With such an expression then it becomes clear that the volume increase, given by $\Delta V/V \propto \epsilon t \sigma^n$, is obtained from the cavity growth relation $\mathrm{d}v/\mathrm{d}t \propto \sigma$ and is in accord with experimental results when $n \approx 3$ (Needham & Greenwood 1975). This approach further permits an assessment to be obtained for variation of the time to then fracture with tensile stress and this is deduced as follows.

A given fraction of the area of grain boundaries is occupied when $N v^{\frac{2}{3}}$ reaches a given value which, at fracture, is $N_f v_f^{\frac{2}{3}}$. This value is proportional to $(\sigma^2 \epsilon_f)(\sigma t_f)^{\frac{2}{3}}$ and so, if $\epsilon_f \propto \sigma^5 t_f$, then $t_f \propto 1/\sigma^{4.6}$.

One of the biggest difficulties that has been faced by cavity growth models by diffusion fluxes is in relation to the observation that the product $\dot{\epsilon} t_f$, an approximate measure of the strain at fracture, is roughly constant (Feltham & Meakin 1959). This expression seems to imply that the cavitation mechanism should have an activation energy similar to that for the creep process which is often identical with the self-diffusion activation energy. Such a correlation would imply that vacancies grow by lattice diffusion rather than by grain boundary diffusion and, of the many proposals that have been put forward to reconcile the situation, some suggest that vacancies can only be produced in a grain boundary at the same rate as that permitted by deformation of the grains (Ishida & McLean 1967).

Although this explanation seems an entirely plausible one, it does not always appear necessary because steady pursuit (Morris 1977) of the separate factors controlling nucleation and growth of cavities in copper has revealed that the growth of individual cavities, taking into account temperature, follow a relation of the form $v \propto \sigma t \exp(-Q_g/kT)$. Additionally, the formula for nucleation, $N \propto \sigma^2 \epsilon$, has a negligible temperature dependence, at least over a range where the rate of grain boundary sliding does not differ substantially from the rate of overall deformation.

Taking again the criterion that failure takes place when a given area of grain boundary is occupied by cavities, this condition may be written

$$(\sigma^2 \epsilon_f)\{\sigma t_f \exp(-Q_g/kT)\}^{\frac{2}{3}} = \text{a constant.}$$

Where creep life is related to the secondary creep stage, so that $\epsilon_f \approx \dot{\epsilon} t_f$, and the creep rate $\dot{\epsilon} \propto \sigma^5 \exp(-Q/kT)$, where Q is the activation energy for creep, then the fracture criterion becomes

$$\sigma^{\frac{8}{3}} \dot{\epsilon} t_f^{\frac{5}{3}} \exp(-2Q_g/3kT) = \text{a constant.}$$

Substituting for σ, we obtain the condition

$$\dot{\epsilon}^{\frac{23}{15}} t_f^{\frac{5}{3}} \exp(8Q/15kT - 2Q_g/3kT) = \text{a constant.}$$

Now the activation energy for creep, Q, is similar to that for self-diffusion under the conditions considered (Perry 1974) and the activation energy for grain boundary self-diffusion $Q_g \approx 0.6Q$ (Gibbs & Harris 1969); thus the exponential term is approximately $\exp(2Q/15kT)$ which shows that the fracture condition is relatively insensitive to temperature. This is in agreement with most observations, except where the temperature is too low for grain boundary sliding or sufficiently high for grain boundary migration since these aspects are not incorporated in the analysis. It also follows that the approximate measure of the strain to fracture, ϵ_f, is roughly constant. It was noted that $\epsilon_f \approx \dot{\epsilon} t_f$ and the above analysis indicates approximately that, at fracture, $\dot{\epsilon} t_f^{\frac{25}{23}}$ is constant.

It follows from this that the elongation to fracture is not strongly sensitive either to temperature or to the level of tensile stress. This is in accord with many experimental results and follows from the theory of cavity growth by vacancy diffusion where strain is necessary for cavity nucleation.

The question next arises of the possible extension of the analysis to interpret the influence of more complex stress systems on creep fracture. It is noted that the rate of vacancy flow to a cavity is dominated by the principal tensile stress and, where a hydrostatic pressure P is superimposed, this stress is $(\sigma - P)$. More precisely, the term $(\sigma - P - 2\gamma/r)$ must be used since $2\gamma/r$ may not be negligible when compared with $(\sigma - P)$. It is also assumed that the same stress term governs nucleation. In contrast, the secondary creep rate is governed primarily by the shear stress components σ_s which is thus not strongly dependent on P and may be taken to be some fraction of σ. Thus, the fracture criterion now becomes

$$(\sigma - P - 2\gamma/r)^2 \epsilon_f \{(\sigma - P - 2\gamma/r) \, t_f \exp(-Q_g/kT)\}^{\frac{2}{3}} = \text{a constant.}$$

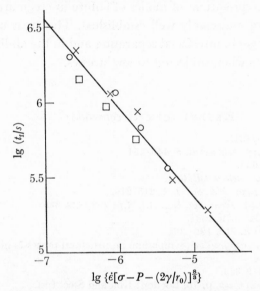

FIGURE 4. The relation between the time to fracture t_f in seconds and the parameter $\dot{\epsilon}[\sigma - P - (2\gamma/r_0)]^{\frac{8}{3}}$ for copper with grain size 230 μm at 500 °C, where σ is the applied tensile stress, P is the superimposed hydrostatic pressure and $(2\gamma/r_0)$ is the stress required for cavity nucleation, taken to be 5 MPa. $\dot{\epsilon}/s^{-1}$ is the creep rate. The circles represent points for $\sigma = 20.7$ MPa; the squares represent points for $\sigma = 24.2$ MPa; the crosses represent points for $\sigma = 27.6$ MPa; P has values between 0.1 and 13.6 MPa. The graph has a gradient equal to -0.60 (Needham & Greenwood, to be published.)

At present, studies have only been made at constant temperature and so, neglecting the temperature term when substituting for ϵ_f, the criterion can be written:

$$(\sigma - P - 2\gamma/r)^{\frac{8}{3}} \dot{\epsilon} t_f^{\frac{5}{3}} = \text{a constant.}$$

Since $\dot{\epsilon}$ has been found to depend more strongly on P than originally suspected (Needham & Greenwood 1975) it is not possible at this stage to make further substitutions, but some experimental evidence supporting the above equation is presented in figure 4.

It is tempting to pursue these arguments to explain all situations of fracture at high temperatures. They have the merit that they can distinguish between the effects of different stress systems and particularly between the deviatoric and non-deviatoric stresses. There appears to be quite strong evidence, however, that the theory cannot be applied in all or even in the majority of practical situations. The reason is probably that even in the case of pure metals there is a small but finite threshold stress that is necessary for vacancies to be produced at the grain boundaries, but in solid solutions, and particularly in materials with precipitates (Harris 1976), the threshold stress may become quite substantial. At this point the theory of cavity growth by vacancy condensation must be linked closely with deviations from the theories of Nabarro–Herring and Coble creep. A number of theories have recently emerged to suggest why the action of grain boundaries as vacancy sources can be inhibited and these eventually must be expected to relate closely to the prediction of creep life.

Thus the situation concerning cavity growth is not fully resolved but there are some clear indications of the directions in which further work may be most profitably pursued.

Conclusions

Although the proposed interpretations of modes of failure in creep are far from complete, a number of features now seem reasonably well established. The areas of greatest uncertainty appear to lie in the initial stages of interfacial separation and in the ability of grain boundaries in alloys to produce vacancies when subjected to low stresses.

References (Greenwood)

Ashby, M. F. 1969 *Scr. met.* **3**, 837–841.

Balluffi, R. W. & Seigle, L. L. 1957 *Acta metall.* **5**, 449–454.

Barnes, R. S. 1960 *Phil. Mag.* **5**, 635.

Barnes, R. S. & Mazey, D. J. 1958 *Acta metall.* **6**, 1.

Burton, B. & Greenwood, G. W. 1970 *Met. Sci. J.* **4**, 215–219.

Chen, C. W. & Machlin, E. S. 1959 *Trans. met. Soc., A.I.M.E.* **209**, 829–835.

Coble, R. L. 1963 *J. appl. Phys.* **34**, 1679–1681.

Cottrell, A. H. 1958 *Trans. A.I.M.E.* **212**, 192–203.

Cottrell, A. H. 1961 *Intercrystalline creep fracture* (Symposium on structural processes in creep). London: Iron and Steel Inst. and Inst. of Metals.

Dyson, B. F. 1976 *Met. Sci.* **10**, 349–353.

Eborall, R. 1961 *Structural Process in Creep*, p. 75. London: Iron and Steel Inst.

Feltham, P. & Meakin, J. D. 1959 *Acta metall.* **7**, 614–627.

Fleck, R. G., Taplin, D. M. R. & Beevers, C. J. 1975 *Acta metall.* **23**, 413–417.

Garofalo, F. 1965 *Fundamentals of creep and creep rupture.* New York: Macmillan.

Gibbs, G. B. & Harris, J. E. 1969 *Conference on interfaces, Melbourne.* Melbourne: Butterworths.

Gifkins, R. C. 1956 *Acta metall*, **4**, 98.

Gifkins, R. C. 1959 *Fracture*, pp. 579–627. New York: Wiley.

Greenwood, G. W. 1963 *Phil. Mag.* **8**, 707–709.

Greenwood, G. W. 1973 *Creep life and ductility* (Conference on microstructure and the design of alloys, Cambridge). London: Inst. of Metals, and Iron and Steel Institute.

Greenwood, G. W. 1974 *Diffusion creep and its technological relevance* (Conference on the physical metallurgy of nuclear reactor fuel elements, Berkeley, p. 53. London: Metals Society.

Greenwood, G. W. 1976 *Grain boundaries as vacancy sources and sinks.* (Conference on vacancies, 1976, Bristol) London: Metals Society. (In the press.)

Greenwood, J. N., Miller, D. R. & Suiter, J. W. 1954 *Acta metall.* **2**, 250–258.

Harris, J. E. 1965 *Trans. met. Soc. A.I.M.E.* **215**, 471.

Harris, J. E. 1973 *Met. Sci. J.* **7**, 1.

Harris, J. E. 1976 *J. nucl. Mater.* **59**, 303.

Harris, J. E., Tucker, M. O. & Greenwood, G. W. 1974 *Met. Sci.* **8**, 311–314.

Hopkin, L. M. T. 1957 *Nature, Lond.* **180**, 808.

Hull, D. & Rimmer, D. E. 1959 *Phil. Mag.* **4**, 673.

Hyam, E. D. & Sumner, G. 1962 *International Atomic Energy Symposium, Venice*, p. 323.

Ishida, Y. & McLean, D. 1967 *Met. Sci. J.* **1**, 171.

Kelly, D. A. 1975 *Acta metall.* **23**, 1267.

Leckie, F. A. 1978 *Phil. Trans. R. Soc. Lond.* A **288**, 27 (this volume).

McLean, D. 1963 *J. Aust. Inst. Met.* **8**, 45.

Morris, P. F. 1977 (To be published.)

Needham, N. G. & Greenwood, G. W. 1975 *Met. Sci.* **9**, 258–262.

Needham, N. G., Greenwood, G. W. & Wheatley, J. E. 1975 *Acta metall.* **23**, 23–27.

Perry, A. J. 1974 *J. Mater. Sci.* **9**, 1016–1039.

Raj, R. & Ashby, M. F. *Acta metall.* **23**, 653.

Ratcliffe, R. T. & Greenwood, G. W. 1965 *Phil. Mag.* **12**, 59–69.

Ritchie, S. M. 1970 Ph.D. thesis, University of Sheffield.

Smith, E. & Barnby, J. T. 1967 *Met. Sci. J.* **1**, 1–4.

Speight, M. V. & Beeré, W. B. 1975 *Met. Sci.* **9**, 190.

Stroh, A. N. 1954 *Proc. R. Soc. Lond.* A **233**, 404.

Stroh, A. N. 1957 *Adv. Phys., Phil. Mag. Suppl.* **6**, 418–465.

Sykes, E. C. & Greenwood, G. W. 1965 *Nature, Lond.* **206**, 181–182.

Taplin, D. M. R. 1973 *The hot fracture story* (International Golden Jubilee Symposium, Banaras Hindu University, India).

Weertman, J. 1973 *Scr. met.* **7**, 1129.

Williams, J. A. 1967*a* *Acta metall.* **15**, 1559.

Williams, J. A. 1967*b* *Phil. Mag.* **15**, 1289–1291.

Williams, J. A. 1967*c* *Acta metall.* **15**, 1564.

Woodford, D. A. 1969 *Met. Sci. J.* **3**, 220.

Zener, C. 1948 *Fracturing of metals*, pp. 7–8. Cleveland: A.S.M.

Discussion

P. E. EVANS (*Joint University and U.M.I.S.T., Metallurgy Building, Grosvenor Street, Manchester M1 7HS*). I feel that Professor Greenwood's hope that creep in metals may throw some light on creep in the mantle may not be *directly* realized. However, there is no doubt that the study of creep in metals has provided useful ideas on creep in ceramics. In turn I see ceramics as intermediate, both in terms of complexity and in terms of our understanding of their mechanical properties, between metals and rocks. I personally tend to consider ceramics as model rocks and I am sure that the study of creep in ceramics will increase our understanding of creep processes in the mantle.

H. H. SCHLOESSIN (*Department of Geophysics, University of Western Ontario, London, Ontario, Canada*). Is not the situation of stress concentration for faceted bubbles with sharp edges as described here similar to that of crack propagation by internal pressure in silicon–iron observed by Tetelman & Robertson (*Acta metall.* **11**, 1963)? There the internal pressure was generated by preferential accumulation of hydrogen in cracks. Stable crack propagation can be obtained if the accumulation of hydrogen by exsolution from the surrounding matrix keeps pace with the pressure drop due to the extension of the crack volume.

There is evidence from laboratory studies and actual rocks and minerals in the field that bubble nucleation and growth can occur under considerable hydrostatic pressure in the presence of partial melting. For example, basalts drilled from several hundred metres below the ocean floor show high concentrations of bubbles, usually thermal segregation vesicles or decorated vesicles. Their sizes vary from several millimetres to below 1 μm. As shown by scanning electron microscopy the walls of some decorated vesicles exhibit glassy linings which are riddled with multitudes of spherical bubbles all equal in size. Because of the higher solubility of gases in liquids compared with solids and, for most gases, decreasing solubility with decompression both liquid and solid phases become supersaturated with gases and it is thus possible for bubbles to nucleate during crystallization of partial melts under considerable hydrostatic pressures. The mechanical effects of gas bubbles may play a significant rôle in the ascent and emplacement of magmatic liquids. The large volume of gases indicated by high concentrations of vesicles in oceanic basalts gives evidence of very strong effervescence.

G. W. GREENWOOD. Dr Schloessin enquires about the possible similarity of the stress concentration for faceted bubbles with sharp edges and that which has been shown to arise during crack propagation by internal pressure in silicon–iron.

I should feel that there are some similarities and some significant differences. The similarities may arise because in both instances there may be a continued accumulation of gas in the bubbles or the cracks and their growth can be influenced if the gas pressure is maintained at a sufficiently high level. Nevertheless, there appear to be some significant differences. In creep processes, failure almost invariably occurs at grain or interface boundaries rather than along cleavage planes and so the shape of bubbles can be further modified by grain boundary tension. A still more important difference between the work mentioned on silicon–iron and observations on creep fracture is that diffusional movements appear to play a major rôle during creep and the tendency to form equilibrium shapes of bubbles can become predominant. Where bubble growth is controlled by diffusional processes, the tensile stress component is redistributed in a way that can be calculated and can become a maximum across grain boundaries in regions midway between adjacent bubbles.

The information that Dr Schloessin provides of bubble nucleation and growth in actual rocks and minerals could have close analogy with some forms of creep fracture in metals. It would be interesting to apply some of the theories that have been developed for the latter to the laboratory studies and observations that have been made on rocks and minerals to allow some quantitative predictions to be made of macroscopic behaviour.

The effects of pressure can be readily incorporated into theoretical treatments and the effects caused by crystallization and other phase changes should be amenable to interpretation to some extent.

I was interested to learn of the importance of gas precipitation in materials of the Earth and it seems that, as is now well appreciated in metallic systems, the influence of the gases may be all-important in some situations.

R. W. CAHN (*School of Applied Science, University of Sussex*). The tendency for pores to form at a grain boundary under uniaxial stress is a function of the inclination of the boundary to the stress vector. Is it feasible to enhance fracture resistance under creep conditions by controlling grain shape in such a way as to reduce or exclude the boundaries which are orientated favourably for pore formation?

I believe that creep ductility can also be enhanced by resorting to a very fine grain size. This might need to be stabilized by means of a disperse second phase. Would the presence of such a dispersion be likely to generate so many extra pores that this strategy would prove counterproductive?

G. W. GREENWOOD. It is quite feasible to enhance the resistance to creep fracture by controlling grain shape and size. It is generally true that cavities form most readily on grain boundaries perpendicular to the maximum principal tensile stress and that they have the most deleterious effect when they occur on such boundaries. Further, when cavities are situated on grain boundaries of large area, then they present the most serious limitation to creep ductility. Thus there is the immediate possibility of using materials with a microstructure consisting of elongated grains where only grain boundaries of relatively small area are orientated nearly perpendicularly to the tensile stress.

These comments also link up with Professor Cahn's second question. Creep ductility can be enhanced simply by decreasing the grain size of a material and it may be thought that a material with fine equiaxed grains might be entirely adequate. There is, however, the allied problem that fine equiaxed grains often result in material of reduced creep strength and so a careful balance must be made in alloy design to reach a suitable compromise between creep strength and ductility. From this point of view, grains elongated in the direction of the tensile stress, but with a relatively small area of cross-section perpendicular to this stress, may have the best combination of properties. Such structures, however, may still require second phase particles for stabilization. The presence of such particles can cause a nucleation of pores in the material and so reduce the creep ductility, but a good deal remains to be learnt in this area since not all second phase particles seem capable of nucleating pores.

Phil. Trans. R. Soc. Lond. A. **288**, 229–234 (1978) [**229**]
Printed in Great Britain

General discussion

G. L. ENGLAND (*Civil Engineering Department, King's College, London WC2R 2LS*).

Some aspects of creep in concrete

First I wish to thank the organizers for allowing me to offer a few comments in relation to another widely used engineering material which exhibits creep behaviour, namely concrete. This material exhibits a number of features which are similar to those described already for materials such as rocks and metals, while at the same time offering some distinctly dissimilar features. For example, the dominant state of stress in concrete structures is compressive and in consequence it is the compressive creep which is of importance in most practical situations. Concrete exhibits deviatoric and volumetric creep and thus differs from many other materials which show zero creep dilatation. The creep rate is related linearly to stress and continuously decreases with time under load; it increases with increase of temperature. Creep occurs at all stress levels: there is no threshold stress level which must be exceeded for creep to occur.

The creep of concrete has three levels of importance:

(1) The creep behaviour originates at the submicroscopic level in the gel and hydration products of the cement phase of the material, and here water plays a dominant rôle in determining the magnitude of creep observed in experiments.

(2) Creep at the macroscopic, or engineering, level is observed as an integrated effect which represents the behaviour of a heterogeneous material consisting of both active and passive material constituents. The observed behaviour is taken to represent an isotropic property at the engineering level.

(3) In the engineering structure, creep, although isotropic, may be influenced by environmental factors such as humidity and temperature and may thus become a non-homogeneous property throughout the volume of the structure. When this happens gross departures of behaviour from the predictions of normal elastic analyses are encountered. Experiments conducted on concrete beams and portal frame structures have indicated that supporting actions not only change with time due to creep, but will in many cases exhibit a change of sense, whether they be bending moments or forces at the foundations.

Experimental creep data reveal at least two separable components of creep strain. One is of the viscous flow type and the other reflects a delayed elastic strain response to changes of stress. The first is temperature-dependent and may conveniently be normalized with respect to stress and temperature for use in analysis. At elevated temperatures it dominates over the delayed elastic component, which is essentially temperature-independent, and for this reason the latter component may be either ignored as a significant creep component, or simply included with the normal elastic strain response by modifying the elastic modulus. After normalization the strain rate equation may be written

$$d\epsilon_c/dt = \sigma f(T) g(t), \tag{1}$$

where $f(T)$ and $g(t)$ represent respectively the normalizing creep-temperature function and an ageing function of time, and σ is the applied stress.

At constant stress and temperature equation (1) represents the creep rate and its variation with time. A comparable but simpler equation may be developed when the normalized creep strain itself, $\epsilon_c/\sigma f(T)$, is taken to represent a pseudo-time parameter, t', say. It then follows that the creep rate, $\dot{\epsilon}_c$, with respect to t', has the form

$$\dot{\epsilon}_c = \sigma f(T). \qquad (2)$$

In other words, the creep rate is related linearly to stress, to some function of temperature, $f(T)$, and is constant with respect to t'. This last property permits considerable simplification to be achieved in the formulation of analytical techniques in structural analysis for creep of concrete in non-homogeneous situations. Conversion to real time is made at the end of the analysis by reference to the normalized creep–real time curve for the material.

Use of the formulation of equation (2) and its corresponding equivalent representation in three dimensions has allowed calculations to be made in terms of the rate at which work is done on the structure by the external loads undergoing displacements caused by creep. The power $\dot{W}$ associated with the external loads is equated to the rate of change $\dot{U}$ of stored energy in the material and the power $\dot{D}$ dissipated due to creep. Then

$$\dot{W} = \dot{U} + \dot{D}. \qquad (3)$$

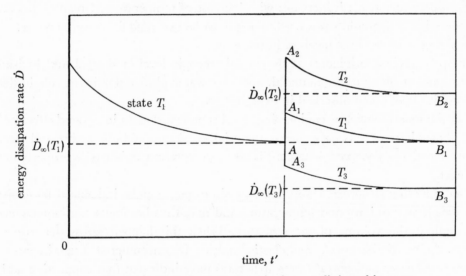

FIGURE 1. Variation with time t', of power dissipated in creep.

The two quantities $\dot{U}$ and $\dot{D}$ represent the integrated sum of corresponding density quantities in the structure; they thus vary in space and time. In addition, there is an interchange between the stored energy and dissipated energy, even when the applied loads are constant in time. Analyses indicate that a preferred long-term solution exists and that this corresponds to a condition in which the power dissipated in creep is a minimum with respect to the pseudo-time parameter. At this time the stored energy becomes a constant in time but does not represent a minimum as in the case of a wholly elastic system.

Further observations indicate that it is possible to obtain by direct calculation the limiting state of stress in a structure undergoing non-homogeneous creep by taking advantage of the knowledge that the energy dissipation rate due to creep alone in the steady-state stress condition (see England 1966) is a minimum with respect to all variations of stress from the true state (see England 1968).

An interesting feature of the rate at which energy is dissipated due to creep is shown in figure 1, where $\dot{D}$ is plotted against t' for concrete subjected to sustained temperatures and loads. At point A, it is assumed that $\dot{D} \approx \dot{D}_\infty$, the steady-state energy dissipation rate and that at this time changes to the system are introduced which alter the state of internal stress. Two cases are considered:

(*a*) A self-equilibrating system of stress is introduced without change of temperature. The resulting response of $\dot{D}$ is as shown by the curve A_1–B_1.

(*b*) A change to the state of non-uniform temperature is introduced. This causes a change to the internal stresses and alters the local creep behaviour of the material in a non-homogeneous manner. The subsequent behaviour with reference to $\dot{D}$ has two possible forms. These are shown by the curves A_2–B_2 and A_3–B_3. In each case the new preferred state for $\dot{D}$ is of smaller magnitude than the value of $\dot{D}$ immediately following the change of temperature, irrespective of whether this preferred state, denoted by points B_2 and B_3, lies above or below the point A for the initial set of conditions.

The behaviour illustrated above forms a basis for the understanding of concrete at the engineering level and allows the laws of mechanics to be used in the formulation of new theorems, which in turn, permit reliable engineering predictions to be made by direct calculation in areas hitherto examined only by historical step-by-step forms of solution.

In conclusion I leave by asking this question: is it likely that an energy dissipation formulation, to the flow and recrystallization problem of creep, can lead to a smooth steady-state creep rate as observed in experiments and to suitable predictive methods of analysis for other materials discussed at this meeting?

References

England, G. L. 1966 Steady-state stresses in concrete structures subjected to sustained temperatures and loads. *Nucl. Eng. & Des.* **3** (1), 54–65, also **3** (2), 246–255.

England, G. L. 1968 Time-dependent stresses in creep-elastic materials: a general method of calculation. *Conference on recent advances in stress analysis, March* 1968, session 2, paper 1. London: Royal Aeronautical Society.

A. M. NEVILLE (*Department of Civil Engineering, The University, Leeds LS2 9JT*).

The materials considered at this meeting seem to fall into two categories: those which undergo creep under a tensile stress (mainly metals) and those which creep under a compressive stress (principally rocks). Now, there exists a material which creeps under both these stresses, indeed under all practical states of stress, and which can be tested in a like manner in the laboratory: concrete. Tests on creep of concrete in compression and in tension and measurements of recovery of creep after removal of either of these stresses are helpful in elucidating the mechanism of creep.

An important feature of the creep of concrete is that it occurs at all temperatures, at least in the practical range of something like -50 to $+400$ °C, and under any stress; there is thus no threshold stress or temperature, although creep is of course a function of both these variables.

We do not have a satisfactory understanding of the mechanism of creep of concrete but we know that at temperatures below about 80° C the presence of adsorbed water in the cement paste is necessary. Strictly speaking, it is only cement paste that creeps, the aggregate not doing so at stresses which are practicable in concrete. So, although concrete is a two-phase material, the major part of creep deformation occurs not at boundaries between phases or between grains but within the cement phase.

Creep of concrete under cyclic stresses is of interest not only for structural design purposes but also in the study of the phenomena involved in creep and recovery; the activation energy approach has been very helpful. Work on this has just been published (Neville & Hurst 1977; Hurst & Neville 1977) as a supplement to a book (Neville 1970).

Finally, I should like to thank the organizers of the meeting for the opportunity they gave me to look at other materials. I am sure this is good for all of us.

References

Hirst, G. A. & Neville, A. M. 1977 Activation energy of creep of concrete under short-term static and cyclic stresses. *Mag. Concr. Res.* **29** (98), March.
Neville, A. M. 1970 *Creep of concrete: plain, reinforced and prestressed*, 622 pages. Amsterdam: North-Holland.
Neville, A. M. & Hirst, G. A. 1977 Mechanism of cyclic creep of concrete. *Proceedings of the Douglas McHenry International Symposium on Concrete and Concrete Structures*. American Concrete Institute.

D. C. Tozer (*School of Physics, University of Newcastle*).

Creep studies and planetary problems

Although my contribution to the discussion was made in response to the first paper of the meeting presented by Professor Weertman, having now listened to the rest of the discussion I should like to broaden somewhat the scope of my original comments to cover what I regard as a serious omission. My comments to Professor Weertman were concerned with a particular misconception of the boundary value problem, involving the creep behaviour of *in situ* planetary material, that is posed by Earth and indeed all the planets. It is very important that one's thoughts about planetary creep be guided by some preliminary understanding of the way this boundary value problem is posed, for without it I have noticed that any discussion of 'The creep of the Earth' is likely to be replaced by a description of observations properly subsumed under the heading 'Laboratory studies of creep in rocks or minerals'. In turn these will generate a host of unanswerable, if not irrelevant, questions about the 99.9999...% of planetary material that is not accessible to similar observation. Let me say at once that any criticism is not intended for those who study rock creep for its intrinsic interest, but at the extremely naïve views that some of them entertain about the relevance of their results to planetary scale problems.

As a group, engineers are probably far more aware than most geophysicists that quite unforeseen problems of material science can arise when one is trying with all one's skill simply to scale up a prototype system by a factor of ten in linear dimensions, so I may well be preaching to the converted in saying that I would profoundly distrust the reliability of using results obtained on a system that was, say, 10^8 times smaller than the one of interest *even if I knew that size was the only respect in which the two systems differed*. Of course, there is absolutely no hope of settling even this latter point for a planet, but some very careful experimenter may well ask whether such distrust is justified or even helpful when, unlike some he could mention, he is at least producing 'hard data' about Earth properties. To answer this, it is necessary to examine exactly what has happened when someone claims to have produced hard data about a macroscopic property of matter. Although the experimenter may believe he has observed and measured, for example, a viscosity, what he has done in fact is to use a solution of equations containing the concept of a viscous continuum to connect limited ranges of two quantities that he interprets as a rate of strain and stress. He will not waste much time testing whether the fitted value of viscosity is a truly intensive quantity because he is usually convinced before he starts the experiment that there is a viscosity loitering inside his specimen waiting to be measured. Such imagery is quite harmless until one asks the question: what is the relation between stress and rate of strain for a specimen of the same material n times larger (or smaller)? The 'hard data' man with his realist views about material properties will have no doubt that the answer is to be found by assigning the same viscosity to all points of the new system's interior and, of course, provided n is not too big or too small this often works. However, by using the matter-of-fact description of the experimenter's activities, one can see that what is really involved in this question is whether the equations whose solution was used to connect the observables 'stress' and 'rate of strain' in the original experiment can be dynamically scaled. What makes me so pessimistic about using laboratory creep data in planetary scale problems is that their substitution into the appropriate equations of motion leads to equations that have no scaling laws. In other words, the empirical data are already telling us that a different relation will hold between the same pair of observables if the size is changed, and that the concept of property data that are 'hard' in the sense of being applicable to a material, whenever or however it is found, is no better than a dangerous half truth. Perhaps to reassure those geophysicists who think I am proposing to throw alway all laboratory data, I should say that I see this difficulty mainly in connection with the dynamical rather than the static properties of materials. While the latter can probably be safely seen as labels attached to a material whenever it turns up, the dynamical properties must be viewed with the equations to which they belong as forming an abstract model connecting the observables of systems. In this abstraction many will recognize that we are already using a systems approach to dynamical problems on a laboratory scale and, in fact, what I am suggesting is that it is only the realist misapprehension about material properties that makes people think they can do otherwise in a planetological situation. However, for the same reasons that engineers find moderately scaled models useful, if not perfect, I have much higher hopes of finding an integrated view of dynamical processes in different planets than of also explaining laboratory phenomena with exactly the same theory.

Although there are a number of planetary observations of deformation on a large scale in whose interpretation a creep resistance in the form of an effective viscosity has been assigned to the interior (Cathles 1975), the most basic problem involving creep uses the concept of a system undergoing an internal heat transfer process to connect observations of surface heat flow and the large-scale secular deformation of near surface rocks. The heat sources are visualized as only very slowly changing, temperature-independent, radiogenic sources distributed throughout the interior. This problem is more basic in the sense that if, instead of treating effective viscosity as an adjustable parameter (as is done in the other problems), we introduce the assumption that *in situ* planetary material also has a very temperature sensitive creep resistance, we may show that within wide limits of choice for such creep functions, the mean values of an effective viscosity controlling large-scale deformation are quite closely regulated by the heat-transfer process itself. My comment to Professor Weertman was made because he assumes, as do so many others for purely historical reasons, that it is the temperature rather than the effective viscosity that is directly regulated by the planetary heat-transfer process. Those who are further interested in the self regulation of planetary creep resistance can pursue the matter through a recent article of mine (Tozer 1977), but I would make the following comments to illustrate the importance of some preliminary understanding of this boundary value problem before getting too deeply involved in discussion of the creep process. It is the effective viscosity $\sigma/\dot{\epsilon}$ for large-scale deformation that is directly regulated by the heat transfer process, and I may add that the regulated value of *ca.* 10^{21} P† agrees satisfactorily with the other methods of assigning a viscosity

† $1\,\mathrm{P} = 10^{-1}\,\mathrm{Pa\,s}$.

to the deep interior. Whether $\dot{e}$ is fixed by Newtonian or non-Newtonian steady state creep processes is an academic question insofar as the observations of surface movement do not require any close commitment about it. However, I believe it is true to say that the heat transfer process will regulate conditions in which Newtonian processes are collectively going to make a significant contribution to the total creep rate $\dot{e}$ of the deeply buried planetary material vis-à-vis processes like power law creep that imply an infinite effective viscosity as $\dot{e} \to 0$. The second point concerns the attempts to use so-called deformation maps. While there have been a number of efforts to put the régime of *in situ* planetary material deformation on such diagrams, the results are very misleading because the boundary value problem is not the one of applying a homogeneous stress or strain rate to planetary material. What happens approximately in the heat-transfer problem is that if any material is too cold to deform the creep resistance is decreased somewhere else to accommodate for it – hence the rather loose talk of 'rigid 'plates to describe the movements of cold, near-surface material.

Lastly, I would say to those geologists who hope to determine conditions inside the Earth by examining the microstructure of odd bits of rock they find on the surface that they will always have to face the problem of how representative their results are on a much larger scale. At the moment and again largely for historical reasons, such questions are largely brushed aside by assumptions of spherical symmetry, but an understanding of the self regulation of creep resistance inherent in the heat-transfer process for such a large self-gravitating object will soon convince one that intrusive and/or volcanic events only occur *because* a tiny fraction of Earth material can be brought to a very exceptional state of low creep resistance by deformational heating associated with the heat transfer process.

References

Cathles, L. M. 1975 *The viscosity of the Earth's mantle*. Princeton University Press.
Tozer, D. C. 1977 The thermal state and evolution of the Earth and terrestrial planets. *Sci. Prog., Oxf.* **64**, 1–28.

0508

971
.3
541
May

Mays, John Bentley.
    Emerald city : Toronto visited / John Bentley Mays ;
photographs by Richard Rhodes. -- Toronto : Viking,
1994.
    xxi, 355 p. : ill.

    Includes bibliographical references (p. 347-355).
    07465866   ISBN:0670853569

    1. Architecture - Ontario - Toronto. 2. Toronto (Ont.) -
Description and travel. 3. Toronto (Ont.) - Social life
and customs. I. Rhodes, Richard. II. Title

    2051   94AUG30   06/he   1-00635414

John Bentley Mays

emerald city
Toronto Visited

Photographs by
Richard Rhodes

VIKING

VIKING
Published by the Penguin Group
Penguin Books Canada Ltd, 10 Alcorn Avenue, Toronto Ontario M4V 3B2
Penguin Books Ltd, 27 Wrights Lane, London W8 5TZ, England
Penguin Books USA Inc., 375 Hudson Street, New York, New York 10014,
U.S.A.
Penguin Books Australia Ltd, Ringwood, Victoria, Australia
Penguin Books (NZ) Ltd, 182-190 Wairau Road, Auckland 10, New Zealand

Penguin Books Registered Offices: Harmondsworth, Middlesex, England

Printed and bound in Canada on acid-free paper ♾

Canadian Cataloguing in Publication Data

Mays, John Bentley
Emerald city

ISBN 0-670-85356-9

1. Toronto (Ont.) - Description and travel.
2. Toronto (Ont.) - Social life and customs.
3. Architecture - Ontario - Toronto. I. Title.

FC3097.3.M38 1994   971.3'541   C94-931421-8
F1059.5.T684M38 1994

All the texts in this collection first appeared, in shorter and considerably different form, in *The Globe and Mail*. Grateful acknowledgment is made to
Thomson Newspapers Ltd. for permission to reprint material which appeared
first in *The Globe and Mail*.

All photographs appearing in this book were taken by Richard Rhodes.

to Erin
city kid

ACKNOWLEDGMENTS

The earliest versions of all these pieces appeared as "Citysites" columns in *The Globe and Mail,* and I must acknowledge the guidance of my senior editors Katherine Ashenburg and Karen York, from whom the idea for this weekly feature came; William Thorsell, editor-in-chief of *The Globe and Mail,* an unfailing source of ideas and encouragement; and the several editors on the arts desk who helped hammer my thoughts into publishable form. Little here is exactly as it was in the newspaper. For helping me in this work of rethinking and rewriting, I am indebted to Jackie Kaiser and Meg Masters, the book's editors at Penguin. Lee Davis Creal, my agent, has been an enthusiastic advocate and adviser at each step. I am also indebted to Mary Adachi for her fastidious and helpful copy editing of the manuscript.

My special gratitude goes to Richard Rhodes, who created the remarkable portfolio of photographs found in this book, and to David Olive, whose acute commentary on an early draft of the manuscript nailed many an imprecision, pointed me down avenues of thought hitherto unnoticed, and saved me from committing more than one error of fact and judgment. As for the mistakes that survived David's scrutiny, and any others, I take full responsibility.

Throughout this book, I have sought to acknowledge every source, personal and literary, in the text itself. At the end of the volume is a list of books and articles which provided facts, or informed my thinking on each topic, or opened my eyes to a site, style or place I had not noticed. There is also full bibliographical information on works cited in the text, for readers who wish to follow my paths deeper into the forest of cultural forms and meanings. Absent from the notes, however, are references to a handful of indispensible reference works always at my elbow during the writing of this book, and constant resources of fact and guidance: Patricia McHugh, *Toronto Architecture: A City Guide,* 2nd edition, Toronto: McClelland & Stewart, 1989; Eric Arthur, *Toronto No Mean City,* 3rd edition (revised by Stephen A. Otto), Toronto: University of Toronto Press, 1986; Robert Bothwell, *A Short History of Ontario,* Edmonton: Hurtig, 1986; and Alan Gowans, *Styles and Types of North American Architecture: Social Function and Cultural Expression,* New York: HarperCollins, 1992. If there is one group of writers whose works I have plundered while grievously underacknowledging throughout, they are the city reporters of *The Globe and Mail,* whose stories, throughout our 150 years of publication, add up to an extraordinary archive on urban occurrence and process. The useful phrase "Depression Modern" has been borrowed from the title of Martin Greif's *Depression Modern: The Thirties Style in America* (1988).

There are also a number of people whose names appear never or rarely in the text and notes, but whose contribution of collegiality, friendship or personal example should not go unrecognized. Thus, my special gratitude goes to Anne Collins, Richard Handler, Antanas and Snaige Sileika, Adele Freedman, Anita Fonda, Gianni Vattimo, Anne Gibson and Ken Winters, Adrienne Fonda and Jean-Phillipe Finkelstein, Larry Richards, Robert Harbison, Robert Everett-Green, Anne and Robert McPherson, Stephen Godfrey, Marie Day and

Murray Laufer, and the Venerable the Archdeacon of Trafalgar.

Thanks, also, to Laura and Antonio Bechelloni, Michael Valpy, David Sobel and Susan Meurer, Diana Birchall, Andrew Lipchak, Jack Diamond, Philip Johnson, Witold Rybczynski, Dennis Reid, Miriam Pretty, Ivar Kalmar, Cheryl Rief, Roald Nasgaard, Scott Jones, Diane Burchmore, Jane Walker, the Reverend Jane Watanabe, Beth Potter, Allen Moore, Robert Hollands, the Reverend Philip Hobson, the Reverend Brian Freeland, Eberhard H. Zeidler, Michael McMahon, Terry Fenton, Dr. John D. M. Griffin and Dr. Cyril Greenland, Randy Sorensen, Peter A. Gabor, Richard Stromberg and other researchers at The Toronto Historical Board, the staffs of the architecture collection and the John P. Robarts Library of the University of Toronto, the Metropolitan Toronto Reference Library and the library of *The Globe and Mail,* the University of Toronto Bookroom, Pages, Ballenford Books and the Bob Miller Bookroom; and the staff and bookshop of the Canadian Centre for Architecture. I am also grateful for the many critical and informative letters and phonecalls from readers of the original articles.

Two people, however, deserve my gratitude more than any others, because of their unfailing love and loyalty: Margaret Cannon, my wife; and Erin Anne Bentley Mays, my daughter, to whom this book is dedicated.

John Bentley Mays
Toronto
September, 1994

CONTENTS

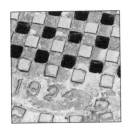

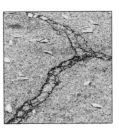

emerald
city

INTRODUCTION

Finding the Emerald City

I arrived in Toronto, by bus, on a blazing day in August, 1969. Though I had been living only twenty miles across Lake Ontario, in Rochester, New York, I imagined Toronto to be a quaint old fishing village with a large and famous university settled improbably in the middle of it. This misperception—not less absurd, because unquestioned—is not unusual among Americans, who tend to have odd ideas about Canada in any case. Upon stepping out of the bus station on Elizabeth Street, I was astonished to see, almost at once, the dark towers of the Toronto-Dominion Centre rising into the white-hot summer sky.

So I went to my nearly bare flat in the north-Toronto neighbourhood of Deer Park, spent one night there, and set out the next day on the first of many pacings-off of the city, just to find out what sort of place I had come to.

My route that first day out took me on a wander southward down the broad declining slope of the lake's ancient, higher shoreline, towards the city and the Toronto-Dominion towers, on the horizon. But instead of heading directly downtown, I drifted south-westward, through the old village of Yorkville—thick with hippies in lovebeads and tie-dyes, and with suburbanites in town to shop Yorkville's boutique circuit—then across the Gothic campus of the University of Toronto, and thence into the narrow streets west of Spadina Avenue, among the gamy-smelling, thronged vegetable stands of Kensington Market.

I ended that first day in Toronto on the hot, dusty western-waterfront grounds of Exhibition Place, where the 1969 edition of the Canadian National Exhibition had just opened.

I had loved state fairs since childhood—and had not yet learned of Toronto's peculiar anxiety about *this* fair—so it seemed like the most natural thing on earth, to take those spectacular rides churning against the sky, and plunge into the sweaty human surge teeming up and down the Midway. My best memories of that day are of the dense mix of delighted screams, rock and funk, barkers' barks and pig-squeals, bingo calls and laughter from the canvas-top beer parlours, the pneumatic hiss and gearbox clatter, all hanging like a bright, deliriously toxic cloud over Exhibition Place.

Only years afterwards, I found that here, on this site of rides, spectacle and façades, of fanciful pleasures and of architectural memories—dreams of Crystal Palace and Brighton Pavilion—*here,* and not in the city's more obvious centre, in the skyscrapered financial district around King and Bay, Toronto had begun.

The only reminder of that foundation visible today is an inconspicuous monument standing on the south-west corner of Exhibition Place, where the petting zoo and kiddie rides are installed during Ex time. Near that plinth of stone, in the winter of 1750, French carpenters and masons under the Marquis de la Jonquière built a small wooden trading post, which was called Fort Rouillé by its builders, in honour of the current colonial minister in Paris.

The population of Jonquière's little settlement, in its decade of life—the fort was burnt by the hopeless French after the British defeat of Fort Niagara in 1759—never numbered

Ferris Wheel, Conklin Midway

more than about fifteen souls, mostly soldiers and kitchen help. It appears that no Christian priest ever went there. But though small and short-lived, the memory of the little palisaded camp lingered in Toronto's folk imagination and fireside tales for more than a century. And it seems to have cast a certain spell on the ground that has never quite gone away.

Though ideal for both, the grounds around what Torontonians called "the old French fort" would never be used for farming or habitation. Rather, the place would remain forever martial and jovial—a zone of tentative dwelling, offering only temporary solace, and standing as an emblem of the grand ambitions of nations and peoples, like any military post on far frontiers. So it was after 1787, when the British bought the place from the Mississauga Indians, having decided to make it part of the military compound of Fort York. And so it remained after 1878, when the Victorian garrison turned the vacant site over to the Canadian National Exhibition, as its permanent home.

Industrial and agricultural fairs, festivals of modernity that sprang into existence in Europe and North America with the coming of industrialization, had been held in Ontario as early as 1820, but only on an itinerant basis. By the 1870s, however, the Ontario railroads had created the arteries necessary for the rapid pulsing of people into Toronto and out of it, and the city's growth created the centre of gravity for the province's accelerating economy. The Exhibition acquired its first permanent home on the site of the old French fort in 1878, and presented its first edition the following year.

From the beginning, the Exhibition, like the burnt fort that preceded it, was a representation of a distant empire—not a political realm, but the metaphysical imperium of modernity, radiant with promise of abundance and wonders. The thousands drawn from their Ontario farms and small towns by the magnetism of those early fairs came not only for the show, but for a taste of this coming apocalypse of modernity.

And they entered through an appropriate monument, the Princes' Gates, a tribute to the victory of industrial democracy, and the imperial British loyalties of the common man.

It was at dusk that day, standing alone in the Princes' Gates, looking east along the lakeshore towards the Modernist tall buildings at the city centre's present heart that I got a first hint of Toronto's peculiar conflict over its own urbanity, its struggle with modernity. To this day, on the mental map of the city I carry in my head, the Gates is the dividing line between these two options for a city style Toronto cannot choose between: small-town British Victorianism, with cliquish fidelity to class, the old school tie, the anachronism summed up by the persistence of an agricultural fair in a metropolis of millions; and the American, hard-driven high Modernism symbolized by the downtown towers, with all they express of the yearning of at least some Torontonians to escape the parochialisms of the past, and enter full secularity.

This conflict, between the weight of tradition and the weightlessness of Modernity, is a theme in this book, probably because it is the central theme in my life. I was born into a Southern American family with deep roots in the land and the traditions of living on the land; yet from the age I knew anything, I knew I would be cutting those roots, little by little, and breaking loose from the South and whatever destinies it held for me. It was not possible for me to know then that the project lying before me was Modernity itself, unfinishable, in ways incomprehensible, at times unendurable. But even in the darkest hours of loneliness, I have never imagined the initiations, the understandings, of the Modern condition to be anything other than the only intellectual and moral tasks ultimately worth embarking on, and carrying through to the end.

Which brings me to the title of this book, and its source— not in the book, by the way, but in the movie version of Frank Baum's *Wizard of Oz*. As a child, during the war years, I saw it again and again, while never then (nor now) understanding it. The whole film is Dorothy's quest to get back to Kansas. But since Dorothy had the supreme good fortune to be picked up by a tornado from a drab black-and-white farm and set down in a prismatic country of endless delights, adventures and surprises, what could she possibly find to do, once back on the farm? Why would *anybody* want to go back to Kansas after catching a glimpse of the delightful and busy Emerald City?

The best thing to do was, or should be, (I believed from childhood onward) to keep riding the tornado for as long as possible, in hopes of being dropped into some desirable Oz. So it was that I came to Toronto in 1969, to teach for three years at York University and then, so I thought, to catch the next tornado out of town. It took me some time to figure out that this was to be no short stop, but a home-coming, and the discovery of the Emerald City I had been looking for since I was high enough to look over the cotton on my father's plantation.

This is a reporter's book, and an account of coming to Modernity, Toronto's and my own. It is not history. Though it is about Toronto places, the book will, I hope, be used as a guide to discovering and thinking about any place, in any city. It is hence for any stranger in metropolis, including those born in one; and especially for anyone who has ridden, as I have, what E. M. Cioran called civilization's whirlwind from agriculture to paradox.

Thinking Places

July, St. Clarens Avenue

THE PORT INDUSTRIAL DISTRICT

Wherever I've lived, I have always had a thinking place.
The first was a shaded gulley, lined with blackberry
bushes and wild roses, water-gouged into the red Louisiana
dirt at the edge of a cotton field near the house in which I
spent my earliest years.

More recent thinking places have included a sunny hillside
in North Carolina; the dapple-shadowed dirt under a bridge
over the Genesee River in Rochester; a cold room, open to the
sky, high in the stone stump of an ancient Irish church's ruined
belltower.

So it was that, not long after coming to Toronto some
twenty-five years ago, I soon found myself a thinking place.
It's in a desolate, declining downtown zone known as the Port
Industrial District, just east of Toronto's inner harbour, and
within clear sight of the skyscrapers at the city's centre of gravity. It is bounded on the north by the cement colonnade bearing up the Gardiner Expressway, on the south by Lake Ontario
beach park frontage—now deprived of its once-panoramic
view of the lake by the long artificial finger of the Leslie Street
Spit—and, to the east, by a sewage-treatment plant.

Most of the district today is a low, dead-flat version of
what's now called an industrial park, traversed by straight
broad avenues—laid out to serve much grander buildings than

the ones that actually got built—and pierced by murky rectangular tanker slips. Apart from the port facilities, little rises far above the district's level damp floor of landfill, except huge mounds of road salt shrouded in black plastic sheeting, tall cylindrical tanks built for chemicals and heavy oils, and the poetically boxy, angular hulk of Ontario Hydro's Richard L. Hearn thermal generating plant. Most of the off-street parts of the district are paved by oily, rubble-strewn earth or dotted with forgettable buildings, survivors from the final years of Toronto's Industrial Age.

In 1969, cargo tonnage at the Port of Toronto attained a record level of 6.3 million tonnes, and has been declining ever since. What's left is a strange landscape, the home of tanks and concrete emplacements and ageing factories, cranes and chimneys and silos; and a temporary stop for Great Lake tankers, lying still and low in their slips. Only one strip of the District, running parallel to the lakeshore on a long ridge, the last memory of an ancient sandbar, is still wild, thick with poplar, sycamore and willow and with dense, prickly undergrowth, and busy with marsh and shore birds.

The district is not appealing aesthetically, nor should it be. Urban thinking places shouldn't be tourist attractions, lest the sightseers make you forget the reason for being there. Or, much worse, lest some earnest, environmentally and historically conscious government—municipal, provincial, federal—decides to turn it into a park. Which decision, of course, would bring in its train the baleful works of "improvement": paved walkways, good lighting throughout, well-maintained toilets, and lots of finger-wagging laws about proper human and pet behaviour. As matters stand now, it's not park; and within this wilderness of trees and rust are many places to be alone and think, into which one can disappear completely from the crush and rattle of urban existence into a shady nook and consider life, fate, some specific decision or the world—all within a ten-minute drive of the hectic middle of Toronto's financial district. If not picturesque, this industrial district is graced with a subtle melancholy that is peculiarly right for thinking things through.

To get a sense of what I'm talking about at its most enchanting, you need only cycle or drive down there late one afternoon, just before night has begun descending on the incandescent city.

Find your thinking spot—a sun-warmed concrete slab beside the inner harbour's Eastern Gap, or a log alongside a damp footpath snaking through the marshy dense foliage near the shore, or some crumbling concrete wall near a battery of oil tanks—and watch the ghost-mist from the vanished marsh upon which the district was built rise over its ramshackle buildings, and the pools of oily water gleam with the first cold light of the rising moon.

At such a lonesome moment, the mind may drift towards the things of which ruins have always made people mindful: mortality and the brevity of life, the decay of what Dame Rose Macaulay, in her book *Pleasure of Ruins,* calls "the stupendous past." In the presence of the rusting bones of Toronto's once-vivid port, one naturally inclines to think about the futility of worldly ambition—individual ambition, surely, but also that of the civilization to which we've fallen heir.

Tumbledown castles, vine-enshrouded monasteries made eerie by screech-owl and bat, fragments of once-mighty temples raised to gods no more adored, have long occasioned reveries of this sort. Rusting petroleum tanks, little boarded-up cinder-block boxes slowly tiring and crumbling, junked trucks littered here and there like fallen, ancient animals in the evening mist—these, too, speak to us of the world and ourselves at dusk. The ties with a powerful element in Toronto's past, as incarnate in its artifacts, has been cut; loading cranes have been left to go still and die, concrete grain elevators demolished, once-magnificent factories swept away. The connection of such built objects with more traditional ruins, I believe, is in the poignance we feel whenever witnessing a great dream—in the case of the Port Industrial District, the tough, ambitious idealism of heavy-mechanical capitalism—slowly falling victim to the impersonal forces of time and obsolescence that have overcome all cultures of the past, and will overcome every person in the end. "A monument of antiquity is never seen with indifference"—so wrote Thomas Whatley in 1770. "No circumstance so forcibly marks the desolation of a spot once inhabited, as the prevalence of Nature over it."

However delicious such reflection may be, it has the side-effect of dulling one's curiosity about the hard history and real archaeology of a site. It's the uneasy sense of that anaesthesia that prompted Henry James to call his own seeking after ruins "a heartless pastime; and the pleasure, I confess, shows a note of perversity."

Had I not been assigned a weekly *Globe and Mail* column about city spots, I'd probably be still indulging in the pleasures of my favourite ruins and wastelands, and still as ignorant about them as ever. But among writing's many gifts to the writer is a dissatisfaction with impression and reverie, the commonest of temptations leading to horrible journalism. You don't have to spend much time typing away, before the hankering to get past the Jamesian "perversity" sets in, and you want the facts of the site. I was fortunate to find, early in my descent into the obscure history of my thinking place, the illuminating work of Toronto architect Jeffrey Stinson, entitled *The Heritage of the Port Industrial District.*

Like a good art critic, in Baudelaire's famous definition, Stinson is "passionate, partial, political." Whether his topic is a tank farm and hydro pylon, or a popstand on Cherry Beach, or the long row of trees lining Unwin Avenue—the paved road just north of the lonely, thicketed ridge where the best thinking places are—Stinson writes with pleasure, attending to details dilettantes easily overlook. As we might expect from a student of the recent proletarian past, Stinson cautions against the unthinking knock-down of old industrial buildings, and rises to indignation at the snobbery embodied in the tired but still-entrenched Victorian Brick school of heritage, with its fixation upon the noble residences of our dead betters, and its aversion to the industrial operations our ancestors built and worked in. Only yesterday, a right-thinking urbanist like Stinson would have shunned those heavy-industrial belchers of smut and leakers of poisons in our midst. Their few fans (including me) had to go to Buffalo in order to get the whiff and feel of naked industrial might, and the human culture generated by it. But as Toronto companies have vanished or fled to cheaper climes, leaving their dilapidated homes behind, this non-nostalgic sort of urban scholar has arisen, concerned with local building's immediate past, and devoted, as earlier heritage folk were not, to remembering the great industrial age *here,* as it slips away into long twilight.

If nostalgia obliterates history, or at least dulls the desire for it, Stinson's pleasure is in the precise and definite. As we learn from his report, done in 1990 for the Toronto Harbour Commissioners, the first Europeans to take up residence on the original town site of Toronto looked out on a large near-lagoon, separated from Lake Ontario by the Toronto islands archipelago, and, just east of this calm sheet of water, on to a spacious marsh lying behind a long sandbar. The source of this wetland was the Don River, which gathered its waters from the high forested plateau and lower lakeshore terrain north of the lake, and carried them south, ever more broadly and slowly, finally emptying them into the great marsh.

By Victorian times, after the city had put down lasting roots in the stiff glacial clay by the lake, Torontonians had taken to escaping the summer's heat at cottages on Fisherman's Island, as the sandbar dividing marsh from lake came to be called. There, for some years anyway, they could enjoy respite from Toronto's blistering late-summer heat, the delight of the long sandy beach, the stalking of the marsh's innumerable waterfowl.

Left Turn, Commissioner's Road

But even before the turn of this century, the wastes generated by Toronto's burgeoning industries and population had already taken a near-mortal toll on the Don River system as a whole, and the marsh in particular. The once-luxuriant fen had thickened into pestilential sludge. So it was that the Toronto Harbour Commissioners was summoned into existence by growing protests from the public against this festering stew—as well as by calls from industrialists for new land near downtown, and better marine facilities than those afforded by the inner harbour up to that time. In 1911, the commissioners were handed 1,385 acres of polluted marshland, and the daunting job of cleaning it up.

Which they promptly set about doing. By the end of 1912, city council had approved the commissioners' plans and by 1914, the work had begun: draining and filling the marsh with sludge, sand and dirt; engraving deep, broad shipping canals; constructing docks and wide paved roads, storage and docking amenities, connections to railways tracks. At the same time the renovation proceeded, a strip of public beaches and parks and bridle paths a thousand feet wide was being developed along the four miles of District lakefront, bringing into existence North America's first harbourfront reconstruction to take into account a city's need for both profit and fun.

Stinson's style of archaeology is notable for its sensitivity to the varied uses people have made of this place. Activities the author catalogues range from the mass-production of shells from scrap metal by an army of workers during the First World War to weekend cycling, and the serious back-seat sex that's still an important use of the dead-end roads of the District year round, and the production of a modern industrial port culture, joining humans and machines into a vivid whole.

Though a preservationist at heart, Stinson is also level-headed. He does not want the District frozen in place, turned into a theme park of Toronto Past, or an open-air museum of industrial archaeology. At the same time, he warns against any plan that does not mandate an adequate time-period for considering the possibilities of reuse offered by sturdy, workable abandoned buildings. Stinson does not hide his anxieties about developers who have exquisite, immaculate, computer-generated visions of a "perfect, money-making future" that would simply displace the untidy industrial zone with soaring condominiums and manicured parks.

Barring an unending depression of real-estate values in Toronto, it's probably only a matter of time before the suburban-biased Metro politicians and the land developers do indeed strike a deal to arrest the decay, purify the soil and put up bright new office blocks and condominiums and public housing on this site, so temptingly near downtown, and so blessed with uninterrupted, breathtaking views of the skyscrapers.

When I think of that, the perverse Jamesian resident in my ruin-seeking soul returns. I think of the loss of my refuge—which is a need, not just a diversion—and of the sites of thought to be lost by my anonymous compatriots, the many other lonely twilight walkers I have sighted on my visits to the Port Industrial District. Gone will be a certain free balm for the soul's inner noise, now so easily and quickly available on those broad avenues, where pools of waste water so beautifully catch and set shimmering the glow of our skyscrapers in their oily mirrors.

THE JOHN INGLIS PLANT

In November, 1989, the last washing machine rattled off the assembly line at the 108-year-old John Inglis plant on Toronto's Strachan Avenue, just north of Exhibition Place, just south of King Street West.

The 650 workers who still had jobs at the end of the factory's decades of decline exchanged phone numbers and farewells over coffee, then punched out for the last time. And for the first time in more than a hundred years, the sprawling nine-hectare site suddenly became what it has been ever since: a dead zone of weedy gaps and unpeopled buildings, some vast and imposing, others mean, and yet another space of gloomy quiet in the once-mighty manufacturing district west of the downtown skyscrapers, along the lakefront railway tracks.

Only one mechanical object continues to animate the eerie stillness of the site: a high electrified sign adorned with the Inglis logo, and treating both inbound and outbound commuters on the Gardiner Expressway to a robotically flashing message of moral uplift. (Sample exhortation—it changes every day—: "Speak ill of none, but speak all the good you know of everyone.") A connoisseur of melancholy urban sites, willing to take an afternoon to kick around Inglis's rubble-strewn parking lot, will find much worth savouring in that scatter of structures. The red-brick Victorian Romanesque of an old jail incorporated long ago into the complex, for example; and the main plant's titanic modern pavilion, its horizontal walls made of glass framed in steel and reinforced concrete.

A forlorn majesty suffuses the site, confirming—as only once-great abandoned factories can—the truth of *sic transit gloria mundi.* Not even the tall sign with its moralizing electric message breaks the spell. It is a poignant reminder of the theory, dear to the hearts of Victorian colonial industrialists of the sort who founded Inglis, that an allotted burden of the rich was to raise the Common Folk's ethical standards and self-respect, by example and by education.

Yet along with eliciting regret for well-nigh vanished capitalist humanitarianism, the Inglis sign reminds us of the people who were the objects of this concern, the workers who toiled at this and the other industrial emplacements once so common in Toronto's downtown. Inglis produced indelible memories, sad and happy, it's worth recalling, as well as washing machines and such. It was a site of desire and narratives, the secret histories so easily dispersed when plants shut down, and of which the extant buildings, for all their impressive emptiness and wistful grandeur, never whisper a word.

"No matter how many new factories are built, no matter how many new jobs are found, the culture in the plant cannot be re-created," Toronto labour historian David Sobel and researcher Susan Meurer have written. "Workers' culture is the mortar between the bricks, the stuff that holds workers together, the salve that makes the hardness and roughness of daily work in a factory bearable."

The quote comes from what its authors call an "intelligent scrapbook"—a compilation called *Working at Inglis,* and underwritten by United Steelworkers of America local 2900, intended to preserve in words and pictures what traces remain of workers' culture at Inglis. A specialist in applied technology and communications, Sobel says that the project gave him the unexpected opportunity "to understand what a closure is all about." The project is partly an act of remembering, partly resistance to the forcible destruction of the past strongly linked to closings. As an automotive plant manager told Sobel: "Historical information about employees is non-value-added information, so we destroy it." Burrowing through the heaps of old ads and in-house publications and photos of Inglis workers toiling, boozing, dancing and striking which have survived loss and active suppression, one discovers the richness of that experiential tapestry woven at Inglis over the course of a century. The loom was the physical equipment itself, on which the working culture was formed. Inglis was always a zone of continuously changing technologies and ceaseless transformations of metal into myriad useful objects, from the day it opened at its present location in 1881. Early on, it made the engines and pumps for Toronto's 1883 massive expansion of water distribution; later it produced factory turbines and home water heaters, shells and howitzers, and, during the Second World War, 186,000 gas-operated Bren guns.

And forever easing the human hardships and hard work at Inglis was the salve of working-class culture, the chief interest of Sobel and Meurer. As many as twenty thousand people toiled in the plant during the era of Bren gun manufacture, but also fell in and out of love there, became enduring friends on the assembly line, and, after work, went over to the Palace Tavern on King Street to plan strikes and company shindigs, to cry the blues, to tell stories or just have some fun. Many of these workers were single young women, drawn from Canada's Depression-crushed towns to Toronto's war-inflated industrial operations. Many soon found their way to the Inglis Girls' Recreation Club, where friendship circles, "socials," cabaret performances, dancing lessons and other pleasures provided an antidote to the loneliness of city living and the monotony of the assembly line.

Now the charm of such memories is perilous, simply because nothing can be more effortlessly sentimentalized than the cultural life of a vanishing work force—especially when its artificiality and ideological function are forgotten or conveniently ignored. A bit reluctantly—both are leftist romantics at heart, even though they know better—Sobel and Meurer do remind us that the sedatives for homesickness provided by the long-vanished Recreation Club helped "reduce absenteeism and prevent (out-of-town immigrant) workers from returning...home," and kept the workers' energies focused on the industrial process.

Such strategic "culture" is something many urban wage-earners nowadays may have a hard time recognizing. Instead of taking anodynes provided by the plant, the post-industrial workforce provides its own, in our choices of distraction, movies and television shows, travel destinations, home-improvement schemes, consumer products. As surely as the Recreation Club kept the newly urbanized Inglis girls in line, so do expensive entertainments keep the informational proletariat of the late twentieth century—*us,* that is—mobilized, persuaded, and properly disposed towards management. All moments in the history of industrialization have had characteristic methods of shaping a reliable work-force—methods which, like Inglis's working culture, seem especially liable to be forgotten and disregarded when the process requiring them shuts down or is superseded.

Much of the value in a separate project, based on a heap of brief silent films Sobel unearthed while rummaging through the National Archives of Canada in Ottawa, lies in its unadorned recollection of this ideological grease, and how it has been marketed and applied. Though documentary evidence about these pictures is scarce, the basic information about their intent and target audience can be deduced from the seven shorts compiled by Sobel's Labour History Images Group in a video package called *The Moving Past*. All were made by the government of Ontario's now-disbanded Motion Picture Bureau, between the First World War and the Crash— a period notable for the remarkably rapid influx of rural folk into Toronto's industrialized workforce, and the intensification of mining activity in Ontario's north. *Her Own Fault* (1922), about the deportment of young female employees at Toronto's Gutta Percha Rubber Co., is aimed at young women new to both city life and the mechanized workplace. *Life In Mining Camp* (1921), on the other hand, is a direct appeal to out-of-work Toronto labourers—veterans returned from the war in Europe to recession and hardship—to forsake the big city and head north, where the pay is good and so's the grub. *A Story of Stone* (1924) is a lofty hymn to the nobility of stonework, from dusty quarry to sculpted quoin, from the building of Solomon's Temple to that of a neo-Gothic government office tower at Queen's Park; and seems to have been made to lure older school-kids into construction and associated trades.

But however various the audiences they were shot and edited for, however different the cinematic styling of each, all seven films gathered in *The Moving Past* nourish an attitude of compliance and conformity, satisfaction with one's place in the swiftly changing army of industrial producers, and admiration for assembly line and the rationalization of work.

The agenda is easiest to see (and comes across funniest) in *Her Own Fault*. This little melodrama about two newcomers to industry shows the rise of delightful Eileen, who leaps out of bed early, does calisthenics at morning "rest period," is not ashamed to wear glasses as she mass-produces shoe heels, and enjoys embroidering and chatting with the other girls during lunch break. She is humble, industrious, contented, and she spends her evenings in healthy recreation, such as canoeing with her friends.

Eileen's demonic opposite is Mamie, who gobbles her food, wears spike heels and is too vain to wear glasses (even though she needs them). Because she's "too nervous to rest" in the evening, Mamie flings herself each night into orgies of dance-hall shimmying. In the end, as the caption tells us, "both girls get what was coming to them": Eileen, a promotion and, it's hinted, a beau; and dissolute Mamie, tuberculosis and a pink slip.

The other films are more subtle, and some are beautifully crafted. Both *Silver Mining in Ontario* (1919) and *Making of an 8" Explosive Howitzer Shell* (1918) present heroic, pulse-quickening portraits of complex production. The brilliant film about the Howitzer shell, by the way, is less documentary than avant-garde rhapsody, a jagged song to precision military technology and—the true object of worship—the robotized workplace.

Sobel's works of recollection address (though they occasionally become complicit with) the tendency of contemporary academics and some white-collar labourers to romanticize Inglis beauty pageants, the passion of its union's strikes and other aspects of disappearing working-class culture. They do not encourage as much critical attention as they should to the facts and ideological lubricants of post-industrial work-life.

At their best, however, these projects remind us of the odd way historic methods of mind-moulding slide into oblivion, and they rebuke a trend among urbanists and the architecturally interested public towards a purely formal appreciation of former plants.

They are Parthenons, our Colosseums, the architectonic climaxes of our industrial civilization's physical mark upon the world, and, as such, sites saturated in thought and memory. While Ruskin would doubtless be horrified at the use to which I'm putting his words, factories finely illustrate the dictum, pronounced in his essay on memory, the sixth of *The Seven Lamps of Architecture:* "There are but two strong conquerors of the forgetfulness of men, Poetry and Architecture; and the latter in some sort includes the former, and is mightier in its reality." So while not opposed in principle to swooning at the rusty front gate of a dilapidated vacuum-cleaner plant, or admiring in timeless aesthetic bliss a harbourside grain elevator whose day is over, I agree with Prince Charles—who would, like Ruskin, be scandalized to know he'd been put among the factory fans—that "when a man loses contact with the past he loses his soul. Likewise, if we deny the architectural past—and the lessons to be learnt from our ancestors—then our buildings also lose *their* souls."

The sort of research being conducted among former Inglis employees can provide a useful corrective—or addition—to the formalist forgetfulness afflicting current industrial preservationism. The ideal candidates for this salvation are retired work-environments which appear to have been sanitized by a neutron bomb—every brick and mullion and spandrel intact and subject to inspection and criticism, but devoid of memory, of *thinking,* and of every trace of the cultural fabric people spin around machines in the architecture of production.

Of course, a dead factory is very often a thing of remarkable beauty. Gazing across an immense, silent shop-floor cleared of machines and noise, lit only by afternoon sunshine slanting through curtain walls of dusty glass, anyone with eyes to see would be moved by the clear flow and austere elegance of modern manufacturing space. Yet in all such appreciation for the purely architectural and engineered, there lurks a spirit of pastoralism, of reluctance—especially stylish among architectural connoisseurs at the present moment—to accept the ceaseless ripping down, tearing up and rearranging things that every great and living city is very much about.

Surely, we need it all. An appreciation for the material craft which has given us the industrial infrastructure of modern life, and artifacts to support this appreciation. And also the awareness that abandoned factories, in addition to their often magisterial and tragic beauty, were once sites in which men and women, managers and wage-slaves, bosses and minions, together created durably interesting narratives of conflict, technique, human agency and human cooperation.

CANADA MALTING

"Our eyes are constructed to enable us to see forms in light," wrote Le Corbusier in his famous 1923 tract *Towards a New Architecture,* then decreed: "Primary forms are beautiful forms because they can be clearly appreciated."

To illustrate his point, he littered his pages with photographs of Canadian and American grain elevators, and thus touched off a fascination with North American silos and such—the functional spin-offs of industrial process—not dead among urban visionaries to this day. The buildings of Le Corbusier and other pioneers of heroic Modernism "were explicitly adapted from these sources," architect Robert Venturi has written, "largely for their symbolic content, because industrial structures represented, for European architects, the brave new world of science and technology," while buildings referring to the masterpieces of the past did not.

Heavy production facilities still cast a spell, though less as mentors than as victims. Perhaps it is in our nature to want to spare victims, including architectural ones. In any case, the affection for North America's decrepit, abandoned and doomed industrialization is today as fervent as ever, at least among the watchers, guardians and students of built form.

Toronto's port facilities, once the pride of our inner harbour, provide an excellent example of the dynamics of this desire.

With the decline in Great Lakes shipping from the end of the Second World War into the 1960s, the grimy industrial emplacements on Toronto's inner harbour were doomed. Recognizing this, in 1972 the federal government, hand-in-glove (as usual) with developers on to a good thing, stepped in to "save" our waterfront—i.e., knock down almost everything in sight, and sponsor the "renewal" of the former industrial zone as an upmarket condominium market and recreational and "cultural" fantasy-land.

Whereupon the second sort of concern with the older architecture kicked in, inspiring urban archaeologists, architectural historians and the heritage people to join forces in efforts to spare examples of vanishing port industrial culture. They saved very few reminders of Toronto's historic orientation towards the Great Lakes marine highway; so today, much of the passion is focused on the Canada Malting complex on Bathurst Quay, one of the only defunct industrial plants on the inner harbour to have so far escaped the wrecker's ball.

The harbourside complex is not old. The silos were put up in 1928 for the storage of barley, hauled in on Great Lakes ships, for transformation into malt. In 1944, the plant capacity was doubled, to meet increased demands for booze and other malt products, and buildings continued to go up until 1961. But even though Canada Malting is no antique, the cracks and pocks and water stains on the grey concrete cylindrical shells, the ignored fine brick cornices on the germinating house, the unchecked shrouding by vines of a *Moderne* office building in the shadow of the tall silos, even the burned-out boat slowly rotting on the grounds, all make this ruin an affecting artifact, and help explain the outlay of time that has gone towards saving it.

In early 1988, only weeks after the factory's November, 1987, shutdown, Toronto City Council struck a committee representing a number of keen public and private interests, to look at possible futures for Canada Malting. In September, 1992, some forty architects, artists, urban planners and historians gathered in Toronto to ponder the future of the former storage and processing facility, then the property of the federal government, and since deeded to the City of Toronto.

In the late summer of 1993, city planners issued their most specific call to date for plans to "restore, reuse and maintain" the complex. By that autumn, a number of proposals had come in, most obviously unfeasible, a couple—one, to turn the silos into recording studios, classrooms and other facilities for musical production and study; another, to make a kind of *son-et-lumiere* out of it all—were felt by city officials to have promise. Everyone is united in a belief that the silos should be saved, that is; some people even have ideas that might work. Nobody, to my knowledge, has come up with anything that would save the taxpayers the expense of just forgetting about Canada Malting altogether. Demolition is estimated to be a $3 million job, according to planners, and would sweep away one of Toronto's best examples of an original North American building type. The other would be mothballing, which would cost about $1 million for starters, consume some $38,000 a year in maintenance, and give the taxpayers nothing in return, except an old artifact, continually in need of face-lifts and cosmetic touch-ups, to look at.

While, in principle, I favour the recycling option, I am not optimistic that an economical reuse for these buildings will ever be possible.

Now, before anybody pops a copy of the Toronto Historical Board's *Silos Can Be Re-Used!* into an envelope and posts it to me, let me say I've got a copy already, and find both the THB's two examples of redeployment wanting.

Exhibit A is the Quaker Square Hilton in Akron, Ohio, the handiwork of Curtis & Rasmussen Architects. Finished in 1980, this project involved the conversion of thirty-six former oats silos and related buildings into a hotel and commercial and retail mall, three restaurants and four nightclubs, all under the same roof. Unlike any such fun paradise that might get built on Bathurst Quay, however, the Hilton is downtown, not on a windswept jut of landfill, separated by an expressway from Toronto's bright lights—though, to be sure, the advantages of a downtown location in Akron is offset by surroundings so depressed and depressing that few Hilton visitors venture outside its maze of shops and clubs and rooms. Sources both inside and outside Akron's hotel business have described the Hilton to me as an island resort in a sea of nothingness. But what tourist or business traveller would ever want to stay on isolated Bathurst Quay, with myriad hotels in the very midst of downtown Toronto's shops and cultural attractions?

In any case, the mid-1990s is not the time to raise the hotel option. According to John Hamilton, spokesman for the Metropolitan Toronto Convention and Visitors Association, "there is no room in Toronto for new hotels." The city now has 15,414 rooms downtown, and 32,030 Metro-wide, with a steadily declining average occupancy rate over the past three years—a glut, says Hamilton, caused by massive overbuilding in the 1980s.

Exhibit B is La Fabrica, the Barcelona home and office of architect Ricardo Bofill, his "freethinking lay convent, dedicated to work." The old cement factory on which Bofill and his colleagues performed their genuinely delightful magic, between 1973 and 1975, was a disused group of short silos and extensive underground passages and surface structures, set in a dry, rugged Mediterranean landscape begging for the oasislike gardens eventually planted there by the architect and his associates. Now, a highly inventive and extremely rich architectural partnership might be able to turn Canada Malting into a faintly similar fantasia—though the cold, flat lakeshore setting, the dank silos and oily waters of the lake, will never add up to Catalonia.

The city's 1993 call reminded the interested citizens of what could be put on the Canada Malting site under then-current zoning regulations, and urged them to think "community centre," "public playground," "elementary school," "day nursery," and other programming of a "cultural" or "artistic" nature—everything, that is, except a much-needed infusion of heavy, imaginative private capital. While artists in need of cheap studios and parents in need of day-care facilities will likely find these prospects exciting, the city's lack of interest in attracting serious investment to Bathurst Quay should alert Toronto taxpayers to the absurdity of encouraging idealistic folk to make outlandish proposals to a municipality already strapped for cash, for a general area already much damaged by piecemeal planning and short-sighted development.

At the time of this writing, no decision has been taken. But since large private investment is not likely to be forthcoming, I cast my vote with great reluctance for a possibly viable scheme that surfaced briefly during the 1992 think-session.

It would involve the takeover of the silos by the Archives of Ontario for use as the central storehouse for about 200,000 cubic feet of documents and historic materials now scattered among warehouses throughout Metro. The cost of converting plant to archive has been variously estimated by federal and provincial offices at between $80 million and $100 million, while a new building on some other site would come in at between $40 million and $80 million, plus land and site preparation costs. Whatever the exact figure might turn out to be, retrofitting would probably be more expensive at the end of the day—and, given Ontario's dim financial prospects, a go-ahead for a huge capital building project of this sort should not be given without the gravest consideration. At the present time, that consideration would almost certainly result in the nixing of the project.

But, were this interesting pipe dream to come true, at least a good use would have been found for the factory's thick, solid walls and the silos' adjacent office buildings. Architectural preservationists would get the industrial fabric intact, without windows punched in it and with few other rude external changes. The province would have its long-needed central storage facility. A forgotten place would again become a place that thinks, that recollects and provides room for recollection. And all of us would see the humane resolution of yet another bit of business left dangling after twenty years of our waterfront's thoughtless devastation and transformation by governments and private interests.

On the Land

Holiday Inn, Toronto West

THE SHAPE OF THE CITY

A couple of autumns ago, after Toronto caught its first snowfall of the season, I rang up Environment Canada's Ontario Climate Centre to find out what they could tell me about this debut.

The weatherfolk there seemed faintly perplexed by my call. After all, everybody who knows Toronto knows the annual rhythms of cold and snow, which determine so much of what we do, and how and when we do it. As climatologist Bryan Smith confirmed, the first serious snowfall had come that year around the same time as always, the middle of November. And, as usual, it was a trifle: just 63,000 litres of frozen water—which sounds like a lot of water to me, but, said Smith, was nothing to get excited over—descending on Metro's 632 square kilometres. The fall began before dawn, and had stopped nearly everywhere before noon.

Throughout the city that early morning, in Forest Hill palaces and Etobicoke ranch-styles and Scarborough bungalows, Metropolitan Toronto's millions enacted the first rites of winter. Among them: locating the car-window brush, and discovering you forgot to replace last year's broken one; and rummaging inside the hall closet for overshoes and the kids' boots, devoured by the darkness deep behind the coats on the last day of slush the spring before. And remembering how much longer than usual it will take to get to work or school.

These are not weighty matters. Winter in Toronto is not as dangerously cold and stormy as it is on the Canadian prairies, or as windy and wet as the Maritimes. It does not endure as long as winter in Montreal, nor is it, say lifelong residents, as long and severe in Toronto as it used to be. Nor do we get the very heavy accumulations of snow that regularly paralyze nearby Buffalo and Rochester. To dwellers in this city and climate, winter is merely an inevitable fact of life, like aging, to be got through as gracefully as possible.

Because it is neither especially mild nor particularly snowy, Toronto's winter is no source of joy, except to skiers. But to those of us who don't ski, and who grew up in warm climes where "snow" was a kind of toxic plastic fuzz sprayed from a can on Christmas trees, the first Toronto fall of the season brings an ineffable lift and delight, and a novelty most born-and-bred Canadians never know.

Of course, this enchantment will wear off halfway through the second snowstorm. By March, I'll be disliking the threatening skies and blowing snowflakes as much as anyone else who drives or walks around town. But throughout the first snowfall's early hours, the city seems to lie under a spell, everything transformed.

An air of great beauty—solemn, subtle, softly calm—graces the town's streets and structures. All the usual bright colours and sharp, tall shapes of the urban cityscape have been softened over night, as though wrapped in cloud. The top of the wooden fence outside the bedroom window, the angry brow of an ecclesiastical gargoyle I pass each morning, even the round lids of the wine barrels stored in my neighbour's backyard—so many dark, ordinary things we seldom notice—all seem strangely vivid, as though heightened by sweeping strokes and dense scrubbings of luminous silvery chalk.

Driving my daughter to school in this first wintry weather, I enjoy glimpses of the deep ravine that cuts south from St. Clair

Avenue West near Spadina Road. It's just a gloomy winter gulch lined with dry, sharp trees, rasping against each other in the autumn wind, until the snow comes. Then it becomes an abstract etching composed of ricocheting, criss-crossing black lines, the bare undersides of branches sketched on the greys, whites, silvers of snowy backdrops and shroudings.

Most Torontonians, I suppose, find snow a kind of lovely nuisance. If thought of at all, it's the only substance that can provide a delightful sport and, after pollution by salt-bearing humans, rot your car. I cannot say precisely when snow became more interesting to me than just that, though it seems to have been during a time of much reading about the underground art of our stone-age European ancestors. Those people who so splendidly decorated the famous caves of Spain and the south of France lived in a cold world only gradually returning to general plant and animal inhabitation after a glaciation of great expanse and devastating force. Prehistorians naturally give scant notice to the ice sheet, since nobody and no animal lived on it, or survived its crushing advance. But snow is wonderful and fearful, especially when it stops being the ephemeral thing Torontonians experience, and begins to add up and up—as it has done on this continent, every few thousand years, for a very long time—becoming a tool of almost unimaginably grand devastation and transformation. One can live happily forever in Toronto without giving mind to the glacier that once bore down upon this site. But nobody can understand the cityscape without acknowledging the vast powers which created it.

The landform over which Metro sprawls today was crossed by the Wisconsin glacier—the fourth to crush our territory in the last million years—and rendered lifeless by this advancing sheet of snow-driven ice more than 110,000 years ago. Then, about 13,000 years before the present, as the most recent glaciation was ending across North America, the layer of ice melted away in our neighbourhood, leaving the Great Lakes behind as a memento, and an unspeakable mess of frozen mud, rubble and clay. After the titanic weight of ice, the next principal instrument of landscape formation was the enormous volume of melt water released by the retreating glacier. We have this water to thank for the creation of Metro's treasured pattern of sharp, deep ravines, and the division of its site into two more or less distinct levels.

North of the steep drop-off running lengthwise across the city between St. Clair Ave. West and Davenport Road is a high, well-drained plateau, rising gently as it backs away from Lake Ontario, and much scarred by valleys and gulleys. The downtown skyscrapers and all other structures built south of Davenport stand, in contrast, on the low, swampy bed of post-glacial Lake Iroquois, still a deep catch-basin for the glacial runoff less than 12,000 years ago. After the waters had fallen below the clay bluffs and rocky beaches of Lake Ontario, the great forest took root where it could, among the beaver ponds and marshes on the old lake-bottom. The early explorers who passed this way were struck by the thick stands of walnut, sycamore and chestnut edging Lake Ontario; de Lamothe Cadillac, in 1702, viewed these hardwood stands as "so temperate, so good and beautiful that one can justly call it the earthly paradise of North America." Cadillac knew nothing of the glaciers, of course, since their existence had not yet been discovered. The land, however, remembered the high, ancient lake on whose bottom the beautiful forests stood. Its steep, long shoreline is balefully familiar to anyone who's ever skidded in an icestorm down the abrupt decline of Avenue Road (or Yonge Street, or Dufferin) south of St. Clair.

To imagine the titanic and enduring force of those endless winters 18,000 years ago is to start recognizing what snow can do, which is infinitely more than send you into a fender-bending skid. Fortunately, life is short enough that no one now living has to think about the next ice age. For Bryan Smith, my informant at Environment Canada, the more immediate worry is global warming. If it continues, he warns, this gradual rise in world temperatures could loose the waters now frozen in the polar ice-caps, causing a rise in sea levels, and thus drowning Vancouver, New York, New Orleans. In sharp contradiction to griping Toronto snow-shovellers, Smith says that annual snowfall in Toronto has in fact been *decreasing* over the last several years.

But that there will eventually be another ice age, erasing all traces of our passage and again reconfiguring the face of the land, is virtually certain, and also worthy of contemplation, according to Canadian scientist E. C. Pielou, author of *After the Ice Age.*

Pielou is not a popularizer. Her book is a heavy-going technical study done for fellow specialists, not interested eavesdroppers like me. But you don't have to be a scientist to understand one well-argued observation in this book: that the current epoch of Canada's geological history is *glacial,* meaning that for most of the last million years massive ice sheets have lain over huge areas of the continent—though every now and again we get a brief melt-away, such as the present one. The interglacial warm spells are always short; and this one is due to end soon, Pielou believes, slowly allowing the return of winters that never cease.

The next glaciation will begin with an almost imperceptibly slow back-up of the date of each year's first snowfall. Symmetrically, the spring melt will begin later, and the winter's snow will run off slower.

Then, after a great many years, somewhere far to the north of Toronto, summer will gradually close down to a brief interval between snowfalls, then be extinguished altogether. The snows will never stop, and the white ground cover will never melt. And thus the glacial age will begin in earnest, with the inexorable compression of underlying snow by new snowfalls on top, locking up almost unimaginable volumes of water in a steadily growing ice sheet. Meanwhile, our more southerly summers will be disappearing, gardens will no longer grow, the beautiful hardwood forests of the Toronto region will die off as tundra conditions again come gradually to prevail.

Now few Torontonians will still be around by the time the enormous build-up of snow begins to push the ice sheet southward, crushing the low hills of Muskoka and Haliburton, bulldozing the abandoned towns and villages into oblivion. No one will be left to witness the approach of the grinding ice, wreathed in eternal, swirling snows, on the edge of our doomed northern suburbs. But if our distant descendants, hundreds of generations hence, will have found refuge in some warmer clime, the architectural trace we and our ancestors have made on this land will disappear forever.

I am not trying to ruin anybody's day with all this talk of disaster and disappearance. It's just that living fully and mindfully anyplace, I believe, involves giving thought to all the rhythms we move within—the personal ones, from birth to death, but also the historical ones, preserved and recalled by the artifacts of architecture and urban planning, art and writing and music. And the far grander cycles, as well, which leave their evidence inscribed in the clay and dirt, the topography and scenery, on which Torontonians have dwelt, worked, made love and brought forth children, gardened and worshipped and built for the first two centuries of our fragile urban prevailing.

LIVING WITH WHAT IS

Given the remoteness of the next glaciation, it's enough for now to think about the land under our feet and weathers over our heads, and how these mutable facts shape even our intimate, simplest decisions. Like the time we leave the house for work, for example—later than usual on an icy day, so as to make sure the road-crews have salted the roads descending

Near Bathurst Street, January

down the old Lake Iroquois shoreline, where no up-and-down roads should ever have been built. Or the exact spot in the soil you plant your clematis. This glorious high-summer vine, by the way, likes Toronto's sunny summers; but it will never understand if you thrust it unceremoniously into the ground without first loosening and treating the tough, sticky glacial clay, everywhere just below our paper-thin topsoil. You must carefully make a place for any clematis you want to grow. You've got the last advance of the ice sheet to thank for that.

In fact, the way one gardens is perhaps the best indicator of how well he or she has accepted the great, but also the minutely specific, natural framework in which living and working are done here.

Until recently, I was not a gardener, and despite some study and practical learning, am still an ignorant one. Not so my grandmother and Aunt Vandalia, in whose rambling house and ample grounds I lived and played for a few childhood years. They gardened as ferociously as their unswerving Victorian commitment to a deathly pale complexion allowed. And with the help of a handyman to do the hard work, they kept the grounds of their comfortable Edwardian house ineffably beautiful with exquisite old roses, wisteria, common honeysuckle and other blooming things, aristocratic and common. It was perhaps these early memories—a curious mix of recollections about fragrances and colours at twilight, fireflies among the night-blooming blossoms, bee-stings and thorn-pricks, blue bottles of poisons in the shed, the scatter of petals on the grass after a storm—that left me with an enduring delight in idly leafing through gardening books, and a permanent belief that someday I would get my hands into dirt.

I eventually did so a few years ago, after settling into the reconstructed tool-and-die factory I call home. The only available open space for a garden was a large third-storey deck, which faces south and, on sunny days, is gloriously bright and warm. The dirt, hauled up the stairs bag by bag, was dumped into insulated wooden planters I designed and had built; and into the dirt went vines, small trees, shrubs and annuals. And shouldering up out of the ground or gushing from unpromising stalks and trunks came much beauty—not exactly what I'd hoped for, and nothing I dare take credit for, but a sobering lesson, as things turned out, in the variously shy, pushy, mischievous and pampered personalities of plants. Contrary to what the stern greenhouse people told me, by the way, the clematis variety Lady Betty Balfour loves its place aloft, and is not a bit snobbish about having to look at a toilet factory's backside and a laneway teeming, in summer, with muscle cars, boom boxes turned up to max.

My decision to create a garden came at almost the same time as the assignment to write a column about city sites; and, though I didn't realize it at first, my deck garden was to become a particularly subtle, intimate tool for learning Toronto, the weathers and winds, spectra of light and atmospheres specific to this place, and what it means to live, and live *here*. While I spend considerable time with my plant books, at least in winter, I am only interested in facts that help me refine the instrument of discovery my garden has become. To make a garden of salad greens in Toronto strikes me as time-wasting and even faintly strange, given the ready availability of locally grown, cheap greens at the grocers throughout the summer months. Nor do great formal gardens, with their immaculate lawns and intricately orchestrated plantings, have any appeal. Of the other sorts of stately European gardens—the rationalist Italian or English Romantic ones, for instance—I have read something, but apparently not enough to make me go out of my way to visit one.

My earliest model—indeed, the garden that got me started in earnest—lies just beyond the edge of Toronto's eastward reach, on the country property of gifted city friends who are together creating a garden very much attuned to our soil and bugs, breezes and droughts and downpours. It has been organized with a casually theatrical sense, but without ostentation. Turning off the town road into the driveway, the visitor in mid-June is immediately saluted by a line of rose bushes, their abundant blossoms bright and alert as an honour guard. Arriving at the house, one is greeted by an informal platoon of more frank, sociable blue and yellow flowers blooming by the door.

But it's in the garden proper—the array of flowers low and tall, noble as roses and ordinary as daisies, falling away from the conservatory window down a gentle, shaded slope—that the real enchantment lies. The uncareful observer would sense little evidence of the assiduous planning that came before the blooming—simply because the florals and non-flowering plants seem so much at home there. The ground cover of low plants ranging in colour from acrid green to a burnished silver, the taller flowers, those translucent dabs of white and yellow watercolour on a rough ground, cascades of roses on the garden wall, the gradual opening of the view through the trees towards the farmer's field beyond: it all appears effortless, without rigour.

Until you sit a while under those tall trees, considering the visual music of the garden's layout. Melodic transitions take place, from a wall of climbing roses which stop the eye with a barrier of matronal opulence, through an open conference of large golden blossoms nodding on tall stems, to the sunshine of that very different world of growing things, the field.

When I asked the lady of the garden to explain how it all worked, she gladly obliged. It's a planned and carefully tended project, she told me, but not so perfectly organized that it becomes an abstract exercise, or a kind of botanical travelogue of some other clime or place. It's deliberately not meant to transcend its site on the depleted soil of old Ontario farming country, with the trying, tiresomely unpredictable winters and abruptly hot summers. It's not supposed to turn into a fantasia of ye olde England, or anywhere else. Rather, this garden is to be both beautiful and tough; a haven, to be sure, but also a kind of vegetative architecture that never forgets what and where it is.

Every garden embodies some kind of symbolism of human dwelling, some suggestion of a strategy for living. To this viewer, the garden of our friends is about survival and grace under pressure. The pressure is great and unceasing, from critters of four to innumerable legs, from disease, from blustery summer storms and snapping winter cold and blasting, intolerant sunshine year-round. The beauty here has been hard-won; but, in good southern Ontario fashion, the flowers that have survived adversity bear their triumph modestly, as though nothing could have been easier.

Some day the bugs and blights may win, because our friends will move along, and newcomers without any interest in this green building will take up residence in the shaded house. Until then, however, the present tenders expect to go on minding their plot, as one aspect of the quietly defiant attitude towards life and work they share—though they do not talk about gardening in so hifalutin a way. That manner belongs to us Toronto-dwellers, out for the day, and out of the usual urban power-game, involving mastery and defeat.

We do have such a hard time with things that will not bend to our will. Rivers, for example, always seem to be in the "wrong" place. Therefore, they must must be bridged, redirected, dammed, or channelled through concrete pipes. Hills must be amputated or swept aside altogether to make way for roads. Swamps are almost always "wrong"; they must be filled. And if a feature of the natural world is simply too big to "correct"—the Great Lakes, for example—it still must be subjugated to human use, as a power source, or merely a sump for our poisons. The project of Western urbanism, which our branch of the human family abruptly embarked upon eight thousand years ago in the Middle East, has never been minded just to let things be as they are. Our work has always been one of eliminating the "wrongness" of Nature, and making it conform to our notion of "rightness."

Which is exactly why gardening under the unsteady weathers and in the bad soil of Ontario is instructive, and a good way to learn grace and patience where we really are. We cannot correct Nature; we can only change it a little, into the "Nature" which Aldo Rossi defines as "the artificial homeland which contains all the experiences of mankind." Yet it forcefully remains itself, neither Eden nor pure Culture. My friends' plot is certainly no Garden of Eden. It's just Ontario as it can be, given time and a certain devotion to tending and letting be, and knowing the difference between the two.

DWELLING AND THE GROUP OF SEVEN

Beshorted in cabbage-green Bermudas picked up at Sears on the way out of town, equipped with a gym bag full of books that really ought to be ploughed through someday—say, the novels of Edith Wharton—and armed with a vat of bug-off, I joined the tense, sluggish flow of cars northward on the expressway one recent summer Friday. Three hours, two pitstops for the kid, and one milk-shake spilled all over the back seat later, we'd reached Fairy Lake, to which a family friend had kindly invited us.

The lake is a sheet of water sheltered by low hills, near Huntsville. The cottage itself, which Anita has been renting for the summer since time out of mind, is a charming storybook house built in the 1930s, nestled among throngs of unattended blue and golden perennials planted long ago. I was bewitched by the hammock slung between tall pines on a broad lawn sloping down to the shore. Three days after touchdown, not one book had been opened, there was sand in my shoes, and cottage country had worked its spell, melting away city knots I didn't even know I had.

At Fairy Lake, there may be more to that spell than just warm sun and sparkling blue water. The lake was given its curious name by the first European settlers in the area, who had heard from the native people that spirits haunted the waters. Early lakeside dwellers reported strange voices in the night, and mysterious lights skittering over the surface. Whether or not the lake's magic is otherworldly, it surely touched the imagination of Canada's first important modern artists. Tom Thomson, the extraordinary Canadian landscape painter, dropped by on one of his sketching trips into Ontario's north woods. A result of that stop was an oil, done in 1912, called *Fairy Lake*. Both Arthur Lismer and A. Y. Jackson, later founding members of Canada's nationalistic Group of Seven landscape painters, visited Thomson at Fairy Lake, and may have sketched with him there.

As often happens when the art-minded venture into Ontario's near north, thoughts turn almost automatically to an interesting puzzle of Canadian culture. When we look out the cottage window at, say, a sunset smouldering over the sparkling waves, *exactly* what are we seeing? Is it Nature, redolent with raw, primordial beauty? Or are we seeing a scene so subtly colonized by our experience of landscape paintings, that we can't tell the difference any more? Is that really a glorious sunset? Or did we learn the meaning of *glorious* from a painting by some member of the Group of Seven?

After all, the visual education of Ontarians, and perhaps all Canadians, is saturated with Group pictures, whether in the real or in reproduction. Each year, some forty-five thousand school children are bussed down to the Art Gallery of Ontario to look at some of Canada's most famous landscape paintings. Those who grow up to become gallery-goers will certainly be seeing that art, in settings ranging from special exhibits to permanent displays, every year, probably forever. And even those who never set foot in a gallery after high-school graduation will never quite escape the calendars and posters, postcards and greeting cards, or the bottom-line fact that Group works are among the most highly valued Canadian artworks, predictably making auction-room news a couple of times each season.

There's no scientific way to tell how much of Fairy Lake is nature, and how much is art. But dozing on the dock through a Saturday afternoon, with my toes in the waters, I made certain common-sense observations that may have a bearing on the case.

The day was muggily hot, hazy and still, with a thin grey scrim covering the sky. The hills across the lake were dull olive-green. A tiny, rocky island between this side and the far shore was an angular blackish brown sherd against the gun-metal water.

By that time, I was no longer surprised to catch myself thinking: "It's just not a Group of Seven kind of day"—a day of stormy lights, of water lashed by summer storms or made irridescent by brilliant, failing light. What is odd is the way Fairy Lake, overcast and calm, tends to disappear, or become strangely remote. The lake doesn't look much different from Lake Ontario on a dog day afternoon.

In reality, a muggy day in Muskoka during August is not unusual. Despite the regional tourist hype, the stuffy weather farther south often creeps up north of Orillia, into the rocky lakes clawed open by the ice centuries ago. This condition may be disappointing to those who have fled Toronto for what they believe will be cooler climes—but definition, not disappointment, is what I'm talking about here. A hot, becalmed Fairy Lake seems to disappear from our mental atlas of "cottage country sites" at the precise moment it no longer evokes the nationalist, nativist, vitalistic mystique of Group painting. It comes back, however, as soon as the wind whips up, the clouds become ragged and the water begins to lash the boathouse.

There is nothing really peculiar about this common experience. The visual codes of painterly wildness, freedom, urgency, the primitive and such have become the basis of our visual doctrine about what's really cottage country and what isn't—supplying, that is, all that's missing from the *real* Muskoka spectacle of endless electrified, plumbered, septic-tanked little lakeside houses, each with its gas barbecue and each readily accessible by paved road. But our eyes forget things slowly, and the codes of popular art most slowly of all.

Even so, the staying power of the Group's ways of seeing the world is a remarkable fact. Notable recent attempts to re-vision the Ontario landscape have been made by a number of contemporary Canadian painters. But whatever their interest to art-world professionals—and the interest is often keen—these attempts to debunk or displace or revise the Group's visual myths haven't worked. At least not on the level of those forty-five thousand school children who are guided through the AGO each year, or for the millions for whom Canadian art *is* Group art, and Group art is the only true national art. It should come as no surprise that every weekend between Victoria Day and Thanksgiving, *Tout* Toronto hits the great road north, to see—actually *see*—a visual world constructed by a few magnificent and wonderfully memorable paintings done at the far side of the twentieth century.

But did the Group of Seven know both ethos and public *too* well, and address their nation a little *too* cannily? They were certainly in a professional position to do so, since almost all the future group artists earned their living as commercial artists in the Toronto studio of Grip Ltd., using its offices as a kind of clubhouse in the years leading up to their 1920 launch. The Grip touch is certainly to be found in many a slick, illustra-tional canvas and sketch by group members, early and late. Many a finished Group canvas seems more cartoonish than painterly, more akin to drawings in popular field-and-stream magazines than alert to the venerable traditions of landscape painting. And the work often seems dated by its lazy display of gimmicks and flourishes borrowed from popular design trends such as Art Nouveau or Art Deco; see, for example the streamlined styling affected by Lawren Harris when trying to be very inspirational to us all, and the stiff, furniture-like *Jugendstil* of Thomson at his least inspired.

But having said all that, we can still be impressed and moved by the most restless, thoughtful work of the Seven. Indeed, I suspect that we may yet have much to learn from the Group of Seven, despite the fame of these artists and the scholarly effort expended on their art over the past three decades. Perhaps the day has come for an exhibition that will chart the relation of Canada's best-known images in painting, not back to antecedents in high art, but out to the actual day-to-day operation of Toronto's Grip Ltd. and the industries of mass communications, advertising and north-country tourism in Canada during the first three decades of this century. By placing the group's work in the context of consumerism in the Edwardian and postwar Dominion, even the images of the North we know best might be coaxed into disclosing their origin in *Toronto's* urban understanding of itself, in the nostalgia that city-folk often develop for rural zones, the oppressions and bitter hardships of which they no longer have to endure.

Binding and
Loosing the Waters

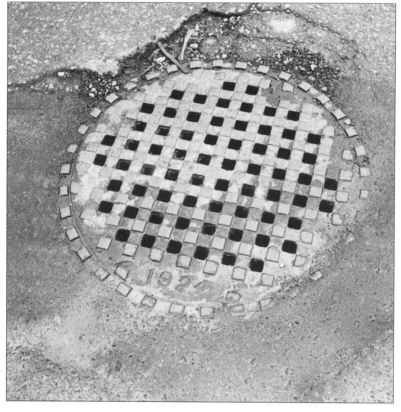

Manhole, 1924

LOST STREAMS

If we think of vanished urban beauty, the ghosts of foolishly demolished buildings—remembrances of photographs, old drawings—are the first things that pass before our inner eye. We think less readily, for some reason, of the myriad ancient streams and brooks and little rivers, all but the largest of which—the Don River and the Humber—were beneath the concrete a century ago.

On a city map drawn up in 1884, we find that Mount Pleasant Creek, just at the northern edge of the urban advance, had already been pushed underground as it approached the Don River. Rosedale Brook was still running free through its deep ravine, and thence into the Don—though a little way upstream, it was already inside brick tunnels and under earth. This creek was open only north of Yorkville, a respectable suburban village in decline before its annexation by Toronto in 1883. By the turn of this century—like so many formerly clear downtown streams, providing delightful banks for picnics— Castle Frank Brook had become a sluggish, stinking ditch, and so was banished under Rosedale Valley Road where it still flows through its conduit. Indeed, a close look at the 1898 city map turns up very few streams still above ground anywhere south of Bloor Street, the venerable dividing line laid out by Toronto founder and governor John Graves Simcoe in the

1790s between Civilization and Beyond, and still our most famous cross-street. Russell Creek, which provided water for cows grazing below Bloor alongside dusty, rural Bathurst Street is gone, doomed to burial, along with the others, by the powerful surge of Toronto's population in late Victorian times.

Among the more lamentable of these disappearances is Taddle Creek, which rises from springs north of St. Clair Avenue and to this day feeds a pond under 250-year-old oaks in Wychwood Park, a secluded residential estate on the escarpment north of Davenport Road and west of Bathurst, then empties south under the bus barn of the Toronto Transit Commission. As recently as 1886, Taddle Creek sparkled in the meadows, shaded hillsides and tangled ravines of Queen's Park, future site of the Ontario Legislature. This much, we know from a charming painting of that year called *Barbara and Alice, Queen's Park,* at the Art Gallery of Ontario, a work of Marmaduke Matthews, who incidentally founded and planned Wychwood Park. In it two young ladies relax by the rivulet, now deep under the park. (The boring flatness of Queen's Park was inflicted during a revision of its original picturesque appearance, which had been created by landscape architect William Mundy in the 1850s.)

From Queen's Park, then part of the University of Toronto and still owned by it, Taddle Creek went underground, flowing through sewers or meandering through basements beneath "new" downtown, west of Yonge Street and far west of the old town of York—though at least the name of the stream has survived, if not the sight of it, in the name of a short, dead-end lane on the University of Toronto campus.

On this map of 1898, at least one remnant of the vast network of streams draining Toronto can still be detected: a short stretch of Garrison Creek, still wandering freely north of Bloor Street westward as far as Ossington Avenue. That twisting line, no wider than a silk thread on the large map, is all that remains to remind us of the creek that once supplied the Imperial

Grape Spill, Emerson Alley

legions stationed at Fort York with an easily defensible, navigable source of abundant fresh water—hence its name—and also supplied a good reason for centring the fortification there.

In the boom years before the First World War, Toronto did its best to erase the streams from the cityscape, for two reasons. The first, and more commendable, had to do with the stinking hazard so many of them had become. The second reason, and (from a modern perspective) less noble one, was the dogged tendency of these wandering waters to obstruct Toronto's historic insistent drive to remain what it had been at the very beginning, a city laid out on a Euclidean grid, every angle not a snitch less than ninety degrees, and topography be damned.

But the habitual walker has little difficulty finding and following the rainwater's old paths. No matter how urbanized, old landscapes have stubborn ways of remembering, and reminding, despite the encasing of their waters in cylinders of stone, brick and concrete. An otherwise inexplicable dip or bend in a street with no reason to dip or bend, for example. Or the damp chill breathed up from underground into a sunken, lawned park at dusk.

Sometimes very early in the morning, before the city fully awakens and long before the rush hour makes such hearing impossible, I walk to a sewer grate at the bottom of a ditch near my house, and listen to the whisperings of an old stream, now imprisoned deep under concrete and asphalt. I try to imagine what it might have been, to see it bearing away the cold melt under vaults of birch, oak, aspen and sycamore boughs, a thousand springs ago.

THE DON RIVER FORGOTTEN

It's no pleasant stroll, whatever the weather, to hike the lower Don River, in its final stretch from Cabbagetown, southward to its ignominious ending in a port industrial ditch south of the Gardiner Expressway. The inhuman din of the Don Valley Parkway makes conversation softer than a shout impossible. A few ducks churning the turbid waters and the odd fish are just about the only wildlife left to remind us of what creatures roamed the valley in Toronto's earliest days—lynx, bear and fox—and the annual salmon runs that made the river famous. A few lovely willows and other trees are all that remain of the forests that graced the Don flood plain.

Virtually everything else one sees is the result of human handiwork, evidence of the human drive to control and organize—visually abrasive, surely, yet deeply expressive of the values that drove Toronto from village to metropolis. One is immediately struck by the linear geometry of it all. The gentle meander of the Don in its final approach to Lake Ontario was strictly straightened more than a century ago. Parallel to its present concrete channel runs a straight stretch of railroad, with bridges and hydroelectric wires spanning both river and railway crossing at right angles.

But for all the visual rigidity, a visit to the homely Don still has a certain pleasure—if only because, unlike the numerous lovely Toronto streams running, almost forgotten, under city streets, the Don was just too big to bury. One has to admire that about a river, even if it is today jailed in a grid of steel and concrete.

If a lively group of urban activists and Toronto city bureaucrats has its way, however, the Don might yet get a parole from its long imprisonment. Instead of being made to dump its waters into a port channel by Lake Ontario, the Don would once again be allowed to end serenely in a broad lakeside marsh busy with wetland wildlife. The portion immediately to the north of this marsh—the part straightened in Victorian times—would remain straight, the Don Valley expressway would remain intact, but the lower valley would be green again. Farther upriver, in the vicinity of Rosedale, the stream would once more be allowed to spread and puddle, nourishing marshes and the wildlife that lives in them.

Such were the recommendations of the city-sponsored Task Force To Bring Back the Don, outlined in its $80,000 report released in 1991. The nobility, practicality and visionary power of this document have been widely and justly hailed, and its recommendations were widely reported.

As the introduction explains, this proposal is "to start thinking of the lower Don Valley as a place in itself, not as a gap between places"—a place with memories of its own, and a peculiar monumentality. One could imagine few notions more foreign to Toronto's prevailing spirit of urban modernity, which strait-jacketed its principal river system in the first place. Three years after the report's release, not surprisingly, little movement has taken place on the political front towards letting the Don become the old river, and active city site it could be.

Yet the task force soldiers on, guiding would-be converts to the cause through the traffic noise, and the mud and garbage alongside the Canadian Pacific tracks parallel to the Don's final passage, and doing other good works of reminding. On a Sunday morning not long ago, for instance, I found them wiring up some two hundred black-painted plywood silhouettes to a rusty fence running along Bayview Avenue and the CP tracks, south from Rosedale Valley Road to just above Queen Street. Designed by artist, writer and task-force activist Marie Day, and fabricated with the help of a dozen or so kids recruited from the nearby Regent Park public-housing project, the cut-outs represent wild mammals, fish, fowl and assorted slinkers and creepers that once called the Don Valley home, and some of which, astonishingly, still do keep a claw-hold on this wasteland of rotting concrete, scraggly trees and incessant din.

The bears and moose and other big animals seen in outline on the fence lumbered away north almost two centuries ago, when the millers and brick-makers took over the valley of the thirty-eight-kilometre river for their factories. (The mammoth had vacated the premises some time before, though not without leaving a skeleton or two behind. Salmon left the Don in the middle of the last century, while the brook trout held on until fifty years ago.)

Despite all the industrial waste and street mess dumped into the river, however, great blue heron can still be seen majestically lifting into the air from the valley's pools. Foxes, too, have their holes in the clay embankment, and are said to raid the chicken coops at the valleyside Riverdale Farm from time to time. And not many months ago, writer Pat Ohlendorf-Moffat reported in *The Globe and Mail* her discoveries of "sunfish and perch...creek chub, white sucker, blacknosed dace and longnosed dace" still negotiating the dirty waters of the lower Don.

The task force's reason for putting up the silhouettes, members told me, is to remind commuters, cyclists and others passing through the valley of its past richness as a natural habitat, and to summon up a vision of what the valley could be again if restored to even a shadow of its former wildness and beauty.

Nobody wiring up pieces of plywood by the Don thought he or she was going to save the world, or even the river, by this action alone. But doing that was at least something; and the task force thereby served notice on the politicians that it is not a group to be easily daunted, even though its options are now limited to small gestures. Mark Wilson, computer consultant and infectiously devoted chairman of the group, speaks fervently of the "reconnection of the people of the city and the Don, the heart of the city," from which we have been barred by fences, railway lines, the expressway. He foresees "a restored natural landscape, a wilderness in the city"—no pipedream, in his view, given the gradual abandonment of the river's edges by the industries that once hemmed it in.

The decline in Toronto's industrialization—or, more accurately, the shift of the city's prime economic force from mechanized object-production to cybernetic information processing, electronic banking and publishing may in time allow the river to come back, and the citizenry to return to a refreshed valley. All that, of course, remains to be seen. But no one who witnessed the wiring-up of those fragile animal silhouettes would have come away unmoved by this small pledge to the waters, and to the renewal of our awareness of their healing powers.

GARRISON CREEK

Garrison Creek can lay fair claim to being Toronto's most famous waterway never seen above ground, at least by anyone I've ever met except Roy Wood, of whom more presently.

Its system of forks, gathering rivulets and ravines, focusing around Bloor and Christie Streets to Fort York and finally emptying into Lake Ontario, had been turned by early Victorian farmers and workers into a stinking latrine and garbage dump long before the Old Queen died. Also, as Toronto's most conspicuous natural feature between Bay Street and High Park, Garrison was doomed to be a "problem" for city planners. The more or less organized westward urban sprawl reached the creek around 1850, building up rapidly after that, and leaving many a skew around this "obstacle," in the form of kinks, twists and wrinkles in street pattern. But the residential developers of the latter nineteenth century seemed happy enough to build on the soggy trash that had by then clogged whole sections of the creek, even though almost anyone with a scrap of engineering know-how could have predicted the outcome: a lot of nice Victorian and later brick houses today slowly tilting and listing and sinking into the sub-surface muck. The result can be seen in those streets, north of Bloor, that look like rows of crooked teeth.

Like many an ancient victim of urban modernity—the Don Valley is another example—Garrison Creek became much cherished by urban historians, naturalists, and community activists in the early 1990s. A residents' group in the neighbourhood of Trinity-Bellwoods Park, under which the creek's waters silently slide today, is writing its history, from the glory days as the strategically vital water-source for Fort York to its current ignominious state. The formation of a Garrison Creek Historical Society is under way, I understand, and tours of the area are conducted by John Harstone, a local historian.

In the spring of 1993, I got wind of perhaps the most ambitious and intriguing plan so far inspired by this invisible stream. It was called the Garrison Creek Community Project, and, if the modest resources being sought are soon found, will be carried out by Toronto landscape designer and horticulturist Terry McGlade, architects James Brown and Kim Storey, and ecological activist Whitney Smith. The Project's initial goal is simply to create public awareness of this topographical, ecological and historical fact, which links so many ethnically, culturally diverse communities along its path—to educate, in other words, and to animate. Longer-range objectives include architectural and natural-history studies, and, eventually, the unbinding and resurrection of at least part of Garrison Creek from its brick-lined grave.

I hope Roy Wood lives to see it.

Long ago retired from schoolteaching, Wood is the only person I've ever met who remembers Garrison Creek when it flowed freely above ground. One sunny, crisp autumn day in 1992, this bright, elderly gentleman took me on a guided tour of the old trace, where he played and splashed as a boy.

Travelling by car, we commenced our trip at St. Clair Avenue West just east of Oakwood, following the old track of the creek along steep, twisting streets. The brief descent took us down the precipitous shoreline of old Lake Iroquois, down to Davenport Road, at its foot. Once south of that thoroughfare, just before twisting over the site of the former sand quarry known as the Christie Pits, at Bloor Street West and Christie, Garrison Creek made a sharp bend, crossed Ossington Avenue under a wooden bridge demolished long ago, and flowed through the Wood family's rhubarb patch. (Like so much of the west end developed shortly after the beginning of this century, the area is now a zone of small-scale workers' housing.)

Making that short southbound trip, one could easily imagine the old creek, which slowly gathered its waters on the plain overlooking the dense forest on the Lake Iroquois lakebed. Then, abruptly hastening, it bore its wild waters down over the edge, carving the gullies remembered today in the twists of Mount Royal Road and other small, winding streets on the escarpment. At the bottom, it broadened and slowed, picking up more waters from forest tributaries, finally gaining the weight to gouge a broad, sandy trough that can still be traced in the densely populated urban landscape between the Christie Pits park and the lakeshore.

By the time Wood came along, the burying and shrinking of Toronto's streams was, of course, far advanced. He recalls that, when he was small and Garrison Creek flowed openly across the Woods' truck farm, it was only a skinny rivulet. Except, that is, during spring run-off, when it briefly remembered its ancient task of gathering melt water and early rains from atop the escarpment, and bringing the torrent tumbling and splashing down through the middle of the farm. You can't have that sort of wild thing happening on a good farm, on the north-west edge of a rapidly growing city. And so it was, in 1913, when Wood was about seven, that the last visible trace of the old creek was put away underground.

While enjoying Roy Wood's reminiscences, I could never quite push from my mind the unsettling thought that this was probably the only person I shall ever meet who actually *saw*

Garrison Creek—diminished, but still flowing under the sun, as it had since Toronto emerged from the cold waters to receive the returning forests.

Some time after my visit with Wood, I did see Garrison Creek, or what's left of it, the only safe way one can nowadays: suited in orange coveralls, rubber gloves and boots and a hard hat, trussed up in a web of straps resembling rock-climbing gear and outfitted with an emergency oxygen tank, and being lowered from the daylit world, through a small manhole, into the eternal darkness of the century-old sewer through which the creek flows today.

As I dropped down the hole, Salvatore Pasquale, an inspector for the city's public works department and our guide, was waiting for me at the bottom, lethal-gas detector at the ready. Coming down into the nether gloom after me, with his tape recorder, was Jeffrey Kofman, host of the CBC rush-hour radio show which had set up this descent, and Richard Stromberg, an archaeologist with the Toronto Historical Board. Each of us was equipped with a flashlight designed not to spark an explosion of undetected gas when switched on.

Seeping into the blackness from only five metres above us, the strong morning daylight swiftly dimmed, turning blue as it sank, becoming merely a ghostly pale glow at the bottom. Along with light, we had left behind all the sounds of the city. Down there, one could hear only the sounds of water: the loud, distant rush from a slaughterhouse down the line, the nearer trickling from small lines running into the main sewer, and the sloshing of our high boots through the shallow run-off, hardly higher than the steel-reinforced toes of our footgear. If the humidity was stifling and somehow insidious—I experienced an unfamiliar tightness of the lungs for two days afterward—there was no foul odour. Even the slaughterhouse effluent turned out to be clean water. I was disappointed to see no rats scampering away from us (or making ready for attack), nor evidence of any other living creatures, except some wispy strands of spider webs dangling inside the narrow pipes feeding into the main sewer.

In fact, there was nothing to see, except the lights of our flashlights on the sewer wall, and the structure itself: a perfectly round tube, eight feet in diameter, built of double-layered brick between 1885 and 1892.

The trickle under our feet was all that remained of Garrison Creek, once navigable by canoe, Richard Stromberg told us, as far north as Bloor Street West, which bridged the waterway. Increasingly fastidious about urban sanitation, Toronto's Victorian forefathers decided as early as 1884 that this "open sewer" and "absolute nuisance"—so-called by a Toronto mayor in his inaugural speech—had to go; and go it did, down into the dank, eerie tunnel through which we groped our way.

Many of the artifacts of civil engineering—bridges, dams, hydro lines, and so on—forcefully and eloquently express the Modern spirit, without meaning to be anything other than unassuming, workaday things. Garrison Sewer is a piece of such engineering, and deserving of special praise, even if its beauty was unintended.

The Victorian immigrant masons who laid the bricks of this tubular fabric did so with the care for detail and precise pointing one might expect to find on the façade of a Rosedale mansion, and hardly in a place of utter darkness where the round walls are meant only to serve, not to be seen.

Deep under our city—a city of bricks, since Toronto's earliest urban builders did not have the stone so readily available in other parts of the country—there is a masterwork of Toronto Victorian brickwork, too handsome to ignore, far too dangerous to visit under any but the strictest supervision, too historic to forget, yet destined to be forever unseen under the city's skin of concrete and asphalt.

THE DON REMEMBERING

In the late spring of 1992, plans for the huge, troubled Ataratiri mixed-use project, slated to rise on thirty-two hectares of polluted, expropriated industrial land east of downtown Toronto, finally died on the drawing board.

As it happened, the death blow to this visionary complex of houses, schools and community facilities was administered by the Ontario government. But for some time, several powerful forces had been mortally threatening to knock the pins from underneath the dream structure. Two of them had origins in the doings of humans: the steep slide of Toronto real-estate prices in the early 1990s, and the soaring costs of rinsing the poisons from the soil.

A third had to do with the lay of the land, and the fact that the site picked for Ataratiri—a roughly pie-shaped wedge broadening eastward from the crossing of Front and Parliament streets, and showing its flat side to the Don River—lies on the Don's ancient flood plain, at the edge of Lake Ontario. What had not been foreseen was the near-impossibility of keeping the river from soaking the soil each spring, flooding basements and underground parking garages, and standing above ground in puddles everywhere. Although corseted in concrete, its final meander towards Lake Ontario straightened out and banked up long ago, the Don, we find, is still wild enough to wreck the dreams of idealistic folk whose plans did not pay it proper respect and take it into proper account.

As I walked the rubble-strewn vacant lots and demolition sites of this district on a blustery cold spring day shortly after the end of Ataratiri, pacing its long, straight and nearly deserted streets, the term "flood plain" took on new, faintly menacing meaning for me. There was no sign here of the rises and ravines and the gulleys of ancient streams that give the old lakeshore topography of Toronto much of its character. The marshy, muddy ground is nearly flat, remorselessly smoothed and scoured by brimming melt water from the ten thousand spring thaws that have come since the glaciers withdrew.

If factories and warehouses and industrial enterprises of various sorts were once thick on the ground here, many had gone, leaving only ghostly oblong shadows of brick dust on boggy fields. Other buildings remained. Some of these survivors stood abruptly isolated—the empty, ponderously pillared warehouse, for one, now a pedestal for an enormous Toshiba sign, beaming its message at the Don Valley Parkway's inbound commuter traffic. Some stood together in the interestingly impure architectural ensembles typical of industrial zones—finely detailed Victorian red brickwork, dumb postwar cinder-block fabrication in the barracks manner, an office building from the 1930s, all tumbled together as the enterprises on the streetscape artlessly multiplied.

Bought up by the province and vacated some time ago, almost all were dark, quiet and ominous, like skulls. About the only living structure in the district is the red-brick building at the corner of Front and Cherry streets, deep in Ataratiri country. The earliest part of this romantic architectural fabric was stacked up in 1859 to house a public school. When the residential area it served was displaced by industry, the old school was turned into an tasteful, small hotel for business travellers known as the Eastern Star. Today, it is home to some small businesses, and to the Canary Restaurant.

Driven off the bleak mud and cracked sidewalks by damp wind, I found a good cup of coffee and a place to warm up in the Canary. But I could not get the eerie music of the spring breeze out of my ears, or stop thinking about the patient surrounding flat-lands, which had soaked up so much public money and defeated so many well-meant plans, and were now waiting under the leaden skies of late winter for the rains and run-off that were due in Toronto, in not so many days.

Humans certainly had played a role in this desolation— but it is easy to dream, while sipping coffee at the Canary, that the wreckers had really been the secret agents of the river, which once owned the land and now was taking back its own. Nor was it hard to imagine what the Ataratiri site would look and be like, were both province and local idealists to surrender, and let the Don get it all. Instead of the present wasteland of rubble and dead buildings—which surely has its own quiet and poignant beauty—we would have a deep, damp forest just touching the eastern edge of downtown. It would be flooded in the spring, and city folk would be driven from its paths. But the little stands of poplars, sycamores and other lowland trees would be annually renewed and invigorated, and begin to spread, gradually embracing the ruined buildings, while healing the wounds inflicted on the ground by the factories. As we were reminded by the Ataratiri cancellation, the river will let us use its ancient flood plain, but on its own terms. Perhaps it is time to accept this reality, and give the land back forever to our silent, powerful partner.

Tales of the Pioneers

The Pavillion, Allan Gardens

Unromancing Fort York

In full swing during the spring and summer of 1993, Toronto 200—as the city's official bicentennial jubilee called itself—was mostly an embarrassment and a bore. Such "heritage" extravaganzas always are. Across the town, we got the usual dull unveilings of plaques. In April, around the anniversary of the 1813 American sack of Toronto—or York, as it was known then—costumed and bewigged actors representing the British 8th Regiment celebrated this defeat by drilling, marching and blowing off muskets. I can't tell you why this disgrace was thought to need a commemoration, though I am certain it didn't. And in August, lest we forget the 200th anniversary of the arrival by founding grandees John and Elizabeth Simcoe, along with some colonial poobahs and a musical band of oom-pa-pahing Queen's Rangers, we got a pageant at the harbour featuring a sailing ship and, in the odd words of a press release, "costumed reenactment troops."

What never occurs to the bureaucrats who plan such tourist carry-ons is that history is more interesting than costume. A brush with the real history of a Toronto site, its special wrinkles and local textures and how they came to be that way, might actually pique new curiosity in otherwise indifferent folk among the Metro area's 4.2-million inhabitants. Among the better opportunities to do so is "Historic Fort

York," the lakeside military installation Governor Simcoe came in 1793 to carve from the wilderness.

Thanks to the Toronto Historical Board, which runs it mainly as a tourist attraction, Fort York is a veritable Santa's workshop of the "heritage industry," with blunderbusses booming, cookies in the oven, the works. But beyond the marketing of the Fort, there is a military and architectural history of high interest. Inside its earthen enclosure—the foundation of long-vanished log palisades—is a scatter of eight buildings. None date from the very earliest days of the fortress, an outcome partly intentional since Governor Simcoe meant what he called "comfortable Barracks of log work" to last no more than seven years. The survivors, however, all hold considerable architectural and historical interest. Included in this ensemble are two handsomely ponderous blockhouses built in 1813, instances of expert log construction on the uneasy British-American frontier, three one-storey brick barracks raised in 1815, and a thick-walled stone powder magazine from the same year, skilfully ventilated and furnished with spark-proof copper and brass fittings. Taken together, the Fort York structures provide an excellent early vignette of colonial military architecture, and are notable for that reason alone.

Why they exist where they do, and why they should be visited, may seem a mystery nowadays. The shoreline lies little less than a mile south—pushed there by landfill over the last one hundred years—and the little fortification itself, once on a low headland overlooking the lake—is virtually invisible behind the immense concrete colonnade bearing the Gardiner Expressway, smokestacks and a dusty cement-block factory, and billboards meant to catch the eye of high-speed commuters with ads for TV sets. The stone wall running round the site is a make-work project of Depression-era vintage. Its abrasive industrial context seems to heighten its character as a fantasy-redoubt of staunch Englishry, perfumed with memories of red-jacketed Christian soldiers living the hard, chaste and loyal life in defence of British North America.

Tourists have become visibly upset upon learning that red-jacketed Jews served along with stalwart Anglicans at Fort York, and that the wives of officers occasionally lived and kept house within the fortification. (Married officers more commonly lived in the nearby town of York.) These tiny facts are arresting in a way mulled cider and cookies fresh from the oven, staged bayonet attacks, cannons booming and so forth never are. But if such fact is what's wanted, there is no better resource than *Historic Fort York 1793–1993*, a readable and profitably illustrated book by Carl Benn, Fort York's curator of military and marine history, and one of the few serious memorials to remain after the bicentennial hoopla had come and gone.

Credit for the founding of Toronto, we learn from Benn, should go to the Americans, indirectly at least. By 1793, the revolutionaries to the south had again become restive about not having grabbed *all* British North America when they had the chance. Whereupon Colonel John Graves Simcoe, lieutenant-governor of the British colony of Upper Canada, fearing attack by U.S. troops on his more or less indefensible forts at Detroit, Niagara and Kingston, decided to build a battle station and administrative centre far up the wild north shore of Lake Ontario. The desolate spot later to become Toronto was an easy choice. It featured a defensible harbour, with only one entrance in those days, a good strategic water supply in Garrison Creek, and a healthy distance across open water between itself and the United States. Simcoe quickly threw up some log bunkers at the entrance to the harbour—the site of present Fort York—and ordered the first plans for a civilian town and port to be constructed about a mile east, in the relative safety of the enclosed bay. On August 17, 1793, Simcoe christened his tiny settlement York, since he believed in giving his foundations names resounding of England.

The colonial administration of British North America turned a deaf ear to Simcoe's pleas for a heavy military build-up of Fort York—meaning that there was only a tiny force here and a tiny town in 1796, when John and Elizabeth Simcoe departed York forever. Only 241 souls lived here. In 1810, there was a single brick house in the town.

If it hadn't been for the ongoing belligerence of the Yankee rebels against the Lord and His Anointed King—my revolutionary ancestors, that is—one of Simcoe's successors might well have abandoned the boggy frontier outpost at York, and moved the capital of Upper Canada to a more civilized spot, Niagara or Kingston, say, or—Simcoe's preference—London. Abandoned by its legion, Fort York and its attached town would then have surely vanished, and millions of people would not now be living in bungalows and apartment blocks on land once covered by dense forests, and later large farms. There would be no commuting, no high-rises, no SkyDome. Just to mark the spot, perhaps, some government might have put up a little theme park, of the "ye olde" sort the heritage people love. (Chicago got lucky, too. The muddy location of the Americans' Fort Dearborn after 1804, Chicago didn't amount to much until the later nineteenth century, when it took off for the same reason Toronto did: the continental railways.)

But as it happened, the Americans did remain feisty, and greedy for Canada; President James Madison was particularly keen on establishing a trade monopoly covering all North America's fantastically abundant natural resources. Such idealism had no room in it for a European outpost, so, on June 18, 1812, Washington declared war on Great Britain. Not that the officers, civilians, colonial bureaucrats and military planners at York could do much else than stay huddled in the mud; the official shorting of York's defensive capacities had seen to that. When the Americans got around to attacking York, in April, 1813, they had little trouble taking the town, which they looted and terrorized before taking their leave in May. In retaliation, British troops burned Washington to the ground in August, 1814. John Strachan, Toronto's first bishop, later told

Thomas Jefferson that the United States capital got just what it deserved. Shortly after the catastrophe, the British began strengthening Fort York—the earliest extant buildings date from this time—which, by 1814, were imposing enough to make an American naval unit think twice about attacking, then back off.

Although Fort York is advertised as a relic of the War of 1812, its most heavy use as a military base came in the later nineteenth century. It was abandoned as an official military fortress in the 1930s. By that time, the only action its troops had seen were in skirmishes against other Canadians. The first was in 1837, when the garrison defended Toronto's Anglican colonial élite against republican-minded Upper Canadian merchants, industrialists and farmers, trouncing them. The 1837 rebels were suppressed, banished or executed, and then transformed into figures of textbook myth. (For some reason, Canada has a tradition of turning its traitors into heroes.) The second engagement took place in 1906, with violent Hamilton streetcar operators on strike.

After the War of 1812, the United States would never again try to finish its incomplete seizure of all British North America. But military strategists in London and in Canada did not know that, and could not count on the eternal international peace Canadians now take for granted. As recently as the 1860s, defence planners considered using the Toronto fort to help save this country should the huge armies of the Republic and the Confederacy decide to stop fighting each other and join forces against Canada. But by the 1890s, Fort York had become militarily obsolete, and so played no part in the last known plans laid, just before the First World War, to counter an American invasion. (The idea was to have Australian and Indian troops attack California if the U.S. launched an aggressive action against Canada.)

What remains of Fort York isn't much to look at, which is why Carl Benn's book should be carried along on every visit.

The sober, thorough text is supplemented by nineteen maps and seventy-seven reproduced photographs, drawings and paintings, enabling any visitor to envision this volatile site, with its built forms rising and changing and disappearing— walls destroyed in 1916 to make way for streetcar tracks, an encroaching slaughterhouse, barracks quickly constructed and quickly demolished, obsolete eighteenth-century cannon put into service as fencing, the gunboats and frigates now utterly gone. Too, these pictures can kindle in the imagination moving pictures of the men and women who lived, served and died here, and laid the tiny foundations of what would become, by a curious historical irony, not a mere memory but a megalopolis.

BLACK CREEK PIONEER VILLAGE

Black Creek Pioneer Village is a theme park nestled behind hedges and artificial hillocks meant to mask the fast surburban traffic in its crime-ridden environs on Metro's north-west fringe.

A self-described simulation of a "typical nineteenth-century crossroads settlement in southern Ontario," Black Creek comprises forty-odd houses, stores, barns, utility buildings and other edifices, most rescued from ruin elsewhere in Ontario, trucked in and set down along dirt streets winding picturesquely through a fifty-six-acre landscaped tract. Among the oldest structures at Black Creek, and the only ones standing where they were built, are the 1816 log cabin of farmer Daniel Stong, his piggery and grain barn, both built in 1825, and the more comfortable house Stong put up for his family in 1832.

Roblin's flour mill, the most imposing industrial structure in the "village," and complemented by a charming pond and mill-race, is a fake, built to resemble an 1842 facility in Prince

Edward County and to house its internal machinery. Many of the other buildings, however, are genuine artifacts, giving Black Creek the potential of being a useful architectural set-piece. There's the Emporium, built in 1856 at Laskay, Ontario, and the circa-1860 blacksmith shop from Nobleton, all looking, we are supposed to believe, as they did when put up.

No book can be found among the doodads in the sprawling gift shop to supply a sense of these dislodged structures' social and cultural contexts. But it would appear that few buildings here, apart from the Stong piggery and such, had much to do with pioneers or pioneering. The Emporium probably did sell local produce; but its very existence, and what was supplied by it—imported wines, molasses, and tropical fruits such as oranges—suggests the existence of a widespread, efficient system for distributing both basic and luxury goods throughout Upper Canada by the 1850s.

It follows that numerous other features of Victorian industrial culture—pornography and prostitution, popular magazines and penny dreadfuls, pervasive religious doubt, mass-produced consumer goods of every sort, the dream of upward mobility—were also rapidly making their way into the Ontario outback. There's no sign of any of that here. What we have at Black Creek, instead, is neither a "pioneer village" nor a reproduction of a rural Canadian community, but a picturesque heap of fakes and architectural relics of Ontario's continuing commercial and industrial modernization.

Such matters seem to be of little interest to the Metropolitan Toronto bureaucrats who created and operate Black Creek Pioneer Village. The idea appears to have been to make, not a place where our understanding can be quickened, but a benign opium den of escapism into "the way we were," and really weren't. As visitors pass through the toll gate, says the glossy guidebook, "they are at once entering a different world...another age...reminiscent of a time when life was harder and simpler but rewarding."

If that's not bad enough, the prose gets even more queasy-making. "From the first step onto the wooden boardwalk," we read, "time changes. The smell of cooking, the sound of the blacksmith hammering at his anvil, the feel of soft fleece, the taste of fresh whole wheat bread and the sight of crinolined skirts swaying along the pathways, all help to erase the modern world for a short while." As a young female guide in long skirt and bonnet cheerily asked me, as I tried to keep from getting knocked from the plank sidewalk into the mud by a mob of adolescents being herded my way by their exasperated teacher: "It's so peaceful and quiet here, isn't it?"

The answer is *no*. Everything about Black Creek is annoying, from the musty museological atmosphere of the "village" itself to the junk-crammed boutique and fries-and-burger outlet masked by sentimental, schmoozy exaltation. But Black Creek can be a delight for the connoisseur of ironies and human folly. The buildings so diligently hauled to Toronto and carefully put here to evoke nostalgia for a "harder, simpler" time, for instance, were abandoned precisely because the society they served was much harder, and made obsolete by Ontario's swift industrial modernization between late Victorian times and the end of the Second World War. They were left to rot where they stood, because most people with good sense got out of those cruelly isolated "pioneer villages" as soon as they could, and moved into Ontario's industrial towns, for a crack at the jobs being multiplied astonishingly by new industries and enterprises.

The 1890s saw the first dramatic moves of Ontario farm folk to the towns and cities; by 1911, the urbanized population had surpassed the number of people still living in the country. In 1921, 58 per cent of Ontario's 2.9 million people were city-dwellers; and, by 1961—a year after Black Creek opened to the public—a mere 8 per cent still lived on farms. The people who streamed into the cities did so to join the industrialized, mass-political life the curators at Black Creek

Pioneer Village want to help us escape.

Every year, almost a third of a million people heed the call, and come to the "village" to recollect a sanitized, sentimentalized, utterly pleasant version of the wretchedness their ancestors left behind. This is a common human foible, and is not uncommon even among those of us who have undergone the traumatic leave-taking of the country and the embrace of the city, and should know better than to be nostalgic. Every year, untold millions of North Americans pay to "go back in time" to another architectural or archaeological fiction of the same sort—some "native encampment" or "frontier town." Nobody needs a survey to prove that this kind of fantastical tourism is hugely popular with urbanized mass populations.

I am sure of two things in this regard. First, that the Victorians who answered the siren song of urbanity, and forsook the miseries of farm work or the deadly constrictions of small-town life—indeed, everything Black Creek tries to make seem appealing—cannot be blamed for leaving. (Why don't these nostalgia factories ever include touches of real rural Victorian life? A child dying of typhoid or cholera, for instance, with a doctor and mother looking on helplessly?) Second, that Black Creek Pioneer Village and places like it are only helping the urbanized population avoid the consequences, rewards and complex moral questions involved in embracing urban modernity without reservation. One is not born a city person, after all. Urban sophistication, devotion to city liberty, are things that must be learned, patiently and carefully.

This learning is unquestionably difficult. Evidence: the thousands upon thousands of people fleeing the city via expressway every weekend, bound for the carefully regulated "hardness" and "simplicities" of cottage, chalet, anywhere but here; and the hundreds of thousands who each year visit Black Creek Pioneer Village, and perhaps believe that this self-induced amnesia is in fact a kind of recollection of paradise lost.

WESTON

At first, the property manager of the Weston apartment tower thought I had dropped by to write something about the bodies.

The corpses of two people, dead in an apparent murder-suicide, had been found the afternoon before in her high-rise building, a block or so east of Weston Road, and the place was busy with the comings and goings of investigators. She wasn't talking to reporters. It took a few moments to convince the lady that I had little interest in the bodies, which I had not even heard of, and considerable interest in her flat-topped metal and concrete block, towering improbably over the low-rise, rambling mumble of beautiful old houses, a church, seedy stores and weedy empty lots just north-west of what was once the town centre of Weston, when it was a town.

That once-upon-a-time, by the way, wasn't very long ago. Weston surrendered its official civic identity only in 1967, when it joined the Borough (later City) of York, itself a visually indistinguishable zone in Metro Toronto's westward expanse. As usually happens to villages overwhelmed by expanding large cities, the little one, Weston in this case, becomes a romantic victim, and the subject of myth and legend. The older citizenry of such towns seem particularly susceptible to the idealization of the past, and the twisting or more subtle editing of the facts to fit memories, real or wished-for. Those who live there invariably use prejudicial words like "gobbled up" or "buried" to describe what has happened to places like Weston.

Towns that have utterly vanished are, of course, not likely objects for such poignancy, simply because no architectural traces remain for the preservationists to draw their wagons round and fight for. Weston, in contrast, is still adorned with distinctive sites built in better times, excellent old houses, variously rustic, august and quaint, on deliciously secretive dead-end roads and fine maple-shaded residential streets, as gracious as any in rural Victorian Ontario. Too, Weston can boast Metro's most noble local library in the Romanesque Revival style, graced with gleaming mosaics around its stained-glass windows—a work of 1914 built with largesse donated by U.S. philanthropist Andrew Carnegie—and a history going back to the earliest British enterprise in this region.

Quite enough of old Weston remains intact, that is, to keep the local history and preservation industry busy, and quietly militant. Examples of this polite defiance includes the local historical society's *Pictorial History of Weston,* launched in 1981 on the occasion of the centenary of the ex-town, and the six mimeographed reports of The Town Project, a group of students pledged "to preserve the flavour of the village in 1882." In none of these publications do we find references to the Modernist slab high-rises so typical of Weston Road's streetscape today, or the large assembly-line operations that gave Weston economic cohesion during its final days of independence. The concern is, as always, with that intangible thing, "community spirit." "With the inflow of newcomers to the town, and the growth of apartment complexes," says a Town Project flyer, "the community spirit may be forgotten in the rush of development."

For a taste of what's meant by "community spirit," we have the archival photos gathered into the *Pictorial History.* This refined and busy Weston of myth and half-memory is captured in an 1869 snap of Prince Arthur, son of Victoria, turning the sod for the new railroad line; in a 1907 team portrait made when the Weston Lacrosse Club bagged a trophy; in a 1911 tableau, showing the Presbyterian congregation parading staunchly through the snow, like good soldiers of the Lord, preceded by billowing flags, to open its new Sunday school. We are supposed to be moved by the recollection, from 1924, when the village turned out to watch Miss Fisher, the school principal, march her charges to the fairgrounds down by the Humber River.

In the pre-metropolitan tale sketched out in such texts and compilations of images, Weston of old was a simple community of good folk of solid British stock, ever marching shoulder-to-shoulder, whether for Christ, a new railroad, the Orange Lodge, the Crown, the advancement of lawn bowling, or abstinence from liquor. It was certainly not the kind of place in which people were found murdered in twenty-eighth-floor apartments, or spoke foreign languages in the Jumbo Save, or worshipped strange gods which Victorian Westonians would have only known about from savage boys' books, or Kiplingesque hero-stories of the Raj.

What I find interesting in pseudo-historical publications such as these—works as numerous in North America as the declining small towns whose citizens produce them—is the pervasive assumption that some terrible gulf has opened between a happily stable, civic-minded past, illustrated therein, and a speedy, restless, incomprehensible present, unworthy of attention and certainly of preservation. Only indirectly or euphemistically—as "the inflow of newcomers" or "the growth of apartment complexes"—can the displacement and disintegration of the village's old Great British establishment be named. "Community spirit" is always another name for power, especially after it has been irretrievably dispersed, or seized by others.

I should quickly make it clear that I have no wish to lay blame on anyone here. Nor do I intend to portray Weston's preservationists as exceptional examples of what is, after all, an ancient and well-nigh universal imaginative tendency. Again and again, in narratives as old as Mesopotamian creation epics and the Bible and the earliest Greek poets, we find history's beginning in a lost, paradisical *illud tempus,* from which we have fallen into the baleful, alienated *hoc tempus,* our prison and destiny and present.

Surely, things have changed in Weston, but not as drastically as all that. The town, it appears from a close reading of

Sports Bar, Weston

the photos in *A Pictorial History*, has always been open to the tumults of modernization, from its beginnings as an lumber milling centre in 1792 until the present—revising itself architecturally and socially, refacing and gutting old buildings that lent themselves to such recycling, ripping down the rest. The mid-nineteenth-century improvement of Weston Road (a key link between burgeoning Toronto and the agricultural communities lying north-west of the city), the building of important bridges across the Humber River and especially the coming of the national Canadian railways in 1856 and 1869, sharply accelerated Weston's population growth and prosperity, and the cultural wrenching that comes in the wake of such change. Unlike other small towns besieged and occupied—if not yet actually annexed—during Toronto's extraordinary nineteenth-century expansion, Weston has always been religiously and ethnically heterogeneous. The process has merely continued: a 1986 provincial study showed that almost half of Weston's inhabitants at that time had a mother-tongue other than English.

Even before it joined the list of Toronto's self-abolished villages and small towns, Weston had lost its innocence of distinctively big-city traits. Metropolitan Toronto's 1960 aerial portrait of this minute region indeed shows the fancily porched Eagle House Hotel, built in 1870, still occupying the best corner of Lawrence Avenue West and Weston Road, the traditional hub of the village. But it also discloses the long shadows of high-rise apartment blocks along the western side of Weston Road, on narrow lots dropping off into the Humber River valley. Aerial atlases of the late 1970s reveal the non-town of Weston still moving right along that path towards full modern urbanism, with ever more conspicuous apartment blocks going up here and there. The Eagle House was gone by this time, its site filled by a cloddish cement mall and apartment superstructure.

And a less lofty survey of the same area, conducted by this

observer in sneakers in the autumn of 1992, tells the same story, but with details aerial photos do not offer. On Weston Road north and south of Lawrence Avenue, we find the hubbub of architectural styles, spread along a tawdry commercial strip, common to all arterial throughways which have remained in active, increasing use. What appears to have been a large Victorian family house on Weston Road, has had a billboard for wristwatches dropped in front of its second storey, and at street level, a pizza parlour installed in what was probably, well, the parlour. A low store-front with Depression Modern touches has been inserted between a tall façade of wackily fancy nineteenth-century brickwork and a squalid building of utterly indeterminate age. And a simple gabled house of the last century has been recently engulfed by a one-storey box sheltering a florist and a "men's shop," leaving only the old house's peaked roof and second-storey windows visible above its modern girdle. Had Weston actually been frozen in place at some point in its history—as its fans seem to wish—its older, distinguished buildings would never have had such interesting, quirky and always instructive perversions visited upon them. Yet these are bits of evidence for a history of demolitions and replacements, of urbanization itself, that remains unwritten, clearly because it is unloved.

To return to the murder building—or King Square, to give the block its proper name—this thirty-two-storey structure was built in 1975, said the manager. Its shadow is the most striking one I saw in the post-1975 atlases, due to the abruptness of the building's full-blown appearance, and its size, grotesquely outscaling the humbler buildings around it. King Square seems to be a monstrous intrusion into an otherwise largely unchanging town plan, however, only when viewed from afar, or from the air, or, lazily, over the rooftops of the much lower commercial buildings along Weston Road. Seen against the actual historical development of Weston, the apartment block is just an especially conspicuous instance of the casually chaotic modernizing that's been going on for what seems like forever.

Granted, you couldn't always browse in a sex-toy shop on Weston's main street, as you can now. Considered as an expressive architectural ensemble, Weston is, in its small way, as rich in the perverse, eclectic, confused spirit of contemporary urbanism as the immense city that took it over after the railways had gone belly-up and the town had fallen on hard times, making it the gobbled instead of the gobbler, and spawning many a devotee to the good old days which, if they did exist, weren't *that* good.

ABOUT WILLIAM GILPIN

This is partly the story of a cottage on Broadview Avenue, but mostly the tale of William Gilpin, who is a curious man. It is also a reminder of the mysteries that gather round every house with the passage of years, waiting for a curious person to come along and patiently untie the knots of history and unhood the shrouds of time.

Speeding down Broadview south from Danforth Avenue, with its panoramic views towards the radiant Toronto skyline across the Don Valley, you don't tend to notice the little old cottage, painted white with green trim. It is set well back from the avenue behind a front garden only gradually being recovered from dilapidation and decades of neglect, and crowded in between two big, muscle-bound Edwardians. If it hadn't been pointed out to me, I might never have appreciated this charming reminder of earlier Toronto, its symmetrical façade still prim, its hipped roof and external chimneys on either side still straight.

Despite the house's demure manner—or, rather, because of it—Gilpin had delighted in it for more than a decade, from the vantage-point of his Cabbagetown apartment opposite, across the Don. He has always loved old things, old stories and romantic songs, he told me over a cup of espresso in his inexpensively antiqued front parlour, and he had been "an archaeologist by inclination" since childhood. He and the cottage were meant for each other—such had been his belief for years—when, in 1992, this tattered survivor of better days went up for sale.

Gilpin clearly had, and has, the urban pioneering sensibility—an acquired trait, by the way—required to repair and live in such a place. A sensibility, that is, disinclined to clear and conquer, and much inclined to save and spare. He cherishes furnishings and ornaments and out-of-date architectural "improvements" that more speedy, self-consciously progressive people would throw out. He is enchanted by traces, however scanty or fragmentary, of the past—which he imagines, with forgiveable incorrectness, to be more elegant and leisurely than the present—and is devoted to preserving each trace, like a talisman enabling communion with the long-dead souls who left it.

But Gilpin, now in his thirties, has also been a rolling stone all his life—working just long enough to save up money for the next wander around the world, odd-jobbing and living gracefully on little—and so did not have a nickel to put down on the house. Nor did he have much time: the lot size was perfect for a duplex, and the asking price was only $218,000, a Toronto developer's dream in the slow housing market of the early 1990s.

But with the good fortune that often shadows gentle fans and aficionados, and with remarkable generosity on the vendor's part, Gilpin managed to piece together the necessary financing, beat out the developers who would have handily knocked down the house in a day, moved in, and got down to work.

The most immediate job was the physical redemption of the cottage from the gradual ruin visited upon it during decades as a rental property. I would not be writing this, however, were Gilpin just your basic Toronto white-painter and fixer-upper, buying cheap this year in order to sell dear three years down the line. What makes his story interesting is his commitment to stay put—not something Torontonians are inclined to do—and patiently reconstruct his house's entire history from the fragments, hints and concealed systems in the shelter itself, and from what pictures, genealogies, deeds and any other documents he can find.

In the kitchen, Gilpin showed me a frayed but still radiant expanse of deep violet printed wallpaper in a flourishing Arts and Crafts pattern, pasted up perhaps in late Victorian times on the wood wall by someone with high style, and perhaps a bit of highfalutin attitude. He discovered it under a stratum of cheap tile. In his cozily fusty dining room, Gilpin takes down an anonymous junktique-store portrait from its hook and handily rips aside a section of modern hardware-store panelling, to reveal a swatch of fine mass-produced wallpaper, circa the Roaring Twenties. Every scrap of wallcovering Gilpin finds, by the way, he carefully preserves and catalogues and researches, in hopes of someday reproducing the patterns, or perhaps even finding long-forgotten bolts of the original stuff stacked in a dusty warehouse.

With the removal of each patch of beaverboard or parquet flooring, another level of the cottage's history, and the tale of the people who lived there, has been revealed. And under each level is another clue on the route backward in time, farther down the rabbit hole which this unimposing cottage has become for the inquiring imagination of William Gilpin.

Of all the clues the owner has uncovered, surely the most intriguing and puzzling is to be found in the main-floor bathroom. There, after pulling away later wall facings, he found a rough wall of square-hewn logs. Gilpin had never believed his cottage was as late as its quaint Victorian front porch suggested—in the 1880s, that is, when the east side of the Don

Valley was annexed by the City of Toronto. His guess, based on a study of early Ontario architecture of the same type, was that the house could have gone up as early as the 1840s.

But finding the log house wrapped inside the later architectural fabric could, Gilpin believes, put the house back to a very early date—perhaps as long ago as the mid-1790s, when European settlement and first construction took place in this part of the world. Peculiarities about its construction, and records and pictures Gilpin has dredged out of public archives, suggest the cabin may have been moved from another location to its present site after 1815, with a mind to using it as the architectural basis for the larger house that was, in fact, built around it. (Part of the cabin's original roof can be seen in the attic of the expanded house.)

The siting was done with "a degree of pretension," in Gilpin's opinion, who makes his point with a sweeping gesture towards the landscape beyond the front window and Broadview Avenue. Indeed, one could hardly have chosen a more propitious spot on the east side of the Don, if what was wanted was a peculiarly Victorian Sublime view of the valley and town beyond. He then heaps the parlour table high with documents, maps, reproductions of old watercolours and photographs, abstracts and wills, branching and twigging genealogies of families who lived here or owned the land, all gleaned from myriad city and provincial sources; then begins going through the stack with exacting concern.

I do not intend to evaluate this evidence here, or further describe it. Perhaps the curious man who owns it does indeed live in the oldest house in Toronto, as he fervently wants to believe. But even if it were possible to pinpoint the cottage's origin and numerous transformations—obviously a matter of considerable interest to local historians and architectural buffs—I would find all that less interesting than the discovery of William Gilpin himself. While I have met only a handful, there may well be hundreds of people like him across

Metropolitan Toronto, who share his sheer joy in the legends and mysteries and odd clues already unmasked in house and home, and the quiet thrill about the many others waiting, behind baseboards and plaster and flooring, to be discovered. I would like to hope there are thousands—if only because each of us needs at least one charmed guide to get us started on this safari of intellect and soul into the hidden heart of where we are.

Pleasures in Places

Fall Yard, St. Clarens Avenue

CHRISTMAS AT COLBORNE LODGE

In an effort to brighten an especially dreary December morning, I dropped by the country house of John and Jemima Howard, known as Colborne Lodge.

The Howards were out. In fact, they've been out now for upwards of 120 years. Nor is Colborne Lodge in the country any more. This eccentric, asymmetric house, begun in 1837 and believed by its proprietors to be one of the oldest examples of the picturesque "cottage ornée" in North America, now stands atop a hill at the south end of High Park, once the Howard farm and now in deep city. The house's bluff-top situation once commanded a fine, panoramic view of nearby Lake Ontario, and it still does—though in between edifice and water now lie a broad hem of landfill and noisy torrents of road and rail traffic which never stop.

Once inside the front door of Colborne Lodge, however, the city and cars seem very far away, as indeed the Toronto Historical Board, which operates it, intends. Hot spiced cider and fresh-baked molasses cookies perfume the air. Candles on the dining-room mantel illuminate a table laden with tarts and cakes, petits fours and puddings. The pleasantly stodgy nineteenth-century decor of the downstairs rooms—all polished dark oak, floral wallpaper, heavy swags, embroidery, patriotic prints—is festive with boughs of pine and red ribbon and festoons of cranberry and popcorn. The Christmas tree stands at military attention in the parlour, bright with quilled snowflakes, beaded eggs and little presents.

Taken together, the Howards' Yuletide decorations are a museum display of what Christmas was for Toronto's middle class, and has almost completely ceased to be. Merry, but not gaudy. Sparing, but never plain; an adult occasion, but without a certain officiousness. Jubilant, but without a whiff of High Toronto ostentation. The word I want to use for the atmosphere in Colborne Lodge is *jovial*—though nowadays this term has lost most of its old associations with King Jove, and fine royal joy, and celebrations high and serious without being in the least stuffy or solemn.

Colborne Lodge is the carefully preserved setting of a fiction created by Sandra Molyneaux, an anthropologist by training and senior curatorial assistant at the Lodge. So how did Molyneaux know what a Toronto Christmas, circa 1870, was supposed to look like?

Some of the sources come from mass-produced images of the period—those popular portrayals of happy hearthsides and Christmas customs that reached their zenith of appeal and jollity during the 1870s in the work of U.S. illustrator Thomas Nast. Then there are the literary descriptions, notably one written in 1850 by Charles Dickens about a contemporary Christmas tree—still a novelty in England at that time, since Prince Albert had imported the idea from his native Germany, and set up the Anglophone world's first Christmas tree at Windsor Castle in 1841. Gorging at Christmas was well established in Toronto—in 1860, *The Globe* ran a seasonal food column hymning the sweets, lobsters, sardines, oysters, wild fowl, grapes and ciders for purchase from a King Street purveyor—though taking the day off to enjoy the feast became official in Canada only in 1867. (William Lyon Mackenzie, editor and rebel and muckraker, docked a copy boy sixty-six cents for taking off Christmas Day in 1839.)

Nobody's stockings have been hung by the Howards' chimney with care, because Canadians did not do that sort of thing in 1870. Christmas cards, too, were new; they had been invented in England in 1846, but would not come into common use in Canada until almost fifty years later. Giving presents to teachers, bosses or anyone outside one's immediate family and most intimate circle of friends, and the paper-wasting business of wrapping gifts, are even more recent "traditions."

What's on view at Colborne Lodge can be looked at merely in a tiresomely grown-up way, as an illustration of seasonal interior-decorating practices in old Ontario. But get as swept up in the anthropology as you will, the decor at Colborne Lodge keeps whispering a question that all the file-cards and scholarly books in the world probably can't answer satisfactorily: What makes this seasonal carry-on work so well, and so durably?

The answer isn't in the strength of something called "tradition," since, we are reminded at the Lodge, the "traditional" Victorian Christmas was itself a pastiche of novelties. Yet at some time around or just after the era represented at Colborne Lodge, the whole heap of stuff known as "Christmas decoration" or "Christmas custom" came together and froze into place exactly as we have it now. Nothing else at Colborne Lodge has worn so well as this instant so-called tradition of holly, wreaths and all the rest. Of course, the taste for the Lodge's kind of Victorian furniture and furnishings comes and goes, but the taste for the Victorian Yuletide ensemble of mistletoe, mincemeat tarts, evergreen and presents and so on is as strong today as at any time in the century since it was all invented.

Despite this novelty, however, the traditions of Christmas have proven astonishingly resistant to the various modern movements in design and architecture, which have profoundly reshaped virtually everything around us and utterly demoted most other instances of Victorian fussiness from their previous ascendancy. Mistletoe is now hung from steel girders in high-tech urban lofts or freshly installed light fixtures, instead of from glittering chandeliers—but it does continue to be hung. Instead of the cut-glass punch cups of yore, neatly laid out on a linen tablecloth, one can now find plastic glasses arrayed on a blond ovoid table of Scandinavian design, or smoked crystal tumblers gleaming atop glass-and-steel wet bars—but what goes into these vessels year after year is still the same old Victorian cider, accompanied by the same old fruitcake and shortbreads.

I am inclined to think that the importance of these dear symbols—which it would feel like a kind of blasphemy to omit—resides in the specious and imaginary link they provide with some vague land of "what came before." This is not a link I otherwise do much to preserve, and have done much to break. Nevertheless, this Christmas as at every Christmas, the wreath of pine boughs and red ribbons will hang on the steel and industrial glass door of the reclaimed and renovated furniture factory I call home. I can do without a front or back garden, I can live with exposed industrial girders, and am happy to have a kitchen fitted out with modern appliances, and a fridge capable of holding a prodigious amount of heat-and-eat. But, without that guilty wreath, it just wouldn't be Christmas.

UNKNOWN GARDENS

Elsewhere, this book is an appeal to curiosity—to the natural intellectual desire, often dulled by the routines of city living, to know just where we stand, and find out what sort of place it is. Elsewhere, that is; but not here.

My topic at this point is the vein-like net of railway corridors. Visiting them is not legally forbidden. Due to an apparent oversight, the federal Parliament left out penalties against trespassing in its most recent legislation regulating the railways.

But this network *is* private property, after all, and peculiarly dangerous property as well.

The most deadly threat is posed, of course, by the rolling stock. Despite their huge size, and the mechanical ruckus they kick up as they move, trains are stealthy. Anyone who has walked railway tracks, as I have, all my life, knows that a train can be almost on top of you before you know it's coming. One has to have an uncommon attentiveness to what's happening on and around the tracks, an habitual resistance to distraction, to keep from getting hit. Then, on the right-of-way, there is many an uncovered pit to break a leg in, and numerous old track ties and iron oddments hidden in the weeds to fall over. Warning signs are, of course, almost non-existent. You are not supposed to be there. And I would urge anyone not already accustomed to them already to stay off railway tracks forever.

That said, I would nevertheless urge everyone to be more alert to these remarkable, much-disregarded facts of urban architecture, which lie everywhere in the city—behind rickety wooden fences at the end of dingy streets of auto-body shops and decaying houses, at the rear of important houses in Rosedale, in the cleft of ravines and on high stone bridges, concealed by street-long factories.

One safe, informative vantage point is from the window of a GO commuter train. Another, good for studying the one-dimensional architecture of railway corridors is any footbridge. My favourite is the old steel Junction footbridge in Toronto's west end, a sturdy little work of utilitarian building from Toronto's late industrial age providing safe passage for pedestrians over the north-south Canadian Pacific tracks, from Wallace Avenue to Dundas Street West. (The newer fashion in avoiding level pedestrian crossings, incidentally, is to run the footpath under the tracks, thereby creating dank, menacing, repulsive concrete burrows.) From the breezy overlook afforded by the Wallace Avenue footbridge, a certain irony of the corridor is apparent at once.

In both directions run the thin parallel threads of gleaming steel, opening prospects as staunchly rational and uncompromisingly naked to the sun as an alley at Versailles. Here, indeed, is a sublime, if unintentionally sublime, expression of one of the nineteenth century's most radically new architectures of transport. Here is the endless metal form that in a generation displaced the river and lake systems' age-old rule over North American trade and travel, and instantly became its era's monument to the new age of mechanized speed.

But if rivers have been routinely honoured by rows of costly houses with picture windows and scenic parks with belvederes, railways see only the city's interesting technical backside. The factories show the railway passage their tubes and tanks, chutes of cascading rubble and racks of shiny car bumpers, tangles and bundles of wires, steam-belching pipes, flapping building exhausts, snappish guard-dogs, junked transformers, the rusting guts of dead trucks. Walkers in the ditch beside the tracks at a certain west-end location see what the street side of a tall brick façade is supposed to conceal: a conclave of skeletal steel titans festooned with wires, all part of a complex and majestic hydroelectric installation its makers apparently thought ugly enough to merit camouflage. So, too, the residents think nobody's noticing—but a hiker alongside the tracks knows of their attachment to clothes-lines, the elegance of their tree-houses, their forbidden backyard businesses (keeping chickens, curing meat in smoke-houses), their past taste in broadloom and sofas, discarded over the fence to decline into soggy pulp.

As an unintended result of its danger and the unpopularity, the urban rail corridor, otherwise so eerily splendid a piece of modern calculation, is also the largest of the city's wild, raw places. In it every sort of garden flower and blooming shrub gone AWOL finds refuge among their feral cousins, the lovely weeds and wildflowers, untamed vines and fruit trees, all come from distant regions into the city's heart with little meddling by any people other than the railway folk.

And people like me. From early spring to the first snow-storm, our dining-room table and desks are rarely without branches of cherry and fragrant blossoms, bunches of wild grapes, cattails and thistles, fragile blue and yellow blossoms cut from broad spills down embankments, and, after freeze-up, selections from the gaunt, rattling skeletons of summer flowers.

The railway officials, quite properly, do their best to keep this botanical invasion at bay, hacking the weeds and striplings and creeping vines back away from the tracks. But some things manage to survive, by luck or just because they're tough. In early spring, the first patches of pushy green can be seen coming up and shouldering aside the dead stalks and grasses along the Canadian Pacific track in my neighbourhood, while nearby lawns are still moribund. The gnarled vines sprawling crazily across the cinder-block backsides of factories are still grey, though scrubby clumps of trees, clumsy thickets and shrubs crowded against fences and walls are thick with plump buds. And before very long, this unfussiest of Toronto land-scapes again is graced with our brief spring's wildflowers, dis-playing their colours under a sky kept wide and open—not for them, of course, but for the trains that have unintentionally created Toronto's sunniest, most savage garden.

ST. JAMES' CEMETERY

The gilded afternoons of October are surely the year's best for rediscovering the special peace and pleasure of city cemeteries. The long days and abundant flourishing of summer are past; the iron-grey skies of winter are yet to come. Yet in this inter-val between, the stone monuments seem to find their perfect backdrop, against the yellow and red blaze of tree and shrub, and green lawns crossed by skipping scatters of flame-coloured leaves.

There are a number of old Toronto burial grounds to stroll through, and one should take them all in. Gradually, one finds favourites; among mine is St. James' Cemetery, just south of Bloor Street East, on a stretch of rather level ground between Parliament Street and the steep drop-off to Rosedale Valley Road.

A simple explanation for the enjoyment such places provide, beyond their scenic beauty, is the sense of unbreakable kinship with the dead they offer, and the gratitude they solicit for those buried there. Now I know St. James' reputation as Toronto's toniest resting place. It is "the Four Seasons of Toronto cemeteries," a reader of this manuscript jotted in the margin. "You pretty much have to be a member of the Social Register to get in." It is true that many of the city's high and mighty await the Last Judgment in this ground, and will doubtless (like the rest of us) have much to answer for.

Like most of those buried in St. James, I belong to the Anglican tradition of Christian practice; the dead here are my kinsmen and kinswomen in faith, to be honoured in death for reasons of loyalty alone. Moreover, if some of the oldest burials are members of the Family Compact—figures of almost demonic wickedness in nationalist Canadian mythology—this colonial ruling group of soldiers and entrepreneurs, presided over by the Anglican prelate John Strachan, has always struck me as a remarkable bunch, courageously establishing British institutions and cultural ideals in a bleak, muddy no-place at the edge of the Empire, because they believed it their duty to do so. Toronto owes a great deal, including its very existence, to the first creators and sustainers of urban civilization on this hostile shore, and to their early successors—men and women buried here, with such names as Jarvis and Gooderham, Ridout and Osler, Baldwin and Austin and Gzowski.

Whether or not another visitor could summon up such an attitude towards the dead of St. James, anyone interested in cities and urban change will be intrigued by the architecture of death here, and its visible symbolism of the great mental change in our thinking about death that occurred in Victorian times.

The surveying and layout of a new cemetery was undertaken by John Howard in 1842 on behalf of the Church (later Cathedral) of St. James, after it had become clear that the downtown burying ground adjoining the church was no longer big enough for the clientele. The bodies which had gone in the ground since the burial site was established, in 1797, were reverently dug up and moved uptown in 1844 to the new cemetery, established in what was then undeveloped countryside, north-east of the built-up town.

Viewed as a moment in the nineteenth-century transformation of cemetery design, St. James is poised almost exactly in the middle. It was laid out by our prolific first city engineer, John Howard, whose work, it should be noted, began after the inauguration of the modish "new" North American graveyard style, featuring ponds, surprising views, shrubbery, Arcadian hills and dells, pioneered by the Boston Horticultural Society's Mount Auburn Cemetery, a Cambridge establishment of 1831. By the standards of such "advanced" cemetery design, St. James was old-fashioned at the time Howard made it. (By 1849, Philadelphia had almost twenty new-style cemeteries, and an issue of *The Horticulturalist* in that year could proclaim that there was virtually no American city of note "that did not have its rural cemetery.")

At St. James, we certainly find no romantic landscape of tiny secret valleys, streets of Greek tombs, few twisting paths, little sorrowing yew or dripping willow—and precious little else to "elevate the mourner's feelings, reduce his sorrow, promote religious meditations and foster a sweet melancholy," as a recent architectural study has described the effects desirable in a rural cemetery. Instead, St. James is characterized by unscholarly plantings of sturdy Ontario shade trees, a generally unmodified flatness of its terrain (at least in its earliest sections), and many more or less unimposing monuments set on a plain grid.

Architectural fashion appears to have held little interest for the planner of St. James, who was merely trying to replicate, on a large scale, the ordinary grid of the English parish churchyard as it had existed from the early Middle Ages through the nineteenth century—though without a church in the midst. In 1861, as the Gothic Revival was gaining ground in all areas of North American design, this unsophisticated English-churchyard look was suddenly reinforced by the construction of the Chapel of St. James-the-Less. Its interior, clearly a one-stop affair, is invariably disappointing. But on its exterior the outstanding Toronto partnership of Frederick Cumberland and William George Storm expended all their romantic love of twilights and shades, and delight in the antique. (Cumberland, after all, knew John Ruskin; and he had only recently returned from England, ravished by a new-found love of the British Middle Ages.)

The result is a perfection of Gothic Revival mortuary building, antiquarian and mindful, and Toronto's most exquisite and sensitively realized instance of Victorian medievalizing. The addition of this chapel, however, does not alter the fact that St. James is *not* a parish churchyard—and that's where lies the interest in the place for the fan of urban transformations. In 1844, when the original graveyard was full, old Christian practice would have dictated that the churchwardens just dig up the old bones, heap them up in storage sheds somewhere on the property—back in England, the church attic had always been a favourite place—and start dropping more bodies into the ground.

This ancient practice is undergirded by an idea that may seem odd to secular urbanites, who do not sense the novelty of a Christian cemetery separate from a place of worship. In a revolutionary departure from the near-universal practice of ancient Mediterranean peoples, who looked upon corpses as unclean and spiritually contaminating, early European Christians cherished their dead, and expressed their belief in the defeat of death's power to break the unity of the Church by keeping the Christian dead as near as possible to living believers. From at least the fifth century, dead Christians were buried in places of worship—under the floor or in the walls—or at least immediately nearby: hence, the evolution of the English churchyard into perhaps the most sociable and delightful meeting-place of living and dead ever created.

The custom continues almost nowhere today. By the middle of the eighteenth century Enlightenment, the old pagan idea of burial outside the gates was making a strong comeback in European thought, under the guise of a "scientific" worry about sanitation. In France first, later in England and in its colonies, the ancient Roman standard of burying the dead beyond the city gates returned, creating—as an architectural historian has noted with a smile—the first modern suburbs. The full-blown expressions of this notion, part of a far more general secularizing and repaganizing of modern thought, are the churchless, romantically rural picturesque cemeteries of the Mount Auburn type, which were only later engulfed by the cities outside which they were originally constructed. St. James is an example of a "modern" burial ground—laid just outside the mid-century bounds of the fledgling city of Toronto—but largely innocent of Rural Cemetery Movement influence.

It did, of course, get a token parish-style church two decades after its inauguration. This fact may cause archaeologists of the distant future, digging up the site of St. James' Cemetery, to scratch their heads over this Christian burial site of the late second millenium. It will look something like other parish churchyards, though the church on the grounds—the little chapel of St. James-the-Less—will seem unusually tiny when compared to the great number of burials around it. Even more puzzling, perhaps, will be the crematorium in the basement of the chapel.

These researchers will almost certainly know that, for most

of the first two thousand years of Christian history, believers strictly avoided cremation, just as they tended not to bury their dead outside the city gates. So how did it happen that Anglicans started cremating? For the sake of posterity, and supposing this book survives longer than its author, here's the answer.

After cremation, like exurban burial, was revived and popularized in the late eighteenth century, thoughtful Anglicans, like other thoughtful Christians, began to worry about this issue. They were pulled back and forth by formidably old, deep customs concerning the disposal of the dead nearby, and by the new, persuasive and plausible theories of sanitation then making the popular rounds. The argument continued until 1944, when the dying William Temple, Archbishop of Canterbury and one of the Church of England's great minds in our century, decided to settle the matter once and for all. He did so in a peculiarly Anglican way—not by issuing a decree with all the august authority of his ecclesiastical office behind it, but by simply ordering that his body, once the spirit had taken flight, be cremated; which it was. The crematorium at St. James-the-Less was installed in 1947.

MOUNT PLEASANT

Mount Pleasant Cemetery is the stylish modern burying ground that makes St. James seem stolid and bluestocking today. It is located on a 206-acre tract north of St. Clair Avenue. Were I not resigned to take my final rest in consecrated ground, I can think of few more tranquil places to await the Great Getting-Up Morning than non-sectarian Mount Pleasant, which opened for burials in 1876.

Nor does Toronto offer a better place to start collecting ideas for interesting markers, crypts and such—inventive markers being one legacy of the Garden Cemetery Movement.

In Mount Pleasant's secret glades and on its sunny hillsides, one can find excellent evidence of virtually every sort of taste in monuments, ranging from the post-Christian and melancholic (exampled by those shroud-draped urns and broken columns beloved by sceptical Victorian mourners) to the defiant (the soaring obelisks and serviceable benches of rationalists, the tall, stately crosses put up by Christians) to the efficiently informative (name and dates only).

Some of the best ideas (and also those to be most carefully avoided) are embodied in the grander monuments. Among the noteworthy mausoleums of old is the Massey family's stout, sturdy Romanesque fantasia, created in 1891 by Toronto architect Edward James Lennox, who also gave us Old City Hall, Casa Loma and other studious memorials to grand-manner styles of the past.

Then there's the imperiously pagan Classical temple wherein the remains of the Eatons, of department store fame and Protestant affiliation, lie forever guarded by bronze lions; and, near it, the smaller, exquisitely proportioned Greek study erected in 1905 for himself by financier George Albertus Cox. (The Cox mausoleum, incidentally, has perhaps the most beautiful pictorial stained glass in Mount Pleasant: a faintly decadent *fin-de-siècle* portrayal of the Risen Christ, floating in opalescent, mystic light.)

It has been a long time of course, since any but the very rich have been able to afford the imaginative design and craftsmanship that went into such masterpieces of funereal architecture as the Massey and Cox mausoleums. Anyway, funeral memorials meant to aggrandize the dead and insist upon their splendour are now out of style, even in bad taste; yet the reaction against this former inclination has not, by and large, been a happy one. The long-term design trend suggested by what's in this burial ground is from monumental towards puny, from ornate towards spare, from exuberantly hand-crafted towards routine, mechanical, trite.

Mount Pleasant Cemetery

Like other cemeteries in which people are still being buried, the area of Mount Pleasant in most recent use is altogether too cluttered by squat, small grey-granite markers with curving crowns and straight sides, all looking more or less alike—"serp-tops," so called, with faint contempt, by monument designers on account of the serpent-like curve characteristic of their crests.

But even if you have decided to settle for a serp-top, there are still a number of lessons to be learned from Mount Pleasant's wealth of monuments. The first, and perhaps most important: make your gravestone as free of lofty words and pious expressions as possible. The older monuments of Mount Pleasant are mercifully short on vaguely spiritual drivel, though as one moves towards the present the twaddle does seem to increase. We are treated to many a dreary bit of Edwardian uplift, along the lines of "the progress of mankind onward and upward forever," or a school-days tag from Shakespeare plucked out of context.

Most people will want to have at least their names put on the rock. But the rule here, again, is: the less said, the better. Few people, I suspect, will want to go as far in the direction of muteness as Toronto art collector and financier Christopher Horne, whose Mount Pleasant monument, by Ontario sculptor John Noestheden, is simply an unimproved boulder poised atop an open bronze frame—just a plain Ontario rock, and no words or names on the plinth. If an unusual statement, the Horne marker is nevertheless a useful corrective, and a reminder of what's good about the best gravestones, old and new, in this "modern" graveyard: their insistence on the heft, density and solemnity of stone, ideal in this picturesque green setting, and a corresponding reluctance to yap on about the dear departed.

Mount Pleasant is, of course, a romantic (if exceptionally large) modern cemetery, intended to rob death of its grimness and give it a touch of the dramatic—an intention its older sections admirably fulfil. The statues and inscriptions, the heroic sighs and pluck made visible in stone, the reminders of all the murders, sinkings, crashes, heart attacks, shenanigans and quiet, noble passings that brought the current inhabitants of Mount Pleasant to their graves among the ravines, hollows and hills seem all to be neutralized by the exuberant landscaping. It should come as no surprise that, unlike St. James, Mount Pleasant has become a favourite place to jog or stroll—bringing the recent history of cemeteries around full circle. Towards the end of the eighteenth century, the dead were banished from the dwelling-places of the living; at the end of our century, urban life would hardly be complete without a gracious downtown cemetery to repair to, to work out in, or merely to enjoy. The picturesqueness of the cemetery has drained proximity to the dead of morbidity and fearfulness—an effect I'm not sure a cemetery *should* have. There is surely health and good sense in being reminded from time to time that we are but dust, and to dust we shall return.

THE EX

Every August, usually on the stickiest, hottest day of the summer, the main gate swings open to the latest Canadian National Exhibition, the modern world's oldest annual agricultural and fun fair, and just about the most gaudy, raucous, sweaty, and low-down-dirty thing we tasteful Torontonians do on a regular basis, in public anyway.

But whatever may go on behind it, the principal entrance of Exhibition Place, known as the Princes' Gates, represents Toronto's most lofty attempt to produce a local item of the "City Beautiful" architectural confectionery popularized by the 1893 World's Columbian Exposition in Chicago. For the North American inventors of our century's best building—plain, businesslike, structurally clear—the fair had been a disaster, and chief designer Daniel Burnham's panoply of white neoclassical fantasy buildings set among artificial lagoons, the source of visual plague. As Louis Sullivan remarked: "The damage wrought by the World's Fair will last a half a century from its date, if not longer."

His judgment would have seemed just to the most forward-looking architects of his time, but to many a builder, I suspect, it would have been absurd. A generation of ambitious American and Canadian architects rushed home from Chicago, their heads busy with plans to put Beauty—the ornamental and grandiose and historic—at the service of civic Boosterism. Thomas H. Mawson swept across the Canadian prairies, laying plans in 1912 for a Saskatchewan Lieutenant-Governor's palace of czarist dimensions, and in the same year envisioning a civic centre for Calgary featuring blocks of enormous white classical office buildings set on a titanic plaza beside an artificial lake.

While these projects in the west came to nothing, Toronto did get its most self-consciously noble piece of city decoration. The Princes' Gates is materially pedestrian, unadventurous in design, and distinctly lacking suitably aristocratic context. It is nevertheless our most wonderfully attractive bit of colonial pomp, show-off and would-be grandeur.

Designed by Toronto architect Alfred Chapman and engineer Morrow Oxley, the "Gates" is actually a single ceremonial gateway, done more or less in the manner of a bewinged Roman arch, but rather more dainty than the heavy old Romans would have abided. The occasion of its building was Confederation's diamond jubilee, in 1927. Its ostensible purpose was to pay patriotic and loyal tribute to the work and know-how that had made Canada a prosperous land.

The gateway is crowned with a flamboyant winged Victory—or, to be precise, a plastic copy of the original, which quickly deteriorated—who stands proudly at the prow of her onrushing ship, a laurel wreath held high for Canada's farmers, industrialists and workers, the true heroes of the nation's modern age. From either side of the gateway stretch complementary colonnades of nine unfluted Doric columns each, tributes to the Canadian provinces that existed in 1927. Both colonnades end in a sort of short curving tower topped by a flourish of Canadian and loyal British heraldry, and fitted out with a lion's-mouth fountain, which occasionally operates.

Unlike a Roman triumphal arch, however, the Princes' Gates is visually most effective when glimpsed from afar and at high speed, like any roadside architecture (such as billboards)—out your car window, that is, while negotiating the broad curve of Lake Shore Boulevard around Exhibition Place.

Stop and examine the structure close up, and the surface grandeur of its styling is quickly deflated. First, there's the matter of the building material, which is not marble but just poured concrete in a couple of complementary tones, one eggshell white, the other gritty beige. And then there's the fact that all the ornamental drama is at the top, to the neglect of what happens down below.

But up along the defining horizontal line of this confection, things are very theatrical indeed. The entablature above the Corinthian columns of the gateway is crisply dramatic in effect and academic in detail. To temper this vigour at the top, Chapman and Oxley fashioned the next level down as a stage for a suite of four pensive allegorical figures, bearing cornucopias and beehives, traditional emblems of plenty and industry. A proper Beaux-Arts monumental fabrication would then have brought the viewer's eye down from this exalted mythological platform, along the serene verticals of the columns, to a dignified horizontal base in the here-and-now world, and, finally, out to the formal garden or promenade which the arch was designed to frame and introduce.

But it's towards the bottom that everything about the Princes' Gates goes wrong. The capitals of the Doric columns representing the provinces, for instance, have been cast correctly and carefully, according to the fashionable textbooks of the day—Chapman learned well his conventional Beaux-Arts architectural flourishes in Paris and New York—but the feet of these columns are merely ugly barrels or cylinders of concrete, without dignity or finesse.

But the real wretchedness starts beyond these pseudo-bases. The Gates fronts directly, with virtually no decent interval, on a busy street feeding cars onto Lake Shore Boulevard. Lying immediately on the inner side of the Gates, on the grounds of Exhibition Place, is a straight avenue which was, until the 1960s, a grand entry-way flanked by trees and fine buildings. All the pavilions of this promenade are gone, along with the trees, save for the beautiful Automotive Building, standing in Art Deco dignity just south-west of the gateway. It is today the only tribute left to Labour and Industry announced by the portal itself. (The handsome Electrical and Engineering Building, on the north side, was demolished and replaced by a parking lot.

The gateway was inaugurated, promisingly, not by an opening-day's crowd of curious visitors, but, as a plaque affixed to the structure tells us, by "a veterans' parade under the auspices of the Canadian Legion & the British Empire Service League for review by H.R.H. The Prince of Wales."

All else in the region is commercial squalour: a jumble of ticket booths and signage, and, worse still, vast deserts of asphalt beyond which rise the tacky façades of the Midway attractions.

Despite the fact that its vision of the nobility of labour in a stately, important architectural ensemble at the east end of Exhibition Place has been lost, the Princes' Gates is surely the most affecting memorial to an honourable aspect of the old Ex that contemporary fair-goers are likely to forget, if they ever knew about it.

Not that the CNE was ever all that serious and honourable. There were weird animal acts and lurid sex shows on the Midway, and all the icky pink cotton-candy you could eat, back in 1927, when Edward, Prince of Wales, and his brother George officially opened the Princes' Gates, and gave the monument its name. The amusements you can buy your way into nowadays are, in fact, distinctly tamer than the really vulgar ones that used to get Toronto's Calvinistic knickers in a proper knot once a year.

But even as the little booths of horrors and tents of wicked pleasures have slowly disappeared from the Ex, so has the fair's role as Canada's grandest annual celebration of its industrial and agricultural progress, and the nation's rootedness in Victorian ideals of loyalty to the Crown and commitment to technological advance declined into insignificance. Only the Princes' Gates remains, to recall this venerable dimension of the Exhibition's long history as a tribute to Empire and excellence, and its role—the role of such fairs in general—in heralding the onset of technological and architectural modernity.

In 1882, the dim night-time alleys between tents and pavillions were turned to day, as Exhibition Place became the first in the world to be lit by the genii of electric lights. The world's first automobile shows were held there, and the first demonstrations of phonograph recordings. Even as recently as the early war years, crowds of CNE visitors were among North America's first people to witness the transfiguration of the human image into electronic signs, as Jack Dempsey and chanteuse Jessica Dragonette faced crude television cameras.

With the electrified fair of 1882, the Exhibition also presented the first instance of the new industrial empire's typical entertainment—not mass warfare, but mass spectacle mimicking the organization of battle-groups. The premiere event was called *The Battle of Alexandria,* and featured bombardments, fireworks, immense painted sets, fiery dragons hovering in the sky and a cast of many hundreds. The spectacles in subsequent years included *The Siege of Peking, The Burning of Moscow, The Last Days of Pompeii*—historical pageants speaking in their content of the empires of this world, but in their form of the coming empire of organization promised by capitalism, catalogues of battles transformed into entertainments, crisis transmogrified into the sustained dazzle of the sign. In *The Burning of Rome,* sheets of fire blazed up, hundreds of actors thronged the market-place before the palace where Nero strummed—then, with abandon, a fight broke out between a coastal fort and an airship, and a collision between a fire engine and an automobile. Already, we find, the spectacle has begun to dissolve historical fact, creating the perverse sidewise slide of signs that characterizes all modern culture, and its sites.

For, indeed, the perverse and the monstrous is the other side of the modern, its demon brother—the subversive side of modern culture that is only discovered, it seems, once the glory of modernist emancipation has been accepted. For the same visitors who came to the CNE to marvel at the revelations of science and technology also came to pay a nickel to peek into the sleazy little tents off the Midway, where JoJo the Dog-Faced Boy displayed his deformity nakedly, the Geeks tore live chickens to pieces and ate them, and the Pinheads, pathetic microcephalic freaks dressed in Mother Hubbards, darted about, chattering imbecilically.

They came for the lurid sex shows, of course, but also for the wonders of primitive movies with titles like *The Laboratory of Mephistopheles*—for tastes of the industrialized empire's promised honky-tonk glitter, allure and horror, sex and the occult, transports of fancy and modern invention at its most wonderful and daring.

Thus for a couple of weeks each year, on the site of the outpost of an older empire, rose this palisaded outpost of the new, in all the optimism, perversity and delight in spectacle of industrialism's infancy.

For me, that delight was always summed up and symbolized best by the Flyer. The 1992 decision by the Canadian National Exhibition to tear down the roller coaster kicked up no fuss. Even the Toronto Historical Board, our town's official worrier about "architectural heritage," did not get involved in one of its usual campaigns of compromise. Nor was anybody else likely to take to the barricades over this issue. The CNE's official reason for dismantling Toronto's only permanent roller coaster was that it just wasn't thrilling any more, or at least not as thrilling as the real stomach-churners being put up at new theme parks such as Canada's Wonderland, just north of Metropolitan Toronto. Only the most fanatical roller coaster aficionado—let alone a less committed pleasure seeker such as this writer, who has ridden the Flyer innumerable times just for the heck of it—would disagree with the CNE's decision to kill off a sure money-loser.

The destruction took place swiftly. By the time the 1992 CNE opened at Exhibition Place, the fusty old coaster was just so many splinters of pressure-treated British Columbia fir—the stuff from which it was constructed, back in 1953—and about a half-mile of crumpled steel track, on which we Flyer veterans experienced so many moments of helplessness, terror and exaltation.

While I never planned to lobby to save the doomed coaster, I did take an early-morning drive down to Exhibition Place after I got the news, just to get a last good look at the Flyer before the wreckers got down to work.

During Ex time, of course, the asphalt wasteland around the Flyer site is thick with the deafening shows of the Midway and crammed with sweaty fun-seekers. The Flyer itself—long ago upstaged by Doppel Looping, the German-built, ideal ride for zero-gravity freaks—was always festooned with glittering lights, like a luxury ocean liner in port, and noisy with the clattering cars and the shrieks of passengers. Late evening was the ideal time to take a ride, since the whip-turns and plunges of the Flyer made the garish carnival lights of the Ex seem passionate, spinning, phantasmagorical.

But in the light of a Toronto spring morning the week the

news came, the Flyer seemed anything but wild and menacing. The waving, soaring lines of its wooden structure rose
from the vast flat expanse of deserted asphalt like a voluptuous
sculpture, its delicate fabric of bolted, white-painted wood
catching the early sun. A roller coaster, I was reminded in that
quiet moment, is architecture, not machinery; it had beauty,
but the rigorous, new beauty of intelligent, functional form
that modern architecture created, and has taught us to appreciate. A roller coaster must be strong but subtly flexible, and
precisely, predictably responsive to the extreme physical forces
it must contain, channel, withstand. Such principles are
expressed in every line of the Flyer, with economical eloquence
and entire seriousness.

The poetry of the roller coaster can be experienced, then,
in two ways, each quite different from the other. One is a
lonely dawn visit, such as the one I just described. The other is
by doing the coaster itself.

The projects of all architecture are to define space, and to
provide a choreography for moving through it. The roller
coaster is an architecture like that of an expressway or an airport runway, intended to abolish limitation, free us from the
staid forces of gravity and the facts of natural obstruction. But
unlike an expressway, which has no real beginning or end,
every coaster creates a precisely predetermined narrative for the
traveller, much as a maze does.

The Flyer, for instance, first assured you with a long, slow,
seemingly harmless turn. Only gradually did the ground begin
to recede, and the fear began to mount, as you were dragged
upward by powerful mechanical forces. Then came the first,
largest fall—a grand, fast, free descent, more exhilarating than
terrorizing—followed by a swooping left turn, a series of quick
rises and falls, and a final, especially harrowing whip-turn and
sudden stop. The repeat visitor quickly learned the sequence
of experiences, though going back again and again never
diminished the pleasure. A favourite short story is no less fun

to reread, just because you know what happens. Coaster riding is among the few architectural experiences with a precise parallel kinship to reading.

The Flyer never was a great roller coaster. It was more like a tale one loves at thirty and has less time for at fifty. But, all the same, I will miss it—especially the vision of its light, sensuous and architecturally sensible lines drawn, as if by a white pencil, against the grey sky of a Toronto dawn.

Modern

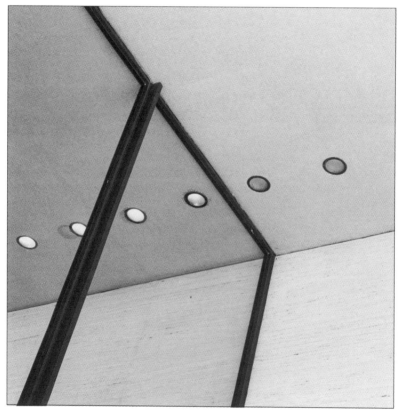

Curtain Wall, Toronto-Dominion Centre

THE CRYSTAL CITY OF HUGH FERRISS

The psalms raised to Toronto journalist and urban activist Jane Jacobs in 1991, the thirtieth anniversary of her famous book *The Death and Life of Great American Cities,* dwelt at length on her opposition to what used to be called "urban renewal," and to her role as North America's most famous officially indignant opponent of urban expressways, the abandonment of the streets and other outcomes of the great postwar love affair with our cars. The crusading tone of *Death and Life* is endearing, and is struck on the first page. "This book is an attack on current city planning and rebuilding," she announces in the introduction. "My attack is not based on quibbles about rebuilding methods or hair-splitting about fashions in design. It is an attack, rather, on the principles and aims that have shaped modern, orthodox city planning and rebuilding."

The enemy was (or at least seemed) clear: the radical idealist megavision in city planning, shaped and popularized among emerging architects and planners in Europe and the United States in the 1920s, eclipsed by the Depression, then brought back into the spotlight and crowned Prom Queen immediately after the Second World War, and given bulldozers to carry her to victory over city messiness. Her quarry, according to Jacobs, was the "organic," makeshift city of scruffy neighbourhoods and casual street life, where presumably workable solutions to urban problems were invented by the folks who actually lived with the problems. In retrospect, the scenario could have been written by Tolkien: real-folk anarchists and small freeholders, facing with courage the robotlike advance of the wicked Prom Queen, and the utopian social engineers in her service.

Nobody nowadays would argue that, by 1961, the record of idealistic planning in North America's cities had become a decidedly mixed business, compounded of magnificent new tall office blocks and of mediocre housing developments, some poorly planned expressways, some superbly planned and executed. Jacobs's portrayal of this complicated scene is caricature, not portraiture. It harped much upon the undeniable ruthlessness of comprehensive city planning, while wilfully shrugging aside the serious philosophical motives in the vision she opposed, and the fervent, even desperate hopes of the young planners and architects riding the bulldozers.

These hopes were rooted in an historic critique of the squalid, crowded industrial city, which had hitherto seemed impermeable, irremediable. The crowded city of the nineteenth century had been breeding grounds for terrible diseases; the great cities of this century, jammed with a poorly urbanized industrial proletariat of new immigrants had been breeding grounds for the even more deadly viruses of revolution and nationalism. To save us from ourselves, in architectural historian Spiro Kostof's words, the great architectural visionaries at our century's dawn wanted to replace all that by "a communitarian utopia beyond regional or national parochialisms," a crystalline, centrally organized "environment consonant with a world free of conflict."

This is not the place to chart the historical trajectory of such visionary megaschemes from their brave beginnings in the drawings of Bruno Taut and Le Corbusier and the young Mies van der Rohe down to the present; and there is hardly a need to dismantle Jacobs's arguments point by point. The successes of Modernist urban planning and architecture that dot and inform Toronto's urban fabric have done that job for me. Yet one Jacobean misrepresentation does deserve debunking: the suggestion that this idealism was an élitist plot, without grounds in concrete historical experience, to smash mindlessly the chummy streets and cozy neighbourhoods of the helpless little folk, huddled in their rabbit warrens within the traditional city (whatever *that* is). Look into the mass culture of the United States or Canada during the 1920s and especially the 1930s, and everywhere—in Buck Rogers cartoons, popular sci-fi movies, futuristic pulp novels—one finds yearning for a clean, monumental, big-city future, and for escape from the limitations of neighbourhood culture. A particularly interesting expression of this street-level longing came my way recently, when a colleague passed along a 1929 book by the popular American architectural illustrator Hugh Ferriss, entitled *The Metropolis of Tomorrow*. Ferriss was no intellectual; but he did understand the international yen for order that had arisen after the First World War, and gave it marvellous visual embodiment in texts, and in his thrillingly evocative charcoal drawings.

His book opens with a description, in word and image, of the spires and masses of the great contemporary metropolis emerging from the mists of dawn. The transfixed writer is plunged into reflection, and imagines himself to be witnessing "some gigantic spectacle, some cyclopean drama of forms…"

Within this architectural spectacle, souls are being formed, lives are being shaped. But how can good souls be formed, he wonders, when so much in the urban environment is still ugly, untidy, dishonest?

Ferriss is appalled by the "thin coating of architectural confectionery," the meretricious "disguises" that line the streets. He is appalled by the jumble of styles, the crowded streets and clumped buildings. "Do we not traverse, in our daily walks, districts which are stupid and miscellaneous rather than logical and serene—and move, day long, through an absence of viewpoint, vista, axis, relation or plan?"

Yet even in the midst of the unkempt contemporary city, Ferriss finds "a hope [which] may begin to define itself in our minds," and he prophesies the arising of a new generation of architects and planners dedicated to bringing this hope to fruition.

The seedlings of Ferriss's hope are, as you might imagine, the first great American skyscrapers. The artist's illustrations of such buildings are reverential and fantastic, not straightforwardly narrative. The Chicago Tribune Building, the Fisher Building in Detroit, the Los Angeles Municipal Tower, the Chrysler Building in New York (here depicted in unfinished form)—these and other buildings, in Ferriss's hands, become titanic, radiant with mystical power and promise. They surge heavenward, dwarfing the streets, and, floodlit, stand out against the gloom of night sky like huge gods adored by the million dwellers of Metropolis.

They are icons of the future city, the scattered hope of things to come. But to what promised land will they take us? Hugh Ferriss whisks us into the modern, mechanical city which he believes must succeed the clumsy, organic one of his own day.

All will be an honest, stern play of materials—concrete, steel and glass—devoid of plaster frills and the ornaments beloved by the academic, historicizing architects. In this truthfulness to materials surrounding us, we city-dwellers shall learn the meaning of truth, and freedom from artifice, contrivance, deceit. Instead of decentralization—a planning goal which "must be dismissed as a mere dream"—the city must be radically centralized, rationalized, transformed into an integrated, compact appliance. (In this celebration of the tight centre, Ferriss is following the lead of the European architectural optimists, from Bruno Taut onward.) To contain and support this intense concentration, the buildings must soar upward from huge bases, becoming objects of geologic scale, like mountains. Through the valleys of this mountain range will surge rivers of expressways, affording free, high-speed movement for all the free citizens of the city's huge population.

Then, in a rather charming touch that betrays Ferriss's typically eclectic American background, the artist pulls back from the European intellectuals' fierce proscription of all historical references, and allows architectural history a literally lofty role. On the highest terraces of Ferriss's gigantic pedestals, Greek Revival temples and neo-Gothic churches rise into the bright, clear air, like shrines on lofty Alpine peaks. But the only deity worshipped therein will be the new Holy Trinity of the future: concrete, steel and glass.

We may smile at the extravagant impracticality of Ferriss's vision, but few people now, I imagine, would find it horrifying. I, for one, find it enthralling. The artist's prophecies are entirely serious, and they are affecting in their urgency and their devotion to showing a way towards a more stable, supportive urban environment. Despite Jane Jacobs's nay-saying, the dreams worked out in Ferriss's head, and the immensely sophisticated mega-urban projects being developed by European thinkers, *were* works of conscience, not drivel, as she has always portrayed them. It is the imaginative richness of this conscience which gave Modernist architectural cosmopolitanism and idealism its early force and popularity, and sustained its European proponents through the tremendous revival of the vicious parochialisms and reactionary politics they despised, through exile, through a second, even more terrible world war. Is it any wonder that such long-frustrated planners and architects felt, after VE Day, that their chance to change urban life and perhaps save us from another such war had finally come—and took it?

THE TORONTO-DOMINION CENTRE

In 1992 the Toronto-Dominion Centre passed its twenty-fifth birthday. By any measure, this was an important date in the history of Toronto urbanism. The original group of buildings was the largest and last commercial real-estate development Ludwig Mies van der Rohe oversaw in his lifetime. It was the anniversary of Toronto's first architectural scandal, remembered as a time of much shock and pall and aghastness amongst the citizenry, who'd never before seen buildings like these in Toronto. The Centre is today, despite a great deal of later development in the same spirit, the most august stamp of the Modernist idea on Toronto's cityscape.

The anniversary was apparently all a big bother to the Toronto-based Cadillac Fairview Corporation, which manages and partly owns the six-building complex, and whose property managers wished I would stop calling them up to ask their plans. For months, they stalled, hemmed and hawed. Finally, in the spring of 1992, a Cadillac Fairview person rang me up to say they'd decided to throw a little public party that June, "with hot dogs and birthday cake and all that kind of stuff" on the plaza defined by the centre's four original structures. The tone of the affair was to be decidedly low-key. In the shaky commercial real-estate market of the early 1990s, I was told, "you don't want to be too lavish. We want to celebrate, but we don't want to throw big bucks at it."

It never happened. The only thing Cadillac Fairview did— and certainly a more useful thing, at that—was kick in support for a city-sponsored scholarly symposium on Mies van der Rohe, which included such distinguished architectural panellists as Phyllis Lambert, George Baird and Detlef Mertins. Even if the management had thrown a party, I suspect few Torontonians would have danced in the streets. By and large, we have always despised the Centre, though, after the first shock, without passion. Mies's magisterial towers of black steel and glass were immediately dubbed "the coffins" by suburban gawkers and downtown enemies. Not that they were the first office buildings in the International Style Toronto had ever seen. Peter Dickinson's Prudential Building, at King and Yonge, had gone up in 1958, John Parkin's Sun Life Building at 196 University Avenue was done in 1961; and there were others. But the Toronto-Dominion Centre offended as others had not, becoming a lightning rod for those wanting to blame *something* for the death of the cozy, low-profile city centre of yesteryear, and the batallions of hackneyed glass boxes that have been marching into our downtown, just behind the bulldozers, ever since.

Indeed, Toronto did lose a considerable amount of charm when the towers went up, and has never forgiven the loss. Gone is Millstone Lane, and it is missed, if only for its charming name. Gone, also, is a Beaux-Arts, columned bank building which Toronto's tribal memory still cherishes as magnificent and noble. Scraps of graceful ornament, tall and well-turned columns, large fragments of façade from these and other structures swept away by Toronto's surge of downtown development do exist, saved by entrepreneur Spencer Clarke, who set them on lawns of his lakeside Guild Inn, in the suburb of Scarborough. Without them, we would have very little material evidence of what was taken away to make way for Mies's buildings.

If unloved by many from the start, the Toronto-Dominion Centre did at least try to be something other than a misliked curiosity. The city's highest restaurant and observation deck and the first underground movie theatre were all intended by developers to make the site a tourist attraction. This new sophistication was quickly made old-hat by the rise of towers yet taller, and the proliferation of submerged downtown cinema buildings; yet, through the early 1970s, the Centre continued to exercise a peculiar fascination, at least over city-desk editors. As late as 1970, two years after the first towers were occupied, reporters were regularly dispatched to the Centre to report on a broken fire sprinkler, the sighting of a racoon on the roof of the Banking Pavilion, and the anxious utterances of naturalists about the mass crash-deaths of migrating birds.

The 1971 blow-out of a glass window from a high floor of a black tower even prompted a high-profile city safety investigation. There were hints of forthcoming revelations from T-D Centre employees about more glass about to blow. But whether no more windows fell out, or nobody ever came forward with the promised horror stories, or the newspapers just lost interest, I do not know. By the early seventies, in any case, the Centre was well down its slide into the state of taken-for-grantedness it was to enjoy at least until the summer of 1993, when a thirty-nine-year-old corporate lawyer employed in the Toronto-Dominion Tower jumped playfully against a floor-to-ceiling glass panel to see what would happen, and was rewarded with a twenty-four-storey fall to his death.

Like most stories about the Centre in the last two decades, the reporting on this unfortunate incident was routine. We had perhaps gotten used to the towers. In contrast, the early newspaper stories—hoots of scorn or ridicule, and tales of things breaking or going wrong—appear in retrospect to have been attempts to come to terms, somehow, with these immense, solemn and strange buildings: some, by dismissing them as brusquely as they seemed to dismiss us; others, by showing that, because they break and get racoons, these menacing blocks of machined steel and gleaming glass are, well, sort of human.

When the original towers were new and not yet overshadowed by the taller Bay Street skyscrapers to come, their purity and severity may well have seemed overbearing, ostentatious, even menacing. At least some also believed them to be a slap in the face to the stuffily comfortable Edwardianism typical of city culture in the past. The towers seemed to be saying: Toronto the Good—church-going, God-fearing Toronto, the

Column, Royal Trust Tower

city of steeples—is no more. If the metropolis needs symbols of its new status, we're the ones: tall, black steel, godless boxes, with nary a whiff of the Victorian spire about us.

Nobody gets exercised about such matters any more, largely, I imagine, because Toronto has accepted the image of itself it glimpses in the mirror of the T-D Centre—an image that is too American for some tastes, and far too imperious and imperial for many Canadians elsewhere. But, by the late 1960s, Toronto had in fact joined the élite fraternity of international centres of finance, industry, investment and development; and it had eclipsed its ancient rival, Montreal, in the race to national economic and urban supremacy. If the exhiliration felt by some about this rapid attainment was called into sharp question during Toronto's anti-development fights of the early 1970s—notably, the campaign to save majestic Union Station from demolition, and the crusade against the Spadina Expressway—the four original structures at the Centre have remained our finest monuments to the sense of civic attainment. They are also intellectually and esthetically compelling, in ways that very few Toronto buildings put up in the Corporate Power Style of the postwar period can match.

For the record, these earliest buildings one sees sited with great precision on the south-west corner of King and Bay streets are: the Toronto-Dominion Bank Tower (fifty-six storeys, officially opened 1967); the flat, single-storied Banking Pavilion (1968); the Royal Trust Tower (forty-six storeys, 1969); and the Commercial Union Tower (thirty-two storeys, 1974), all set on a spacious stone platform elevated above street level.

Walking around them, you sense immediately what early critics loved to hate about these structures: the vast indifference of the project to human scale, however defined; the unrelenting impersonality of the soaring walls of glass and blackened steel; the dramatically vacant intervals among the buildings; the celebration of open space and shunning of cozy nooks; the nearly comfortless stone-lined, glass-walled lobbies (now unfortunately compromised by mediocre fabric wall-hangings). I suspect only the most dogmatic anti-Modernist could fail to find a classic, ancient beauty in the rhythm and consummate grace of the slender square steel pillars holding aloft the roof over the Banking Pavilion, or the rigorous, perfect poetry of black metal, glass and warm beige stone in the ground-level areas of the three original towers. This powerful architecture is also as beautiful at night as during the day, and in this detail the Centre is most radical. No Hollywood-like spotlighting is needed to save it from disappearing after dark—as the designers of futuristic skyscrapers in the interwar years apparently feared—since the walls of glass enable the interior lights of the Centre to become the night sun of the plaza.

Now that Toronto is less cocky than it used to be, the original T-D Centre continues to project immense dignity and intellectual seriousness in the midst of much frivolity and architectural ostentation. It has also been looking faintly run-down of late. One notes with sinking heart, for example, the indifference of the managers, who too often don't bother to change burnt-out bulbs in the magnificent grid of light in the ceiling of the Banking Pavilion, thereby spoiling one of the most sheerly beautiful rooms in Toronto.

If the rigour and purity of the high Modernist building style embodied in the Toronto-Dominion Centre were once daunting, the tastelessness of the designers who have given us the two new towers built into the ensemble since the early 1970s makes such lofty virtues seem downright endearing. In the IBM Tower (1985), we see the transformation of Mies's clear, open entry floor into a boutique-like diversion, with a mezzanine, obfuscating barricades of stone, and shiny stainless-steel staircases.

The most spectacular travesty of Mies's stringent idealism on the site, however, is the Ernst & Young Tower, on Bay

Street just east of the Toronto-Dominion Bank Tower. At ground level, this newest addition to the T-D Centre features the old beige limestone home of the Toronto Stock Exchange (1934), wrapped up in pseudo-Miesian black steel corsetry mixed with references to deluxe Art Deco architecture. This visual gobbledegook goes on for the first several metres of the thirty-one-storey building's rise, whereupon the style abruptly and unaccountably lapses back into the routine Miesian rhymes of black-painted steel and glass the rest of the way up.

This attempt to grind together two great and incompatible agendas of twentieth-century architecture—the radical-utopian and the conservationist—is understandable, if not forgivable, given the avid, eclectic waywardness of corporate architecture nowadays. The Ernst & Young Tower will have done a positively good thing, if it prompts critics to go back and appreciate anew the integrity and beauty of the original three towers and the Banking Pavilion.

Returning to the older buildings, we are reminded that, to Mies's way of thinking, the plain structural elements of steel, stone and glass are not ornaments or images excerpted from the history of building types, and infinitely open to whatever free, witty, imaginative combinations the designer can think up. These very deliberate buildings express no arid logic or functionalism, but rather recall, with splendid clarity, the rational, human mobilizations involved in the making of such tall structures. The industries of mining and steel milling, for instance; the assembly and deployment of workers, the coordination of financing, design, many technologies, and, above all, precise information of every sort. The towers give visual expression to the energies and abilities that have given us the modern world.

Architectural post-Modernism has attempted to bring back into building the wit, inspiration and expression of traditional Western architecture. In the rational-utopian Modernism embodied in the Toronto-Dominion Centre, wit is excluded;

but so is imagination itself, that most durable idea we have inherited from Romanticism.

In this revolutionary view of building, Technology takes the laurel wreath that, for so many centuries, crowned Imagination. That doesn't mean a Miesian building is a simple machine. It does mean that the imagery of such building—the openly architectonic, systematic, non-ornamental ways materials are used and combined, the aloof indifference of the building to context—are grounded in an experience of how modern machines and mechanized social forces look and work, instead of springing from some mysterious play of fancy in the mind of the architect.

I find it interesting to note how unquestioned the idea of Imagination is nowadays, as a way to explain the creative process of building. Many people, who are not inclined to invoke divine inspiration in a discussion of creativity, would be quite speechless if you took the equally mysterious doctrine of Imagination away from them. Yet Imagination is neither a particularly old nor eternal notion. Miesian architecture would have us leave this theoretical novelty behind, and move to a post-religious, post-Romantic philosophy of making things. One is tempted even to call it an architectural spirituality: though its terms would include neither God nor Imagination, the idea of Modern supplies radical and transcendental values for negotiating our position in the world, and for guiding our construction of dwelling.

The towers on the south-west corner of Bay and King are serviceable office buildings; but they are also evangelical statements of a profound idea of our moment in time and history. They call us to resign ourselves to our situation in the wholly technological, post-human yet endlessly interesting world our machines have given us—to leave nostalgia and the sumptuous self-centredness of humanism behind, and embrace the unromantic, creatively alienated present which industrial capitalism has created for us.

No wonder people were upset when these uncompromising edifices in downtown Toronto went up. Yet we would do well to attend carefully to the severe doctrine of creative and spiritual life embodied in them, if only to understand why we are inclined to answer the summons of these buildings with the single, emphatic word *no*.

LESSER MODERN

"If you are ever driving on I-91 through Connecticut, don't miss the Colt Firearms Building on your way to New England diversions."

So begins Ada Louise Huxtable's *Architecture, Anyone?*, a 1986 gathering of reviews done during her long, distinguished stint as architecture critic for *The New York Times*. Not that the nineteenth-century brick Colt building is blessed with any particular magnificence, or historic importance. It's just been the landmark that, for many years, has been telling the author she's gotten to Hartford.

"I find, when I think of it," Huxtable continues, "that I have a set of such landmarks—personal, transient, and indelible—that mark the stages of my journeys and the stages of my life. I wait for these particular places on trips year after year; they are all old friends." I was delighted to discover that Huxtable and I share one of these old friends: the streamlined Bulova watch factory near La Guardia Airport, that, for decades, has been letting me know I've again hit the Big Apple.

Think about it, and you'll probably find that what's true on the long-haul car or plane trip is also true on the short hop across town, or around the block to office or supermarket or mall—though you may not even become aware of your personal landmark until it has disappeared.

Here's an example. Almost every school-day since my tot was marched into junior kindergarten, some seven years ago, I have driven home from the morning drop-off the same way. Though I don't remember consciously having picked it for this reason, the route ensured that I would always pass a small, blockish brick and concrete storage building of some kind, completely ordinary except for a few discs and flutings on its plain façade, and a certain dignity—broad-shouldered, utilitarian—about the relations of flat roofline, door and strip windows. No engineer had thoughtlessly thrown up this building to shelter something. It had been designed by some corporate or municipal wage-slave who could nevertheless draw well, who had a taste for high style, if tiny scope to project his taste into the stuff of architecture.

Then, one afternoon, this unimportant little structure vanished. Until my first pass by its rubble-strewn site the next morning, I'd never realized how much the sight of it each morning had meant to me, or even that it had meant anything at all. The building was always just there, morning after morning: a station on the *via dolorosa* of the cross-town traffic crawl, a signal that I was exactly a minute away from home and peace.

This little loss disconcerted me, and set me to thinking of other personal urban landmarks that I've noted for years, with similar absent-minded appreciation, and without a thought of how quickly they can disappear.

The list is not long. But towards the top of it is surely the derelict steel-framed glass work occupying a narrow triangular property bounded, on the north, by the grim old wall of the Queen Street Mental Health Centre, and unpopulated Sudbury Street, slanting sharply sidewise down to King Street.

On countless trips between home and office, I have driven past this long glass edifice—or sequence of boxy glass structures, apparently added on to one another over time—but hardly ever without experiencing a certain delight. This delight is akin to the exaltation we know in the presence of serious, thoughtful architecture, but is tinged with the regret we often feel about fine, perfectly serviceable buildings that have been allowed to fall into disuse, and slip into decay.

Now the glass shed on Sudbury Street is not supposed to be taken seriously as architecture. It's engineering—built, it appears, just after the Second World War to meet the needs of the mighty Massey-Ferguson farm machinery company, which employed up to eleven thousand workers, and is yet another Toronto industrial titan to have passed recently over the Styx. But even if it's not architecture, the building is, or was, a glittering folk reminiscence of the International Style at its most transparent, utilitarian, explicit. The severe linearity of the glass curtain walls, the crisp right-angle meeting of flat roof and glass wall, the geometric clarity of design, and absence of jerry-building—such are among the beauties we find in the best Modernist factory construction, and also in this obscure, utterly unprepossessing ensemble of glass boxes. Trendy architectural historians hardly miss a chance to scoff at Modernism's claim to have discovered a genuinely timeless, styleless style, a manner of building devoid of ornament yet beautiful, lovely in its response to the basic forces of gravity, light, space and use. Quite often, when passing my dilapidated glass building, I have wondered whether the Modernists had not in fact recovered the great classic architectural truth of the ages.

This landmark on the personal map of Toronto I carry around in my head should not pass into oblivion without at least a notice, and a few words of historical description. Jotting about any such building is never easy, given the general poverty of documentation on industrial construction, and the rapidity and frequency with which they are swept away, *sans* plans or a trace of written history behind. But as near as I can make out, the reinforced concrete edifice at the corner of Shaw and King Street West, from the west side of which springs the glass structure, is a building of turn-of-the-century vintage, and of Chicago warehouse inspiration. It has pillars notched to make them look like masonry, and a cornice that runs all round, presumably to give the façade a touch of class. Neither it nor the glass annex stretching westward appear in any city directory until 1948. In the Toronto fire-insurance atlas updated to 1964, this set of two buildings is called the "north combine plant," and is included as part of the "Massey-Harris-Ferguson" plant site.

Beyond these meagre facts, I have been able to discover nothing. (Not that I have tried very hard to do so. Certain buildings spark curiosity, others, especially derelict ones, ask to be left enshrouded in obscurity; the north combine plant belongs to the latter sort.) But before it was allowed to slide into its present dinginess and dilapidation, this transparent composition of neatly, resolutely joined glass volumes must have gleamed brilliantly in the morning sunshine. And even in its current shabby condition—the structure is doomed, along with the other housings of the defunct Massey-Harris empire, once headquartered on King Street West—its purity and lucidity recalls a recurring dream in Western architecture. That dream, at least as old as the great European cathedrals, is to lighten walls and roof, minimizing the roof-supports, and opening previously dark, cavernous exteriors to sunshine and sky. Perhaps no drive in the project of Western architecture has been more durable. It has often suffered reversals, most recently in the religiose gloom and heaviness favoured by the American Romanesque revivalist Henry Hobson Richardson and his myriad clients and disciples, and in the opaque, lightly decorated wall and indirect or filtered light popular during the *rappel à l'ordre* of Depression-era building.

If, by the early 1990s, cladding whole buildings in glass had become the most boring stutter in the vocabulary of urban architecture, it was not always so. Before the end of the Second World War, it was do-able, even though rarely think-able. Then, in 1948 or so, the Massey-Harris company decided to construct its north turbine plant, and the style picked, probably for its cheapness and utility alone, was that of sturdy glass set in light steel frames, and covered with a flat roof. But even if nobody noticed, and if not many people will ever care, a dream almost as old as Western architecture itself unobtrusively came true on Sudbury Street, and endured, and was still there the last time I drove by it.

CONFEDERATION LIFE

Upon completion late in the summer of 1992, the Confederation Life Insurance Company's national head office, located just below Bloor Street East, became an Instant Landmark.

As opposed to an *historical* Landmark, that is—a red brick mansion or venerable civic building through which schoolkids are herded while being lectured about grizzle-chinned Victorians whose names and accomplishments they forget at once. To become a shrine of *that* sort is easy. All an old building has to do is survive the tsunamis of urban renewal which sweep over old parts of every city from time to time.

To qualify as an Instant Landmark, however, a building has to be, like Confederation Life, a show-off and a likeable bully. It must poke fun at tiresome tastefulness, be impudent, outlandish. It helps, too, if it can be put in a stylish architectural "context." (Sceptics should be aware that neo-Gothic has definitely been in at least since 1987, when Philip Johnson and Raj Ahuaja's medievaloid, spired and piered IBM Tower opened in Atlanta.) It must come flirtatiously close to insulting its hometown's most venerated architectural monuments, thus imprinting itself on the maps in our heads as a downtown edifice we love to hate, or hate to love.

But love it or hate it—and we should definitely view it cautiously—there's nothing in Toronto quite like the eighteen-storey Confederation Life by architect Eberhard Zeidler, Toronto's best producer of comfort-controlled paradises for the era of Late High Consumerism. (Among Zeidler's former projects: the mammoth heaven of consumption known as Toronto Eaton Centre; the provincial government's water playground, Ontario Place; and the sprinkling of fairy dust that changed a dingy cement warehouse at Harbourfront into a glittering shopping mall.) From its enormous footprint—or dinosauric stomp—south of Bloor East's ceremonious strip of insurance companies, the glass-and-marble citadel punches towards heaven, with an occasional dodge-back here, and there, gleaming metal ornaments reminiscent of details from the Batman movie sets, and whimsical slants and angles everywhere. It's a game with eccentric building types, from medieval representations of the Tower of Babel to Buck Rogers, via the pyramidal Romanesque fancy at the Abbey of Fontevrault, France, known as Evraud's Tower.

The tower at Fontevrault, by the way, is remembered in local story as the lair of a robber. Its beaconlike shape was designed to lure lost travellers from the forest, into a place where they could be conveniently mugged. If that's not exactly a savoury association for the head office of an insurance company, who cares? The building is all dance and play. Anyway, serious architectural thinkers I pay attention to had already brushed off this architectural Carmen Miranda before it was finished. So passionate was the feeling that, after my *Globe and Mail* column about Confederation Life appeared, I got a number of anxious letters and phone-calls from professors and builders, who appeared to think my having just mentioned the structure in print was to bless it. Now while I did not then, and will not now, bless the building—everyone can relax—I confess it did set off a train of thinking in my head about how Toronto got flattened, and why we take this condition completely for granted.

For in addition to its gaming with historical forms, Zeidler's snappily eye-snagging structure did give Toronto's knocked-off Modern buildings and the flat skyline they've created a nice comeuppance, with a crown of up-jutting marble fangs and dormer windows, and a high green tower tapering upward to culminate in a voluminously steaming chimney, like the top of Dorothy's pal the Tin Man.

Of course, pitches, dormers, gables and other swell roof details have never disappeared from the single-family house; in Toronto, as elsewhere, the sleek glass-and-steel box-houses built in North America since the 1930s by Richard Neutra, Walter Gropius, Marcel Breuer and others, however exciting they look in picture books, never caught on with residential clients or builders. Nor, before 1945, had there been a great deal of interest in challenging the pieties of traditional historicist building. Toronto's townscape resembled that of any small, staid midwestern American city, with slender steeples punctuating the sky, acres of low Depression-era shanties and slums near industrial areas—and a few cautiously untraditional skyscrapers and monumental public buildings. Toronto was still quite literally pedestrian in 1945—designed, that is, to be viewed at walking, shopping, browsing or church-going speed. Hence, the prevalence of low skyline, and the continued tradition of crowning, hatting, cornicing and otherwise beautifying the façades above the first storey.

Emerging from the Depression and the Second World War with a boom-town in the making, and with attitude and ambition to match, Toronto and its developers lost little time in transforming the Toronto skyline from Victorian steeple-spiked to stylishly Modernist flat. Confederation Life cutely reminds us of this historical fact, simply by upstaging the Modernist drudge-work in its neighbourhood with much uppity, expensive festivity. (From what benumbed, T-squared architectural brain, one wonders, sprang the squat, ugly national headquarters of the Anglican Church of Canada, across Jarvis Street from Zeidler's pistachio sundae?)

In the first decades of the Cold War, the Western cities were ready for the flat roof and geometric outline, for reasons too easily forgotten or ignored. To draw a plain horizontal roofline across the top of an office building could be read—only then, and perhaps never again—as capitalist democracy's reply to several uglinesses. The neo-Gothic monstrosities going up in Stalin's Moscow, for example, and the Nazi romance with old-fashioned pitched roofs and other presumably *volk-ish* architectural recipes. The ideological motive was strong, in ways we find difficult to feel at this distance—so strong, in fact, that it overrode the ancient architectural truth that a flat roof is the worst kind imaginable.

Browsing in an architecture library recently, I came across a recent publication of the British government called *Flat Roofs Technical Guide*. This complicated engineering manual begins with this warning in bold-face type: "Flat roofs should only be considered when the ruse of a sloping or pitched roof is impracticable." (I don't understand why a sloping roof is a "ruse," but never mind.) The manual then follows up this stern advice with myriad illustrations of everything that can, and almost certainly will, go wrong with a horizontal roof—leaking, collapse, warping.

None of this practicality seems to have counted with, and may not have dawned upon, the postwar architectural avant-garde. They, and their ambitious clients, were ready to draw a line in the dirt between themselves and all that had gone before; and the level roofline seemed like something worth fighting for, against the mummery of roofs pitched, gabled, towered or domed. Thus was the traditional roof never even considered for inclusion in the lexicon of styles being developed by the designers of our new apartment blocks and institional and office towers.

Some sensitive observers recollect Toronto's postwar epidemic of high and low flat-topped construction with horror, especially when the epic scale of demolition is factored in. As far as I can tell, this transformation left us with a cityscape visually more healthy, not sicklier; and more interesting than it had ever been before.

The story of the city's revision has been excellently sketched in the Bureau of Architecture and Urbanism's *Toronto Modern: Architecture 1945–1965*, which accompanied an exhibition held at City Hall in 1987. In their introduction, the producers of this show recall the *annus mirabilis* 1948: the year not only of *refus global*, Quebec's famous declaration of cultural independence from past, piety and parochialism, but also the year in which Modernist evangelist Siegfried Giedion's militantly titled *Mechanization Takes Command* appeared, and, by a nice coincidence, the Mechanical Building at the University of Toronto was built. If the book heralded the ascendence of Modernist urbanism and Euro-American avant-gardism in the built world, Allward and Gouinlock's academic pavilion—with its formal rhyming of horizontal volumes, its reinforced concrete and steel bones, its terrazzo floors and stainless steel railing—announced that the idealism of the Modern had arrived in Toronto.

From 1948 through the early 1970s, the Modernist reconstruction of Toronto continued—unrushed, with the deliberate, carefully plotted pacing typical of every project this city undertakes—leaving in its wake a great many forgettable steel and cement cereal boxes, a few masterpieces, and a cityscape and urban imagination transformed.

The catalogue of *Toronto Modern* lists and lauds a number of the great works. Among them: the remarkable new town of Don Mills (begun in 1953), where John B. Parkin, James Murray and other decisive figures in local architecture learned their craft, and the short, eccentric and brightly coloured Anglo Canada Insurance Co. at 76 St. Clair Avenue West (1954)—a controversial choice, to put it mildly. Then there is Parkin's serene, authoritative Ortho Pharmaceutical office building and plant (1955), Peter Dickinson's Benvenuto Place Apartments (1955)—still a paradigm of sophisticated city dwelling—the Toronto-Dominion Centre (begun 1963), of course, and Viljo Revell's daringly different Toronto City Hall (1965).

These are the monuments; the stand-outs from the crowd. The exhibit quietly omitted the embarrassments. But whatever one's discontents with this or that Modernist construction, or reservations about the process as a whole (which included expressways, superblock subdivisions and shopping malls, along with masterworks like Benvenuto Place), Toronto will never again be the place it was before 1945. The change in the city's mind has been wrought, and cannot be undone. As *Toronto Modern* states, "integrated and efficient public transportation, elegant open spaces, and monumental public, corporate and educational buildings are hallmarks of Toronto's ascendency as an international metropolis." The editors could have mentioned the rise of splendid bookshops, commercial galleries dealing in advanced contemporary art, a theatre and experimental music scene, vast record stores, a culture of high fashion—all elements in the economy of sophisticated desire, sexuality and knowledge that matured along with Toronto's measured, pervasive experiment in architectural Modernity.

While we need the occasional reminder of how good Toronto life is, we do not need many more funnily exotic buildings like Confederation Life. But huzzah! for Eberhard Zeidler and his curious beacon anyway, just for inspiring us to ignore Modernism's misbegotten cubes and domino stacks, and cherish Toronto's leap to Modernity, the immediate heritage of us all.

Shopping

Parking Lot, Galleria Mall

PARKDALE

It takes a good reason to get a Torontonian who doesn't live there down to Parkdale's smelly, rundown Queen Street strip. And it takes a strong stomach to stick around long enough to discover the higher cultural pleasures of this ill-famed neighbourhood west of downtown.

I know. I have spent more than one afternoon walking and photographing Queen West between Dufferin and Jameson, all the while avoiding meaningful eye-contact with the prostitutes in hot pants, dodging oblivious mental out-patients weaving along the sidewalks, almost tripping a couple of times over addicts slumped in doorways, and witnessing more human wreckage per metre of street than I've seen anywhere this side of lower Manhattan. But explorers can ill afford to be prim. And no one curious about how Toronto's important Victorian street looked and worked should skip a visit to the Parkdale strip.

As a whole, Queen—from its eastern termination at the R. C. Harris Filtration Plant, a Depression Modern masterpiece of civic architecture, to its less gracious west end just short of High Park—is a retail street unhonoured in Toronto, and even less admired than honoured. It is nevertheless a treasury of neglected architectural and cultural artifacts, displaying remnants of every fashion in commercial architecture to sweep

Toronto from early Victorian times to the late-Modern present and the post-Modern beyond. Apart from the Parkdale stretch of Queen—for reasons I'll get to presently—no nineteenth-century commercial streetscape of comparable uninterrupted length, architectural hauteur or aesthetic variety has survived Toronto's past one hundred years of ceaseless speculative destruction and replacement. Looking along the north side of Queen from the edge of Parkdale closest to downtown—where the eastbound Queen streetcar rumbles down into the stone tunnel under the railway tracks—one sees block after block of sophisticated and often sumptuously turned-out tall buildings from the 1880s, Parkdale's final decade as an independent political entity, and the early 1890s, its first years as a Toronto neighbourhood.

The street level of almost every tall, narrow retail establishment built during this consumer heyday featured a large-windowed shop with a prominent entrance, through which Parkdale's variously heeled citizenry passed in search of fine millinery and humble bonnets, bread and carriage fittings and fresh beef, the latest international fashion as well as heavy-duty clothes suitable for the working man. Retail outlets still face the street, though they are now places like The Wildside bar, countless Vietnamese video rental and doughnut shops, and one discount outlet after another.

Above each shop would typically have been a suite of offices, or a spacious two-storey flat for the store's proprietor, with parlour and large window on the second level, often showing a face of handsome Romanesque Revival brick and stone detail to the street, and with sleeping quarters on the third. Throughout this zone, the street architecture of Queen is self-conscious, ornamented, slow—built and adorned, that is, to be seen and savoured at pedestrian shopping speed, or at most, the speed of a horse.

But the automobile, which created an entirely new way to see the city: namely, as rumps—those of cars up ahead, and of sexually interesting pedestrians—and as a horizontal band of flashy store-front signs, sealed the fate of those fancy finials, upsweeps, knops and other ornaments that once crowned every respectable business establishment worth the name.

Mr. Small, the butcher, knew nothing of automobiles, but did intuit the relation between speed and awareness. So it was that he coiffed his abattoir at 1372 Queen W. with a tall, flourishing cornice, and installed good terracotta heads of cow and sheep on either side of its high bay windows to advertise his trade, and his own importance, and that of street and town. And when, in 1892, the builders of the music school at 1482–1486 Queen West put graceful towers atop a window-fabric of fluid Gothic tracery, they clearly expected their decoration to be a sign of prosperity, high culture and aspiration. Thus architecture, as always, matches civic ambition. Newspapers of the day reflected Parkdale's hype when they wrote of it as "floral suburb" of Toronto, a delicious refuge for those seeking "a refreshing coolness that is lacking in the close and sultry city."

Queen Street played a particularly important civic role in this Parkdale of affluence and ex-urban resort. It was then the principal entranceway for vehicular traffic approaching Toronto from the west—hence its double role, as eclectic shopping promenade serving a local clientele ranging from wealthy to working-class, and a ceremonial, auspicious entrance to the great Victorian city adjoining it on the east. As befitted such a street, Toronto's mainstream architects had a hand in creating the streetscape we see today. "Parkdale was no quaint village," Alec Keefer said as we strolled along Queen together last week. "It always knew itself to be part of the big city."

Though long an admirer of Parkdale, at least at a safe distance, I have Keefer to thank for unveiling the historical and architectural facts undergirding its visual patterns and pleasures. A resident of the neighbourhood, a passionate architectural conservationist and local historian, Keefer has developed every urban connoisseur's knack of simply ignoring the decay, and openly exulting in what's left of imaginative cultural form—in Parkdale's case, "an urban form dictated by this fact: the merchants had to cater to a complete society. They had to provide straw hats, as well as Paris originals."

He singles out two low, connected buildings a casual observer would easily overlook, and hails them as perhaps the earliest remaining structures on this strip of Queen West, dating from the 1860s. He delightedly relates that a broad and gracious palazzo-like fabric of shops and flats was constructed in 1898 as an investment property by the Anglican Church of Quebec, dabbling in real estate at the time. (We assume this origin has nothing to do with the sign reading "Jesus, I Trust You," that hangs from the imposing little Ionic belvedere high on the façade.) He proudly points out a multi-colour brick façade which he and fellow activists discovered and, after a campaign to remove the hateful metal siding which had long concealed it, uncovered.

This remarkable strip of commercial architecture, like many another precious urban survival, owes its enduring presence in our imagination to enthusiasts of Keefer's stripe. Its continuing physical existence, however, is largely due to calamities that turned out to be blessings after all.

The more important of these mishaps was the dethronement of Parkdale from triumphal entry into a dead corner of Toronto by the diversion of heavy, high-speed traffic from Queen onto the lakeshore traffic corridor. Like Weston and numerous other communities on Toronto's spreading fringe, Parkdale had already begun its decline in private riches and urban status before its wholesale engulfment by Toronto. In Weston, however, the continuing usefulness of its principal thoroughfare ensured an afterlife for it, however baleful and fitful; and it supplied a reason to continue demolition and renovation. In contrast, the final shunting of all traffic inbound from Niagara and the United States from Queen—along with the postwar move of shoppers away from street-front specialty stores to malls and chains—deprived shop-owners of any incentive to improve their store-fronts. Both ideological and economic motives for keeping Queen showily up-to-date had thus vanished. So it happened that virtually the whole north side of the strip—the finest commercial row, put on the north side to catch the sun—was left more or less alone as they were at the turn of the century. Less even than Spadina Avenue, a great old retail esplanade always much more proletarian in spirit, the essential architecture of Queen Street has remained largely undisfigured by garish neon lighting, shouting paint colours and metal siding.

Another calamity, this one unalloyed, was Toronto's indifference—at least this is the way Keefer weighs up the history—when it came to keeping Parkdale stocked with public amenities in top architectural styles following the municipality's annexation in 1889. The present public library, for instance, is a low, mean block of bricks, "sucking the life out of the street," says Keefer, and expressing nothing of learning's joy and richness. Among the very few noteworthy newer buildings on the strip is the streamlined police station—now an emergency shelter—put at 1313 Queen in the early 1930s by City of Toronto architect J.J. Woolnough, the same gifted stylist who gave us the magnificent Horse Palace at Exhibition Place.

Then descended the Great Depression, and the subsequent conversion of many of Parkdale's ample, upper-middle-class homes into cheap, crowded flop-houses. Later, after the Second World War, came the destruction of whole streetscapes to make way for high-rises packed with low-renters. More recently Parkdale has suffered the takeover of virtually all remaining detached houses south of Queen by absentee landlords, further contributing to the squalor into which much of the neighbourhood has sunk so deep.

But at the top of this commentary, you will recall, I said urban explorers cannot afford to be prim. I'm ending it with another admonition. Urban explorers cannot afford to ignore Parkdale and the many other high-style Toronto beauties that, through no fault of their own, have slid into urban eclipse and neglect during Toronto's two centuries of radical capitalist development and transformation.

THE GOLDEN MILE

I went looking for the Golden Mile because of its lilting high-Modern name, and because it was one of those fabled places Toronto people pushing middle age mention with reverence. In the 1950s, what would become The Mile—then flat countryside beyond Toronto's north-east edge—had been sold to big manufacturers and their employees as "an oasis of industrial opportunity and harmony. Its booming factories fueled images of limitless wealth." This peppy quote, from a final memoir published by the union local at General Motors' Scarborough Van Plant. Almost three thousand members put in their last shift there in early 1993, and left the once-magnificent engine of local Progress and Prosperity an empty hulk, clean, flat and dead, set back from the broad stretch of Eglinton Avenue East bisecting The Mile's superblock.

The rolling carpets of lawn along Eglinton have gone shabby, and random shacking and shedding and putting up fast-food outlets have ruined the once-dramatic stripes of green and strong lines of horizontal factories paralleling the street. The Canadian General Electric plant still boldly faces Eglinton, though the grand building is an empty shell. Philips Electronics long ago moved its manufacturing of black-and-white TV sets to Taiwan. And no trace remains of developer Avie Bennett's tremendously popular and successful strip plaza that anchored The Mile's western end. It was the essence of The Mile's newness and flash, erected right after West Vancouver's Park Royal, the first shopping plaza in Canada, when the idea of shopping strips surrounded by acres of asphalt parking spots was the hot continental vogue in consumption. By the 1970s, the enclosed mall was leaching the vitality from such strips; and today, where Golden Mile Plaza once stood is a metal shed, sheltering bulk food stores, a discount grocery, cheap clothing stores, boom-box and heavy-metal record outlets, and all the other suppliers of a proletarian clientele on the slide.

When I went, I had in mind the writing of a hymn of praise to this jubilant, instantly famous act of Modernist planning. What I saw made me flip shut my notebook, and forget about ever writing up what I had seen. Until, that is, I happened upon a 1986 column about the site by John Sewell, former mayor of Toronto and sometime *Globe* columnist. It was a happy discovery.

Written on the occasion of the dilapidated Golden Mile Plaza's rip-down, Sewell's story is largely a tribute to the politicians, land developers and industrialists who made The Mile arise and flourish. If The Mile is a ruin today, Sewell reminds us, it is one with a proud, clever past. More district than strip—though strip is what its name makes it sound like—The Mile is the most memorable commercial and industrial superdevelopment to pop out of the serene countryside ringing Toronto immediately after the Second World War. Its creation also marks Toronto's first experience of instant industrial and commercial development as radical therapy for an ailing local economy. The malady of postwar Scarborough township, like many another urban ill, was brought on by loveliness, and a rolling landscape that seemed suffused with infinite opportunity. The green and subtle hills, peaceful farms and deep, shaded ravines made this patch of southern Ontario one of the most lovely places on earth. Indeed, for newlywed veterans and their families, and for waves of newcomers from across Canada and the world, the idea of living in the pre-sprawl expanses of Scarborough, quiet and rural under a cloudless summer sky, must have seemed like a storybook dream.

Eglinton Avenue East

This widespread perception was not lost, of course, on residential developers. Within weeks of VE Day, bulldozers were wrecking Scarborough's barns and farmhouses, and shaving the ground down to a flat plane suitable for the immense housing subdivisions then on the drawing boards. Over the next few months and years, the formerly pastoral township would be carpeted by low-cost bungalows and "starter homes" which, as it happened, became last stops for many, leaving the huge area the visually monotonous, slowly dilapidating scene it is today.

The high-rolling municipality had no crystal ball—and even if the councillors had possessed one, they probably would have ignored the dismal pictures of the future which lurked in its depths. All Scarborough could see were tax-coffers stuffed with gold, plucked from the pockets of the new homeowners riding the tide of economic hope into the subdivisions. But the town's rulers were quickly to learn the fiscal truth that no political entity can thrive, or even survive, on residential taxes alone. So it was, in 1948, that the town government snapped up 225 acres of land at the intersection of Victoria Park and Eglinton Avenue East, and began aggressively courting industrialists to set up shop there. It worked. Soon, fridge and toaster makers, car-part stampers, cosmetics and plastics manufacturers and other big corporate taxpayers had occupied the greensward along "Canada's Golden Mile of Industry," as the tax collector's golden goose had been dubbed. The huge human influx provided the labour power necessary to operate the plants, and generate both profits and taxes. Scarborough was saved.

It was in 1953 that the zone came into its fullest glory, when the first of the forty-two stores in Golden Mile Plaza opened for business. The unroofed strip mall, a revolutionary development in its day, was a hit with the parents of the baby-boomers from the start, and a trumpet-call to all that Toronto had made the grand switch from cozy British high-street shopping to American-style suburban consumerism. Many

Scarborough folk, I am sure, can fondly recall the day in 1959 when the Queen was brought to Golden Mile Plaza, to bless with her regal presence this newest of wonders, strip shopping.

We do not know what, if anything, Her Majesty thought about the Golden Mile. But we do have the musings of John Sewell, who did not forget the thrill of this early episode in Toronto's abrupt jump-up to urban Modernity. His memorial suggests more than mere admiration; indeed, it reveals a real soft spot for the Mile in its heyday: "The Fifties," crooned Sewell "when my generation drank milkshakes at the Honeymoon, and listened to Elvis Presley. We didn't share much with our parents, except perhaps the dream of the Golden Mile on the outskirts of Toronto. Now, in the eighties, dreams could never glitter in the same way."

It was something I did not expect from Toronto's only mayor ever to ride a bicycle to work, to fight the developers like a bobcat before, during and after his term of office in the early 1970s, and serve Toronto's anti-development neighbour-hoodniks as exemplary political advocate and hero. So what do we find in Sewell's nostalgia for "the dream of the Golden Mile," other than that the former mayor is made of the same stuff as us all?

Civic convictions, including the anti-development ones for which Sewell has been so eloquent and principled a spokesman, are formed in the crucible of maturity, urban experience, bad knocks, and revulsion against the damage the selfish and powerful can wreak on the city fabric. But if grown-up experience should lead a man to opinions more measured than those of his impulsive adolescence, he can never afford to forget the music and magazines, the soda shops and back seats, and places like Golden Mile Plaza in which he awoke to sexuality, his soul and adult body and its longings, and discovered the civic stage on which will and desire would be tested against reality. Much of life is spent overcoming the mistakes we made in such cultural environs, and learning from the successes.

So here's to His Former Worship, for reminding us to treasure the sites of our openings to the world, even after we've decided to write out of our political agenda much of what we learned there.

MALL ENDURANCE

Here is your assignment.

You are to drive to one of North America's 2,500 regional malls—defined by retail analysts as a covered shopping centre featuring at least two department stores and one hundred shops, and attracting customers from within a twenty-mile radius—then lock and leave your car on the skirt of asphalt always surrounding such sites.

You are then to stay inside the building exactly three hours—the current average period of a mall visit in North America—and *hate* every single minute of it. Just being bored doesn't count, by the way. You really do have to hate it, from the first to the last minute you are there.

Now that may well be a problem. While pre-testing this assignment at two immense Toronto-area malls—Yorkdale, in North York, and Mississauga's Square One—I found myself becoming crashingly bored long before I could work up a decent hate for these dreamlands of consumerism. That's perhaps because my frequent visits to megamalls are almost always dominated by two specific needs—for the thing I want and for getting my business done as quickly as possible.

Not that I have anything against malls. Indeed, I have much to be grateful for. I am certainly old enough to remember being dragged by my mother from store to store along the blisteringly hot or miserably cold streets of the pre-mall world, and am therefore not keen on seeing this fate visited on any other child. For this reason alone, the abundance and variety, the easy access and quick exit summed up in the phrase "one-stop shopping," which I first heard and witnessed in the 1950s, has never lost its charm for me.

But unlike many other first-generation initiates into this postwar mode of consumer activity and architecture, I never took the common next step for some reason, becoming a browser or grazer. Nor did I ever become a mall layabout. Such persons did not exist during my growing-up years, since, in the 1950s, going to a mall was still a dress-up affair. I do not window-shop, and have never done so. And only rarely, while dashing around a mall on shopping errands, have I experienced what marketing analysts call "the Gruen Transfer." Named after architect Victor Gruen, it's what urbanist Margaret Crawford has described as "the moment a 'destination buyer'…is transformed into an impulse shopper, a crucial point immediately visible in the shift from a determined stride to an erratic and meandering gait." If you are committed, as I am, to picking up your case of cat food, two Jockey Classic briefs and a tube of Polyfilla in the shortest time possible, then the three hours in a mall I've assigned you will be well-nigh maddening.

For Margaret Crawford, those hours would be very hateful indeed. Or so one imagines after reading her essay "The World in a Shopping Mall," which opens Michael Sorkin's recent collection of outcries called *Variations on a Theme Park.*

Crawford is West Coast: an architectural historian and theorist at the Southern California Institute of Architecture. Sorkin is East Coast: an architect, former architecture critic of *The Village Voice,* and now a professor at Yale and New York's Cooper Union. Most contributors to this book, in fact, reside on one coast or the other. From their academic towers beside America's littered, polluted eastern and western shores, these observers gaze out upon a continent blighted by insidious homogeneity, gentrification and rigid social stratification, whole cities of streets empty of once-happy social tumult, and other nightmares embedded in the minds of the current generation of urbanists by Jane Jacobs and her fellow apocalyptarians, when younger and still on the professional rise. And, as is the fashion in academic tracts on North American cities nowadays, the critics' contempt for what they see is cast in the form of anarchist jeremiad, intended to bump us out of our (presumed) complacency and mobilize us into the struggle to save our continent's cities from Baskin-Robbins, Disneyland and the West Edmonton Mall.

The 5.2-million-square-foot West Edmonton Mall—WEM, as Crawford calls it—is her prime exhibit of the hell of simulation and false need about to overtake us all. She is aghast at WEM's eight hundred shops, its twenty movie theatres and thirteen nightclubs, and an artificial lagoon "where real submarines move through...imported coral and plastic seaweed inhabited by live penguins and electronically controlled rubber sharks." She also worries about the malign architectural tactics meant to produce the Gruen Transfer, forcing us to bend the knee to Mammon before we know what we're doing. She is concerned about what's to become of the buoyant intimacy, haggling, personal confrontation and other elements of village-market behaviour being liquidated by the anti-haggling, fixed-price system of merchandising of which the mall is the culmination.

But along with much fretting over the gross manipulation she finds (or thinks she finds) in mall culture, Crawford provides a concise survey of the history of the mall itself, from its rise out of the department-store consumerism launched in Paris around 1850, to its spectacular postwar apotheosis in the suburban American shopping plaza. She also offers a good analysis of the unique combination of applied sociology, marketing know-how, creative financing flexibility that has made the mall the most stunning financial success, per square foot, in the history of building for retail sales. What really makes her essay sparkle, however, is the outrage ever seething just below the surface of the sophisticated historical and cultural summary.

While Crawford's facts are provocative and her condemnations plausible, her essay nevertheless leaves me as incapable and unwilling as ever to work up even a theoretical hate for the big malls I often visit and often use. First, because they seem no more malignant or bewitching than Mr. Wong's corner milk store, the local hardware outlet or any other building designed to shelter the necessary acts of consumption and exchange. The Gruen Transfer can happen in little shops as easily as in great malls, though it need not happen at all, if you keep your mind on what you are doing. And as for getting distracted by all the goodies in the stores, which Crawford finds awful: gimme a break. Getting seduced by alluring things, being lured onto unexpected paths, entangled by unpredicted passions are all parts, if potentially dangerous parts, of being incarnate and alive.

My second reason for not hating malls has to do with a certain hesitation about stern moralizing against anything so successful with the public at large. At bottom, Crawford's attack is snobbish, and driven by an élite nostalgia for a kind of hectic working-class street culture which nearly all urbanized North Americans disliked before Margaret Crawford or I were born, and which most have happily abandoned since then.

My last reason springs from the almost certain doom of malls and mall culture. What's the point, after all, in getting worked up about an architectural type that, some fifteen years ago, had attained numerical saturation of its market, and may now be enjoying the Indian summer of its short life?

No urban architectural formations are as liable to sudden change as those shaped by desire, consumption, longing and satisfaction. As street shopping lost its appeal for the middle classes, and strip shopping died after that, so the department store—the indispensible key to the success of every regional mall, and consumer retailing itself for more than a century—appears set to follow suit. For every $100 spent by Canadians in stores a decade ago, $9.53 went into department stores tills; nowadays, the figure is around $6.60, and headed down. Doing the damage are the chain-linked specialty monster houses, such as Toys 'R' Us, Business Depot, Aikenhead's hardware, each standing vast and stuffed with things on its own parking lot.

If department stores are crucial to malls, and department stores are vanishing, then what architectural mode will desire-supply take next? Is the titanic warehouse-outlet the next stage in the history of consumption, or, like the strip plaza, a short-lived glitch? I do not have the answer, though such are the questions ambitious young architects should certainly be asking themselves. And learned urbanists of the sort represented in Sorkin's interesting anthology might more profitably spend their time dreaming up new civic and suburban configurations for the more chaste, pennywise shopping of the 1990s and beyond, rather than beating the expiring horse of mall culture.

Eaton Centre

Once, while strolling through Toronto Eaton Centre's multi-level galleries of shops, I caught a glimpse of myself in a shirt-store mirror, and beheld the enemy.

The suburbanites are causing one sort of blight. But it's downtown-dwellers like *that*, I thought of the middle-aged visage in the mirror, whom many a critic believes responsible for everything that's wrong with the urban core.

We have forsaken the street. We avoid festive collisions with persons from other socio-economic groups.

We urban traitors do most of our consuming in the city's bright, climate-controlled, constantly policed shopping tunnels and multilevel city-core malls. While benighted suburbanites may be forgiven for their mall addiction, educated urbanites will never be acquitted of the crime of abandoning the corner store, and flinging themselves into the ecstatic gleam of Eaton Centre. We take refuge inside burglar-alarmed, camouflaged, anti-street-oriented "stealth houses" (as a hostile architectural critic has called them)—homes deliberately concealed from the attention, and even knowledge, of our neighbours.

I confess. I live in a stealth house. I shop in downtown malls. I walk between the tall buildings through pedestrian tunnels and bridges, even when the weather is nice. I am one of the enemies of the people who, in the view of Trevor Boddy, is deforming the contemporary city from "a zone of coexistence, of dialogue, of friction" into an ugly, self-absorbed network of "monoclass, monoform, and decidedly monotonous hermetic architectural archipelagos," i.e. malls and enclosed shopping corridors.

In Boddy's fact-rich contribution to *Variations on a Theme Park,* the Ottawa architectural historian and theorist delivers the ominous news that "under the guise of convenience, we are imposing a middle-class tyranny on…downtown streets." He ends by calling upon us to "resist the temptation to fancy ourselves the new Medici, with our continuous sealed walker's highways to art gallery, shopping centre, health club and other splendid palaces of refuge…We must quit the splendid surroundings of our new bridges and return to the streets, with all their hectoring danger, their swirling confusion, and their muddled vitality."

Specifically, "we" would-be Florentine plutocrats must abandon Eaton Centre, Boddy's Toronto example of these hateful "palaces of refuge."

Curiously, the author seems unaware that the genesis of the downtown enclosed mall is no novelty, but merely the latest move in a century-old trend to intensify and focus city-centre pedestrian movement and consumption. Be that as it may, however, "we" aren't about to give up Eaton Centre, because "we" like it very much. Each year, some 42 million people wander among the Centre's 1.6-million square feet of enclosed retail areas—that's a crowd about ten times the size of the Metropolitan Toronto region's total population—because (1) they are crazed or (2), as I believe, they like what they see for sale there. The 14.5-acre site, extending along Yonge Street the entire distance between the cross-town thoroughfares of Dundas Street West and Queen Street West, is middle-brow and middle-class, with mid-priced items for every middle-range lifestyle. That is, it's got no low-down, dirty video stores or bargain warehouses; nor does it sport a Holt Renfrew outlet in which the ostentiously affluent can flaunt their affluence. It's got Roots and Gap Kids and Shopper's Drug Mart and Birks and Tip Top Tailors, and a self-proclaimed "classical" record store featuring Franz Schubert and Barbara Streisand side by side. For a megamall that's middle-aged—sixteen years old is getting up there, as such palaces of refuge go—the vast building is as spiffy, clean and neat as a north Toronto *en suite* bathroom.

Quite apart from its enduring vogue, and despite the hostile, slit-eyed glances it gets from Trevor Boddy and other self-appointed defenders of "downtown vitality," Eaton Centre should be cherished for its peculiar architectural importance.

That's not saying we have to love it. Most of us don't. But the behemoth is Toronto's grandest commercial real-estate statement of the styling called, with attractive frankness, Brutalism. In its twilight during the decade of the Centre's design and construction (by Toronto's Bregman & Hamann, and the Zeidler Roberts Partnership), this thuggish Anglo-American architectural manner marshalled exposed ductwork and elevator machinery, obvious structural steel and showoffishly pre-fab surface coverings and massive poured concrete, all to one end: the celebration of mass democracy's victory over ideological scruples, right or left, and its stylish embrace of godless materialistic consumerism. Now think what you will of the Brutalist gospel—it has always seemed pedantic and exhaustingly optimistic to me—Toronto Eaton Centre still constitutes an unforgettably forceful expression of its final, faded popularity.

Perhaps because I find shopping there practical, and the galleria's apocalyptic architectural message interesting, I am not able to view the Centre through Boddy's jaundiced eyes, as an instrument of anti-urban evil. And while agreeing with many of his useful factual observations, I cannot share his horror about mall-and-tunnel architecture's transformation of my hometown's population into what he calls "a wholly disaggregated series of social, racial, class and sexual subtypes, without the possibility of contact, divided by occupation, bylaw, habit, or default." As it happens, I downright like the separations and distances and alienations which all metropolitan life and building increasingly, generously afford, and which Toronto offers more amply than perhaps any other Canadian city. (Really, Trevor Boddy should get out of Canada's tidy, officious, inflation-proof capital more often, fly down to Toronto, and get a stomach-full of real big-town street life—*then* tell us how thrilling it is.)

Anyway, if mingling with the motley throng is what the Eaton Centre shopper wants to do, it's available right at the north-east door. Twenty-four hours a day, that corner of Yonge and Dundas has more hectoring danger, swirling confusion and muddled vitality than any other intersection in Toronto. It's a constant, colourful near-riot of punks, grunges, passed-out addicts, teen dropouts waiting to score some dope or sell their bodies for a fix, evangelists noisily attempting to win souls for Jesus with conundrums and gimmicky magic tricks, faux-poor suburban boys and girls in "Rock Against Fur" T-shirts skipping school and looking for sins to commit, Salvadoran street bands and boom-boxes blasting out Stompin' Tom to lure gaudily track-suited American tourists to the cheap mirrored-sunglasses pushcart.

Handed the prospect of being plunged day after day into that or any other kind of hectic Boddyesque mixing, I'd take permanent confinement to my stealth house in a second.

SCENES FROM THE UNDERWORLD

Most city folk know the labyrinth beneath the city, or at least short stretches of its twelve-odd kilometres of connected pathways, forking and winding through the spacious lobbies and food courts, and basements of tall buildings between Front Street West and Dundas Street West, and connecting an estimated fifty office towers, six hotels, five subway stations, our principal commuter and inter-city rail station, and more than a thousand shops, restaurants and other amenities.

For those who do not know it, and want to know how strange reality can be, here's my favourite route. The outing begins with a descent from street level at the network's northwesterly end, the recently spiffed-up thirties-style lobby of the old Gray Coach terminal on Bay Street. From there, the track continues eastbound through the shops of The Atrium on Bay. After the Dundas subway station, one goes south across the bottom levels of the Eaton Centre—a realm of blue jeans and cookies, and a huge waterfall cascading over an escarpment of stone—then under Queen Street, through an opulent grocery store deep beneath The Bay (formerly Simpson's), west under the Thomson Building, then generally south again below the tall bank towers at King and Bay to Union Station, and the end.

Now you will soon find this trek is not as reassuringly straightforward as I've made it sound. There has been a recent proliferation of directional helps, the so-called *PATH* markers designed by the Toronto firm of Gottschalk & Ash. But signage in the underworld remains confusing, and, despite improvement, the maps available at variety stores are still hard to follow. Toronto's underground, like some other places in the world—Venice comes to mind—may require living in, or at least a thousand walk-throughs, to understand thoroughly. Anyway, if you're like most urban wanderers, you enjoy getting lost, just to see what happens, and where it gets you.

To be sure, tracing this system of subterranean trails, with their shops and banks and even law offices and dental clinics, is to be bored by the gleaming sameness of it all, made uncomfortable by the humid atmosphere, certainly overloaded by the incessant, enveloping consumerism. Yet such vagabondage is fascinating, if only because it reveals how completely this new urban architecture overturns the connotations tunnels have had in Western culture until quite recent times.

Until this century, most underground structures—tombs, cellars, mines—have been the most fearful of places, connoting death and dire straits and hell. They were unnatural: tyrants maintained secret tunnels to escape through, concentration-camp inmates dug illegal burrows for the same reason. No one went willingly into an underground place, unless engaged in work, or in some dark, secret mission or conspiracy, or to escape catastrophe, such as bombing or tornado. The underworld was a place of death and the dead, of pirates and gangsters and monsters. The only terrifying place on the grounds of my otherwise unfearful childhood home was the storm cellar—a dark, dank haunt of spiders and unspeakable crawlers reserved for our use in case of violent storms. (The Roman catacombs may have been hallowed by their use as refuges by persecuted Christians, but this has not made them a place you want to stay in long.)

The modern, more benign architectural history of the submerged passageway was heralded some five hundred years ago by Leonardo da Vinci, who sketched out an imaginary city under which vehicular corridors would run. But like the airplane and other schemes by this Renaissance genius, the underground city was destined to remain unrealized until modern technology made it possible. London's Inner Circle

Museum Station

steam railway, put into operation in 1863, appears to have been the first subsurface transport system, and the sixty-two-mile network of freight tunnels burrowed under Chicago, licensed in 1899 and opened in 1909, may be the first true underground passages for individually operated and propelled vehicles. And throughout this century, architects and theorists—Le Corbusier, the Swiss engineer Max du Blois, the Italian visionary Sant'Elia and many lesser figures—have dreamed of dissolving the traffic jam by channelling the vehicular flow along ramps under or over habitation levels.

None of these precedents or notions, however, seems to be a direct ancestor of the contemporary subsurface *pedestrian* passageway. This structure, I'm inclined to think, is a rather straightforward result of the desire by postwar developers of tall buildings to maximize the commercial value of their deep foundation structures. Skyscrapers rise, after all, from posts sunk into the earth, not from platforms on it. So you open the otherwise unused intervals among the under-earth pylons, making the space bright and various and accessible, fill it with shops, make what short links as may prove necessary between your basement and adjoining ones, and market this network as escape from cruel and insufferable weather, which nobody had ever before thought cruel and insufferable. In any case, it is certain that Toronto's first underground shopping concourses came into being in the mid-1960s, along with the first Modernist office towers; and the former has spread at the same pace the latter have risen towards the sky.

The largely uninterrupted flow of Toronto's passageway—from tower lobby, down escalators, into retail zones, back up to food fair, back into office complex, and so on—was not, however, a feature of the system at the beginning. Behind that change stands the ambition of Toronto's Reichmann brothers, Paul, Ralph and Albert, for a while the most successful private real-estate developers in world history. And behind the Reichmanns' innovation is a story worth telling.

It all began around 1970, when the brothers decided to put up the world's tallest bank tower and a new home for the Toronto Stock Exchange on a seven-acre site at Bay and King Street West. The complex was to be known as First Canadian Place. No sooner had they embarked on their project, however, than they found themselves, plans in hand, frowning over a table at a squad of young, reform-minded and equally hard-headed City of Toronto planners and architects, pledged to the preservation of downtown breathing room and open corridors for public movement and amenities in Toronto's core.

"We wanted First Canadian Place to be a place where people would want to be," Ron Soskolne, then a planner sitting on the city's side, told me. "What we got was five acres of public space, and a new spaciousness in the lobby. It marked a change from the formal purity of the office tower to mixed use in a large development." If today the towers of the Toronto-Dominion Centre, rising from their low pedestal just across King Street, remain Toronto's finest instance of purist commercial real estate, the entry areas and the shopping mall of First Canadian Place—despite the monotony of all the white marble slapped up on every surface—have the expansiveness and readily available services which would characterize all the best Reichmann projects thereafter.

The Reichmanns got a winning formula from the guys across the table, and Soskolne, fellow planner Michael Dennis, city housing commissioner in the 1970s, and Toronto chief planner Tony Coombes eventually got top jobs in the Reichmann's Olympia & York Developments Ltd. For Soskolne—now with Reichmann International, successor to bankrupt O&Y—his switch-over to his former opponents was a natural development, not treason. "I was inherently a promoter, not a regulator," said Soskolne. "I much more enjoy making things happen instead of keeping them from happening." As for the urban values he upheld when working for the city, "they have changed hardly at all. Any of the projects we've done have those characteristics. I have learned a lot about development from the feasibility side, but the essential values are still valid."

These values can be summed up in the overworked phrase "public space," and the overworked word "access."

Access is guaranteed for those targeted to benefit from it. Sophisticated city executives and workers who want the on-site fashion and sports shops you find at First Canadian Place, the tailors and *chocolatiers,* dry cleaners and dentists and deluxe toyshops, situated along bright boulevards, both above and below ground. People who want quick access, without a trip through the slop and chill of a Toronto winter, to luxuriously appointed restaurants for serious lunches, to fast-food joints that are just fine without saying so, and to coffee nooks in pools of quiet, outside the stream's main current. People for whom the good life is a matter of proximity to subway and record stores, to ticket outlets and art exhibitions—all the things available in First Canadian Place's 500,000-square-foot shopping mall, or in nearby tunnels and basements.

The ambiguous nexus of overground and underground at First Canadian Place is as good an example of "access" as Toronto can offer. One recent morning, after the riptide of inbound commuters had slowed to a trickle and the traffic in the white and beige expanses of the shopping concourse had slowed to a grazing pace, I found a table and a coffee in a fast-food place overlooking the Bay Street entrance to the complex. My high vantage-point, on a level above the entry level, provided an impressive view of the corridor below, and the large glass-encased Bank of Montreal branch beyond it. Instead of being hidden underground in a catacomb sharply separated from the official entries of the building—like the shopping precinct beneath the slightly earlier Toronto-Dominion Centre, across the street—the restaurants and stores of First Canadian Place have been comfortably knit into the very fabric of the towers' multilevel public areas. The flow of space and joining of interior volumes are smooth, continuous, integrated with dignity—eloquently communicating the late-Modernist idea of erasing the formerly rigid boundaries between public and private, work and pleasure, production and consumption.

But just how public *is* public? Certainly, anyone could walk in through the doors, at least when the doors are open, which isn't all the time. But would just anyone do so?

The omnipresent Carrara marble, the swank decor of the shops and grand curtains of glass, the luxurious (if now dated) lobby light fixtures and other interior appointments, along with a dozen other not-so-subtle signals, say *keep out* to the homeless, the jobless, the penniless urban wanderer. This absorption of more and more "public" spaces and retail amenities into office towers and their underground components has recently come under heavy fire from urbanists and architects, for a number of reasons. One has to do with the presumably undesirable establishment of ever-larger exclusionary zones in the city's heart—exclusionary, that is, to the poor, homeless, derelict. Another criticism, closely allied to the first, has to do with the creation of similarly disagreeable territories of homogeneity, where people only see mirror images of themselves, and are spared the sight of the unlike.

Though ruefully, I find such a critique wanting, because it hangs on a feeble scaffolding of *non sequiturs* and wishful thinking, not the girders of urban fact. Poverty will not be eliminated by putting difficult distances between sophisticated city people and their pleasures and recreations. Nor will most of us willingly go very far out of our way to meet *les autres,* unless it's in our best interest to do so—in which case, such an encounter will happen. Until a way to alter this tendency of people to seek the company of others like themselves is found, the mixed-use tunnels will probably continue to attract working city folk, who seem destined never to be as idealistic as the urban visionaries among us would like.

I am concerned, however, about a certain metaphoric loss to our language these submerged malls will likely bring in their train. If developers keep providing vast subterranean pathways as utterly benign and non-sinister as those under Toronto, future generations may not understand what writers of the last two centuries have meant when they spoke metaphorically about "underground men," the "anti-fascist underground," and the like. Trodding the washed marble tiles under Toronto's financial district, moving through those immaculate and almost shadowless corridors, one finds none of those characters typically associated with the undergrounds in legend and story—sexual desperados, outlaws, mad hermits, wild boys who rule whole terrifying tracts of the dark world, and dangerous monsters, like the Minotaur, whom we descend into the underworld to fight.

Instead of all that, we discover respectable, well-dressed professional people, come down from the towers to shop in the thousand stores of that underworld. Terror has been banished from the experience.

But the person who goes through the tunnels may come back with one peculiar memory—of just how the system erases one's traditional sense of the city core as a rhythmic ensemble of level streets, right-angle intersections and high edifices. Twisting and turning among the roots of tall buildings, the particularity of each structure above ground cannot be sensed. Little other than a change in the pattern and composition of the floor tiles signals the passage from the basement of one building to that of another. Everything tends to meld into a continuous spectacle of shopping and regular movement, of sparkling glass and light travertine wall-facing, of broad-leaved tropical trees and upmarket fast food, all backed by the faint, incessant hum of the electrical and mechanical devices which share the corridors with us.

With light and music and pleasant finishings, the planners and makers of the urban culture have created a new, very popular structure. But they have also begun banishing from the city its ancient cloacal darkness, its space of the sinister, perverse, untamed—the scintillating depths which have inspired so much fine modern fiction, poetry and cinema, from Dostoyevsky to Derek Jarman. But this vanquishing may be inevitable. As long ago as 1965—at the start of the great pedestrian underworlding—the magazine *Progressive Architecture* printed an article by American architect Malcolm Wells called "Nowhere To Go But Down." His title has proved prophetic of what was to come; to walk Toronto's underground system is to feel oneself in the future of cities, and the slow turning of urban experience from the natural rhythms of day and night into something like unending day.

Suburban Idylls

Mall Path, Scarborough

Revisiting Mississauga

Though I'd never before seen it in print, I must have heard a story like Andrew McAlpine's, published in *The Globe and Mail* just after Christmas, 1992, a dozen times.

His tale was about a recent, wintry drive by his childhood home, where his parents lived twenty-five years before moving to the Ontario village of Fergus. Memories rushed back from 1963, when Andrew, aged seven, decamped with his family to a new tract house at the end of a remote, idyllic cul-de-sac, out among apple orchards and strawberry patches and deliciously inviting country vistas of suburban Mississauga.

The idyll was short-lived. Andrew's mother soon found herself bored silly. And the boy himself lived there long enough to witness the burial of the orchards under concrete and asphalt and lawn, and the building of "more houses and shopping malls and high-rise apartment blocks and glass office towers." Returning to suburbia after a decade abroad, McAlpine found it all "as awful as I had feared. The mindless patchwork of banal and ugly speculative development that passes for a planned city has continued its cancerous growth." The question that came to mind as he motored down the cul-de-sac, noting the present-day residents, was whether the dream his family bought thirty years before was theirs as well.

I would expect not. The suburban family-folk I know

personally give quite practical reasons for living out there: more house for the same money, big yards for the kids, a non-specific sense of security, and the good, long distance living there puts between the secure fortress of domesticity and the downtown towers they earn the mortgage money in.

Except for a brief time many years ago, I have never lived in suburbia, so I cannot say how much of what suburbanites, past and present, say is true, and how much delusion. But that the original inhabitants of Toronto's postwar sprawl were dreamers is beyond reasonable doubt. If the tract-house advertisements and "modern living" supplements in magazines for the years between 1945 and 1955 tell us anything, the dream—or commercial come-on—was about freedom and fresh air, a healthy young, white-collar Dad coming home from the office, Mom hard at work among her delightful children and miraculous appliances. It was about stability, and a few square feet of ground to call one's own. It was about predictability, security, solidity.

I am inclined not to hate postwar suburbia, if only because it delivered on its promise, at least for a while. Too, I can just remember the conditions under which McAlpine's parents may well have grown up in—the scarcity of the war years, the abandonment of children to whomever was handy while Mom went to work at the bomb plant, the absence of men and the constant worry that those absent men might never return from the conflict in Europe or Asia. And I know more than I wish to know about families demoralized and embittered, even ripped to pieces by the Great Depression. Given the miseries of the years between the Crash and VJ Day, it is no wonder that developers found the marketing of the tract-house dream easy in the postwar period. It was not a meretricious or absurd dream. So what went wrong?

The current wisdom puts the finger on greedy, speedy developers, money-grubbing townships wanting to up the tax base, and utopian planners slap-happy about the power of bulldozers to make real their concrete New Jerusalem, express-wayed, malled and bungalowed. All these ingredients were present at the beginning, at least in the ill-planned suburbs. (There were others which were *not* ill-planned.) But a recent book about sexuality and city planning called *Gendered Spaces,* by urbanist Daphne Spain, raises the possibility of yet another factor in the reported unhappiness that descended on the sub-divisions: the obsolescence of the social ideals suburbia's endless rows of neat, secure ranch-styles were trying to re-create.

Spain's book throws interesting light on, among much else, the boredom of McAlpine's mother in her paradise of Tupperware and gadgets. For almost a hundred years, she tells us, North American house-builders and their middle-class clients had been inundated with books insisting on the "home" as the physical expression and enforcer of social stability. All these books "glorified the wife's role in creating a calm and peaceful retreat for her family." It was the "faire ladye's" lot in life—the ridiculous spelling is from one of these tomes—to preside over a parlour furnished in "simple, elegant and harmonious style," and to provide her husband with a "sanctum" worthy of the "master of the house." Meanwhile, the tide of reality was running another way. Since the mid-nineteenth century, women had been steadily winning more and more rights—to vote and work, to keep their earnings, to own property. Most women, it appears, liked what they got. A labour research team has discovered through recent interviews, for example, that women who spent the war years making machine-guns in a Toronto plant again and again recalled those years of financial independence as the best in their lives.

Then, in 1945, Rosie the Rivetter, Molly the Machine-Gun Maker and myriad other Canadian and American women were abruptly demobilized, and displaced by returning male veterans, then subjected to a stunning mass-media campaign aimed at turning them into the women depicted in those Victorian architectural handbooks: serene housewives, busy consumers of new commodities, and contented, fertile queen-bees making cookies while hubby was out earning the cash. Many young wives went along with this vision, and tried to make it work. And it all seems to have worked, for a time, at least for some people. But a century of gradual emancipation from rigid expectations of what women should and must do had done its irreversible work. Even if it was, is and always will be desired by many men and women, stability in gendered roles will probably never again be sustainable for very long. It was the fate of Mrs. McAlpine, and that of millions of other women, to come of age at precisely the moment when the old, old human hope of snug security was briefly rekindled in mass culture, and their destiny to live long enough to see it snuffed out, this time perhaps forever.

THE STRANGEST HOUSE ON PARKHURST BOULEVARD

Shady under its old hardwood trees, East York's Parkhurst Boulevard is a comfortably modest street of family homes, none grand but none particularly mean, arranged side by side on narrow lots with mid-sized aprons of lawn fore and aft.

The small brick houses, with their neat flower beds out front, seem to have been grown from the seed of country cottages; they imply coziness without crowding, the rural hearth without any of the heartbreak of urban living. There are no signs of institutional life (churches, offices and so forth) on the street. Rather, the word *family* is brought immediately to mind—just the word itself, free of sociological or critical qualifiers such as nuclear, extended or whatever, and comfy with connotations of poignance and warmth.

I cannot say whether all the people who dwell on Parkhurst enjoy warm, comfy family lives. But if they don't embody such values, the nostalgic visual rhetoric of their street surely does, with plain conviction if also with rather too much emphasis on social uniformity, or, to use the nicer word, *community.*

Along with everything else on the street, the house at 160 Parkhurst, built about 1948, conveys this message, with its warm brick front, the petunias nodding by the door, the stately maple on the lawn and the white picket fence. But not much human kindliness was ever generated under this house's gently pitched roof. Walk through the tall white front door—as I did with Jim Balmer, manager of operations for the East York Hydro-electric Commission—and you find yourself on a red-tile floor, standing among huge, loudly humming and whining metal boxes. Into these boxes, explained Balmer, come 13,800 line-volts of electricity, supplied by Ontario Hydro. The power is reduced inside the house—known in the business as the Parkhurst Sub-Station—then sent on for further reduction by small transformers affixed on hydro poles along the street, and thence to the local residents' toasters and TV sets.

The building at 160 Parkhurst, then, is a house, but hardly a home. It just looks like one. Or, more to the point, it is an instance of modern technology (without which suburbia would not be possible) resolutely disguised behind an explicitly anti-technological mask.

And it's not alone. Around 1987, Toronto artist Robin Collyer, who lives in Willowdale, became fascinated with these non-homes of suburbia and has since photographed some forty of them, which are only a fraction of those built. Collyer's subjects have ranged from the Toronto Hydro installation at 555 Spadina Road, elegantly disguised as a Forest Hill mansion in the Georgian style, built about 1951, to ones which amount to little more than a skimpy lawn and tacky bungalow façade dropped in front of a clump of buzzing transformers.

During the same three-year period, Collyer has also been taking pictures of "monster houses," a suburban blight of more recent vintage, which, in the photographer's view, embody urban values precisely opposed to those of the transformer houses, and much less desirable. Monster houses show "no consideration for the character of the neighbourhood," while the concealment of transformers within home-like structures was a "quite honourable decision" which showed "sensitivity to the neighbourhood."

Since this phenomenon of "sensitivity" happened right across Toronto during its postwar suburban building boom, it is clear that builders and home-owners were giving the hydro authorities the same message, which, in turn, became a remarkable architectural fact. But what exactly motivated it? What was it that Torontonians in the postwar era didn't like about transformers? To say people thought them "ugly" and therefore wanted them camouflaged explains nothing, since it doesn't make clear how these ruggedly handsome, tough items of technology came to be perceived as ugly, hence undesirable.

You'll get an idea of how that happened if you take another walk along Parkhurst Boulevard, this time thinking about its history.

Like that of countless other suburban streets in Toronto, the architecture of Parkhurst is an artifice created in response to trauma. As the first houses began to go up, the Great Depression was still a vivid, terrible memory. The last houses were completed in the years immediately following the Second World War. The architects of Parkhurst showed how well they knew their clients when they crafted a street of small, old-fashioned homes redolent with memories of happier times.

The result was a kind of hospital disguised as a street of homes, instituted for the healing of souls damaged by some fifteen years, first, of economic breakdown (which hit the borough of East York very hard) and, next, mechanized warfare. In the closely related styles and closely aligned sizes of the houses, the street speaks of a uniform commitment to an idea of family life from the earlier twentieth century—the benign, companionable patriarchy which succeeded the more rigorously hierarchical Victorian one. By offering a continuation of the suburban building styles of twenty years before, the builders proclaim the street's continuity with the pre-crash, pre-war world of the buyers' childhood. There was to be no place here for reminders of disaster. In a therapeutic setting such as Parkhurst Boulevard, a bald reminder of the modern network of industrialized electric power would indeed have been out of place—hence, its designation as undesirable, and its concealment behind an ideological façade.

Since the Second World War, Toronto has been a thoroughly modern city, yet its urban modernism, like Janus, always faced two ways. One view was towards the future—towards mastery over urban and human disorder by means of the strict, serious command of space and planning, and by technology and industrial power. The other has typically faced the other way, preferring a nostalgic play with masks, façades and the known styles of the past.

Any Toronto suburb presents evidence of this kind of nostalgic modernity, which, since the building of suburbia itself began, has a kind of visualized anguish for lost roots and interrupted traditions. Normally, and as often as possible, expressions of the two modernities—soaring American-style office tower and English story-book cottage—are kept from meeting publicly in the same place, on the same street.

Occasionally, however, the technological fact and architecture of roots are required to share a street. Hence, the transformers disguised as houses, in which Toronto has proposed an interesting solution to the enduring conflicts between ideas of what it means to be modern, urban, free.

DON MILLS

In 1953, with the Convenience Centre ready to go up on the bull's-eye of the 2,058-acre mixed-use project at the intersection of Lawrence Avenue West and Don Mills Road, and with houses, schools and churches (coordinated in colour, materials and design) already arising on the lanes of its meticulously planned residential quadrants, Don Mills was ready to receive its founding families.

Those first Ozzie and Harriets, for whom this innovative town of up to thirty-five thousand souls seven miles north-east of downtown Toronto had been projected on the Ontario countryside, found themselves in Modernity's postwar Emerald City, radiant with the New—new cars, new babies coming along, the dads' new careers opening up, new houses to outfit with gleaming novel appliances, new neighbours to meet.

Their immediate vision is commemorated by two relief sculptures still affixed to Glen Gordon Court, a modest garden apartment emplacement near the Convenience Centre. On one panel is stalwart father, happy baby and serious little girl; on the other, dutiful mother, healthy boy—and, naturally, the family pooch. Given their preoccupation with making this vision real, few of the Don Mills pioneers of '53, I suspect, had much time to ponder what their town would be like forty years on—and, even had they taken a moment to do so, few could have believed that this experiment in Modernist social engineering would be more fascinating in 1993 than ever before.

True, Don Mills is today a place less refined and astute than the paradigm of Modernist probity decreed in the early 1950s by Macklin L. Hancock, the gifted Harvard undergraduate whom Toronto financier E. P. Taylor picked to mastermind his giant real-estate scheme.

Urban jumble long ago nudged its way into Don Mills' green belt, intended by Hancock to quarantine the self-contained community of commercial, residential and industrial facilities from precisely this sort of encroachment. The low horizontal lines of architect John B. Parkin's Convenience

Centre (completed in 1955)—an historic work of mall construction—has been broken by the recent erection of an office tower, and its serene autonomy smudged by isolated fast-food and mini-mart agglomerations with no architectural reference to the commanding central structure.

Worst of all, some condominiums in hideous French Provincial and neo-Georgian flavours have muscled their way into the small, well-spaced parks between the older flat-topped concrete and brick buildings around the intersection of Lawrence Avenue and Don Mills Road.

What's remarkable, given the project's compromised bounds and disarrayed centre, is the dedication with which the quadrant residents are keeping together the crisp Modernist coherence of their instant neighbourhoods, and the distinctive poetics of their streetscapes.

If the odd maverick has been demolished or turned into a garage, the single, open carport—the domestic showcase of postwar consumerism and mobility—is still the standard place to find the family auto in Don Mills. One can find the occasional monster house, loathsome pink and frantically ornate, its cavernous multi-car garage crowding the street line. But for the most part, Don Mills folk have rebuilt and remodelled modestly, and with consideration for the relatively small scale, style differences, simple lines and material harmonies—beige flagstone, tan and grey brick, dark brown wood trim—dictated by the original plan.

Dictated, by the way, is hardly too strong a word for what Macklin Hancock, his assistant Douglas Lee, and his backer—E. P. Taylor's Don Mills Development Corporation—effected on that countryside once just beyond Toronto. Don Mills was not to be just another monoform subdivision, or merely another add-on to Toronto's suburban supernova. The project was to be a new creation, its every detail and distinction strictly supervised, its amenities and houses constructed by the best architects available, and carefully directed towards the technological and architectural satisfaction of human desire, as understood by Hancock, his Modernist teachers at Harvard, and the now much-despised social and urban theorists of mid-century.

The free spatial movement afforded by cars—the analogue to the free social mobility so highly treasured by these theorists—was understood as a basic need and pleasure in the Don Mills model of human nature and urban culture. Hence the scarcity of sidewalks—a presumed "fault" balanced by the extensive system of discreet off-street paths—and the complete absence of corner stores and local hangouts within Don Mills' residential precincts. It was expected that one would cruise over to the Convenience Centre, when in need of such services. In this model, similarly, the need for articulated visible order, wide and regular spacing of houses, and the sharp curtailing of visually jarring variances and innovations, is taken for granted.

Anyone inclined to dismiss as boring and sterile the result of such thinking should spend a day roaming the pleasant residential quadrants of Don Mills. The street, dedicated to cars, is each sector's central architectural fact. But, because no street is straight, the eye is always being led onward, round yet another turn in the road, both by the gentle bending of pavement over the rolling landform and by the broad strips of neat lawn bordering each thoroughfare. Residents, by and large, have honoured the designers' intent when it comes to these continuous green street margins. Only rarely does one find a front yard transformed from oblong of lawn into a garden, thus "individualized."

And though the original houses are closely attuned in shape, size and colour—a changelessness of programme perhaps meant to recall the changeless virtues of grace, good taste, good form—the progress of the house line along the street is restrained from tiresomeness by its graceful curving with the street, and by the uncrowding that allows each house to *rotate* with respect to the passer-by, revealing various curt angles as the observer moves along. (Le Corbusier, whom the anti-Modernists blame for every urban ill short of flat tires, issued a refreshing challenge to modern planners in the 1920s, to see if it were possible to "impose an architectural character on the winding street." Macklin Hancock obviously picked up the gauntlet, and won the battle.)

If Don Mills didn't work out quite as its planners hoped it would—the expected mix of persons of differing means and ages did not materialize, and Don Mills people have remained stolidly middle class, and ever older and greyer—this most famous of postwar Canada's new towns is still a monumental rebuke to Modernist idealism's numerous enemies, and an instance of the moral seriousness that characterized the best Modernist urban thought and architectural practice in its heyday, some forty years ago.

THE TROUBLE WITH SCARBOROUGH

Retired accountant Tom Abel and his neighbours, founders of a pep squad called Friends of Scarborough, are "fed up" with their Metro city's reputation as a crime-ridden, decaying and centreless bore. The tiny group's goal is to "build pride" among the 500,000-odd dwellers in the vast east-Metro suburb. "We have a lot to offer," proclaimed *West Hill News,* a Scarborough community paper in late 1993. "Scarborough is the most ethnically diverse city in Canada. A lot of good things are happening…but more needs to be done to foster a positive attitude."

The crusade of the Friends and of *West Hill News* is doomed, of course, simply because the Metro member-city is a crime-ridden, decaying centreless bore. Between 1988 and 1992, Scarborough's muggers, rapists and murderers increased their business by almost 40 per cent. This visually inert expanse of bungalows, shabby strip malls, mass-designed schools and factories is regarded by many Torontonians—with justice, given the prevalence of ethnic gangs—to be our town's likeliest launch-site for any future race war. Prostitution and drug trafficking are on the rise, right along with poverty. Apart from the Scarborough Bluffs, jutting in rugged grandeur into Lake Ontario, I can think of few reasons for any non-resident to go there—unless, like me, you happen to be fascinated by the phenomenon of postwar suburbia, its successes and failures.

I do not wish Scarborough ill. Nor do I share the cool hatred for the suburbs oozing from much contemporary writing by academic urbanists and architectural theorists. Indeed, I am convinced that it is time to show mercy on those whose highest earthly hopes were embodied in bungalow and lawn, comfortable conformism, freedom from ideology and the liberty to consume in a leisurely fashion.

That having been said, Tom Abel and his comrades should take a long look at the historic errors made in the creation of the place and state of mind known as Scarborough.

Nowhere in this suburb do these errors find more eloquent expression than in the so-called centre of this political unit, which in fact has no architecturally designated focus or edge. Even that official "centre"—the ensemble of buildings situated south of expressway 401, between McCowan and Brimley roads—is multipolar, disparate, centrifugal. The generic names of the streets in its surroundings suggest, not specific site, but slow drift through a dictionary of names of nothing: Estate Drive, Production Drive, Corporate Drive, Grangeway. The visual force giving a sense of location, presence, to all else was supposed to have been Toronto architect Raymond Moriyama's Civic Centre (1969–1973), an extravaganza roughly circular in plan and theatrically angular in elevation. Sprawling northward from the Civic Centre is Scarborough Town Centre, a regional shopping mall linked to the municipal building by a row of huge steel arches.

What makes this otherwise boring place interesting is the innocent exuberance with which its planners embraced Modernist development doctrine as a cure for what ailed Scarborough after the patient was already moribund, and got the medication wrong to boot. If by the late 1960s the city was feeling all the psychological and social ouches of aimless sprawl—so the thinking went—the thing to do was create *ex nihilo* a centre to give focus where there had been none. (The success of Don Mills is largely a result of the anticipation of sprawl problems, particularly ones caused by vast over-sizing of uncentred "neighbourhoods," and coordinated attention to them before they arose.) And if the rising incidence of gang conflict and plain fear was due to factional alienation in the huge, cheap-built residential sectors, then civil harmony could be attained by providing a forum in which persons from all classes, races and tax-brackets might gather to celebrate the one transcendent sacrament remaining in secular culture: the consumption of commodities and public services.

The complex that resulted from this meditation is anchored by its two principal structures. The enclosed mall features imposing department stores (Sears, The Bay, Eaton's), between which are strung numerous little outlets for Canadian and international retail chains, such as Taco Bell, Japan Camera, Cineplex Odeon theatres, Baskin-Robbins, Pro Hardware. Like virtually every shopping mall, it is nowhere and anywhere—the topographical equivalent of the raceless, genderless, non-sectarian and hence ideal citizen which secular humanism would have us be.

For such a soul was Moriyama's Civic Centre erected. While the smiling attendant did not appear to understand the quite simple question I asked on a recent visit—where and when the lecture by the project architect would begin— she was quick to point out the wares on sale at her information desk: Scarborough ballpoint pens, Scarborough scarves, Scarborough baseball hats, tickets to Canada's Wonderland. In

Fire escape, Scarborough

fact, the Civic Centre advertises almost as many products as the Town Centre: an Italian cultural festival; innumerable "educational opportunities"; social services for persons with any need, deficiency or ailment, or none; a programme of regular brisk walks and health lectures among the Town Centre shops, co-sponsored by city and mall management.

In this strange non-place of amusements and consuming and servicing, we have an excellent instance of the unlimited hedonism deployed as a cure for the disorder of non-planning, non-thinking. Government is presented no longer as the restrainer of desire, but its coordinator and satisfier, and the provider of "culture." In this view, however, it also becomes the parody Theodor Adorno called "a manifestation of pure humanity without regard for its functional relationship within society." Which is only an abstract way to name the hectic round of Scarborough's (and every suburban government's) ethnic singalongs, minglings, festivals and other unreal events. Burdened with such deeply flawed idea of governing, and a hopelessly wrong-headed idea of culture, is it any wonder that Scarborough can't understand what is going wrong with it? At the Civic Centre, every politician and speaker at the Moriyama event—a homage to the architect on the twentieth anniversary of his building—drivelled on about what a "people place" Scarborough is. If there's something wrong, well, the citizenry and politicians just have to work harder to make it a "people place." That is, a place devoid of *real* people, and populated only by quiescent phantoms of Adorno's "pure humanity."

Into the vacuum created by the failure of any government to make constraint, restriction and the enforcement of sensible ethical standards its first duty, brutality and injustice are always quick to rush. So far, three hundred bannings of young people, most of them Filipino in origin, have taken place in the mall during the last year. A young Filipino was hauled by guards out of a hamburger line, charged with trespassing, and told to sign a form saying he would never come into the mall again. Another Filipino, a Taco Bell employee, has alleged harassment by guards as he tries to get to work.

Apparently, the problem is that some Filipino youths hang out—gasp!—use naughty words—shock!—and dress flamboyantly. They are *not* vessels of the "pure humanity" Scarborough would create. Or so I gather from the warnings throughout the mall. "Loitering makes shoppers uneasy... Running, littering or swearing in the mall upsets other shoppers... Disguises unnerve shoppers." That is, in the course of just being teenagers these people threaten the compulsion to consume, while it is precisely this compulsion, suburbia's real law and hope, that lies at the heart of its tragedy.

TROUBLE IN NORTH YORK

Some years ago, during an informal chat with Mel Lastman, North York's eternally electable mayor, I inadvertently called His Worship's vast suburban turf a "borough," which it indeed once was. Whereupon the usual cheery demeanor vanished from Lastman's face. He fixed upon me a stern gaze, and firmly, very deliberately, said: "It's the *City* of North York. Not borough. *City.*" The mayor was appalled.

On a Sunday afternoon in early December, 1992, Lastman was appalled again. This time, the occasion was a customary holiday ceremony on North York's civic-centre plaza, which the mayor has graciously allowed to be named after himself.

It seems that about a thousand Greek-Canadians had turned up in front of North York City Hall to stop the hoisting of a flag by about the same number of people from the transnational Balkan region of Macedonia. In the shoving and shouting match that ensued, the mayor suffered a bloodied shin. He also suffered an even more severe blow to his optimism about life in Canada.

"I'm a nervous wreck," Lastman told a reporter the day after. "I saw hate for the first time in my life yesterday, real hate. Here were people with their eyes bulging out and their mouths frothing.... When people come to Canada, they've got to leave their hate at home." Alas for North York, which proclaims itself "The City With A Heart."

I cannot explain, and do not pretend to understand fully, the historic feud that led to that Sunday's shoving in North York. And if Lastman's complaints in the newspaper stories about the tussle called to my mind that trivial *faux pas* I'd almost forgotten about, I mention the two incidents together here not to suggest their equivalence, which would be absurd and tasteless. Rather, I do so to call attention to the widely held, flawed notions of urbanity suggested by Lastman's comments, both then and now—notions that, it seems, still pervade political culture in the gigantic suburban terrain north of the City of Toronto.

When Lastman gave me a comeuppance over the "borough" remark, it came as a sharp reminder of how attractive and sophisticated-sounding the word "city" had become to politicians in Toronto's suburbs, and how readily these chieftains bristle at the suggestion that their jurisdictions are—as many a hard-core urbanite deeply believes—occasionally interesting but largely misbegotten superblocks of tacky boxes somewhere north of beyond.

Of the five boroughs that joined Toronto City to bring Metro into existence in 1953, all but East York have proclaimed themselves "cities" in name. Since their incorporation into the titanic Metro conglomerate, a number of these entities have rushed (or are rushing) to become cities in fact, with impressive city halls and other public amenities, large libraries, galleries and theatres, clusters of soaring, gleaming office and apartment towers, and other emblems of high urban aspiration.

But Lastman's dismay and astonishment after the showdown between Greeks and Macedonians suggest a curious lack of realism, at a couple of levels, about the sparkling city life he and other suburban mayors seem to want so dearly.

Real city life is, after all, a matter of continual conflict as well as glistening tall buildings, a matter of ethnic passion as truly as it is a commitment to wheelchair-accessible theatres. More doggedly and longer than newcomers to the United States, recent immigrants to Toronto treasure the national chauvinisms and ancient grudges they brought with them in their trunks and bags. This reluctance of the recently arrived to join hands with others, even tangentially or nominally, and thus create a homogeneous "Toronto citizenry," is not their fault. As an immigrant myself, I can testify to Toronto's myriad ways of letting the newly landed know—even if they come from another Canadian city—they are allowed and tolerated, but not really welcome. More than any large city I know, Toronto is clannish to the core, characterized by tightly circumscribed ethnic and professional networks and old circles of friends, and by little open psychological space in which any new centres of real political or social power can be established. The message is: stay with your own kind. The only alternative is to ingratiate yourself with an existing power network—a nearly impossible task for many an immigrant—or be so brattish you cannot be ignored. If you have no kind to stay with, and are not able or inclined to be ingratiating or conspicuously obnoxious, then stay alone and out of the way. Don't look for a glad hand here.

Everyone who lives in Toronto knows and, most of the time, abides by the codes of this frigid civility. Except for the odd visit to an "ethnic" restaurant, we *do* stay with our own kind. Perhaps this is a reason Americans, on holiday from their disintegrating cities, find Toronto such a peaceable and agreeable place. What they do not appear to realize is the extent to which this "peace" is a result of extraordinary self-policing and self-segregation.

Pandemic in the satellite-cities of the Metropolitan Toronto complex, and perhaps especially in North York, is the fantasy of being the Great Welcome Wagon which Toronto City certainly is *not*. In this suburbia, the vision of true democratic urbanity would prevail, mediating conflicts instead of ignoring them, and bringing people together into a commonwealth of difference, celebrating rather than fearing variety.

Or so we are to believe, if we subscribe to the politicians' twaddle, and their architecture. The great public square, with its entertainments and opportunities for meetings of the citizenry, is an inevitable part of this notion's substantial rhetoric; and Mel Lastman Square is an excellent example of how it's all supposed to work. Located on Yonge Street just north of Sheppard Avenue, a belated architectural cluster installed at the centre of absolutely nothing, the North York City Centre is as eloquent a witness to this dizzy urbanism as we are likely to find anywhere—one in which historical memory and the strife such memory engenders are supposed to be eliminated by unrestrained consumption, opportunities for self-improvement and unlimited distraction.

If the sumptuous Romanesque bulk of Toronto's Old City Hall speaks to us of solemn encounters between wayward citizens and socially superior, strict magistrates, the North York Civic Centre is supremely about citizenship as a kind of permanent state of bliss. The welcoming, open City Hall, with its smiling receptionists, occupies one side of Mel Lastman Square. On the south side is the Board of Education, and facing it is the city's central library. A theatre lies next to City Hall, and just north of the library is a hotel, and a marble-lined shopping galleria, its glass roof opening towards a slit of sky defined by the sheer glass walls of office towers.

Elsewhere in the complex one finds a swimming pool, offices of dentists and doctors, and a city-operated parlour for weddings. Outside, in Mel Lastman Square, we have a band-shell and an ice rink, an amphitheatre and a box office—satisfactions for almost every need and desire, short of religious or sexual ecstasy (or a fine bookstore or superb restaurant), one could imagine.

It could be argued, and Lastman is obviously prepared to argue, that members of a free and privileged society should behave themselves better than the Greeks and Macedonians did that Sunday in Mel Lastman Square. He was shocked. But how seriously should we really take this shock? Human nature being the potentially horrific thing it is, I should think the mayor would prefer a little hooting, pushing and uproar to the kind of thing that goes on in places where no public social space is provided for dissent—in the rural districts of the world, for instance, where death squads, barn burners, lynch mobs and other furtive, elusive actors have always been the main performers in the rituals of difference. Other than the mayor's scraped shin and jangled nerves, little real damage appears to have been inflicted during the Greek-Macedonian melee.

What the politicians in the new "cities" want, apparently, is neither country-style marauding nor Toronto-style coldness, but apolitical Eden. In the long history of the city, nothing has ever been quite so unreal, so remote from the true fact and face of urbanity, as the "centres" of these suburbs longing to be cities. Big-city reality has a way of invading even the enchanted gardens of the suburban politicians, of which fact I submit the faint scar on Mel Lastman's leg as Exhibit A.

Concrete Dreams

Column, Gardiner Expressway

CAR ARCHITECTURE

Officially, it was just a proposed highway that died that summer's day in 1971, when the Ontario government decided to swing its bulk behind a grassroots coalition of environmentalists and neighbourhood-minded folk, and stop the southward thrust of the Spadina Expressway towards the heart of Toronto.

But as many on both sides of the issue realized at the time, the more important casualty of the decision was a certain idea about Toronto's future. Its devotees called this idea Progress. To its detractors, what was stopped was an ideologically driven programme of therapy that would kill the patient, by destroying virtually all the things that make cities worth living in. Since 1971 until now, Progress has been an idea suppressed in Toronto. And so it is that this remains perhaps North America's only metropolis with expressways that abruptly stop, and empty their load onto ordinary streets—the Allen Road, as the Spadina Expressway is known today—and soaring ramps, like those at the eastern end of the Gardiner Expressway, that suddenly lose their surge, go limp and sag to the ground in the middle of nowhere.

And as intended, comfortably antique neighbourhoods approximating the anti-Modernist visual ideal have survived downtown—at least in the minds of the aging urban professionals who took up the populist slogans of Stop Spadina, and occupied and gentrified the Victorian houses of Cabbagetown and the Annex. The massive internal migrations of well-heeled citizens during the astonishing real-estate boom of the 1980s, incidentally, did more to destroy the social fabric of "traditional neighbourhoods" than the Spadina Expressway ever would have. But the visually homey streetscape is as important to the architecturally savvy as to the upwardly mobile. It's also the principal criterion of what's a neighbourhood and what's not. A number of mobile Torontonians I know staunchly claim they live in "vibrant neighbourhoods," even though they don't know the names of the people next door. Nor do these educated, middle-income invaders, of which I am one, spend much time wondering publicly where the indigents, addicts, students and artists who once found cheap housing in these zones have gone.

Before the 1971 death of Spadina, the vogue for high-speed modernity had enjoyed a career in Toronto quite as remarkable as it had anywhere else on this continent, reigning virtually unopposed in the planning departments and the minds of theorists after the Second World War. The soaring expressway was democracy visualized, social mobility given infinite physical tether, the decisively right use of heavy technologies after a half-century of brutal misuse. In the sacred name of Progress (or using it as an alibi), civic planners and private real-estate developers wiped out city blocks and nearly the whole of the large warehousing district just west of downtown, raised office towers in the avant-garde Euro-Modern style from the scraped-down ground, blitzkrieged the urban fringe with subdivisions, constructed regional shopping plazas ringed by parking lots. The halt of the Spadina in 1971 did not, of course, instantly stop this upheaval, so conspicuous in the buckaroo days of Modernist bulldozing. But it did reveal—even to us beneficiaries and admirers of the Modernist ethos and *élan*—that the quasi-religious belief in the power of wrecking balls to set us free was, like most other optimistic gospels, fatefully vulnerable to error, and open to loss of credibility.

Few Torontonians under forty nowadays, and not a great many over that age, require much convincing on this point. The tight, comprehensive weave of expressways, conceived by Modernist planners and brought to full realization in many an American city, is a bad thing; such is general conviction today, almost never publicly contested any more—more a matter of sincere sentiment, that is, than a talking point in the politics of the urban future. It suffuses the contemporary literature on urbanism, like a dogmatic truth that should be self-apparent to all but psychopaths and the benighted.

In common with most other fellow-citizens, I believe Ontario's decision to halt Spadina, and the dramatic expresswaying of Toronto, was the right one—not because it would have ruined Toronto, but because it would not have solved anything. Or so I believe today, some twenty-five years after the fact. It was a sentiment that did not spring spontaneously from my heart, in which the love of expressways had taken root long before. Growing up in the rural American South, I was trapped in a small community by narrow, pocked roads that seemed to have been designed with my imprisonment in mind. Then, just as adolescence struck me, the Interstate Highway came through—a huge gash in the red clay of the Louisiana hills, later filled with concrete stretching away to the far horizon. The first spring the expressway was open, the din of trucks and cars that fell on my ears sounded like a song of freedom. I wanted to go where those cars went, and I did. To be sure, the first sizeable city down the line was pretty pokey when it came to civilized pleasures. But to me, it was Babylon and Paradise, all rolled into one; and the expressway got me there.

Even today, when I know better, I still feel primitive excitement when whipping along an expressway, or banking hard

Parking Lot, Queen Street East

along a ribbon of concrete at a basketweave interchange. Nor is such mystical regard for expressways, I suspect, uncommon among those who hit adolescence thirty years or more ago. I recall a newspaper article written by Toronto sociologist Arnold Rockman in the mid-1960s—I read it some years later—in which the newly completed Don Valley Parkway is praised as "sculpture in motion." In a 1991 interview, Rockman recalled his excitement upon driving on the Don Valley Parkway for the first time, the admiration he felt for "the way it conformed to the landscape." The article was, in his words, "a futurist rhapsody" of a sort, he believed, could not be written in the 1990s.

If not the expressway, at least what flies along it *was* being rhapsodized in 1991, though neither Rockman nor I knew it. Dutch architect Moshé Zwarts was writing these words, destined to appear in a little anthology called *Architecture Now:* "I propose that starting from tomorrow we all privately and publicly declare our love of cars. For our profession, this would mean having to treat cars affectionately in our designs. I say this because I dare to recognize that the car is an almost ideal form of transport...unrivalled by any other vehicle in the amount of freedom it offers...apparently affordable for most of us...environmentally friendlier per passenger kilometre than the train." The problem, in this view, is not the car, but the failure of architectural imagination in accommodating our stubborn allegiance to it—a failure that is nearly universal, except in one spectacular instance: the invention and development of the expressway. While not the only tool in the kit of transportation planning, the high-speed, limited-access highway has proven itself to be an extraordinarily versatile one. Now we need others—many others, so far unimagined—that, like expressways, will enhance the pleasures of metropolitan choice and movement.

THE GARDINER

Admirers of urban form and transformation can never afford
to doze, since there is simply no way of telling just when and
where a city will start telling stories you never heard before.
Such was the unexpected lesson that came my way on a wintry
day, when I was trying to get a good look at Fort York.

It's easy to get a view from inside what's left of Toronto's
Georgian imperial defence, such as it was. A magnet for
tourists and school tours, the fort is open to the general public
most every day. The hard part comes when the would-be spec-
tator tries to find an instructive standpoint *outside* the structure, in the gritty downtown clutter of factories, expressway
ramps and truck yards, smokestacks and tall electric signs that
has long hemmed it in, shutting it away from almost any long
view.

It was during a search for such a viewpoint that I found
myself slipping and sliding on icy, mucky rubble beneath the
traffic deck of the Frederick G. Gardiner Expressway, where it
billows upward and soars between Fort York and the lakeshore,
just west of Bathurst Street. No scenic prospect presented
itself—until, that is, I gave up trying to get a useful topographical glimpse of Fort York and abruptly noticed the sub-
limity of just where I was.

Few sites more forsaken lie this close to Toronto's busy,
dense downtown mountain-range of glass. Overhead, the wide
steel belly of the Gardiner's traffic level lies like a flat green
snake on a series of tall, water-stained concrete brackets.
Underneath spreads the expanse of loose gravel, some of it
used as a gathering place for trucks, some of it the dusty yard
of a factory in which big cement blocks are fabricated.

One hesitates to use the word *beautiful* of such a forbidding place, though the word fits the bill. There is strong visual
surge and power here: in the dignified rhythms of the expressway's tapered reinforced-concrete supports, marching away

into the distance like an immense Baroque colonnade, in the tough muscularity, in the ensemble of cement factory and rumbling trucks. There is a gruff beauty here that the swank towers nearby can't touch.

In penning this hymn to the Gardiner and its industrial aesthetic, I mean no insult to those who fought hard and successfully to stop the total expresswaying of Metropolitan Toronto, or to people for whom expressways are the very incarnation of urban menace. But the battle *is* over—really and forever. The opponents of freeways can sleep soundly at night, and need not fear the presence of one aficionado of the grand idea and the massive architecture they stopped. Indeed, had the concept of salvation by expressway so much as a flicker of life left in it, the Metropolitan Toronto government would probably not have dared inaugurate, in 1991, its new archives building on Spadina Road with the handsomely fair visual and documentary exhibition called *Concrete Dreams*.

Drawing on the extensive holdings of pictures, maps and other resources conserved by Metro council, archives exhibits manager Michael McMahon, historian Rosemary Donegan and photographer Jim Miller assembled a lively, engaging evocation of Toronto's period of heroic urban Modernism (roughly 1950 to 1970), and a memorial especially to Frederick "Big Daddy" Gardiner, first Metro chairman and our Robert Moses (urban planning czar of New York City), a massive schemer and bully who was the power behind the plan to drop a comprehensive net of interlocking expressways on Metro. (Gardiner is regarded with affection by Torontonians. His motto: "You won't leave your footprints on the sands of time by sitting on your ass on the beach.")

If Big Daddy and his gang on Metro Council were hardly Modernist visionaries and original theorists, they did believe sincerely that any city, to be great, must have expressways; and they very much wanted Toronto to be great. The crumbling, crowded neighbourhoods sullying Toronto in 1945 were disgraces reminiscent of the Depression, and were best done away with. The future expansion and glory of Metropolitan Toronto, Big Daddy's baby, lay in blitzing the blight and stinking small-industrial zones, and raising from the rubble a new town of high-rises and parks, populated by well-to-do cosmopolites working in shiny, tall buildings that would say *cosmopolis* to the world. His agents were to be, and were, the newest generation of planners and developers, who believed, also with undeniable sincerity, that the path to urban peace and private wealth would have to be cleared by bulldozers.

In this scenario, the cure for the city's image problem was a radical dose of demolition, followed by big spoonfuls of high speed. As McMahon put it, Space (the neighbourhood) was to be conquered by Time (the car). Significantly, official documents from the early 1950s, in which plans for the new elevated lakeshore expressway were enunciated—it was later dubbed the Gardiner—made much of speed and volume of traffic flow, but said virtually nothing about the homes and streets and buildings of historical interest that would be swept aside to make way for this river of heightened mobility. The latter did not exist in Big Daddy's dream. Fort York, for instance, turned out to be in the way of the lakeshore expressway. The solution proposed by Gardiner was simply to move the fort. His plan was given solemn blessing by city engineers, who proclaimed that pulling the proposed highway around Fort York would involve a "sacrifice (of) design standards, resulting in unsafe driving conditions." Such was the tenor of thought in those days, at least among bureaucrats, politicians and real-estate capitalists devoted (for various reasons) to the "urban renewal" taking place in American cities. At its worst, the project was driven by greed, naked political ambition, foolish chauvinism. At its best, such Modernist super-planning was driven by compassion and idealism, honourable civic boosterism combined with holy hatred for Depression-era slums and for what they were doing to the miserable souls condemned to live in them. But Modernist urbanism was never what Jane Jacobs, as recently as 1993, was still denouncing as "fraud" and intellectual treason. Nor do I believe anything more noble than the usual hesitancy about American influence was at work in the late 1950s, when the Toronto citizenry and press forced the imperious Gardiner to leave Fort York *in situ*.

BURYING THE GARDINER

No document more eloquently conveys the romantic mood of current planning orthodoxy than *Regeneration,* the 530-page final report of former Toronto mayor David Crombie's Royal Commission on the Future of the Toronto Waterfront, released in May, 1992. It is suffused by the piety that cars, and the architecture they make necessary, are both horrid.

To the surprise of many observers, however, the report stopped short of calling for the wholesale knockdown of the Gardiner Expressway. "Shying away from the enormous cost," lamented urban activist Colin Vaughan in *The Globe and Mail,* "engineers and planners were timid about the removal of this elevated eyesore that blocks access to the city's lakefront." But, we learn in another newspaper story, if there's no simple solution to eradicating this "wedge between Toronto-area communities and Lake Ontario," the commission "suggests that burying part of it would make the Gardiner less of a barrier without increasing traffic woes in the process."

While it did not strike me as odd that Toronto activists, such as Colin Vaughan, would use this occasion to bash the now-heretical view that limited-access elevated highways are balm for gridlock, other early responses to the report were stranger. It struck me as odd, for example, that anyone could refer to the Gardiner as an "elevated eyesore," without qualification. Those who rode it in the early days remember it much differently. Back then, the expressway was illuminated by parallel lines of fluorescent tubes recessed into the guardrails on either side. Coming from the direction of Scarborough, one came onto the ramp at rapidly accelerating speed, climbed into the sky, the equivalent of some three storeys in only fifteen seconds, and then was guided off into the distance by ribbons of light. Ahead lay the flashing Consumers' Gas flame, and the shining city. Nobody complained then about the expressway's ugliness, nor should anyone do so today. It is Toronto's largest and most spectacular monument to the tragic utopianism typical of Modernist culture as a whole—and, at an instantly more mundane level, still a good, fast way to get downtown, thus worthy of preservation on both counts.

Nor do I sympathize with the claim, very often heard from its critics, that the Gardiner is somehow impeding our access to the lakefront, like the former Berlin Wall, an insuperable obstacle between landlocked Torontonians and a realm of forbidden delights we can now only dare dream about, or see on television. Is it *really* paradise over there, among the Florida-style condominium towers that East York's wretched bungalow-dwellers, imprisoned behind the Gardiner Expressway, can only lust after from afar?

Such a bizarre vision of the glories of living beside Lake Ontario—the vision informing much of the wildly inappropriate tropical architecture at Harbourfront—caused even David Crombie to marvel. The designers appear to have simply overlooked the fact of our bone-chillingly damp winters, Crombie told *The Globe*'s Craig McInnes. "Most of the buildings along the waterfront in the public realm are built as if they are in the Mediterranean. If you think about it, that's crazy. We have winter here every year."

What exactly, then, is the Gardiner blocking our access to?

Nothing. You can drive or walk or take the light-rapid-transit car under it or over it at many points.

And what does one find in this wondrous Land of Oz south of the expressway ramps?

Well, for the discerning few, there's The Power Plant, a public art gallery which often displays contemporary Canadian and international art of outstanding quality. There's also a couple of other interesting cultural facilities—a dance theatre, a workshop for making and displaying crafts, a forum for literary readings and so on. For the tourists, there's deluxe shopping at Queen Quay Terminal—though less deluxe, and a lot pricier, than similar shopping north of the Wall. You were supposed to be able to stroll for miles along the alternately blisteringly cold and blazingly hot waterfront, if that's your idea of a good time, but the condo developers have put considerable swatches of the harbour off limits to ordinary folk.

Once upon a time—indeed, for a couple of years after I first arrived in Toronto—the harbour still possessed a certain wonder. It was a run-down Great Lakes port, past its best days, which had never been very good; but still a port, with ships and sailors hailing from all the world. The cracked concrete docks and corroded cranes and little tin service buildings were mute reminders of the era when the Lakes provided a great marine highway from the Atlantic into the very heart of the continent, linking Toronto to Cleveland and Chicago, to Montreal and London, Capetown and Leningrad with the strong cords of trade. Then the cords snapped one by one, the marine traffic declined, and the docklands became Harbourfront: soaring, boring condo towers and hotels and well-kempt parkettes and promenades, all nakedly exposed to the winds that sweep inland from the lake most months of the year.

The winds do die down in May, and Toronto experiences a seasonal miracle: the fleeting moment when the waters of our cold, dangerous lake briefly sparkle brilliantly, and all seems balmy. This moment is usually cut short by the regular summer die-off of small lake fish, whose silvery, stinking corpses mass in the boat slips. If I dislike Harbourfront, and never go there except on business, it's perhaps because I was not born in Toronto, and therefore never knew the rough romance of its zenith as a port town. Nowadays, when I peep over the Gardiner, all I can think of is the good job this mammoth expressway is doing, just by keeping phoney Miami out of sight.

THE MOTEL STRIP

If you've always meant to do a dirty weekend in a cheap room down in the Etobicoke armpit known as the Motel Strip, and just haven't gotten around to it, take heart. There's still plenty of time.

That's despite the 1993 decision of Ontario's highest zoning power that put an end to twenty-five years of wrangling among builders, politicians and citizens over what should be done with the notorious fifty-acre lakefront site at the west end of Metropolitan Toronto, and gave the green light for the bulldozers to move in. When they do, the motels will go, and in their place will rise gleaming condominium towers, a hotel, upmarket retail stores. The idea is to attract, as a *Globe and Mail* reporter put it, without a trace of irony, an "influx of stable, community-minded residents." The scheme also includes the construction of another of those waterfront parks Toronto planners keep wanting us to have. (When will the planners realize that anyone of sound mind will not want to frolic beside the cold waters of Lake Ontario? And, for that matter, haven't they figured out that Torontonians don't like to frolic *anywhere*?)

The reason why the window of opportunity for sleaziness remains open is the flat Toronto condo market, which is not expected to pick up until the mid- to-late 1990s, if ever. But if I were you, I wouldn't put it off too long. Nobody will shed a tear when the Strip's famous culture of sex and drugs, and the refuge it's historically provided for Bonnie and Clydes on the lamb, falls victim to the full-scale gentrification now on tactical hold. When the wreckers start to move, nobody is going to stop them.

Speaking of tears, a motel operator burst into them one cold day late in 1991, when I told her that I found her establishment's streamlined, luxury-liner styling and *haut-moderne* glass-brick touches all quite interesting, and surely worth a hail-and-farewell in *The Globe and Mail* before it all gets knocked into oblivion. The distraught lady's fear, I learned, was that I would write something which would get her motel designated a protected historic site—when all she wanted was *out*.

No such designation has taken or will take place, with or without my help. But some note should be taken of the fifteen-odd boarded-up or feebly operating motels on the Strip before they slip away beyond the edge of city memory. Apart from a similar row of one-night-stand motels on Scarborough's Kingston Road, this one-kilometre stretch of Lake Shore Boulevard West is Toronto's last substantial preserve of an architecture specific to the Great Automobile Age. It was during this period—launched by the high-speed expressways of the 1940s (including the Queen Elizabeth Way, Canada's first controlled-access road) and brought to an abrupt halt by the oil crises of 1973 and 1979—that the popular worship of the car-as-sex-symbol reached its zenith. The motel styling on the Strip is the sacred architecture of this cult of the fast, sleek, daring, hedonistic.

Before this moment, the North American automobile traveller could expect to find accommodation in an old-fashioned hotel. For those who preferred, however, on the outskirts of every city there was always a row of huts called "tourist cabins," cheap digs which nostalgically connoted pioneer dwellings, and recalled slow, purposeful journeys across great expanses. (There used to be such a place in Scarborough called Wagon Wheel Court.)

Since the 1950s, however, the temporary shelter of choice has been the "motor hotel" chains—low stacks of prefab hotel rooms, without room service, virtually identical in design coast to coast, typified by Holiday Inn and Howard Johnson's. These motor hotels have always been camouflaged as apartment buildings. On occasion, their architecture merges seamlessly with that of recent hotels, so that one cannot tell the difference any more. It's all the same beige broadloom, easy-care plastic wall coverings, uniform windows that will not open, Formica table tops. The inner environment suggests "home away from home," by creating mediocritized, generic versions based on the domestic bedroom of the moment.

These mass-replicated environments—along with fast-food outlets, among the world's first such items of commercial architecture—were an instant hit with the mobile public; their success was also an excellent instance of capitalist industry's ability to meet a genuine need with speed and considerable technical sophistication. This need on the part of travellers was begotten by the notorious unevenness and unpredictability of overnight accommodation during the Motel Strip era. Beyond not knowing whether the bed would collapse, or you'd find roaches in the bathtub, there was always the possibility of a chance collision between motel cultures: the older one, involving hookers and the clandestine rendezvous and next-door noises very hard to explain to a curious six-year-old; and the newer ethos of the happy, typical nuclear family on holiday.

The Brady Bunch, in other words, would *not* stop off at the Motel Strip, hence its decline—but also its value for recollecting a distinctive architecture and unique style whose day is past. These buildings draw their aesthetic and connotations from neither the pioneer sod-house nor the Home, but from the Car. Like '59 Caddys, most are long, low, brightly coloured, easy to slip in and out of quickly. The sleek red stripes of the Strip's Cruise Motel echo the chrome stripes on the automobile. The smooth wrap-around design of the boarded-up Lake Edge is reminiscent of the wrap-around windshield of its postwar epoch.

Both car design at the time—roughly the mid-1940s through the 1950s—and the classic motel architecture on the Strip express the new, permissive erotic style which came into the open in North America during the early *Playboy* period. The tourist cabin was still, after all, a cabin, a kind of home, often set among trees. The Strip motel of the 1940s and 1950s, on the other hand, takes its cue from the back seat. It's a row of bedrooms parallel to a row of parking spots. The motel acknowledges the new mechanical sexiness by dropping the discreet lobby, the seemly vestibule—the separation of sex and street characteristic of stately hotel design—and offers unencumbered instant passage from the erotic zone of the car to that of the bed.

If the car overcame the distances between cities, it also eliminated an interesting new kind of interval, between the motel and the rest of town. It was commonly known in the small Southern American town I lived in that a local pillar of the community, when his wife went to visit her sister in Dallas, always met his girlfriend in the motel in the next town along the highway, just over the Texas line, five miles away. He imagined his meetings were secret, of course, and my buddies and I thought we were the only ones who knew. As I later learned, everybody in town knew, though no one talked. Like the car, the motel was a zone invisibly cordoned off from proper small-town morality, in which one could easily imagine getting away with fantasies and acts prohibited in town. (Hence the understandable surprise of televangelist Jim Bakker and Jessica Hahn, and Jimmy Swaggert and his hooker, when their strip-motel trysts were discovered and frowned upon by contributors.)

The classic modern motel, such as those we find in Etobicoke, does not pretend to be a home away from home, or a cozy guest-house or anything other than what it is: a partner with the car in the uniquely mid-twentieth-century culture of high-speed travel, fast and novel powers and sex.

CONCRETE REALITIES

Until the early days of 1993, city critics could credibly claim to know the exact death-date of architectural Modernism, and its lofty dream of mass cleanliness, urban rationality and salvation by skyscraper and park.

All that idealism bit the dust, so the old story goes, in April, 1976, in St. Louis, when city authorities dynamited the giant Pruitt-Igoe housing project. Hailed as a model of progressive urban renewal when put up in the early 1950s, by the early 1970s Pruitt-Igoe had become a cluster of horror-houses, raddled with violent crime and abandoned by all but gangsters, rats and a few victims too poor or frightened to move out.

While acknowledging the historic importance of the spectacular demolition in St. Louis, Modernism's steel-and-concrete vogue has certainly continued, enjoying new significant anniversaries. I here propose to add one to the calendar: a new one, just as spectacular, and accurate not merely to the month or even day, but to the very minute: Friday, February 26, 1993, at 12:18 p.m.

This was the moment a powerful bomb exploded in the parking garage under one of the two immense towers comprising New York's World Trade Center, killing seven people and injuring about one thousand in the skycraper above and the rapid-transit station below.

In addition to the human harm inflicted, the blast knocked out the 110-storey building's communications systems, air-conditioning apparatus, fire and smoke alarms, and 2,721 toilets—virtually all the unseen or discreetly tucked-away equipment which made this titanic structure convenient and safe for the 50,000 people who came to work in it every day.

The St. Louis and New York stories are linked, interestingly, by two facts. Both Pruitt-Igoe and the World Trade Center became important architectural events only with the help of high explosives—safely, professionally set off in St. Louis, detonated with deadly intent in New York. And, by a curious irony, both architectural settings—the St. Louis development (1952–1955), the Manhattan office tower (1962–1975)—sprang from the quintessentially sub-modernist mind of U.S. architect Minoru Yamasaki.

For those of us who've lived much of our lives in high-rise apartment buildings, Yamasaki's St. Louis fiasco has never seemed as significant as anti-Modern critics have made it seem. Some of the thousands of housing blocks cobbled together from concrete slabs, steel beams and prefab windows and doors during the last forty years were bound to fail, for reasons of structural faultiness, or because of the sorts of people who live there. Pruitt-Igoe was among the flops. The clean, handsome, completely prefab thirty-storey Toronto block near High Park, in which I rented a well-proportioned, inexpensive apartment for sixteen years was a success by any standard. I could have lived there all my life, had renting not become uneconomic, and had we not simply run out of room.

All we learn from the recent history of apartment construction is that some modernist megastructures work, some don't. But the 1976 lesson wasn't grim enough to slow down developers from carrying right on with the erection of tall modernist residential and commercial buildings—though without the dolling-up with perky hats, glitzy brooches and marble high-heels popular nowadays.

Once its full significance sinks into the minds of architects, developers and urban planners, the New York bombing may be recalled many times as an instructive event, even an upturn in the hopes for the tall building.

If decorative beauty in an edifice meant to be used by more than a couple of hundred people is optional, an underground parking garage is currently unthinkable. True, since their introduction parking garages have been dangerous places, especially for unaccompanied women. But the person who touched off the bomb under the World Trade Center brought us face to face with the fact of how dangerous they can become, instantly, for up to fifty thousand people at once—and how relatively harmless. The explosion did not blow out the perimeter columns of Tower One, bringing down the whole edifice; the blast damage was largely confined by various structural barricades to the garage itself and to underlying structures. While it did reveal the vulnerability of a great building's electro-mechanical, electronic and safety systems to so powerful a detonation, it also demonstrated the sturdiness and containment powers of a reinforced concrete parking garage, properly buried and technically isolated.

My optimism is qualified by only one fact: that the underground garage built as sturdily as the one below the World Trade Center is rare. We may never again be able to trust entirely this benign, useful building-type, created by and for the car, as we once did. What security systems are now in place to prevent a fanatic or lunatic from driving into any subterranean parking garage, under any residential or office tower anywhere, with enough explosives to make his point?

Be that as it may, these structures will endure, we will continue to use them, and New York will not affect our practical decisions to park underground. Nor will parking garages rapidly lose their practical interest, or their air of novelty. From time immemorial until the invention of the subway, the vehicles of transportation were always kept above ground, in a stable or (in any case) stablelike, free-standing building. At least that's where my great-grandmother kept her buggy and horses, until my great-grandfather turned the stable into a garage for his new car, the first ever seen in Palestine, Texas, and a gadget which attracted great public curiosity.

The descent of car barns underground, City of Toronto architect Robert Glover speculated in an interview, may have begun with "basement parking" in domestic or small office structures. While it's hard to say just when the true underground "parking garage" came into existence, the earliest instance I've found of an office tower with parking facilities built into it is Chicago's Pure Oil Building (1924), with the Title Insurance Building (1928) in Los Angeles a close runner-up. But never mind. "Going underground," says Glover, "was the only way of maximizing parking without losing open amenity space."

And, of course, it still *is* the only way. We shall probably never see the end of underground parking. But while I don't know about you, I'll never again feel quite the same about driving off the street, down the long, winding ramp, past the robot that burps out my ticket, and into the concrete-lined semi-dark under the belly of the city.

Streets

Curbside, St. Clarens Avenue

FORSAKING KENSINGTON MARKET

If the millions of people who live in North America's cities don't appear to care that the casual bump, shove and commotion of traditional big-city street culture is dying, Spiro Kostof cared very much. This outstanding American architectural historian, who died young in 1991, wrote about cities past and present with rigour and verve, and with a handsomely broad command of his subject. But he also did his scholarly work with evangelical militance, as though commanded by an angel to stand at the gates of the city, defending what was left of untidy, organic, unpredictable urbanity from the rational Modernist barbarians and planners camped without the walls.

We learn from a headnote in *The City Assembled*, his final testament, that Kostof died on the barricades, revising and sharpening his arguments almost to the last. It is in this lavishly illustrated book that Kostof paid his final, most passionate tribute to the old-fashioned street, and thundered forth his most dire prophecies about its future, and our own fate without it.

"In the past," Kostof writes—exactly when and where this *past* was is not specified—"the street was the place where social classes and social uses mixed. It was the stage of solemn ceremony and improvised spectacle, of people-watching, of commerce and recreation...both school and stage of urbanity..." He suggests that all this goodness has been diminished or destroyed by our recent inclination to "keep our own counsel, avoid social tension by escaping, *schedule* encounters with our friends"—his emphasis—"and happily travel alone in climate-controlled and music-injected glossy metal boxes." Change your ways and revive your streets, warns the author, lest they become "the burial place of our chances to learn from one another, child from bagwoman and street vendor from jock; the burial place of unrehearsed excitement, of the cumulative knowledge of human ways, and the residual benefits of a public life."

Kostof's is a moving appeal to the urban soul; it is undergirded, as well, by perhaps the most venerable of all perennial visions of urban utopia. Writing in 1913 from a point of view similar to that of Kostof, C. J. Cameron declared that "from almost every country in the world the immigrants come to Canada like the magic assembling of a hundred constituents to form a chemical compound. ...Canada is the vast laboratory of grace in which God is fashioning the final man. The final race will not be any one nationality, but will be composed of elements from all races."

In every view informed by such apocalyptic pictures, the street is the "laboratory of grace" *non pareil.* With all the respect due Kostof, however, we are not likely to change our ways—simply because most of us have willingly chosen the new anti-street alienation, and have no intention of reviewing our choice.

If the word "street" has a certain charm to it, our scepticism and apprehension about this urban setting is revealed in the adjectival uses we make of the word. "Street people" are those unfortunate souls with no place to sleep. "Street kids" are teenagers who have descended into the hell of drugs, violence and prostitution that "street culture" has become in every large city. To be "street wise" is not to be wise at all, but merely clever at defending oneself against the predators who currently dominate "street culture." The gaudy, most sordid sections of Yonge Street, the sites of "street culture" with a vengeance, are not what Kostof had in mind. In search of some remnant of his dream-streets in Toronto, he would certainly have ignored Yonge and gone to Kensington Market.

Out-of-town visitors enjoy being walked through its narrow lanes and streets to savour the spirited open-air commerce now vanished almost everywhere except here, in the densely peopled, hectic retail and residential district of downtown Toronto south of College Street and east of the high Victorian brick façades, from the 1880s, lining Spadina Avenue. The Market is no neat theme park of Shopping Past. It is living, bruising, sweaty history. Elbowing your way through the mob on a sunny Saturday morning is probably as tough today as it was in the early 1920s, after Jewish merchants had established a thriving open-air market with "a shtetl atmosphere," an historian has affectionately recalled, on the narrow sidewalks of Augusta and Baldwin, Nassau and Kensington.

Every wave of immigration since then has left its mark, its particular commercial trace—jobbers and kosher butcher shops, West Indian roti joints, Portuguese cut-rate clothing stalls, Vietnamese restaurants, the bakeries and spice stores and greengrocers perfuming the air, brightening the street on the greyest winter day. And especially the earlier immigrants— Jewish and pre-war Italian—brought their political passions, leaving the area (as playwright and journalist Rick Salutin has written) "a centre for every kind of opposition and alternative to the sober Upper Canadian mainstream." The last struggle of anarchist Emma Goldman was waged there, on behalf of Italian anarchists, you tell your guests; and there she died, in 1940. As recently as the late 1960s, Grossman's Tavern was a hangout for black American draft dodgers and a drop-in centre for elderly Communists, while guys down from the

University of Toronto double-dog-dared each other to lead the way into the Victory Burlesque (formerly a Yiddish theatre), where you could see some of the dirtiest hip-grinding west of Times Square. (It's now a Chinese movie house.) The stench of fresh fish and a hundred kinds of cheese mingles with loud socializing in a dozen languages under the awnings of the open-air marts, and we and our visitors are charmed, especially because that's not the way we have to provision or entertain ourselves each week.

But that's just the point. Very few immigrants not forced by poverty to do so continue living in Kensington Market voluntarily. Some people still call Kensington "the Jewish market," and Jewish it was, from the turn of the century to the 1940s. At that point, prosperity, education and assimilation opened the way out, and the Jews sensibly took it. Ukrainian, Hungarian and Italian immigrants flooded into the Market after the Second World War, and by the 1960s Portuguese from the Azores had become the local mercantile ascendency. While the campuses of North America were ablaze with protest in those years, the Market's fame as a hotbed for political and social dissent had evaporated. The postwar immigrants are now following the Jews into suburban enclaves, and being replaced by East and West Indians, Koreans, Filipinos, Latin Americans, who will soon do the same. No amount of hand-wringing and nostalgia about what Toronto is losing with each flight will make one person stay, or anyone come back.

To my knowledge, the residents of the Market today are almost all recent immigrants too poor to do otherwise, university students playing poor, and a handful of sophisticated urbanists who, for some reason, *want* to live among the Market's rats, racket and odours. There is something touching about Spiro Kostof's decision to include in his superb final book a cheer for the busy old street, and a heartfelt appeal to those who've deserted all that for backyard barbecues and family rooms, private decks, *scheduled* encounters and life in "climate-controlled and music-injected glossy metal boxes." May Kostof rest in peace, undisturbed by urban humankind's firm rejection of street culture, and may he remain in everlasting ignorance of our complete absence of guilt or remorse about having made the decision we did.

NOT FINDING MAIN STREET

The front-page news stories about Toronto's rampage of May 4, 1992, was dominated by the facts of what actually happened, musings about what caused it, and speculation about the ultimate social and dollar costs of it all. There was much public curiosity about the event, simply because it was exotic. Unlike U.S. cities similar in size and ethnic diversity, Toronto has rarely witnessed a widespread social blow-out of any sort, let alone a major uprising of citizens. The last memorable outbreak of equivalent civil violence here was, I believe, in 1837.

Be that as it may, the otherwise exhaustive coverage of this affair overlooked at least one aspect: the aimlessness of the route of both the peaceful demonstration and the violent display that followed. Toronto, it revealed with unusual force, lacks a Tienanmen or Wenceslas Square, an open place of celebration and conflict which Torontonians identify as uniquely our own, and a traditional crucible of our history. Nor could one escape being struck by Toronto's peculiar lack of a proper main street, along which a parade or violent demonstration could move with spectacular effect.

The affair began at about 4 p.m. on that spring Monday, as a peaceful gathering by about five hundred people in front of the United States consulate, an unimportant New Deal edifice huddled behind ugly anti-terrorist cement barricades on lower University Avenue. The demonstration had been called to protest the weekend shooting of Raymond Constantine

Lawrence, a young Jamaican-born man, by Toronto police, and also the recent acquittal of four white Los Angeles police officers in the videotaped beating of Rodney King.

While there may have been some symbolic reason to protest in front of the U.S. consulate, no real public effect could have been expected. That's because University Avenue is a grand and sociable boulevard only at first glance. On closer inspection, it reveals itself to be a failed main street, grandiose in concept but dead in fact, where people neither live nor enjoy themselves. If, long ago, University was a spacious thoroughfare lined by stately houses, today it is just an especially broad boulevard fronted by government buildings, insurance companies, financial institutions and hospitals. The employees who work by day in these buildings were all packing up their briefcases to go home by the time the demonstration began, and were tucked into the family room by the time it became ugly. Virtually the only people left who could possible appreciate the spectacle in front of the U.S. consulate would have been those sick or dying in the hospitals. A less promising street for a demonstration of outrage could hardly be imagined.

The crowd, still more or less peaceful, next made its way north to the mid-town intersection of Bloor and Yonge. The demonstrators then sat down, blocking what traffic there was for about an hour. Again, fact and perception collide: for if Bloor and Yonge is famous for marking the principal interchange of the Toronto subway system, its importance as Toronto's main vehicular crossroads is a thing of the past. The intersection only *sounds* important, and even then only to Torontonians. Like declaiming on near-deserted University Avenue, blocking Bloor and Yonge should have been recognized for the hopeless exercise it is. The net of wireless communications, particularly the traffic and news reports available on car radios, makes it easy for any motorist simply to take another route. And in any case, most sensible drivers heading out to suburbia or into the heart of downtown avoid Yonge Street at all times.

The trashing and looting began in earnest after the mob started milling around the toney shops near the intersection, and drawing in teenagers who had not taken part in the demonstration at the consulate. The untidy gaggle of demonstrators and mere vandals next started rambling down shabby Yonge Street, yet another of Toronto's main streets *manqué*.

Now Yonge Street *is* famous, because *The Guiness Book of World Records* has declared it the longest street in the world. Built early in the city's history to serve as a northbound escape-route from maurauding Americans, Yonge now measures 1,178.3 miles from its intersection with Front Street in Toronto to the bridge at Rainy River over to Minnesota. It also sports a handful of refined and important architectural monuments, especially along its lower blocks. But Yonge south of Bloor is, for the most part, gaudy and raucous, day or night— a showcase for male and female meat for sale, a stew of rock record stores, sex shops, electronic outlets, sleazy fast-food outlets, a hangout for drug dealers and destitute teen runaways. Could anyone realistically imagine that storming Yonge Street could have the effect of storming the Winter Palace, or the Bastille?

After milling and mobbing all the way back down from Bloor to Queen, the crowd then found itself at New City Hall. The great plaza in front of Viljo Revell's curving, embracing towers structure is perhaps the nearest thing Toronto has to a Winter Palace courtyard, though the effect of spaciousness has been largely negated by the clutter of bandshells and skating rink and peace monument and a Henry Moore sculpture— hardly a suitable or desirable place for a full-tilt demonstration.

Whatever was left of the peaceful element of the demonstration was then completely swept aside, and a more violent mob began moving back up Yonge. Now there is venerable Toronto tradition of crowds taking over Yonge Street for a night of merry wildness on special occasions, usually victories in sport. What happened here, however, was a perverse twist on this tradition: an angry rampage of trashing cars and looting shops, with the hooligans fighting the police, bystanders and, it seems, each other, all the way back up to Bloor, where the complex series of events finally fizzled.

The habit of remembering such civil upheavals by the names of city sites—Chicago's Haymarket riots, the storming of the Winter Palace, Watts, and so forth—is old and stubborn. It is still with us, despite the fact that organized demonstrations are now everywhere staged mostly for the benefit of the mass media, not to summon the local citizenry to arms or to frighten an offensive despot, holed up in his castle, into quick abdication.

Because the 1992 riot had no urban focus, any more than Toronto does—because it had no real main route or geographical aim—this almost unprecedented occurrence is doomed to be quickly forgotten, and will certainly not colour the meaning of Yonge, University or Bloor in popular imagination and urban mythology. Within days, police had downgraded its importance from "race riot" to "giant swarming," and mostly the work of teenagers greedy for the jackets on display in the Giorgio Armani shop. The drunken post-game looniness on Yonge Street will doubtless continue unto the ages of ages—while Toronto will have again absorbed and buried a memory by the sheer fact of its curious lacks of focus, ghettos, "festering zones of poverty," even *neighbourhoods* in any sense except that used by real-estate agents.

The End of University Avenue

Just as the century began, Edwardian Toronto's prosperous and powerful took a long look south, at the urban transfigurations under way in Buffalo, New York, Boston and Chicago, and they very much liked what they saw. In those transportation and commercial hubs, akin to what Toronto had become, public power and private wealth were combining forces to make their cities as magnificent in visual form as they had become in financial and commercial fact. Displacing the low downtown jumble were long spacious boulevards and plazas, from which gracious residential streets branched. Grand theatres, heaven-defying steeples, parks and monuments were being deliberately sited and built to impress and delight, and to advertise urban prestige and glory. Here were cities ablaze, as Toronto was not, with the angelic fires of Progress and Refinement, sweeping away the old and stuffy to make way for "The City Beautiful."

When Toronto's plutocrats and politicians looked back at the old home town, they did not like the miserable buildings and crowded blocks they saw. In 1905, The Guild of Civic Art, inspired by the 1893 Chicago world's fair, had called for a major revamping of Toronto in accordance with City Beautiful notions—but nothing had come of it. "For me it is not a question of the city beautiful," said financier Sir Edmund Walker the next year. "It is just a question of practical common sense. Do we really believe in the city of Toronto?"

Had he been in the audience that day, architectural historian William Dendy would surely have responded with a resounding *yes*.

Dendy, who died at age forty-five in 1993, exulted in the cultivation, public intelligence and self-confidence of so much built in Toronto during its first two hundred years. That was the easy part. More remarkable is the sturdiness of his interest in Toronto throughout the writing of his prize-winning 1978 book *Lost Toronto: Images of the City's Past,* and the arduous updating of this indispensible work in the last months of his life.

Dendy was a connoisseur of lost chances. It bothered him to think, especially, about the doomed proposals for University Avenue which sprang from pre-war City Beautiful enthusiasm.

In a lovely drawing by Earle C. Sheppard, done for City Council in 1929 and reproduced in Dendy's book, we are shown a breathtaking proposal for the southern termination of University Avenue. The great street would broaden just south of Queen Street into an immense plaza, to be called Vimy Circle, focused by a tall, slender war memorial. As though paying deep homage to the fallen heroes of the First World War, refined and dignified civic and commercial buildings—none too high, all appropriately conservative and martial—would line the curving edge of this urbane *place*.

Traffic continuing towards the lakeshore from Vimy Circle would be angled slightly eastward, along a wide radial street intended to terminate, at Front Street, in another focused opening, to be called Britannia Square.

As things turned out, however, neither the Circle nor the Square ever got built, though the street did—producing one of the most unattractive conclusions of a great city street imaginable. Today, the broad part of University Avenue simply fizzles south of Queen, narrowing into an ignominious "extension." From the north start of this tightening, University Avenue continues southeastward stingily to Front Street West, where it twists and dives into a loathsome tunnel under the railway tracks, disgorging its traffic at last onto the waterfront.

This is not to say the last passage of University has been deprived of all architectural punctuation. What's wanted, of course, is a splendid exclamation point, such as the proposed Britannia Square. What the southbound motorist gets instead, just before the plunge into the sewerlike tunnel, is just a smudged comma, in the form of Citibank Place. Though only nine years old, its silvery twenty-storey façade at Front and University is already stained and shabby—unfortunate even (or especially) in a building that was as charmless as a public toilet fixture on the first day of its occupancy.

Anyone brave enough to get out of the car and walk around this confused, hectic tangle of crossing streets will see that Citibank Place is hardly the only offence against eye and good taste. But before the architecturally well-advised Torontonian hurls this book into the dustbin, I hasten to acknowledge that, yes, the intersection *does* feature two of Toronto's great twentieth-century buildings, the Royal York Hotel and Union Station. As William Dendy pointed out, both were briefly considered for deployment as framing and focusing elements for the magnificent Cambrai Avenue proposal. This thoroughfare would have begun its straight northward passage at the portico of Union Station, concluding above Richmond Street in a vast open circle to be known as St. Julien Place, in memory of the Battle of Ypres. Had Cambrai Avenue been built, Dendy wrote verily, "Toronto's downtown would have had a focus and grandeur it sadly lacks." It could also have provided a highly imposing initiation into urban sophistication: a grand and thoroughly urbane street, lined with elegant towers and gallerias, fine shops and exquisite restaurants, all dedicated to the highest civilized pleasures and pursuits for the some seventeen million travellers and commuters who arrive in Toronto via Union Station each year.

As matters stand, the arriving passenger steps from the train, descends through an indifferent passageway, and is deposited into a low-ceilinged, meanly appointed arrivals room located directly below the station's justly famous Great Hall. The traveller can then come up directly from the swarming hall onto Front Street, as from a basement flat, to be greeted by a welcoming committee out of nightmare, comprised of an escarpment of giant buildings, all staring down coldly upon the hapless newcomer.

The alternative is to make a beeline for the network of man-made caves fanning out northward underground from the Station, leading to hotels, shops, subways. But whatever you do, however—scram in a cab, flee into the underground maze, jump on a subway—the only appropriate response is panicky retreat.

I realize that to utter negative thoughts about Union Station is to invite arrest on a blasphemy charge by Toronto's architectural preservationists, who, more than two decades ago, fought long and successfully to save the Station from the wrecker's ball.

The nub of the problem I'm pointing out is that all the architectural ceremony of Union Station is, and was originally designed to be, reserved for people who are *leaving* town. Arriving by car at the Front Street portals, the traveller passes though a magnificent colonnade into the Hall, with its fine tiled barrel vault overhead and everywhere the luxurious rhythms of warmer and cooler stone. One passes to the trains from this triumphant (though hardly pompous) space down a brief, luxuriously appointed ramp guarded by tall pillars topped by flourishing Corinthian capitals, the city's last bouquets for its *departing* guests. Does Toronto *always* have to save the best until last, the climax for the conclusion?

THE BURYING OF THE WIRES

We seem prepared to go to almost any length to flaunt certain accomplishments of modernity—soaring towers celebrating technical and financial mastery, cars conspicuously loaded with chrome, or compact to the verge of ostentatious minuteness—while relentlessly hiding others. In this latter category, I'm thinking of almost everything that has to do with electricity. The visible net of communications technology (telephone, television cable strung overhead), for example, and the physical apparatus of energy supply. (The CN Tower is, technically, an exception, being the tallest communications mast in the world; but its function as a TV and radio mast is incidental to its presumed usefulness as a city status-symbol, an emblem to be stamped on endless souvenirs, a stalk for a revolving restaurant from which nothing can be seen except water and Buffalo, and so forth.)

Most hateful, apparently, are the street wires, the burying of which was briefly a hotly debated public topic in Toronto early in the 1990s. Until then, I had never thought of hydro and phone wires as either homely *or* lovely. In fact, I'd never thought much about them at all. City councillor Howard Levine had been thinking about them a great deal, and come to the conclusion that they are things we should all give grave thought to indeed.

"If we live in an ugly environment," reasoned Levine, in a utopian-Modern (or Ruskinian) mood, "it stimulates ugly social patterns. A beautiful city is important." I suspect that most people are sympathetic with the sentiment, even if the behaviouristic social theory that produces the equation "ugly wires equals ugly acts" is, of course, quaintly false. But even if they disagree with Levine's philosophy, no Torontonians will take to barricades to stop Ontario Hydro's enormous programme of burying its wires beneath the sidewalks.

Now underway, the great interment will cost taxpayers an estimated $1.5 billion and take twenty-five years to complete. The occasion of this project is a massive changeover, for which Toronto Hydro has argued persuasively, from the 4.16-kV (kilovolt) grid, in service since early in this century and now approaching its limits of service, to a more capacious and otherwise up-to-date 13.8-kV system. According to the official Hydro proposal, the new high-voltage network need not be buried, though aesthetic considerations about poles "out of proportion to the street" and about "industrial-type hardware... not fitting for residential communities" have combined with practical considerations to make burial the chosen option. (The vulnerability of exposed wires to Toronto's occasional savage windstorms, which, to my mind, is the best reason to put the things underground, appears to have been only a minor consideration.)

Not every wire in the new 13.8-kV distribution grid, incidentally, will be put underground. In industrial areas—around

Toronto Harbour, in the Eastern Avenue wasteland of junk yards, and around west-downtown's crumbling, vanishing factory district, among others—the new cables will still be strung between and from tall poles, as before. But on many arterial and commercial streets, around public parks and along what are being called "historical streets"—mostly in the vicinity of Queen's Park and in "olde York"—the concealment of transformers and conducting wire is intended to be total. Street lighting will be provided, in all districts where burial is complete, by lamps mounted on stalks or walls.

As for residential streets, Hydro is going halfway. The transformers, and the primary 13.8-kV lines serving them, will be put under the sidewalks, inside tiny burrows excavated, for the most part, by remote-controlled mechanical moles. Each house on a given street will be served by a slender bundle of secondary, reduced-voltage cables strung from an eight-metre pole. The effect, then, will be to reduce the overhead clutter in residential neighbourhoods, but not to return them to the appearance of premodernity altogether.

Unless, of course, one is prepared to pay for it.

Under Toronto Hydro's so-called "local option," residents of each urban area will get the choice of eight-metre poles at no extra cost (beyond the increased hydro rates made necessary by the city-wide conversion), or total undergrounding at a cost of between $4,000 and $5,500 extra per property, payable by the owner. Complete burial will require approval by two-thirds of all property owners in a given area, and will then be mandatory for all. Thus, having hydro wires on your street, or not having them, will become, for the first time in the history of urban electrification, expressions of neighbourhood status.

Once they're gone underground, the swags and festoons of wires now overhead on every street, and criss-crossing in loose weaves at most important intersections, will probably not be missed, if only because we have not been much aware of them for most of a century. Only at the odd time when they get in the way—when we're trying to take a snapshot of the factory we live in, to send back to Texas cousins who can't believe we really would live in such a place—do we actively dislike them. Long past is the day when the incising of those lines against the changeable Toronto skies was a moment of excitement on a given street or in a given neighbourhood—the very herald and sign of modernity, coming to pierce the dark of pre-electric night with steady brilliance.

Removing the wires will not make us beautiful people, as councillor Levine thinks; indeed, it will give new visual expression to the turn of attitude towards the city long underway—from a certain delight about our machines and electric devices to a downright dislike of the things, and a will to rid ourselves of the very sight of them.

CHRISTMAS LIGHTS

Now while we're on the subject of electricity and streets, and ridding the latter of the former, I wish to register a complaint about the one electrical custom I don't like, which is the ritual of Christmas lights. Our single strand of outdoor bulbs, manufactured to cheer the hearts of passers-by, lay heaped in a chair in the bedroom all through the last Christmas season, a sad example of the unstrung and the unlit.

Blame it on the dragging economy or rumours of an impending jack-up of the provincial sales tax or whatever: most people I know have had problems revving up the old Yuletide cheer the last few Christmases, and a number have failed, as we have, to put up our outdoor lights. No one should conclude that I and my crowd are grinches; we aren't. I will advance an alternative theory of our behaviour presently. To be fair, it does seem that most other Torontonians except us have been merrily stringing their estimated ten million lights, inside and out, around Metro. (The estimate came from Noma, Inc., a Canadian company and the world's largest manufacturer of Christmas lighting.)

This last Christmas, few households in my downtown neighbourhood were content merely to put an old-fashioned glowing orange candelabra in the window. The more usual style has been increasingly Las Vegas post-Modern. Galaxies of big bulbs in nail-polish colours, set on rapid-blink, in the trees. Strings of twinkly green or poisonously red lights outlining fence, porch railing, roof. Whorls of brilliant white pinpoint bulbs around evergreens, programmed to light up one after the other at blinding speed, like casino marquees.

The festoons of Christmas lights which bedeck, year-round, a local front garden shrine of the Immaculate Conception was enhanced for the holiday season with a mini-Niagara of gently luminous blue bulbs. Less pious decorations included innumerable glowing plastic Santas, Rudolphs and such.

But, in any Christmas season, to motor along the wide streets of Toronto's outer suburbs is to find mere front-yard showoffishness replaced by a war of spectacles, with everybody on a street aggressively trying to outdo his neighbours. Vastly long snakes of twinkling bulbs in the trees surround the ranch-styles in delirious fantasias of colour and light, flood-lit mechanized reindeer rock stiffly back and forth, while Santa's arm wags back and forth on rooftop or lawn. What is more wondrous? The army of glowing gnomes in Yard A, or the full-sized, flood-lit tableau of the Holy Family in Yard B?

Hoping to feel noble about my abstention from doing an outdoor display, I rang up Ontario Hydro to find out how much money everybody wastes, province-wide, on the Great Christmas Light-Up. To my disappointment, I found it's very little, comparatively speaking: only about $21,000 an hour on top of the hourly $1.3 million that Ontarians would ordinarily be spending on electricity at that time of year. If Hydro statistics cannot be used to make us feel thrifty, and virtuous in our concern for the environment, those who dread the onset of Christmas—the decorating and partying and shopping, as well putting up outdoor lights—could take comfort in the notion that we may well be avant-garde, yea! the very Spirits of Christmases Yet To Come.

It should be obvious that what is dreaded is not Christmas itself—which means nothing to those who aren't Christians, and something wonderfully important to those of us who are—but certain "traditional" elements that have come to be associated with it. The enforced conviviality of obligatory parties, for example. And the shopping in bad weather or in crowded malls. And the virtually unavoidable overeating, over-drinking, overloading on chocolate. And lights. While pigging out on Christmas Day is perhaps a custom of high antiquity, the pre-Christmas party at office or school appears to be an invention no older than the 1920s. And, says a spokesman for Noma, outdoor Christmas lighting only goes back to the latter 1940s. Thus, the exhaustion some of us feel at the very idea of Christmas may not be a sign of impending Scroogehood or grinchness at all, but just a sign that the latter-day novelty of Christmas lights may be going the way of other fads, such as the rib-crushing corset and the boogie-woogie.

LANEWAYS

A history of the residential laneway, which nobody has gotten around to writing, would make colourful reading; and the Toronto chapter would go farther back than you might imagine. Leader Lane, in the heart of downtown, is thought to be from circa 1850, while Grand Opera Lane dates from around 1874. The oldest alley of which the City has official knowledge—this information from public works planner Julius Keddy—is a century old, and, unlike most of its counterparts

Emerson Alley, Morning

throughout the metropolis, has a name, Mincing Lane. I only recently learned that it was called that after the ancient London street. Until then, I had wondered whether the *mincing* had to do with what you do to meat, or from the way tall men walk when attired in spike heels; I considered no third option.

The practice of putting paved lanes between rows of back gardens endured into the early 1930s, principally in Toronto's working-class residential neighbourhoods. By the thirties, of course, its principal original uses—the removal of kitchen garbage and night soil from privies—had been rendered antique by the installation of underground sewage pipes. But architectural habits die hard, and zoning habits even harder. If you do have a lane out back, it's fairly certain you've bought into what was and may still be a workers' neighbourhood. (The notion of the residential laneway as mews, for the discreet movement of horses, groceries and servants, is Old World; though a grungy version of this older idea can still be seen in the commercial alleys, full of dead lettuce and soggy cardboard boxes, running parallel to market thoroughfares.) By the same token, the more upper-class the neighbourhood, the rarer the lanes. Some can be found among the late Victorian brick houses of gentrified Cabbagetown, but only a few in comfortably middle-class north Toronto or in generally posh Forest Hill (on the really posh *east* side of Spadina, anyway).

As it turns out, Keddy said, the overhauled circa-1915 factory I live in lies in the heartland of Toronto's 233.6-kilometre system of laneways. It's that stretch of town south of St. Clair Avenue, between Christie Pits and the Junction, transformed quickly from vacant cropland into a tract of industrial plants and workers' houses shortly after 1900. Apart from a complete overhaul in ethnic makeup—from British, to a current mix of Portuguese, Italian, Asian and Yuppie—it remains the densely populated, unspiffy and unstylish territory it was *ab origine*.

Like every other planned city site, the typical lanescape in this district is in constant, often contradictory architectural process. The outhouses that once stood at the alley-end of the properties have long ago disappeared, but the rickety old wooden garages that replaced them are falling down, and sturdier brick ones are everywhere under construction. Many of these newer buildings are not used for parking cars. In garage-looking structures behind my factory one can find a little auto-body shop, a meticulously well-organized welding operation, a tiny furniture factory, a junk dealership, a storage facility for Volvo parts—the infrastructure of that display of individual initiative known as the "underground economy." After some older garages have collapsed, however, homeowners tired of mowing and weeding have mercilessly paved their back lots, while others have plowed them up for the spring planting of tomatoes and grape vines.

This physical environment is the stage for volatile live action of the alley, the urban theatre which is too little known, and too little appreciated. For the kids, it's a winter battleground for snowball fights, and a summer street-hockey arena. For our Sikh neighbours, it's a place to gather in tight bunches and chat quietly, even on very cold days. For the tubby, ugly Siamese tomcat who lords it over all the other cats in the area, it's turf to be guarded with fang and claw, and, when he's lovesick, where he howls for his passion of the evening.

We dwellers in the factory know that spring has come when musclebound guys pull a sputtering car into the lane, put some heavy metal rock on the boom-box, plunk down a styrofoam cooler full of iced beer, and spend a Saturday afternoon ripping out the engine's guts. Or when the car-washing begins, spattering the pavement with frothy suds. We know it's nightfall in summer, when all the little kids abruptly stop bashing each other and vanish indoors, and the lanky pimp in a Kermit-green tracksuit slouches to his post up by the fender-mender shop, and settles in to do business.

And we'll know it's September, and nearly the end of another summer of gardening, when a broad streak of transparent ecclesiastical purple appears in the middle of the lane's grey concrete pavement one morning—the evidence of grape-dregs, clandestinely dumped by our wine-making neighbours into the storm sewers in the dead of the night before.

Gardens

Fence and Rose Bush

GROUNDS AND FRONT GARDENS

Only the wealthy few in any metropolis have what can properly be called *grounds*. Almost everyone else has unwalled yards, fore and aft, with a walkway or drive of varying width running perpendicular to the street between one house and the next.

While I have neither grounds nor yard, most of the people in my district of west-end Toronto have front and back gardens. Almost every one has been endowed with a specific character, giving the neighbourhood a character entirely different from the uniformity attempted in suburbia, with its regular, uninterrupted lawn-strips and "foundation plantings," or in the upmarket zones, where gardens have been reduced to ostentatious monotony by imperious designers. Low, inconspicuous fencing is common, and an important aspect of a streetscape rhythm of enclosure and disclosure, what's mine and what belongs to everybody, or what can belong to both me and everybody, to varying degrees and at different times. Most front gardens around here are clearly laid out to be beheld by passers-by, and not strolled in by anyone, including the owner—a crucial departure from the aristocratic European garden, designed for the owner's eyes only—while the front porch overlooking it is arranged as a quasi-public area, a proletarian belvedere, for neighbourhood gatherings on summer evenings.

A few of the small front yards in my territory have been allowed to go weedy and scruffy. Such a fate is rare, because universally disliked. The more usual attitude taken by my neighbours is one of idiosyncratic attention towards the front patch as an infinitesimal state under absolute monarchy. The experience of the streetscape, then, is one of pleasing, wild variety, as one might have found in nineteenth-century *Mitteleuropa*—an archipelago of minute principalities, each with its own currency, cuisine and customs.

One front yard will be a strictly uninflected rectangle of grass, sown and grown with exquisite care, and chemically encouraged to resemble Astroturf, each blade clipped to precisely the same length. Another will be laid out in beds of red and white petunias, alternating with a pattern as regular as houndstooth, and set in rigidly straight rows from which, one imagines, it would be certain death for a petunia to wander; while yet another will be dotted with annuals selected, it appears, to illustrate the heraldic colours of the owner's home country.

A front garden around the corner lies almost invisible under an overwhelming scramble and tumble of pink roses, while the one next to it is a square slab of smoothed, uniformly grey concrete, perforated in its exact centre by a small hole stuffed with lurid impatiens. Nearby is an exceedingly neat, faintly tragic garden of devotion to Our Lady of the Immaculate Conception, whose sacred image, bedecked year-round with Christmas lights, stands in a finely constructed brick shrine—tragic, because its creator clearly wishes his front garden to be a great place of piety and quietude, a refuge not unlike Lourdes, while his property is small, and abuts one of the busiest streets in the area.

If there is one frequently recurring plant in the various front-yard arrangements, it is a kind of astonishingly abundant, climbing rose with screaming red blooms, giving many a porch the appearance, in high summer, of having been hit with rocket-launched bombs filled with nail polish. Whatever one makes of the contribution to beautiful urbanism by this strange rose—a bit of genetic fallout, surely, from the same progressive postwar moment that gave us Tang and Teflon—virtually everyone seems deeply concerned about the effect of their little front yards. I should add that many people on my street eschew all spectacle, making quietly eccentric, flowerful or fruitful yards, lovely and visually melodic despite (or perhaps because of) their makers' angelic ignorance of "good design."

Now I have nothing against good garden design—the real and inventive thing, as preached and practised by the likes of Gertrude Jekyll or Vita Sackville-West. It's just that good garden design can rarely be found here, and most rarely of all in the pretentious, predictable, unfelt deployment of flowers and shrubs one sees on the front yards of one carriage-trade street after another. Such plantings express no joy, declare a lack of abiding care, and reveal a ballooning surplus of image-anxiety on the part of the owners.

My thoughts on front gardens were not prompted, however, by unhappy drive-bys of the front yards of the downtown rich, or by affection or disaffection for the way my unrich neighbours utilize their little lawns, but by the surprising discovery of what little information is available about the history of this commonest site in Toronto and every other North American city. It happened when I reached for the architectural dictionary I keep ever at my elbow—a fat, recent book abundantly endowed with entries on everything from aedicules to zoophoria. In it was not a word about yards, most ubiquitous of North American urban forms, their development or the stage they provide for the enactment of the rites of metropolitan culture. It was like finding no article about God in a large, authoritative dictionary of Christian theology. The yard appears to be to contemporary architectural historians what toilet activities were to our great-grandmothers: universally known to exist, though something absolutely not to be mentioned.

So far, I have been able to find only one reason most of Toronto's working-class tract houses, quick-built on farmland, between the 1880s and the First World War, have front and back yards and narrow corridors between; and it is a legal one. The British Parliament's Public Health Act of 1875, urged on legislators by the squalidly crowded, disease-ridden slums which blighted Britain's cities during the Industrial Revolution, set strict standards for intervals separating domestic dwellings, in the name of health and well-being. The result of this law was the development of a model for urban housing throughout the Empire, which, in turn, gave us pre-suburban Toronto, i.e., most of downtown constructed before air-conditioning and central heating. The standard working-class house of this period is a detached structure, or, perhaps more commonly a semi-detached one—an architectural peculiarity in North American building history, very common in Toronto—with large windows to allow free ventilation, small front "gardens" to provide distance from the dusty street and unenclosed by walls, to assure proper air and light—a novelty, coming from a European culture much given to walling up gardens, yards and farmlands. Some space was required between houses, again to allow ventilation, with long backyards behind each structure—not for growing beans and tomatoes, at that time anyway, but to provide a decent separation between back door and privy.

The death of the privy allowed the transformation of the backyard into a garden for flowers or vegetables or both. Paving and guttering of streets, and the running of storm and sanitary sewers underground, left the unwalled front yard without function as a sanitary interval. But California urbanist Dean MacCannell is probably right when he suggests that peculiar obsessiveness lavished on the front garden in a district like mine may well be grounded in the need of the immigrant labourer—alienated from homeland, expected to obey strange laws and customs, shut out by the native élite—to control at

Lappin Avenue

least *this* much of the land he's come to. Assimilation will undercut the desperation feeding this need; upward economic mobility, and the move on to suburbia, will largely eliminate it. But for now the front gardens on my street are objects of this curiously intense caring—whether what's being attempted is floral fantasy or salad-growing, the obsessively cute piling of rocks to create a miniscule Hobbit shire, or, in the case of a yard directly across the street from my factory home, a primly maintained grassy shrine of the Lord Buddha.

ESSENTIAL URBAN GARDENING

In 1891, a young man working in a Manchester factory put an ad in a local newspaper, asking for help in planting a window-box three feet long and ten inches wide. He knew nothing about growing things. Would someone kindly give him advice?

Fortunate was that man, for the ad turned up on the breakfast table of a portly, myopic, middle-aged English-woman, known to her friends as Aunt Bumps, who decided at once to take the boy and his box firmly in hand. "The post brought him plants of mossy and silvery saxifrages, and a few small bulbs," Aunt Bumps recalled later. "Even some stones were sent, for it was to be a rock-garden, and there were to be two hills of different height with rocky tops, and a longish valley with a sunny and a shady side."

Now she who so promptly answered the working-boy's appeal—the lady known outside her inner circle as Gertrude Jekyll—was accustomed to gardening on a somewhat larger scale than a window-box on a working-class street. Her home turf in Surrey covered fifteen acres, and kept a half-dozen workmen busy year round. In the remaining forty years of her very long life, Jekyll would design, directly or in consultation with fifty architects, some 350 gardens, large and small, for a galaxy of wealthy clients, including the Duke of Westminster, Prince Alexis Dolgorouki and Vita Sackville-West.

But, as she knew perfectly well, and as snobs never learn, creating a wonderful garden has very little to do with money or grand space. It has everything to do with rigorous attention to both detail and ensemble, a thorough grasp of plant structure and behaviour, reverence for the *genius loci* expressed in the suite of lights and shadows that informs the shape of site, even the tiniest window-box, making it unique.

A passage on gardening, and gardening tips, may seem like an odd addition to a book on city sites. My reason for putting it in is simply that no city site is so common, or numerous, or various than the garden, created under the particular constraints of city living; and no city site is less complex or various. On it converges every sort of desire, longing, hope and love of pleasure, and into its creation goes peculiar and intimate knowledge, disciplined by restrictions the country-dweller does not have to face. This, then, is very much a personal story; but it is also an experience, I believe, that is not unlike that of a million other people who, for some reason, have taken it into their minds to make a garden in the midst of metropolis.

Fortunately, I discovered the example of Aunt Bumps before building the boxes on my third-floor deck garden and putting my hands into the dirt, thus saving myself from doing what I usually do in practical matters, which is put cart before horse. Instead of rushing out and buying a lot of nice-looking plants, I first amassed a small library of botanies, practical growing guides, works on the history and philosophy of gardens, and handbooks on successful container gardening, then went to work, studying and drawing and planning. The principal difference between gardening on a deck and on the ground had to be dealt with immediately—the absence on rooftops of any possibility of dialogue between the given and the constructed, that is. Up there, *everything* is constructed, even the dirt—or "dirt," as this fluffy ersatz compound should properly be called. My first idea was to create a belvedere from which to contemplate the housetops, chimneys and laneways to the south, with birches framing the roofscape and focused on an old-fashioned black metal watertower in the distance. After the first summer, however, I decided to turn the deck into a *hortus conclusus,* or monastic garden—an enclosed dead-end for meditation and retreat, since the view definitely did not turn out to be as interesting as I thought it would be.

Never mind that all my schemes have gone awry, and few of the plants behaved exactly as the books said they should. Aunt Bumps's advice is still sound: the plan comes first. Much of the value of doing things this way is psychological, for I might not have been able to summon up the courage to make my first garden in the only place I had—a bare oblong of outdoor deck opening south over the rooftops and backyards of my neighbours—without a detailed strategy in hand.

Jekyll brought me the benefits of the great change that took root in the British garden in the 1860s and 1870s, the years she was coming of age. The older, exclusive ideals—the outdoor botanical museum, the stiffly geometrical planting, the carefully disposed rhythm of copse and field, the garden as intellectual exercise or stylishly novel showpiece—were already under attack by garden designers flying the flags of wildness, the romance of colour, the irrational, the natural and intimate.

She would eventually give such ideas some of their most glorious and influential expressions. As a young woman, Jekyll was aware of this horticultural revolution afoot, and would shortly begin contributing the first of hundreds of articles to the journals promoting it. She perhaps could have struck out as an independent leader of it, because her comprehension was brilliant. Instead, this deeply mid-Victorian soul for many years devoted herself to the traditional womanly crafts of her age and place, from textile design to embroidery and water-colour.

It was while taking these steps into art that she drifted effortlessly into a circle of romantic writers and artists including John Ruskin, with whom she shared the worship of Turner, and William Morris. And there were the Jacques Blumenthals, who hosted musical evenings, and the Duke of Westminster, Princess Louise—Victoria's daughter and a talented artist—the dashing amateur watercolourist Hercules Brabazon, the painters Edward Burne-Jones and Lord Leighton, the odd Anglican prelate and many another genteel and cultivated persons.

From our late-modern overlook, Jekyll's social scene seems (as the title of a book about her puts it) like an endless glide through "gardens of a golden afternoon," *English* in the tea-and-crumpets sense that would die in the trenches of the First World War. Yet this gentle cultural loam nourished in young Jekyll a blossoming interest in fine hand-crafted design which would finally find its fulfillment in one of the great creative friendships of modern times. This relation began in 1891—the year of the Manchester window-box—when she met an ambitious young architect named Edwin Lutyens. Aided in his rise by the socially well-connected Jekyll, this last of the great English country-house architects returned the favour by commissioning her to do more than one hundred gardens for his projects.

One wonders what Gertrude Jekyll would find to do nowadays. As the gardens of Toronto's poshest neighbourhoods testify, the rich want show, not subtlety—a big visual impression, not art, invention, fugitive delights. And most of us non-rich are too busy making a living to spend the years of patient experiment and rigorous self-education that it takes to create a garden on the large scale. So we borrow the concrete ideas about plantings and such from Jekyll and her heirs, just to kick-start ourselves—working out designs with a colour wheel, studying up on heights and textures and colours so as to avoid making a mess, and so on. Only later do we begin to find that gardening the Jekyll way is as much a spiritual discipline as a hands-on craft. The joy of it is in learning how the sunlight of each summer hour falls on every leaf, petal and twig; of noting how an autumn shower changes the colour of one's late-blooming clematis and hydrangeas, of learning to listen to the fragile soul of the world, speaking in every garden's ever-evolving, evanescent beauties.

Not that many serious plantspeople outside cities would consider my deck-top boxes of dirt a real garden. Perhaps in self-defence, my use of the word is promiscuous. I am prepared to allow the word to be used of an island of hot-pink impatiens on a flattened, poison-green suburban lawn, and also my neighbour's tiny plot of exuberantly abounding salad greens. A clutch of potted petunias on a twentieth-storey balcony is a garden, and so is a capacious estate planned for a cowed client by a tyrannical landscape architect and maintained by a battalion of weeders and trimmers. So is a row of shrubs, or a line of proletarian annuals plunked down each May to mask an unsightly concrete foundation.

Some private gardens are actually quite public, like the little patches in parks allotted by the city to apartment-dwellers. Others are (or are meant to be) secret, like the secluded patch of tomato and marijuana plants I discovered a summer or so ago, while scouting an abandoned railroad right-of-way for wildflowers.

Now that I've defined *garden* so broadly that it includes whatever you do—from gently putting in *Cyprididium insigne* to growing pot on the sly—you'll probably be wanting some practical tips. I don't have any, except the obvious ones. Just because you like a plant is not good enough. It must fit the plan. Learn how to refuse, firmly and finally, cuttings spontaneously proffered by friends trying to be helpful. Their helpfulness will turn out to be your downfall. At the same time, develop polite strategies for getting your friends to give you cuttings that will fit your plan. Be brave in your retromania, since it is a grace given to every plantsperson who works and prays. Above all, *think*. And be imperious. Every really good gardener is.

What counts is the development of a correct attitude. And this is my reason for recommending to all city plantsfolk *V. Sackville-West's Garden Book*. If notorious for the oddity of her marriage to Harold Nicolson—she, lesbian; he, gay; the marriage, idyllic (most of the time); the children, smart and charming—the late Vita Sackville-West was also probably the best horticultural journalist ever to write for a newspaper. The *Garden Book* is a selection of weekly columns contributed to *The Observer* between 1947 and 1961. After its publication in 1968, this volume went through ten subsequent impressions in hardcover; and, since coming out in paperback in 1987, it's been reprinted twice. Read twenty pages, and you'll understand the reasons for its durable enchantment.

Just remember that Kent isn't Canada; and few of us have the barrow-loads of money, the land, time or willpower to create anything like the world-famous gardens at Sissinghurst, the country seat of Sackville-West and her diplomat-husband. What counts in this book, apart from the marvellous writing itself, is the columnist's exemplary stance, which any city gardener will do well to mimic, even if his or her garden is, like mine, just some insulated boxes on a rooftop.

Central to the Sackville-West position is an unflinching ruthlessness, coupled with an unbounded delight in experimentation. "The true gardener must be brutal," she advises, "and imaginative for the future." Every plant must please. If something doesn't please, for any reason, it must be ripped out and thrown away. Move things around, add, subtract, try anything, but never rest until everything is exactly right. Since nothing about a garden comes out exactly right, you'll be at the job for the rest of your life.

Of course—here I go again—you won't get anything right if you don't have a plan. Sackville-West was an architect at heart, who chose to realize her plans in the ephemeral, unpredictable stuff of plants, with their infinite variety of colour, texture, volume, shape, seasonal variation. Like any relentless designer, she is exasperated when something goes wrong. But every exasperation is just a jumping-off point for a revision of the plan, and the next try. In a fabulous understatement, she reminds us that "one has a lot, an endless lot, to learn when one sets out to be a gardener."

In addition to everything else you have to do, you should learn the sheer botany of it all. You must know exactly what such-and-such really is, when and how it blooms and droops and dies, where it comes from. (Vita has a lovely column about the origin of Bourbon roses on an island off the east coast of Africa.) You must know the minutest detail of every bloom's powers to lure butterflies or repel them, what sorts of bugs will chew it to death, and what house-pets its leaves, lunched upon, will kill.

It follows that flowers must do more than just look nice on a sunny day. They should be interesting in temperament and history, as well as in appearance. Her opinion of campion (*Lychnis haageana*) is that it's "rather untidy," with "ugly leaves" and "shaggy flowers"—but she finds its tendency to come up from seed in unexpected colours amusing, so its life is spared. The attraction of *Narcissus triandrus albus* is its liking for "broken shade, where it looks like a little ghost, weeping."

But she would rip all the ivy off those venerable English walls, because "one gets so bored by its persistent stuffy evergreen." If a terrible snob in private life, Sackville-West is militant about reversing the bad rap given the humble Virginia creeper, after she discovers one in its autumnal scarlet phase, not "glued to red-brick houses," but hung in "great swags and festoons" on a silver birch, and glowing like "wine held up to the light."

So here's a rooftop toast to Vita; and a toast, too, to all my fellow city-dwellers who are planning or learning the beauties we want to spring, someday, from our most intimate city site, be it yard, pot, box, planter, tub, or whatever other bit of dirt to which we have decided to lay total, imperial claim.

ALLAN GARDENS

Newcomers to Toronto are still warned by friends and old-timers to shun Allan Gardens. While I don't remember getting this advice when I settled here a quarter-century ago—not in so many words anyway—the idea of this once-fashionable downtown park as a sordid open-air drug-and-sex bazaar, and a noisy preaching-field for the famously fundamentalist Jarvis Street Baptist Church nearby, somehow got stuck in my mind. Not needing heroin and being certain of Heaven, I stayed away—until one very cold Sunday afternoon in 1991, when I discovered what many another wary immigrant has probably missed for as many years as I had.

It is the charming domed and winged pavilion of glass that stands on Allan Gardens' west side, designed by city architect Robert McCallum, and known as the Palm House. This little structure was put up between 1910 and 1913 to replace a whimsical pagoda of 1878, famed for its concerts, balls and high teas, and which burned to the ground in 1902. At the time of the completion of the Palm House, the posh Edwardian district around Allan Gardens and the Palm House was poised on the edge of decline, which urban decay would inflict in earnest after the First World War and largely destroy by the end of the Second.

In 1910, of course, Robert McCallum did not know what was coming, and so envisioned his building as a gleaming brooch on the ample, upper-class bosom of the neighbour-hood. I have not been able to track down the likely inspiration for McCallum's project, and am inclined to think he had no specific structure in mind. Rather, the Palm House appears to be a little chamber rhapsody on several large, lovely themes current among *fin-de-siècle* French and English builders of canopied and winged conservatories, market halls, exhibition palaces and the like. The lanterned central dome recalls the one Victor Baltard had provided for his Halles Centrales in Paris, though McCallum has given it an exotic taper, perhaps to remind his Toronto viewers of India and Raj. The closest thing to the Palm House McCallum might have known directly was the great glazed tent of the Palm Stove, created by Decimus Burton and Richard Turner in the mid-1840s for the Royal Botanic Gardens at Kew.

The Palm Stove is really an ordinary rectangular greenhouse writ large, with all its angles smoothed into an elongated bubble of glass. It looks like a domed railway car, and seems about to turn from architecture into engineering before our very eyes. Toronto's Palm House does not go so far, nor become so modern. If it glances at the Palm Stove and contemporary pavilions, it also has an air of the old-fashioned garden temple about it. In McCallum's work, the central temple is turned from stone to glass, and outfitted, not quite convincingly, with wings that are really just traditional hothouses.

In the Palm House, there are a number of palms from Africa and equatorial America, some of them surprisingly tall, surging like green fountains towards the dome's clear glass lantern. But strolling through the wings of this sunny palace of plants, the visitor will also find a gallery of cactus, variously sinister, impudent and piquant; a tiny swamp; and splendid orchids suspended aloft in many small baskets, as though enjoying a sociable afternoon of hot-air ballooning. In a sweet, cold gallery, there are banks of white narcissus, those good-natured proletarians of the garden, and beds of pale, neurotic camelia, flowerdom's Blanche Dubois.

One finds no long or panoramic vistas here. Everything is very close, inviting scrutiny rather than admiration. Walking along the Palm House's narrow, meandering flagstone trails, one is usually nose-to-leaf with swags of vine or palm-tree fans, or about to crush splatters of moss underfoot, or bruise fingers of creeper clinging to the path's edge-stones.

While devoted to plants, and rich in everything that destroys artworks—bright sunshine, air that's hot and moist—the Palm House is surely less garden than museum, of a peculiarly old-fashioned sort. The modern museum of art or science is, after all, a place in which the exuberant jumble of reality is tamed, arrayed by type, given an air of order it never has in the real world. The whole idea is to make sense out of South Yemeni tattooing practices, Renaissance easel painting, toed feet in their large but finite variety, and all else serious people become interested in and spend lifetimes studying. The emphasis is on revealing—or, more likely, creating—a system in the apparent chaos of existence.

The Palm House, on the other hand, embodies a notion of the museum as a cabinet of curiosities and wonders, a heap of booty hauled from the world's farthest corners to amaze and delight, and to accustom Toronto's rising colonial middle class to proper respect for the immense expanse of the Empire. It was an idea of both museum and gardening obsolete in Britain and North America when the Palm House was built—though this ideology, enshrined in the midst of late-twentieth-century Toronto, is too delicious to pass over without a notice.

I know no better summary of it than Canon Henry Ellacombe's *In a Gloucestershire Garden,* first published in 1895. The book's twenty-five chapters are columns written between 1890 and 1893 for *The Guardian* by the vicar of Bitton, a village lying under the clement skies of southwesterly England. They deal with matters ranging from the chief floral beauties of February—rather a cruel column for a Torontonian to read—to "the easy cultivation of hardy palms," and how most pleasingly to plant an ancient, ruinous garden wall.

As Canon Ellacombe was penning his articles, he perhaps did not notice his early Victorian concept of the garden as a museum of botanical curiosities was swiftly passing away, dethroned by the new spirit of sheerly aesthetic garden design being spread by Gertrude Jekyll. If the gardening idea coming into ascendence had to do with textures and colours, iridescences and shimmering ambiguities, effect and show, Canon Ellacombe's credo led him to plant less for colour than for botanical or historical interest. Thus, his trees and flowers mentioned in the Bible—palm, fig, olive, "the willow of Babylon," hyssop—for the edification of "garden parties of mechanics, young men's associations, school teachers, etc." He found it "very pleasant" to see growing, side by side, the Antarctic bramble and the Iceland poppy, simply because one is the most southerly, the other the most northerly, of flowering plants. He would grow anything mentioned by the ancient Greek and Roman writers, just because they mentioned it; and he put down a yucca after reading a poem "by the late poet-laureate" which begins: "My Yucca, which no winter quells…"

Toronto's Palm House plantings belong to the spiritual world of Ellacombe, not Jekyll or Sackville-West. Even when mimicking one, a planting in the Palm House replicates no real environment elsewhere, and every one is a fiction. The little swamp in the north wing is a pleasing fiction composed of Egyptian papyrus, sweet flag from Japan, Ontario pond weed. Similarly, the thick jungle vignette under the central dome is a jungle from nowhere—a silver thatch palm from Trinidad, Madagascar's screw pine, poinsettias from Mexico.

As we wander through these fragrant galleries, it's always best to do so in a leisurely late-Victorian frame of mind: charmed by the opulence, sharply attentive to details of pistil and stamen, delighted to practise strolling in this improbable spot, surprisingly reminiscent of the architectural vogue of our own *fin de siècle* for cleverness, the cute, the *quel surprise!*

AND THEN THERE ARE *LAWNS*

Each year in May, everyone in my household wakes up with blinding headaches, scratchy throats and stuffy noses. Soon, we gather round the breakfast table, smiling at each other through the tears of joy drizzling from our reddened eyes.

"'Tis spring at last!" I muse aloud to my gasping wife and coughing daughter.

"From the great aprons of green spread before the ranch-styles of Etobicoke to the long yards behind Scarborough bungalows"—I continue between sneezes—"the plucky buzz of the Toro and the 4.5-hp Noma Brute, and the mighty roar of the 12.5-hp John Deere STX38, arise from the land.

"The folk, both great and small, have again begun their timeless ritual of shaving their grass down to the suburban standard variation of between 2 and 2.5 inches, filling the air of our city with the music and fragrance of their noble industry.

"Soon, out of the metal shed will come the clippers and edgers, fertilizer spreaders, weed whips, poison sprayers, Speedy Weedy Weed Removers, black rubberized lawn edging, the 72-position oscillating sprinkler, lawn food, insect killers powerful enough to kill whole species at a whiff, and herbicides able, like smart bombs, to pinpoint the enemy weed and annihilate him, while leaving the good and kindly grasses to grow—the weaponry, that is, in the unrelenting struggle of Man against the Forces of Nature."

Then, taking the hands of my gagging mate and wheezing child in mine, I say: "If in the beginning there was chaos, we can be sure that in the end there will be The Lawn."

Weeping and uncontrollably choking—obviously so moved is she by my show of faith in Man's Progress Ever Upward—my little daughter asks between lung spasms: "But, Daddy dear and wise, how came this wonder, the Suburban Lawn, to be?"

Passing the family oxygen tank and mask to my wife, now lying half-conscious on the breakfast-nook floor after her most recent asthma attack, and after chasing some antihistamine tablets down my raw throat with a gulp of fresh-mixed powdered orange juice, I sit pensively for a moment, admiring the way the allergic rashes were turning my girl's winter-pale cheeks apple red—then begin my story.

"Once upon a time, in the wicked and dirty place called Europe, from whence we came to this new, free land, there were things known as Walls. Now I know you have never seen one in Toronto, but, believe, they did indeed exist.

"And awful things they were, these Walls, both high and forbidding. They were built of stones by the rich, the aristocrats and other degenerate persons to keep us common folk out on the dingy, grey streets, and to protect all the Lawns for their own use and pleasure. And what miserable things they were, these Walls of stone! Plague-bearing wild creatures, known as 'rabbits' and 'mice' and 'birds'—all extinct now, thank heaven—made their disgusting nests under and among the stones in the Walls, and the rich people allowed even hateful weeds, and flowers not certified disease-free by government authorities, to take root and flourish in the crannies."

My daughter shudders at the very thought, and hacks meaningfully into her paper table napkin.

"But then our stalwart ancestors quit that miserable continent of Walls, and came to this broad, fruitful and free land. After clearing it of trees, Natives and other impediments to Progress, they created Suburbia. There, each man could be a king in his castle, and rule his turf like an emperor. And did those yeoman pioneers, who made Etobicoke, North York and Scarborough what they are today, build Walls around their greenswards? No, not they, these men of democratic vision! For them naught but the Lawn in all its purity would do—the open Lawn, flawless, uniform, free of disease-bearing animals and perfectly green, and free to be seen by all!"

Choking back the sobs for a moment, my little daughter asks eagerly: "And all this wonder began right here in Canada?"

"No, no, my child," is my reply. "Like everything good in this country, the Lawn was invented in the United States. 'Twas in the year 1868, as I recall, that a great American named Frederick Law Olmstead was asked to create, near Chicago, one of the first real Suburbs. And it was he who ordained that all the houses in his Suburb be set back far from the street. All trees were to be uprooted and removed, so that each house would sit in a perfectly green and level rectangle, thereby making it easy to kill all disease-bearing animals, and slash vulgar weeds off at the foot.

"Then an Englishman named Frank J. Scott—one of the few decent men of his decadent nation—was granted a vision of Olmstead's cunning handiwork, found it good, and published a book in 1870 called *The Art of Beautifying Suburban Home Grounds,* in which he offered the Lawn to the world. Like Olmstead and every other person who is careful to include something from all food groups in every meal, and keep the head clear and the bowels open, Scott hated the Wall. 'It is unChristian,' he wrote in his wise book, 'to hedge from the sight of others the beauties of nature which it has been our good fortune to create and secure.' "

With darkened countenance, I go on thus: "And so, my child, you see why we must be ever vigilant. Whether with poison or shotgun, leg-hold trap and caustic fertilizers, the Lawn must be kept green and pure at any cost—else Christian Civilization itself be lost!"

Visiting my wife in hospital later that day, I comfort her with these words: "Sorry I was so wrapped up in our daughter's instruction that I didn't notice that you'd passed out after a windpipe seizure on the kitchen floor. But now our darling understands everything. Family values. Lawns. The importance of chlorinated hydrocarbon pesticides in keeping it all together. Think about it. Isn't the maintenance of our hard-won Civilization worth a couple of months a year in an iron lung, recovering from pollen-induced respiratory failure?"

She can't speak, of course, because of the feeding tube pushed down her throat. But I know the tear running down her cheek says *yes*.

Moral Management

Suburban Crescent, Number Two

999 Queen

In 1989, a young man in a stolen car hurtled down Ossington Avenue, jumped Queen Street, crashed through the glass-fronted main entrance of the Queen Street Mental Health Centre, continued through the lobby and drove ninety-one metres down a glassed-in corridor before the car was stopped by a broken wheel.

Whatever the driver's intentions may have been—one story has it that he was trying to free an incarcerated friend—they probably didn't include a demonstration of how permeable the recent architecture of mental hospitals has become.

But that was one lesson that came out of the caper. Built by the Ontario government in 1956—most of the other structures on the site date from the early 1970s—the building in question fronts Queen Street bluntly, with much glass and concrete pushing out. The lobby is on the edge of the sidewalk, like a store-front. This lobby, and the long corridor which was turned briefly into a speedway, are both at street level, suggesting a smooth continuity between life inside the institution and life outside it. Before the crash-in and the erection of anti-terrorist concrete planters, it was easy to drive in— almost as easy, in fact, as it is to walk in.

Absent are all the standard things about older mental hospitals that would once have made such a spectacular drive-through impossible: the high perimeter walls, the wooded park in which such institutions were frequently set, the flight of steps up to the portico and vaultlike door leading into the vast, mysterious asylum beyond. Instead of being open, *porous,* the structure would have been opaque, solid, resistant to the glances of snoopy passers-by.

Until nearly twenty years ago, as many a Torontonian will recall, the centrepiece of a rural asylum-complex of this sort did stand on the site, just south of the modern building on Queen. That remnant was the Provincial Lunatic Asylum—popularly known as 999 Queen—North America's facility most finely attuned to the psychiatric ideologies current at the moment of its construction. The cornerstone was laid in 1846, in what was then countryside west of Toronto, and the first patients arrived three years later. Designed by the ubiquitous John Howard of Colborne Lodge, it was first acclaimed for its design—said to have been inspired by the National Gallery in London—then later much criticized for practical faults, downright despised by staff and patients after the Second World War, and finally demolished in 1976 to make way for the present scatter of successor-buildings. In its declining decades, the immense edifice was noted for its peculiar and unforgettable stench.

But whatever blame was to be heaped on it in later years, at the time of its opening the Lunatic Asylum was at the forefront of the revolution in asylum architecture and the treatment of mental illness then sweeping North America. Optimism about the curability of mental disorder was in the air. The ill were being unchained from the walls of ordinary prisons, in Ontario as elsewhere. The radical treatments of former times—whipping and starving and purging—were everywhere being rejected in favour of what was called "moral management."

This was first of all a theory of treatment: "activity without excitement, progress and the combination of self-government with appeals to the intellect and sentiments," in the words of a Victorian psychiatrist.

But new systems of treatment required new places in which treatment could appropriately take place. Anything reminiscent of the stinking cellars where the mad had been kept earlier in the nineteenth century would clearly not do; hence the rapid development, at mid-century, of the new theory of asylum design embodied in John Howard's building on Queen. The new mental hospital, wrote U.S. madhouse theorist Thomas Kirkbride, "should have a cheerful and comfortable appearance, everything repulsive and prison-like should be carefully avoided, and even the means of effecting the proper degree of security should be masked." In its classical visual balance, the building should express stability, order, permanence. Its architecture should itself be a kind of therapy, complementing the moral management going on inside.

The Toronto incarnation of these lofty notions was Howard's building at 999 Queen Street West: a forceful expression of what was then fashionable, but soon reputed to be too big, too forbidding. Before it was finally ripped down, historian Tom Brown has written, it had come to symbolize, "for many generations of Torontonians, all the terrors and horrors of the dark and hidden world of the mad." During the debate in the 1970s among architectural preservationists, provincial officials and historians over the proposed demolition, one frequently heard the argument that the worst thing about the building was the fact that it was simply sending the wrong message.

Today, there is almost nothing left of Howard's edifice or its message, except for a gloomy wall running intermittently around three sides of the site, and an old service building tucked away at the back. The general impression given by this large provincial medical facility is less one of asylum than of community college. Most of it has been built in the last twenty years, in the campus style. Low dormitory towers are connected by glassed-in walkways, and bordered by parking lots and tennis courts and grassy lawns. There's even a pleasant student union building—it's called the "community centre"—with a large central lounge, a swimming pool, a coffee shop, a bank.

This is no place of incarceration, or even of treatment, but of education. What's being dealt with are not terrifying illnesses, but ignorance. Those resident or occasional users of this place are not dangerous psychotics, or the compulsory objects of police scrutiny, or people whose personalities are being held together merely by powerful and destructive drugs, but *students,* coming and going as they please through those glass doors and long, open corridors. They will not be there forever. They will graduate.

Or so the architecture of the present Queen Street facility insistently tells us. The reality is different. Reality and architecture usually are different, with architecture playing the role of making rhetorically forceful the myths and theories reality stoutly denies. Like John Howard's old asylum, the new centre is a visible expression more of expectation than of anything real—this time, the fond, recurring hope that the asylum-as-prison is nearing the end of its long career, and that soon all the insane will "graduate," walking out into the world to resume life as productive and happy and creative citizens. The vogue enjoyed by this fantasy among psychiatric professionals in the 1950s helps explain both the extraordinary hatred of John Howard's building, as well as the style of the "campus" subsequently built on the site. The recent discovery of a new family of antidepressants, Eli Lilly's Prozac chief among them, has inspired yet another flutter of optimism among the doctors, who long so badly (and curiously) for their own obsolescence. I should be surprised to find a sensible psychiatrist who really believes the end of forced confinement and life-long heroic treatment is really going to come in our time. But architecture serves as the guardian of foolish fantasies about many things, the junior-college campus on Queen Street being one superb example.

THE BOARDWALK

Toronto has a number of sandy shoreline expanses lapped by the waters of Lake Ontario, but only one known simply as The Beach.

Or, if you prefer, The Beaches.

Once you've decided what you'll call it, by the way, be prepared to stand your ground. The citizens of this east-end neighbourhood are notoriously touchy about what things are called. In 1985, some shopkeepers along its Queen Street getting-and-spending strip decided to have signs put up proclaiming the area The Beaches. As the entrepreneurs should have known, the denizens of The Beach are not folk to be dealt with high-handedly. They organized, they protested—while historians timidly tried to convince everyone that *both* The Beach and The Beaches have long histories in local usages. Residents who'd lived there thirty years had been calling it The Beaches, and thought The Beach a novelty. But, in the end, the loudest citizens' group won. Their turf was, and ever shall be, The Beach.

While perhaps the most pedantic uproar of them all, it was not the first time Beach(es) people had gotten on their high horses, and ridden off to war; nor would it be the last. In 1907, they stopped the greatest rail companies in the land from putting tracks along the lakefront. And in 1988, they again rose up in outrage, on this occasion against the building of a temporary 14.6-metre beaconlike fancy by the distinguished Italian architect Aldo Rossi. Rossi's $100,000 work, his contribution to an international art exhibit, was eventually built on the lower lawn of the R. C. Harris Filtration Plant, but only after anonymous threats of violence against the exhibition organizers and a near-riotous demonstration by Beach(es) activists at the construction site.

There has never been a dust-up about the name of The Boardwalk, though the stubborn Beach(es) residents have been fighting to save it from the lake wind and water since it was officially opened on Victoria Day, 1932. Severe storms sweep off Lake Ontario almost every winter, ravaging and dragging away whole sections of the structure. But walk along it some summer evening, and you'd never know a plank had ever been displaced. The unbroken continuity of wooden walkway along the district's three kilometres of lake frontage—from the Balmy Beach Canoe Club at the east end, to the more recently fashioned Ashbridge's Bay Park—is today the most conspicuous monument to the dogged determination of Beach(es) souls to preserve Toronto's last extensive neighbourhood to front directly onto an unexpresswayed and unrailroaded strip of public lake footage.

While I find the famous grouchiness of Beach(es) people diverting, I find the Boardwalk less so, and seldom go there. I would perhaps have enjoyed the earlier, fragmentary boardwalks more, judging from the surviving photographs and memoirs. Torontonians of late Victorian and Edwardian times seem to have understood boardwalks, not as places of idle relaxation, but as occasions for the costume and ceremony characteristic of Toronto's first great age of middle-class consumerism and the shortened work-week.

One did not merely walk on the extant portions of boardwalk in former times, for instance. One promenaded: if male, in best suit and tie, and if female, in elegant hat and high collar, long sleeves and long skirts, all as part of a continuous unrushed display of pride in the white-collar class and its perks. When picnicking, one did so carefully and uncomfortably—in full afternoon dress, dining off china, never forgetting the show of propriety which was one's duty as an actor in Toronto's rather new pageant of respectability. Or one played; but only as a member of a disciplined, uniformed team, which in turn reflected the disciplined, promenading and uniformed middle-class Beach(es) culture as a whole. Such regulated play was epitomized by the community's boating clubs, still popular at the present time.

The modern store, with its large windows stocked with mass-produced consumer goods, emerged on Queen Street East at the same time. If Queen was the place where Beach(es) people learned the new rites of acquisitive consumer browsing, the Boardwalk was where they learned the rituals of conspicuous leisure, and of free time attained. Recreation, in the contemporary sense of winding down or staying in shape, was not an issue.

By 1932, when the continuous Boardwalk we have now was completed, the cultural ideals and ethics that had brought the promenade into being were already passing into memory. The last time I walked the length of the Boardwalk, they had altogether vanished. It was this discord that struck me most forcefully: the contrast between the lost historic purpose of the Boardwalk as architecture for conspicuous enjoyment, and the use to which it is now put by virtually all its comers.

Today's joggers do not parade; they run along with ears plugged into taped music, seemingly oblivious to everyone else around them, except as obstacles to their onrush. Little groups, twos and threes, walk along more slowly, equally oblivious, talking loudly about such private topics as their lovers and their sex lives, the impending death of parents and so on, as though no one were listening and watching—when the whole point of the boardwalk-as-idea is in fact to listen and watch, and be watched. Granted, this is the age of Oprah, when confession has taken the place of conversation. But *must* people be constantly divulging?

If discourse proper to boardwalk spectacle is gone, the costumed formality of yesteryear has also been replaced. Its heir is what's euphemistically called "informality," but which usually means slovenly dress and swinish manners, pop cans and cigarette butts tossed on the sand.

I recall reading somewhere that the contemporary expressway is an accelerated boardwalk. The analogy certainly held true in the 1950s, when the shells of Detroit cars were designed to declare the power, wealth, suave sexuality, businesslike demeanour or other wished-for traits of their drivers. Be that as it may, boardwalks—like expressways—have lost that richness of spectacle, and become as devoid of stylish showoff as the styling of the cars clogging Toronto's streets in the 1990s.

HIGH PARK

After a long lapse, interest in the public park, that wonder of nineteeth-century municipal design, is on the upswing. The reason, I suspect, is that North American parks themselves are so conspicuously and universally on the skids.

Talk to worried urban planners, peruse recent photographic surveys of parks, page through the books and reports coming off the presses, and you'll find generally the same story. It's about a brilliant social invention—an opening of green and sky in the midst of the industrial city—betrayed by stingy politicians, deserted by the hard-working city folk for whom its bandstands, flowerbeds and playing fields were created, and now colonized by addicts, rapists, prostitutes and other undesirables.

The tone of the reports is usually one of elegy, tinged with hope for restoration of the old ideal of the "people's park." But occasionally there's a dissenting voice. In a recent study, British urbanist Hazel Conway has denounced the public park as a tool of mass social control, designed by the Victorian patriarchy to instil good public manners, patriotism, team spirit, Christian respect for Sunday and other presumably imperialist values in a working class dangerously prone to waywardness, even revolution.

Had Conway been looking for Canadian proof of her thesis, she could have found it writ larger in few places than the 406.8 acres of High Park, Toronto's grandest contiguous green space and its most quintessentially Victorian British rest-spot for the common man. The core is a wildly beautiful lakeshore property donated in 1873 by John Howard, the city's first and gifted official engineer, who left another 45 acres, including Colborne Lodge, to the city in his will. In Howard's day, the park around the lodge was used by its owner for riding and blood sports, the obligatory rites of colonials (then and now) keen to display visible ties to aristocratic Englishry. But when deeding it over to the city, Howard made sure future users would be doing none of that. Perhaps with the precedent of Frederick Law Olmsted and Calvert Vaux's recently opened Central Park (1858) fresh in mind—a development promoted by New York's élite as a hoist for the dilapidated morals of the poor—Howard envisioned High Park as a school of uplift for the slum-dulled working family, and spent the rest of his life, which ended in 1890, planning drainage and changes in the landform, clearing the scrubby brush and proposing "improvements to the site." High Park was off limits to booze and questionable frivolity from day one.

At night, of course, like every other public park in the world, High Park was (and remains) a popular spot for erotic quickies between gay men, and for back-seat and blanket sex between consenting adults and consenting kids. But moral militancy reigned by day, at least until rather recently. In 1913, a playground was established by a women's group determined to provide an "antidote to slothful living." Grown-ups were allowed to stroll politely through the grove of oaks and among the roses on the Lord's Day, and expected to do nothing more vigorous or flamboyant than listen to band concerts. In 1937, more relaxed recreations were allowed on Sundays, though hockey and other team sports, along with tobogganing, were banned on the Christian Sabbath until 1961. We do well not to let our liberalism (or the present-day dogmatic commitment to the casual) be offended by such now-lifted prohibitions, simply because they interestingly reflected the Victorian park's forerunners, in the military parade ground and the field for public ceremonials, and similar institutions, such as Central Park.

Now no longer a training-ground for Victorian social decorum, or informed by any other strong, central idea, High Park is drifting in the same dismal direction as other big-city parks in North America—though drifting not so quickly, this being Toronto, hence slower to inflict destruction upon itself than American cities the same size. As early as 1904, on a return to his native New York, expatriate Henry James was shocked to find Central Park "overdone by the 'run' on its resources...It has had to have something for everybody, since everybody arrives famished." It took the 1985 murder of a homosexual schoolteacher in High Park by young gay-bashers, however—memorialized in Robin Fulford's remarkable play *Steel Kiss*—to make the once-peaceful park *seem* more dangerous, and at least alert Toronto officialdom to the decline of their best Victorian park.

The concern is well grounded. The fear of rape in broad daylight has made women afraid to jog there alone. And apart from the heavier threats we associate with American public parks, petty annoyances plague anyone using High Park. There's the unending rumble of car traffic on the north-south street bisecting the park between the Queensway and Bloor Street—it should have been blocked off year-round long ago, not shut, as today, only on summer weekends—and the difficulty of observing bitterns and herons at the marshy edge of Grenadier Pond without noisy runners galumphing by, terrorizing the wildlife. And there's the inevitable, annoying conflict between the rights of guys who've come down after work for a good knock-down, drag-out game of soccer, and those of drugged or drunk wanderers, of families letting their kids have a visit with the yak in the little zoo or run off some steam before bedtime, and of quieter folk who just want to contemplate the autumnal daylight's slow failing in High Park's beautiful ravines and shadowed copses.

While not yet in ruins, High Park is visibly buckling under the weight of use and the burden of conflicting demands. The salvation of the park from its perilous popularity was clearly on the minds of the City of Toronto's parks-and-recreation people when, in 1992, they handed their political masters a list of proposals for restoring and better managing this matchless resource.

The planners want less car traffic and a freshening-up of the gardens overlooking Grenadier Pond, increased protection for both the park's human users and its rare, vulnerable black-oak savannahs and moist forests of red oak and hemlock. They want people to enjoy Colborne Lodge, and the swimming pool, and the hockey rink. They want dwellers in the apartment complexes north of the park to continue growing zucchini and sunflowers in the tiny community allotment gardens, and they want nature-lovers to be able to stalk peaceably the shy birds that nest along Wendigo Creek.

In fact, the planners want *everything*, only "improved." They can't be faulted for that. High Park attracts millions of visitors each year, and it's the job of the City to see that these people enjoy themselves without undue stress, annoyance or danger.

But even if City Hall were to be enchanted by what was proposed in this report, and came up with the money to fulfil the bureaucrats' every wish, High Park would still be, in my view, doomed to continuing decline. This gloomy forecast has nothing to do with the goodwill of the city or its well-meaning planners, but is based on the peculiar nature of English Victorian parks in general. Though nobody nowadays wants to admit it, these parks worked best so long as they were institutions of official moral management, whose working-class users knew and kept to the rules, dutifully appreciating the exquisite views from well-defined strolling paths by day, refraining from undue noise on Sunday, and confining their sexual ramblings to discreet voluntary encounters at night.

But even then there were vexatious problems that only grew more obvious with time. The English Victorian inventors of the public park wanted their creature to be two things at peace with each other: a field for staid entertainments, and a green chapel for quiet meditation upon nature and nature's God. Almost no downtown urban space anywhere, and certainly not Toronto's largest park (after the harbour islands), is vast enough to serve both purposes. We are gradually learning, as Montreal has learned from Frederick Law Olmsted's splendid Mount Royal park—intended as a sort of church for the adoration of nature—that the North American public park has been beset from its beginning by its inner contradictions.

So far, however, High Park has not sunk to the level of wretchedness notable in some of its American counterparts, and may yet be saved, at least in part. It is bearing the traffic of use better, in fact, than most parks—which is not saying it's bearing it well, or without the inflicting of slow, incremental damage. Crime and the destruction of delicate environments—the old oak groves, the unique blend of forest types—are the most serious symptoms of decline. But what really can be done, short of police-state control, the outlawing of cars, and putting most of its sensitive natural areas off limits to human use? The result, of course, would be the true end of the Victorian mixed-use park in our midst, the death of another idea from the past, and the hastening of High Park's slow decay into a collapsed moment of redemptive urbanism in the city's heart.

BANKING

When banking, my city-born Victorian grandmother always wore hat, gloves and girdle. She could not have foreseen a day when half-dressed and unkempt people would jump out of cars, stick plastic cards into an automated teller on a windy

street or in a convenience store, snaffle up the bills burped out
of these machines—and dare call this *banking*.

Now the notion that getting dressed up to cash a cheque
for twenty-five dollars might be snobbish would have struck
my grandmother as absurd. One dressed importantly to bank,
simply because banking—along with shopping, lunching,
attending Horticultural Society lectures and worshipping the
Divinity—was an important rite of urbanity, requiring one to
dress the part.

While such colourful seriousness has largely vanished from
our increasingly lax social culture, many a dignified Victorian
or Edwardian building still stands, often in the humblest of
neighbourhoods, as a reminder of a certain urban grace and
gravity we lost sometime around the middle of this century.

Vast, costly churches, for instance. Looming, turreted
public schools, tributes to the Elevating Powers of Education.
And, everywhere, banks in many architectural styles, but possessed of unvarying solemnity, appropriate to the respect they
enjoyed in Canadian culture when built, and to the importance people once ascribed to the rites conducted in them.

Look around any turn-of-the-century residential zone in a
Canadian city, and you soon find a bank that illustrates exactly
what I'm talking about. It will be discovered in one of two
forms. The less august of the commonest old bank-types is to
be found on the ground-level corner of a larger commercial
building.

Toronto has many good corner banks; were I pressed to
pick a personal favourite, it might be the Canadian Imperial
Bank of Commerce office at the intersection of Spadina
Avenue and Queen Street West. A work finished in 1903 by
establishment architect George W. Gouinlock, the bank graces
the intersection with an angled entrance guarded by two stout
Tuscan columns and watched over by a Renaissance Italianate
balcony. The deep-cut rectangular windows are similarly
framed by Tuscan columns, but "rusticated"—rude, primitive.

The result is a column that looks as though it were passing up through a series of fat Scrabble blocks. Such stylistic gestures, expressive of stolidity and "ancientness," are repeated across the whole office building's façade, impressing the prospective client with the antiquity and durability of the firm.

The more imposing of the two bank-types is the free-standing pavilion, impressively ornamented, and elevated above street level as a way to insist on the seriousness of the rituals of banking to be done within. In this category, I have many favourites that I never drive past without a nod of thanks: the Bank of Montreal's self-important little Roman mausoleum at the ignominious corner of Christie and Dupont Streets, for instance, and, in rundown Parkdale, the same institution's Greek temple, charmingly mongrelized by two vaguely oriental leaded domes stuck atop it.

For the record, however, Toronto's—perhaps even the world's—most exalted example of a bank branch in the traditional, conservative, free-standing "temple" tradition is neither old nor ponderous. Yet Mies van der Rohe's floating glass and steel banking pavilion at the Toronto-Dominion Centre outdoes all its predecessors, as an expression of banking's seriousness and urbanity. Despite the structure's Modernist formality—or, rather, because of it—my grandmother would have felt completely at ease doing business there.

By 1968, when the T-D pavilion was completed, the automated banking machine (ABM) was already making its way down the technological pipeline, headed for your street and Quik-Mart, and about to doom both the high rites of middle-class banking, and the imaginative architecture designed as a setting for them. Some twenty years after the introduction of ABMs in Canada, the number of bank branches across the country—about 7,600 at last count—is declining, while the 13,000 ABMs in use at present are multiplying like rabbits. Michael Bradley, an analyst with the Canadian Bankers Association and the source of these statistics, sees no end to the services that these machines may provide. We are likely soon to be using them for the processing of mortgage and loan applications, the dispensing of "postage stamps, travel services, transit passes, movie tickets," he says, in addition to coughing up bills and gulping down deposits.

The last time Bradley checked, there were some 14.5 million ABM cards in circulation in Canada. No Luddite, I have one and use it all the time—though with an increasing sense of unease about what this use portends for the future of urban rites. According to the Bankers Association, Canadians used ABMs for some 900 million financial transactions in 1992—about a 20 per cent jump over 1991. That was some 900 million transactions conducted without one person having to deal with another in a civilized manner. Or, to put it another way, that was close to a billion opportunities missed to perform the ordinary courtesies that have traditionally affirmed urban community and helped keep it intact.

But not to worry, says Michael Bradley. Even if we do most of our personal money dealings with machines, the branch and the face-to-face encounter will still have a place in the ecology of advanced capitalist finance: in the purchase of securities, for example, and the retailing of insurance.

Somehow I do not find this very reassuring. First, because any machine that can sell a transit ticket will eventually be able to sell an insurance policy. Second, because even if personal interchanges are still required by law for stock purchases and such, the branch in a less well-heeled neighbourhood, where people have little loose change to put into the equity market, is doomed. My much-liked Roman mausoleum at Christie and Dupont is slated for abandonment and probable demolition—if not this year, then someday not long off.

But the eventual end of all stately settings for city rites, including banking, is something we may as well get used to. Urban experience is everywhere changing from a complex web of publicly enacted social interactions—like going to church or synagogue services, or going to the bank, dressed up—to a diffuse, individualistic, informal and unritualized condition. It all may strike you, as it does me, as a denial of all the richness most of us moved off RR1 to Megalopolis to find. Every time I pass it, the garish back-lit plastic sign announcing 24-hour banking, stuck in the doorway of the splendid old CIBC branch at Queen and Spadina, strikes me, in words my grandmother might have used, as "simply not fitting."

THE ROBARTS

Not that anybody is going to put on a belated funny hat and pop open a magnum of Dom Perignon in joyous celebration upon hearing the news, but I feel compelled to note that 1993 was the twentieth anniversary of the University of Toronto's John P. Robarts Research Library. I do so because the building is the city's most universally despised item of architecture, and also because it is our most celebrated monument to despair over the decline in student respect for law, orderly education and responsible use of resources. Its users have always called it Fort Book. It is certainly more bunker than shrine of learning.

To create the Robarts Library, the university's principal repository of books, maps, microfilm, rare printed objects and other resource materials, the New York firm of Warner Burns Toan & Lunde first bulldozed a downtown Victorian residential block, then formed their structure by pouring an estimated 100,000 cubic yards of concrete on the site. This cement thug today looms fourteen storeys over the houses in what's left of its neighbourhood—an interesting architectural mix of comfortably middle-class domiciles and a few august mansions still left standing on St. George Street, staid Victorian-Gothic St. Thomas' Anglican Church, and an ambitious, largely failed scatter of Modernist academic buildings introduced after the Second World War.

Monstrously bigger than anything nearby, it is also monstrously uglier, and incoherent. Soaring towards the sky are vast snorkels and sharp-nosed ship's prows and defensive parapets, while, at street level, the library is clad with blockish concrete slabs that appear to have been designed to withstand heavy artillery bombardment. Until quite recently, there was no ground entrance.

Crooked exterior ramps lead the visitor up to exposed, bitterly windswept expanses of yet more concrete, the level top of the "platform" on which the library appears to stand. Once through the revolving glass doors, the patron is immediately confronted with more expanses, more streaked, shabby concrete walls, identical to those outside, and the unceasing mechanical rattle and grind of the escalators that hang in the structure's central well. The books lie beyond yet more barricades and cyclopean walls of concrete.

The honesty of the name for this type of ruthless construction—New Brutalism—may be due to the fact that the moniker was meant as a joke. In 1950, so the story goes, Swedish architect Hans Asplund offhandedly invented the term to describe the style of a house being designed by colleagues. The term later spread, via the international network of professional chit-chat, to England, where it was enthusiastically and seriously embraced by young architects.

It appears that there are now two subtypes of Brutalism, one grand, the other mean. The first is Brutalism plain and simple, a term used by enemies to nail the more horrifying ideas and projects of Swiss architect Le Corbusier. Not all Le Corbusier's notions of urban planning and building design are bad, and some of his projects will be thought-provoking for as long as people think about architecture at all. (I am thinking particularly of the splendid church at Ronchamp, France.) Having said that, Le Corbusier's so-called "Voisin Plan," unveiled in 1925—a project calling for the demolition of six hundred acres of central Paris, to make way for a camp of tall

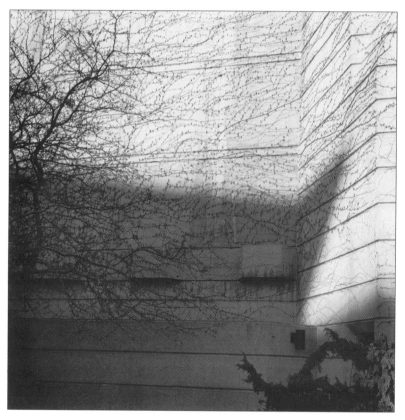

Creeper Vine, Robarts Library

towers X-shaped in plan, and a net of high-speed motor-ways—not only deserves the name Brutalism, but is horrific enough to merit an all-points police alert for the arrest of its mad creator.

He would probably get off with a parking-ticket, how-ever—since in the age of the Voisin Plan megalomanic archi-tectural proposals were thick on the ground, and all the rage. The New Brutalism, principally an English and British colo-nial tendency of the 1950s and 1960s, was "new" only inas-much as it was forced by circumstances to apply the old ruthlessness to sites much smaller than, say, all downtown Paris. During the Second World War, after all, the military had shown its superiority to architects when it came to devastat-ing vast areas. After Hiroshima and Dresden, architects real-ized they would have to be content with depredations on a small scale—the odd neighbourhood, downtown districts, and sometimes, as in the case of the Robarts Library, with just single city blocks.

The common denominators of every act of Brutalism, however, are the cement-mixer, the contempt for context, and the determination to devastate everything on a given site, then build monstrously and afresh from the mud up. British Columbia architectural historian Alan Gowans has given the wickedly funny name *Führerbunkerstil* to this manner of building. Curiously, it is perhaps the only defunct architectural style which brings to mind only melancholy thoughts, tinged with chronic depression. We are reminded, for instance, of the concrete Nazi pillboxes now lying half-buried under Normandy's coastal sands, structures movingly photographed and described by the French intellectual Paul Virillio. And then there are the seemingly indestructible Nazi-era concrete structures still uglifying some streets in Berlin, and, under-ground, disrupting the construction of new utility and metro systems—as though Third Reich nihilism was still at work, silently defying the practical re-creation of its principal city as the capital of a more humane state.

In *Styles and Types of North American Architecture,* Gowans chooses to illustrate, as his best example of New Brutalism, Australian architect John Andrew's Scarborough College, completed in 1965 in a far region of Toronto's eastward suburban sprawl. While I find justice in Gowans's choice, I am inclined to hold out for the Robarts as the best of Toronto's worst.

Yet the real question remains, not which is worse, but why such construction ever took place at all. Fighting my way towards the Robarts against the howling, bitterly cold winds of a nasty Toronto spring, I find myself thinking: How did so repellent an idea take root in architectural practice, spread, and find its way into the heart of the venerable, eclectically Victorian University of Toronto?

But if I understand the rationalizations given by its British theorists correctly, the question answers itself. It's the empirical "truth" it tells about the empirical world that is supposed to give Brutalism its cogency—its unmasking of the ideological obscurantism represented by, say, the University of Toronto's prevailing Academic Gothic. As Gowans explains, Brutalism "corresponded to a fashionable stance among sixties youths of 'brutal honesty,' 'no faking,' 'nothing plastic'…" It was the mirror image of the Victorian Picturesque, a discrediting of humanism in every tough, nasty angle and texture.

Though Gowans doesn't say so, Brutalism was also a quick and dirty way for an architect to gratify, safely, everyone. It pleased welfare-state bureaucrats, who wanted the shelters of postwar mass education built quickly and finished cheaply. It pleased university boards of trustees, by making buildings look like rough military defences of civilization against barbarian hordes of philistine students. And, surprisingly, it pleased the trustees' enemies, rabble-rousing anti-Establishment undergraduates of the 1960s, who thought they were seeing universities finally coming clean about their role as mere industries of knowledge, and assembly lines for information workers.

I intend never to stop mining the lode of some two million books in the Robarts Library, an activity which seems to me very much a part of being a civic, urban human being. And I hope never to lose my sense of revulsion at the wrapper thrown round these treasures twenty years ago—the kind of gut-wrench that reminds me that I still can recognize the architecture of mind control and crowd management when I see it, even when disguised as "truth to materials."

THE SKYDOME

While no important city has a domed stadium—in fact, having one is a sure sign of urban second-classhood—Toronto sports fans love our immense reinforced-concrete bunker, and virtually all the amenities. Well, maybe not the food, widely reputed to be vile—but almost everything else: the sightlines and seating, the weather shield provided by the $90-million retractable roof, the $17-million Jumbotron—a glorified scoreboard operated by a crew of up to eighteen technicians—the handy washrooms and so on.

Torontonians not inclined to be game-goers think of it, if at all, as a gleaming pimple down on the city's backside. Indeed, the SkyDome is a homely and bulky structure, with a bug-like humpiness made only more egregious by the herd of smaller buildings hemming it in, and the soaring needle of the CN Tower next door.

But though not much of a fan (except at World Series time), I hereby distance myself a good nine yards from the building's most dismissive critics, and point out two reasons it deserves fair recognition, if not exactly a twenty-one-gun salute.

The first reason is the sheer effectiveness with which the SkyDome's bulk hides its faintly sinister elegance, both from fans inside and from passers-by. I am not suggesting that architect Roderick Robbie or engineer Michael Allen intended the public to believe the SkyDome to be really as simple as it looks. But the basic message the public gets from the concrete exterior is that the Dome is just another dumb jock of a stadium, only a lot bigger than most.

I got my first glimpse of the truth about the SkyDome in the summer of 1993, on a tour with SkyDome vice-president David E. Garrick. Our winding route, most of it between the outer walls and inner surfaces of the stadium—vast tracts ordinarily off-limits to the game-going public—took us through the semi-darkness under huge moveable bleachers, along corridors washed in cold fluorescent light, through private boxes (rental fees: up to $225,000 a year) and dining rooms and clubs.

But the rooms I found most fascinating were those resembling little skulls, crammed with the building's many unsleeping cybernetic brains, and peopled by almost unmoving men and women, made to seem irrelevant by the flickering banks of computer screens, flashing images relayed from myriad video cameras installed throughout the SkyDome, dials and read-out gadgetry, churning data-processing mechanisms.

Among our stops was the room that contains the electronic unlocking and tracking mechanisms, and the single button, controlling the retraction and extension of the roof's steel parabolic arches along its track-ways. Another was the top-security cell housing the optical and sensing devices which monitor every movement, the operation of every human and mechanical system, throughout the vast building, even every elevator. A glance at computer screens lets you know instantly whether any given lift-door in the entire building is opened or closed.

We naturally visited the elaborate Jumbotron control studio, and the nerve centre of Dome Productions, a deluxe $13-million facility hard-wired to forty-seven permanent cameras positioned throughout the stadium, and capable of transmitting images of a game to satellites in space and thence to everyone on the planet with a dish or cable and a TV set.

SkyDome officials are naturally pleased to hear their building compared to the Roman Colosseum, dedicated by the Emperor Titus in AD 80, and like to press the analogy on visiting reporters. Yet no two buildings are less alike, except in the trivial sense that both are big, and both are made of cement (though the Colosseum was clad in marble). And both were constructed in the round, to create the feeling of community, the experience of seeing both the spectacle and one-self simultaneously, which has been historically required of all sports palaces. Anyway, it's the difference between them that points up why the SkyDome is special. The Colosseum, after all, was a royal propaganda building, elaborately and monumentally ornamented on the outside, while the rounded façade of the Dome is democratically plain and functional.

More important, the Colosseum was merely a set of structurally sound, stepped bleachers for patrons, surrounding a wood-plank stage for the brutal games an architectural historian has called "ritual re-enactments of Rome's humiliation of the nations of the world"—while the SkyDome is principally an intricate mechanism of surveillance sheathed in concrete. Inside are machines for monitoring and sensing and interpreting information gathered from the field, and from every other nook and cranny, and from every electric and mechanical system on the site.

Only a tiny fraction of the information gathered and stored in the Dome—the game happening on the field, and pretaped advertisements—is transmitted to fans on the Jumbotron, and to Canada and the world via satellite. Most of the data flowing silently behind the concrete skin into the monitoring stations is used to secure—or control, inconspicuously—the movement of people through the gates, in the public corridors, in the stands, and to regulate the electromechanical and electronic environment that surrounds everyone. Seen from the perspective of one of its immaculate, equipment-jammed control rooms, the SkyDome seems less like an updated version of the ponderous old Colosseum than a mock-up for an ultra-high-tech prison, more dedicated to surveillance than any recreational facility previously built.

I raise this matter, not to criticize the SkyDome for being so pervasively wired, but to suggest how *comfortable* we have become in the midst of so many watching eyes. It has never been a secret that virtually all so-called "public" space, not just the SkyDome, is now under optical surveillance; or that electro-magnetic information-gathering operates every time we use a credit card, borrow a book, or write a cheque. If we are more aware than ever nowadays that video cameras and sensing devices are aware of *us*—whether we're whipping along an expressway, waiting in a bank line-up, or out on the town— there are two ready explanations.

In the first place, the sensing gadgetry is everywhere. We tend, I believe, to view these machines as mere deterrents to *someone else's* criminal thoughts or intentions, and are disinclined to think that someone, somewhere, is actually sitting there, recording our movements on videotape. With those little cameras oscillating back and forth across the teller area, who, after all, would stick up a bank? Anyway, it seems almost impossible that there could be as many observers as there are cameras—at least until one visits the SkyDome, and witnesses firsthand the vast network of people working the optical surveillance mechanisms.

The second reason we are conscious of this as never before has to do with the trickle-down into popular culture of certain ideas in Michel Foucault's 1975 book *Discipline and Punish,* which treats the modern history of incarceration. Certainly the most notorious and suggestive chapter in this book has to do with British philosopher Jeremy Bentham's Panopticon, first proposed in 1791 as a novel form of prison architecture. The jail would be round, with cells radiating from a central observation centre. From that viewpoint, one guard could keep watch over all the jailbirds. But once the inmates came to believe someone was sitting in the darkness of the watching booth, no one at all would be necessary to keep the prison orderly; and the inmates would then become their own guards. (John Howard's 1838 Toronto city jail, an unfinished structure, was influenced by Benthamite prison theory.)

To the consternation of writers on contemporary urban culture, we citizens of the late twentieth century have become indifferent to the cameras and sensing gizmos almost continually around us, and thus become our own cops. Nor is this phenomenon anywhere more obvious than in Toronto.

But why here?

That glorious night in 1993, when Joe Carter slammed the tiny white dot over the wall, winning the Blue Jays their second consecutive World Series championship, Toronto went crazy. Or as crazy at it goes, which isn't very.

In stark contrast with the four thousand rampaging baseball fans clubbed and gassed into submission before they tore up Saskatoon that Saturday night, the million or so party animals who took to the Toronto streets were real Sunday-school picnickers, hooting and hollering and honking, but doing little damage to property or each other.

Who would have had the situation be otherwise? While competitive with Montreal in many things, sports included, no sensible Torontonian would ever want a Hogtown replay of the riot that swept rue Ste.-Catherine following the Canadiens' Stanley Cup win in June, 1993. We'd be shocked silly if something that ill-mannered happened here.

A reason can be found, I think, by watching the slo-mo replays of Joe Carter's terrific hit. Here is sports television at its most beautiful: Carter, unsure at first about what he'd done, surging mightily off the plate towards first base, then suddenly becoming aware and almost incandescent in that awareness, then flying, free, ecstatic.

Now rewind, play the footage again. Ignore Carter and look at what's happening behind him. Emotionless uniformed

Metro police, previously invisible, swarming out of dugouts onto the field and quickly lining up, facing the stands. A fan has somehow made it past the cordon, and is running alongside Carter when the player rounds third base. There's a rush on the frisky fan by cops—but the camera moves on with Carter, so we do not witness the fate of the fan.

But this exception reminded watchers of the rule. Had even a small number of fans decided to leap the wall and mob the Jays, the police could have done nothing at all to prevent them from doing so. A predictable inner restraint kept, and always keeps, such a decision from being made by Torontonians; and all goes more or less smoothly.

Ditto later on, when the post-game bash got going on lower Yonge Street, consecrated by generations of Torontonians as our promenade of playful rampage. In theory, the doings on Yonge could easily have become dangerous. Were this Detroit or Chicago or Montreal, they quite likely would have. But Toronto isn't, and the doings didn't. "Police who lined Yonge Street in case of a Chicago- or Montreal-style ruckus," we read in Monday's *Globe and Mail,* "seemed to turn a blind eye to those who swigged beer. They mingled with the crowd, as they high-fived or gave the thumbs-up signal to passers-by." Mel Lastman, the everlastingly boosterish mayor of North York pronounced with pride: "When people see us on CNN, they talk about how well behaved we are, what a world-class area Metropolitan Toronto is. It's good for business, it's good for tourism."

Yet once again, the niceness reminded this city-watcher of Toronto's never-slackening sense of righteous force, ever poised for instant deployment against wrongdoers. This comforting thought is gratefully noted by visitors up from the Republic of Fear. The rarity of overt violence by the Toronto police—against white people, anyway—coupled with their omnipresence and unfailing smiles, their firm pleasantness in crowd situations, only makes the atmosphere of dread, and each citizen's duty to behave himself decorously and carefully in public, more keenly felt.

But woe betide the non-conformist in our benign Toronto police state!

The culprit I'm thinking of here is *Globe and Mail* reporter Kirk Makin, who supplied readers with a delightful World Series column called "The Fan," about being just that, throughout the playoffs. Perhaps so the burghers can sleep soundly again, Makin confessed all in his "Fan" the Monday after.

It appears that Makin took a home-video camera to the game that Saturday night and actually used it. He was immediately spotted by the good-behaviour boys, and apprehended. "My moving appeal to the home-town loyalties of an officious security twerp and a Metro Toronto cop fell on deaf ears," Makin writes. "Their steely grip on my arms, I was thrown out of the SkyDome in front of 20,000 revellers. Bad boy, go home."

My only regret about this episode—other than Makin's ouster—was that our reporter did not get caught videocamming *in flagrante delicto* by the roving eye of network TV. It would have happily besmirched the ghastly PR picture of Toronto as a "well-behaved," "world-class area" in American eyes, and proved that even a Torontonian can act really bad when he feels like it.

I am nevertheless happy to report that such a thing did happen during gametime one night in early 1990, just after the SkyDome opened for business.

While engaging to most fans, the baseball game had become boring to a man and woman lodged in one of the expensive, large-windowed hotel rooms overlooking the field—or so we concluded, when they casually doffed their togs and did It, in full view of forty thousand spectators.

This sexual coupling may well have been seen directly by more people than any other sex-act in world history. High historical significance aside, however, many patrons of the stadium reportedly got a real kick out of it—and probably nobody was really scandalized.

Nobody, that is, except SkyDome brass, who reacted with indignant, horror-stricken splutterings to the news media, and the issuance of stringent new rules governing guest conduct in the hotel rooms. "After all," reported *Globe and Mail* sports columnist Stephen Brunt with a smile, "a quickie today could turn into orgies by the all-star break and then what? The moral fabric of the town might be forever torn asunder."

That must not, of course, be allowed to happen. We are very *good* in Toronto. But on behalf of forty thousand delighted spectators, I'd like to thank the randy couple for doing something spontaneous in the one Toronto structure constructed precisely to prevent the unexpected from ever, ever happening.

Moderne *Variations*

R.C. Harris Filtration Plant

THE R.C. HARRIS FILTRATION PLANT

A few summers ago, friends handed me the assignment of showing the best of Toronto to their Parisian cousins. Father was a professor of Italian social history, mother an art historian, teenaged daughter an uncertain university student, pre-teen son continually on the lookout for McDonald's golden arches. I got just one afternoon to do the job, so deciding what to show them was easy. First, whatever it was, it had to be Depression Modern, simply because Toronto is peculiarly blessed with structures designed in this conservative interwar manner, which preceded the onslaught of full-scale glass and steel Modernism after World War Two. Surveying the available *moderne* variations, I had little trouble coming up with a choice: the R. C. Harris Filtration Plant, our Greta Garbo of public architecture, and among Toronto's grandiose expressions of building in the Depression Modern style.

The immaculately manicured site, just east of the Toronto city limit on the Scarborough bluffs, has long been cherished for its panoramic overlook of Lake Ontario. Between 1878 and 1906 a rambunctious amusement park called Victoria Park operated there. Though gone for almost a century, its name survives in Victoria Park Avenue, the long north-south street that ends there. And, to the endless exasperation of vociferous community groups in the adjoining Beach neighbourhood, its reputation as a terrific place to get drunk and party wild and long into the night has persisted in the folk memory of local teen culture down to the present day. The complex of buildings, begun in the years between 1933 and 1935, and completed in the 1950s, was first known as the Victoria Park Pumping Station; it was given the name it now bears in 1946, to honour Rowland Caldwell Harris, czar of Toronto public works from 1912 until 1945.

No lifeless relic of yesteryear's technology and architecture, it remains fully operational, pulling in, processing and pumping out an average 140 million gallons of fluoridated, chlorinated lake water each day, or roughly half the water required by the lawns and begonias, coffee-pots and showers, toilets and kitchen taps and public water fountains of Metropolitan Toronto.

In this array of buildings, closest to the lake stands the tall-windowed pavilion sheltering the huge intake turbines, sumptuously appointed inside with brass railings and brackets, and a wall-mounted panel of great brass-ringed numbers, back-lit to show which mighty pumps are operating. Outside, the limestone cladding bears carved bas-relief patterns of sleek, simplified turbines and waterfalls. Nearby a tall tank of alum, a chemical required in the purification process, rises camouflaged inside a slender tower reminscent—but only vaguely reminiscent, as typical of most Depression Modernism—of a Roman watchtower.

A tall, rounded entrance gives the finely balanced plant administration building the general look of a triumphal arch, fronted by a fountain and by a stately balustraded belvedere from which to view the lake—on a clear summer's evening, one imagines, in formal attire.

The two skylit wings of the filtration hall crown the green slope. The hall's immensely long galleria is among Toronto's most stunning interiors, sumptuously appointed with beige marble flooring and trimmed with brass inlays and fixtures, and lined with costly green marble. Exactly in the middle of this dramatic expanse is a sleek, sculpted tower and bank of dials and switches.

Sun-drenched on a cloudless summer day, the sweep of its buildings drawn hard against blue sky and immaculate green lawns, Lake Ontario at its foot, the site of the filtration plant is one of the grand overlooks in a city otherwise notably short of natural summits. We all love it—party animals, picnickers, sun-bathers, people who just want to sit alone, and gaze at the lake's blue waters and think. My Parisian guests were enchanted, by the way—especially by the interior contrasts between the swank decor and mighty engines, ducts and tanks.

For those interested in the history of building, the R. C. Harris plant holds still more charms, and even a pleasant bit of nostalgia for a modern architectural path that has been abandoned. For this coordinated ensemble of structures comes at the climax, and almost the end, of the great era of stream-lined, subtly decorated, stripped-down-classical waterworks, incinerators, pumping stations, power plants and other public utilities—architectural works which sum up for us the patri-otic, heroic aspirations of R. C. Harris and public servants like him. Their projects were grandiose; but for cities crushed by the Great Depression, as Toronto was, the grandeur of the Victoria Park Pumping Station was exactly what was needed to return to a defeated citizenry the reasonable hope for a decent life. Harris's choice of a crisply modernized Roman shell for his water plant—a late application of academic styling to contemporary building—was part of this architectural scheme "to proclaim the grandeur of the city," as historian Alan Gowans says of all academic architecture, "to help citi-zens perceive themselves as part of it, to make individual lives somehow nobler by being set in relationship to a grand past."

The use of streamlined Classicism for reviving patriotism and public spirit was pervasive in Depression-era Canada and Franklin Roosevelt's America, but was hardly limited to the democracies. Its widespread deployment by populist totalitarianisms in the years between the wars—whether Nazi, Fascist, or New-Deal—was the key force behind the style's sudden obsolescence in North America after VE Day. By the mid-1950s, when the R. C. Harris plant was being completed, its chic styling and all it stood for—strong national revival, government-sponsored public order and uplift, renewed mass politics—was dead. The current moment, and the next twenty years of North American architecture belonged to the presumably anti-ideological International Style introduced to the United States by Mies van der Rohe, Walter Gropius and Marcel Breuer and adopted with enthusiasm by the post-war capitalist élite.

Yet the wheel of fortune never ceases to turn, and the International Style and all it stood for are now things of past affection—to be admired for their ambitions, and studied for the lessons they teach us, but no more to be emulated than the Depression-era styling and ideological scheme of the R. C. Harris Filtration Plant. Let us praise them both, for their expansive public ambitions, their vast intellectual programmes, and the contributions they have made to the built ensemble of urban forms we inhabit.

620 UNIVERSITY AVENUE

To my knowledge there is no corporate site anywhere quite like the one on which the successive office blocks of Ontario Hydro stand shoulder to shoulder on the west side of University Avenue immediately south of College Street. Here, between 1915 and 1975, Hydro built three head office buildings. Not clumsy add-ons or replacements, but three distinct, distinctive buildings, together comprising a remarkable anthology of our century's changing ideals for the architecture of public works. It is a collection also undergoing drastic internal revision, as the Toronto firm of Zeidler Roberts prepares to drop the new nineteen-storey Princess Margaret Hospital inside and behind the two older Hydro headquarters.

Few people will rejoice to learn that the third Hydro office building will remain untouched. I belong to the tiny minority that likes it. Opened in 1975, this thick concave arc of mirroring glass, curving around the intersection of University and College, never fails to remind me of the Horseshoe Falls at Niagara—arc, cascading sheen and all—though I imagine the Falls were not on the mind of the glass-bedazzled architect (Kenneth R. Cooper) who designed it.

You would not necessarily be aware of all the change going on, while motoring up or down University Avenue in rush-hour traffic. In fact, as important downtown dig-ups and knock-downs go, site preparation for the new medical treatment and research facility has been notably inconspicuous, because the work has been mostly veiled by the existing structures the hospital will eventually absorb. The main giveaway that something's afoot is the heavy web of steel girdles and garters holding up, until recently, an ornate Italian-Renaissance façade, the lingering architectural smile of the first headquarters at 610 University, now otherwise vanished.

Commissioned by Adam Beck, Hydro's visionary and despotical chairman, in 1915—five years after the utility started delivering power to Ontario communities—George Gouinlock's Hydro-Electric Building is (or was) a pretty, academic étude on Florentine themes, advertising the renaissance in electrification about to dawn all over Ontario by referring us back to an older, grander Renaissance. We should spend no time mourning the loss of the indifferent building itself, especially since the hospital master-builders have decided to attach the nicest thing about it, its exuberant Italianate mask, to the street-front of their facility.

Luckier than 610 University is the Depression-Modern edifice at 620, Hydro's second head office, destined to survive intact as a hospital office tower. Put up six storeys high in 1935 by Sproatt & Rolph, and topped off by the same firm in 1945 with a tower of ten more storeys, the Ontario Hydro Building is among Toronto's most comely stepped tower blocks. The ascent of smooth, warm stone moves up from the sidewalk along lovely steps, while the lower courses of the building are graced by expressive ornaments of open-handed refinement. Water imagery is everywhere: in abstracted cascades, carved on the streamlined classical pilasters on either side of the entrance; in the shimmering curtain of waving glass brick under which one enters the marble lobby; in wave-forms cast in the horizontal bronze elements which accent and lightly restrain the gracious visual upthrust of the exterior cladding.

It has become customary to emphasize the progressive, upward-yearning aspects of stepped structures, which were first seriously proposed by Louis Sullivan in 1891, an immediate hit with architects and their clients, and, at best, still among the finest results of twentieth-century building art. We are inclined to cherish the soaring sophistication and sculpted styling of such buildings, which have been made to seem only more attractive by the classic postwar office slab, bluntly confronting the sidewalk, or retiring from the street behind windy plaza. Such admiration of a beautiful object from the past is entirely appropriate, though one should not see a proclamation of heroic triumph in the sonorous vertical lilt of 620 University. From the standpoint of architectural history, it dates from the twilight of the expressive North American skyscraper construction in the stepped Sullivan manner, with its judicious ornamentation, street-aligned façade, allusions to past glories of civil engineering in the service of the people. Much of the beauty of 620 University is in its affirmative conservatism, and its skilful visual management of upthrust on a comparatively low scale.

But to understand 620 University fully, something of its immediate historical context should be known. By 1935, when the office building was opened, Adam Beck's dreams of universal, free electrification had been brought near death's door by the Great Depression. Plans for expanding the provincial power grid were stalled, some 134,000 Ontario farms were still without electricity, and the whole idea of "power at cost" had fizzled out. Seen against the storm clouds of dire days, the Ontario Hydro Building seems less like a victory monument than a promise to keep the progressive faith in the midst of evil times. The incorporation of noble touches, fluted columns and dignified, balanced ornaments recalls commitments to what endures, as the water motifs are pointed reminders of the mighty torrents that never cease, even when humans are temporarily unable to harness their immense power.

THE HORSE PALACE

Although railway trains had replaced horses in inter-city haulage as early as mid-Victorian times, it wasn't so long ago that this most beautiful and intelligent of domesticated animals was still the principal engine for moving people and things around town, hence a common sight on Toronto streets. The combustion engines first startled big-city working horses, then abruptly displaced them, leaving only the mounts of Toronto's police on our streets, and those ponies now used for dragging tourist-laden buggies around Olde York on summer evenings. But if the civilian working horse has disappeared from the city, the fascination of urbanites with horses has not. Hence, the electricity that seems to flicker in the cold air down at Exhibition Place every November, as the annual Royal Agricultural Winter Fair gets ready to kick off its world-famous Royal Horse Show.

The first event is usually the $5,000 McDougald Open Jumper. But the official opening event—the $150,000 Crown Royal Cup Finale, often featuring such international equestrian superstars as Ian Millar and Margie Goldstein and Mark Leone—is what we all really wait for, and always find an unsurpassable thrill.

The steeds seen over the ten days of the Royal are, as you might expect, the peerage of the international equine world. Unlike most other revolutions, the one initiated by the automobile quickly eliminated the proletarians and left only the nobles, who now enjoy endless brushing and cossetting by their human servants in luxuriously appointed stalls.

The revolution that swept away the working horse was, of course, long over by 1931, when City of Toronto architect J. J. Woolnough finished building the first suitable temporary quarters at Exhibition Place for these splendid animals. If the excellent swine that turn up each year at the Royal get rather mean little pens, beneath their dignity, and the bovine breeding stock are assigned merely a bed of straw in the drafty, unadorned eastern shed of the Coliseum, the equine visitors and their human valets are quartered in Woolnough's Horse Palace, which may well be the most illustrious and interesting hotel for horses in the world.

With 1,200 stalls in several sizes and designs spread over eight acres, the Horse Palace proclaims the importance of both horse and Horse Show in every graceful detail and dashing line. Woolnough dressed its plain steel box-frame with a smooth skin of Queenston limestone trimmed in copper, bronze and brick, all of it ornamented with stylish equine motifs. A bas-relief frieze of horse heads parallels and accents a horizontal course of windows, full-bodied prancers adorn the dramatic stone curves of the Palace's main entrance. The doors are of glass, decorated with red and black woodwork. And, with a touch of humour we would not expect from a sober sided official architect, bas-relief horses' rears—to be sure, so streamlined one can just make out what they are—grace the exterior entrances to the toilets.

Inside the Palace, we find the same careful attention to

detail. The charcoal-dark elm wood and cream-painted *mod-erne* ironwork of the stalls provide a luxuriant visual harmony, while the sleek, ceremonial banisters on the ramps connecting the vast structure's two principal floors also declare this a palace in more than just name. The lighting, naturally, is neon.

In the late 1980s, the City officially put the Horse Palace under the protective wing of the Ontario Heritage Act. But, as we find in the official memos leading up to the structure's designation, nobody appears to have been quite sure what to call its distinctive architectural styling. Should it just be called Modern? After all, an article in the September, 1931, issue of *Construction* magazine did hail the Palace as "modernity itself." But other writers, responding perhaps more to the building's delightful Jazz-Age ornamentation than to its resolutely Depression-Era horizontal streamlining, have tended to use the term Art Deco, from the 1920s, for it.

In fact, the Horse Palace was built in the midst of the architecturally uncertain period between the Crash, which ended the feverishly exuberant Twenties and took the always costly, high-style Art Deco with it, and the 1933 Chicago world's fair, which was to set the tone for the super-stream-lined, sleekly unornamented nationalist fashion in everything from toasters to skyscrapers in North America for the next decade. Commissioned to make a grand building on the very cusp of two historical epochs—a collapsing era of luxury and display, a rising era of shortage and constraint—the earnest but pedestrian mind of J.J. Woolnough came up with a design of such nice ambiguity that it shouldn't work. But it does work, beautifully—expressing in ornament and line both the passing extravagance of that new, horsey money which brought the Royal into existence in the early 1920s, and antic-ipating, with surprising accuracy, sweepingly stern, horizontal post-Chicago styling at the same time.

Not that the magnificent and expensive horses housed

there care much about architecture, or about anything else except the regal treatment they're used to. But horsey Torontonians curious about architecture may well want to give the Horse Palace a closer look than usual from their limousine windows, as they sweep onto the grounds for the next Royal horse gala.

MAPLE LEAF GARDENS

There are reasons for ignoring Maple Leaf Gardens. Maybe, like me, you're not a hockey fan. In that case, the historical detail that Foster Hewitt broadcast the play-by-plays to all Canada from the Gardens' booth for fifty years is just a fact, not a holy truth of Canadian Identity. And unless you're heavily into nostalgic trivia, the facts that Elvis wiggled, Billy Graham preached, Buster Crabbe back-stroked and manic girls screamed for the Beatles' bodies under the Gardens' roof won't be tidbits to cherish forever.

You may, however, have noted that mammoth, routine modern buildings on the slide—even places as widely venerated as the Gardens (or, say, Exhibition Place)—never age with ruinous and melancholy grace. They just get shabby and faded, and appear all tuckered out, like the down-and-outs hustling spare change on nearby Yonge Street.

Or at least such were my thoughts about Maple Leaf Gardens before I stopped and took a long look at the inside and outside of what I'd been casually ignoring for years. For those who appreciate ambitious modern engineering, the interior of this 16,000-seat arena is the most interesting part of such a look-see. There is drama in the canopy of corrugated metal, held aloft, without internal columns, by four arched ribs bearing all the ceiling weight down on four concrete stumps, one at each of the building's corners. In an era that's brought us the telescoping roof of the SkyDome, the sight of the Gardens' roof won't be a transcendental experience, but that cover is still one of Toronto's most inspiring interior instances of steel's defiance of gravity, and its almost animal will to leap across space—a property of this wonderful modern material noted ever since the Brooklyn Bridge and the Eiffel Tower.

To those more interested in architectural styling than in engineering, and who've never given the Gardens' style much thought, a close look at the exterior of this huge box will come as yet another kind of pleasant surprise. But the finesse of the exterior is hardly news. During the recent controversy over whether the 1931 structure should be named an official "heritage property" by the city of Toronto—eventually it was designated just that—consultant A. M. de Fort-Menares effectively pointed out how the streamlined upsweeps of brick and window elements were joined with strong horizontals of grey concrete and "stylishly modern detailing" to make a distinctive building both staidly traditional and cautiously *moderne.*

But in her report, Fort-Menares also argued against giving the Gardens an exalted place in the story of Toronto's large-scale architecture done between the world wars. Compared with what used to be Eaton's flagship College Street store, a block away—a work of very high quality done by the same designers—the Gardens is characterized by "coarseness of materials," "utter functionalism" in its interior arrangements and "superficiality" of exterior ornament.

And if it wasn't great even on opening day, the building was only made tackier by thoughtless changes to the facade, and by numerous needless uglifications, removals and renovations. Inside, the rich curtains of the top rank of windows, and the huge ceremonial portrait of the sovereign, were swept away to make room for more seats. And outside, there are those hideous panels of mortared stone chunks adorning the entrance to the private Hot Stove Club.

Then we have the more general havoc wreaked on the Gardens in the 1970s, when the bosses decided to clean the building by sand-blasting, thereby blowing away the cladding's distinctive glaze and dulling the porous yellow brick, and making it defenceless before all the city acid and dirt. (The more recent cleaning, completed only a year or so ago, was done with gentler, water-soluble detergents.) The Gardens is more architecturally engaging than it's reputed to be, yet hardly as fine as Toronto's finest examples of interwar architecture. But who cares? When you're talking about the Lourdes of hockey, it's the miracle of winning in overtime that matters to those millions of Canadians glued to their TV sets, not whether the Art Deco styling in the brickwork over the marquee is quite up to snuff.

If originally a sceptic about the Gardens—or, more accurately, an indifferent bystander—I am no longer. After seeing the place, and reading its history, I now understand the broadcast gasp of CBC's Morningside host, Peter Gzowski, upon his learning that Maple Leaf Gardens had been given official "designation as an historic building" by the Toronto Historical Board. "Designation?" Gzowski spluttered in disbelief. "It didn't need designation! It's always been an historic building!" Much the same note was struck, in more measured language, by Toronto historian Michael Bliss in an impassioned 1989 memorandum to the Historical Board, at a moment when demolition was threatened. "Maple Leaf Gardens is one of this country's most historic buildings...the cathedral of Canadian hockey during the sport's Golden Age," wrote Bliss, "a time when hockey became part of the fabric of the culture for millions and millions of Canadians." Being a young country, Canada has few symbols of national identity. "If in 1989 we no longer believe that Maple Leaf Gardens has been a historically significant building—if we are happy to let it disappear under the wreckers' ball—then we might as well give up."

Well, the building did not disappear, as its owners wished and wish it would, and Canada did not give up. And at least one outlander, this writer, finally figured out why the Gardens must never go, so long as there are people alive who remember it in its most happy moments.

Now if you're wondering why a huge, steel-roofed square box in Toronto should get the name Gardens—spelled always with an "s" but always used in sentences as a singular noun—the story goes like this.

It started with the designation of a famous sports facility in New York as Madison Square Garden (spelled without an "s"). Madison Square wasn't a garden either, but it was built on the site of an amusement park called Gilmore's Gardens. In 1912, Toronto borrowed the name, or nearly, when the Montreal firm of Ross and Macdonald built a sports palace called Arena Gardens, on Mutual Street. In a little over five months in 1931, the same firm built Maple Leaf Gardens, for the same ends as Madison Square Garden, but adding the "s" from Arena Gardens (demolished 1989). All of which may explain how gardening and hockey got mixed up in the Canadian imagination.

You always wanted to know that. You really did.

PARK LANE

In a tribute to the brilliant pianist and theorist of technology Glenn Gould, *Globe and Mail* music critic Robert Everett-Green observed that the artist's "playing was a thoroughly modern phenomenon, more akin to the uncluttered lines of modernist architecture than to the Beaux-Arts exuberance of, say, a Rubinstein."

As it happened, Gould spent the last twenty years of his life in a spectacularly messy penthouse atop Park Lane Apartments, a Depression Modern building on St. Clair Avenue West which curiously adumbrates the performance style Everett-Green described. It is a structure notable, not only for the genius who lived there, but for its own linear rigour and its pragmatic clarity. We could only wish that its influence on the street had been greater.

The area I am talking about stands on both sides of St. Clair, between Yonge Street and Avenue Road—once a smart Easter Parade of stylish apartment blocks, now a stern display of important early experiments in Toronto's postwar architectural Modernism. But if the broad street on the south border of toney Forest Hill and Deer Park has been allowed to change from its former chic to operational, the Park Lane, completed in 1938 by Toronto architects Forsey Page and Steele, continues to hold its own as a fine, modest instance of the high *Moderne,* complete with most of the decorative gestures we look for in deluxe apartment construction in its day.

Project architect Harland Steele clad his symmetrical, U-shaped building with dark-brown brick, streamlined and stripped of ornament, other than firm horizontal courses of raised parallel brickwork binding the exterior all round. The entrance is, predictably, a ceremonial piece confected of chromed steel and fluted concrete posts. To enhance the staid drama of this ensemble, Steele let horizontal sweeps of windows move the eye effectively out of the sheltered, set-back entrance, around the inner corner-curves of the structure, to the straight walls of its two wings, fronting the St. Clair sidewalk. Steele's stripe of windows, varying in size and proportion while keeping to the general horizontal programme of the architecture, creates a melodious counterpoint to the flattish brick facing. (The crisp effect of these windows was muted in 1987, when the glass and horizontal mullions, decreed by Steele to echo the brickwork, were replaced by glazing set in vertical frames.)

The happiest architectural feature of Park Lane is the near-universal trademark of 1930s, the swept-around corner. Whether the object being styled was your neighbourhood Trans-Lux movie theatre, a TWA Dixie Clipper aircraft, a cocktail lounge with a sinuously flowing Flexwood bar

Glenn Gould entry

trimmed with opalescent bronze or a Chrysler Airflow sedan, the soft corner was the sign of the deluxe, and of the technologically up-to-date. For the first time in the history of masonry building, and, if only for a moment, the rigidly angled corner was out.

It is difficult to view this apartment block, still neat and smart as a plucked eyebrow, without a twinge of regret. Park Lane stands at the end of the history of opaque, stylishly sophisticated masonry building. In 1938, the style's streamlined external decor and rounded corners, small-scale comfortable living in modern circumstances, were all enjoying their last moments of timeliness. Next would come the era of glass, flat tops and steel frames. Luxurious exterior styling would become a heresy, to be stamped out; and stamped out it was, if the relatively few instances of fine Depression Modern survivals in Toronto are any evidence.

What more appropriate living quarters could have been found by Glenn Gould, this "invisible man" who deserted the concert stage for a life of Garbo-like reclusiveness, than an aloof, exclusive apartment building of the Garbo era? If, as Everett-Green argues, Gould set up "technological filters...between himself and the world," he also chose a domicile constructed as a nearly opaque architectural filter between interior and street—a gesture of construction repugnant to the "democratic" builders of transparent glass and steel slabs who would soon dominate the provision of apartment towers.

Gould was not destined for the revealing eye of the television camera, or for its architectural analogue, the revealing modernist curtain wall of glass. He was born, writes Everett-Green, "for the microphone, and more especially for magnetic tape. His personal distaste for the concert experience, and its potential for accidents, led him naturally toward a technology that invited second thoughts. Editing a bundle of tapes into one performance was not 'cheating,' but putting the studio to its rightful use, which was the creation of an artifact, not an event...

"Gould's instincts led him toward a full-blown ethics of technology. Machines and technology, he wrote, make us better by distancing us from our 'animal response to confrontation.'" As it happened, Glenn Gould lived in an apartment building constructed in the last important style of apartment architecture to propose distance and apartness as existential qualities highly desirable for *wholly intentional* living. At Park Lane, there is simply no question of the free movement of intellectual unequals, of those opportunities for mixing and sociability suggested alike by the windy plazas and cozy, crowded streets beloved by postwar Modernism's theorists and critics alike; any more than there is any tolerance for a dropped hem, or *public* failure of style. Unlike a microphone, Park Lane is an expressive artifact in its own right—declaring in no uncertain terms the distinctiveness of the people who live within from those who live without, and, in its sophisticated and haughty design, their *disdain* for active community with whoever chances to be passing by on the street.

Now we know that in his utopian writings on technology, as opposed to his mandarin performance practice, Gould declared himself on the side of interaction, smudged bounds between artist and audience, and other pieties of the 1960s. "The irony," says Everett-Green, "is that the man who floated the idea of creative anonymity, of blurred boundaries between performer and listener, has been put on a pedestal, and his every recorded note made sacred." But what besides posthumous idolatry could one reasonably expect of an artist who chose to live his last twenty years within *these* stylish and exclusive walls?

High Styles

Storm Windows, St. James Cathedral

GOTHIC

However tenuous or non-existent one's relation to Christianity, and even if you've never set foot in one, I suspect you'd have no trouble singling out a church on a city street. Scale—the fact that the building is conspicuously bigger than what's around it—is important to the idea of churchliness, but not central. After all, some modern, low churches in the suburbs could easily be mistaken for hockey rinks or Legion halls, were it not for the steeple.

The steeple, then, is crucial to the basic package, and so is more or less strict autonomy of the building on its site. Once a congregation is past the store-front stage, only a free-standing edifice will do, preferably as deluxe as possible. It's curious. I could name several good reasons why a new Christian congregation should think about joining forces with a developer on a mixed commercial/residential/ecclesiastical complex, rather than enthrone an independent church on its own lawn.

A few years ago, a Toronto real-estate developer struck a $3-million purchase and lease-back deal with the Church of the Redeemer, at the corner of Bloor Street West and Avenue Road. As the mixed-use tower went up, it got me wondering why the new building wasn't allowed simply to engulf the piquant 1879 church, including its embellished Victorian interior. As things turned out, the church's Gothic exterior was left to stand primly separate and intact, with the new structure rising behind it.

Most people I know, regardless of their belief or lack thereof, would find the thought of integrating an old church with a new commercial development odd. The contemporary idea of what a church should look and be like, a little informal polling reveals, is extraordinarily rigid, even (and perhaps especially) in the minds of people who give little or no thought to what goes on inside these cherished structures. The model appears to be based on the oldest and most conspicuous ecclesiastical edifices people have grown up with, whether or not they attend the services inside.

In Toronto, that means Victorian Gothic buildings which, while venerated, are not terribly old: the Ohio stone and Toronto brick Cathedral Church of St. James (1849–1874), for example; or St. Michael's Roman Catholic Cathedral (1845–1890), or St. Basil's Church (1856-1895), among others. Myriad older and similarly serious secular bits of Victorian architecture have been demolished without public outcry. These churches, however, will almost certainly stand until their brick facings can no longer be prevented from crumbling into dust.

One could argue, I suppose, that a new country needs such pseudo-ancient edifices—theme parks of geographically remote and presumably unchanging spirituality, as it were—to give it an instant "past," even if the past is stitched together from that of another country. Then there's the argument, more attractive to secular moralizers, that such buildings are necessary, since they serve as visible advertisements for conservative religious conviction and unquestioning loyalty to Crown and country, virtues obviously in quite short supply these days. Viewed from such perspectives, St. James' Cathedral is a hymn in brick to the ethical and social values of the Victorian Empire, from its scrupulously correct imitation of the Collegiate Gothic manner to its magisterial and aloof attitude, and to the top of its exceptionally graceful, lofty spire. One can hardly fail to be impressed by evident dignity, and made to remember the good works of the John Strachan, first bishop of Toronto, who is buried under the altar.

William Thomas's Cathedral of St. Michael provides our best Roman Catholic rehearsal of the same tropes: lofty-minded Englishry, expressed in its conscious mimicry of York Minster, imposing reminders of antique craftsmanship, the loyal mid-Victorian "Gothick" thought to be the true architectural style of the Age of Faith, and clearly meant here to declare the political loyalty of Christians who were still objects of suspicion and prejudice by Protestants.

Do these architectural stylistics really express the triumph of Christian faith and order in the wilderness? I am inclined to believe the answer is no. All these model churches, it is worth remembering, as well as dozens of "Gothic" and "Romanesque" ones which went up at the same time or shortly thereafter in Toronto, were built during the mid-Victorian collapse of popular religious faith. For more than a century before the present fabric of St. James' Cathedral came to be, intellectual, then industrial, modernity had been whittling away at traditional notions of transcendence and Divine Providence. Science was making scepticism and the sceptical method increasingly attractive to the brightest minds of the age, and pressing home the point that humankind had created itself by means of technology and ingenuity, and without help from a benevolent Creator.

But if any event can be said to have announced the city's final changeover from a colonial preserve of Anglican attitudes and proprieties to a profane, modern city, it was the provincial act of 1849 which radically secularized the University of Toronto. For John Strachan, consecrated first bishop of Toronto ten years before, the legislation was a catastrophe and disgrace, and also the immediate occasion of his last great architectural and intellectual project. It was to be the founding of Trinity College, upon what he termed the "clear and unequivocal principle" of loyalty to the Church of England. The college was begun on Queen Street West in 1851 by architect Kivas Tully, who understood well the power of the English Gothic to declare a religious institution's stand in the apocalpytic match between Christian Faith and godless Science. It is no coincidence that the ground was broken for Toronto's most important Gothic buildings in the years just before and just after the University Act of 1849.

Looked at against this background of deepening, rapidly spreading doubt and religious indifference, Toronto's early models for churchliness seem to be desperate rearguard actions rather than declarations of victory, and quite open acknowledgments of just how deep a rift had opened between factory and rectory, the culture of technology and the increasingly non-credible culture of religious institutions. For whatever reasons, Western Christianity failed to understand and confront head-on the nineteenth-century technological and intellectual revolution of secularity, and effectively combat its sweeping claims to universal enlightenment and mental emancipation.

But the struggle is hardly over. Indeed, a principled dislike and disregard for technological modernity, and a persuasive critique of it, seem to be gaining ground in large regions of the world—in the West weakly, under the banner of ecological awareness, architectural preservationism and "Green" consciousness generally; and strongly in those areas dominated by reviving Islam. But whatever the future holds, the present state of Western churches has long been one of retreat and decline, and of reluctant abandonment of cultural leadership—though the churches have continued, despite all, to build refuges from the stuff of Gothic mist, mystery and past architecture carefully studied and reconstructed, and assume the role of dispensing Band-Aids to the victims of mass modernization from the safety of the sidelines. The deliberately archaic Victorian churches in our midst, while beautifying the grey, tight urban fabric, also are signs of a disastrous surrender to a secularizing *Zeitgeist* that now seems not nearly so irresistible or overwhelming as it once did.

ROMANESQUE

If every city site expresses a state of mind, a literary genre, and a time of day, then Toronto's buildings in the Richardsonian Romanesque style express the irrational and uncanny, romance, and a mystic twilight—all of it pointedly opposed to the noonday rationality symbolized by the Classical. This is a tale of these places of eternal dusk, where massed dark stone volumes loom, and rounded portals and windows deep-cut into walls of quarry-rough rock conceal cool, mysterious rooms within, where a certain grand gloom, compounded of Dark Age barbarity and a sense of rude Christian faith, hangs heavy in the air.

Toronto has many such buildings. But at the time Henry Hobson Richardson died, in 1886, Toronto's Richardsonian moment was still yet to come; and this consummate Romantic architectural imagery was largely a spent force in the United States, its home.

The judgment of Henry-Russell Hitchcock, delivered in 1936 on the occasion of a tribute to Richardson mounted by New York's Museum of Modern Art, still rings true, if harsh, in our ears: "(Richardson) was not the first modern architect: he was the last great traditional architect; a reformer and not an initiator. Dying when he did, his architecture remained entirely within the historic past of traditional masonry architecture, cut off almost entirely from the new cycle which extends from the mid-eighties into the twenties of the present century."

The "great cycle" Hitchcock is talking about is the one set rolling when U.S. and European architects discovered the lightness and great height that could be achieved by the use of steel and elevators. But if this revolution in materials had already doomed the ponderous, costly Richardsonian Romanesque in America by 1886, the expiring style was destined to leave extraordinary marks on Toronto, in the form of a series of highly important Richardsonian buildings put up around 1890. Thus did Toronto become, without meaning to, North America's heaven of the defunct Richardsonian Romanesque, and the style's great epitaph.

At the top of any list of these buildings stands the rose sandstone Provincial Parliament Building, the handiwork of the London-born, Buffalo-based architectural eclectic Richard A. Waite. Majestically heaped on the south end of Queen's Park and glowering officiously down University Avenue like Queen Victoria herself, the vast confection of semi-circular arches (imitations of ancient Syrian church portals), elaborate carving and massed, quarry-rough stone is best visited at the end of a long summer evening, when the dying day makes the sandstone glow, the deep-set windows seem mysterious, and the gargoyles snarl. It is a charmless edifice, and is little liked by Torontonians today. Architectural guide Patricia McHugh has written off our attitude as "the usual city slicker disdain for the country bumpkin," noting the Ontario legislature's historical association with rural interests. Opinion of the building at the time of its unveiling was rather different. Under the curious banner headline "Legislators in Fairyland," an 1893 Toronto *Empire* story trumpeted forth that the opening of Waite's structure "marks an epoch in the history of the province."

Which, of course, it did. With the inauguration of this gigantic seat of government, the low, little but rich and burgeoning provincial capital at last gained a building of metropolitan stature, a monument to eminence it had never before allowed itself. It was also too good a gesture to let stand alone, which brings us to another of Toronto's great

St. George Street

Richardsonianisms: Toronto architect E. J. Lennox's Old City Hall, begun in 1889.

It appears necessary for anyone who talks about this building to mention knowingly Richardson's Allegheny County Courthouse, in Pittsburgh, and suggest quietly that Lennox got all his ideas from this 1884–1888 "original." True, the Pittsburgh creation influenced Lennox—it was famous throughout the North American architectural community from the day it was finished—but, as we find just by looking at the Toronto building, Lennox was no mimic. Old City Hall is an asymmetrical Wagnerian fantasia on Romanesque chants, done in brown New Brunswick sandstone and a rose sandstone quarried in Ontario, and markedly different in spirit from Richardson's grave symmetry in Pittsburgh. For reasons both practical and aesthetic—he wanted the pinnacle to dominate the top of Bay Street, our Wall Street—Lennox puts his clocktower off-centre, and makes it gracefully slender, almost Italianate. Richardson's campanile, in contrast, is a stolid affair, squatly centred. Too, Lennox's is made even more operatic by opulent carvings evoking weeds underwater, wildly flourishing capitals, and many an exuberant oddity and surprise—all of it, remote in tone from Richardson's late, overpraised essay. In contrast to their indifference towards the undistinguished provincial legislature, Torontonians have always shown affection for Old City Hall, and have stoutly resisted moves to remove or alter it. Lennox's masterpiece was a superbly archaic, ponderous advertisement for a powerful city muscling its way towards continental prominence—and for Toronto, ever uncertain of its importance and identity, such forthright symbols will always have peculiar importance.

The peculiar pleasures of Toronto's Richardsonian Romanesque is all around us, set in place by builders famous and anonymous. We find an isolated, rude scrap of engraved Richardsonian sandstone under the chin of an oriel at the otherwise dull intersection of St. Clair Avenue West and Old

Weston Road, and many a rough-hewn lintel and flush of weedy carving over the video stores and Vietnamese fast-food joints in the once-stylish neighbourhood of Parkdale. But of all our minor Richardsonian knockoffs, my favourite is the Gladstone House, a hotel erected in 1889 by mainstream architect George M. Miller for business visitors to the industrial outskirts of Parkdale, and still operating as a hotel today, long after most former industries in its neighbourhood have disappeared.

Located on what's become a run-down, hooker-ridden strip of Queen Street West, and home to Bronco's Sports Bar, the Gladstone is a sturdy box transfigured by wonderfully phantasmagorical, eclectic ornament. In some places on this tall building's red-brick façade stand quite respectable Corinthian pilasters, with pretty flounces of vegetation at the capital. Other pilasters shoot up and end with a burst of flickering lizard tongues, or, unaccountably, in a rudely hacked stone stump. Carved, winged dragons cling to the underside of deep windows on the tall, squared-off Italianate tower, and lions roar from perches high up the wall.

The scenario spun by this improbable structure recalls an age of energetic barbarian builders, tacking together an architecture of their own from ornaments and elements left them by the winded, distant Empire—an architecture bent on making up by brute strength what it lacks in tasteful reasonableness. The Gladstone, like so many minor Toronto buildings erected around 1890, suggests a culture awakening to itself, and to its future as cosmopolis.

Until near the end of the last century, Victorian Toronto had styled itself in the manner of the distant patriarchate, typified by the Cathedral Church of St. James, to celebrate conservative intellectual and religious values, and loyalty to Crown and Empire.

The Richardsonian Romanesque, on the other hand, speaks of a new order of things, directing our gaze in two directions. One is back in history, to the system of ornamentation and craftmanship of the north-European Dark Ages, when the rough, formerly subject peoples of Europe found themselves free of Roman rule, yet still in awe of Roman culture. The other direction it points is towards the United States, then enjoying its first ebullient moment of prosperity since the Civil War, the conservatism and opulence of which is ably summed up in the costly, grandiose projects of Henry Hobson Richardson. Here, then, was a style for Canada—then emerging into wealthy, ambitious nationhood, solidly in the new American sphere of economic power, and ready for Richardson's Romanesque, which would proclaim the end of its status as a far-flung colony of the Victorian empire, and its launch towards independence.

BYZANTIUM AND ST. ANNE'S

To Toronto's architectural aficionados, St. Anne's Anglican Church—located on Gladstone Avenue, in a district of west Toronto from which most Anglicans departed generations ago—is famous principally for its remarkable interior, decorated in 1923 under the direction of the Group of Seven's J. E. H. MacDonald.

What MacDonald created is a feast of colour and design. Gold floods across the interior of the great central dome, fifty-five feet in diameter, and glints from the dazzling mosaic behind the plain marble altar. Stencilled designs in blue and rose and gold swarm over pillars and walls and vaulting arches. Within the half-dome arching over the altar, painted grapevines coil and spring, separate and join again in an exuberant illustration of Jesus Christ's simple description of Himself as the vine.

Set among these elaborate decorations are St. Anne's greatest treasures: the large paintings executed by some of the best-known Canadian artists of the twentieth century. Surrounding

the altar are depictions of scenes from the life of Jesus—
including *The Adoration of the Magi* by Frank Carmichael, *The
Transfiguration and the Tempest* by MacDonald, and *The
Raising of Lazarus* by Thoreau MacDonald. High in the dome
are F. H. Varley's large, striking portraits of the Old Testament
figures Moses, Isaiah, Jeremiah and Daniel; and a set of four
medallions, depicting the symbols of the four Evangelists, by
Toronto sculptors Frances Loring and Florence Wyle.

The largest of St. Anne's pictures are the oil paintings
which appear in the triangular areas over the four great pillars
supporting the dome. These works portray the four central
events in the great redemptive drama of the Incarnation. First,
there is Varley's nuanced picture of the Nativity of Jesus, the
finest single painting in the church. (Varley has left us a self-
portrait in the shepherd kneeling to the left of the infant
Jesus.) Next comes the *Crucifixion* by MacDonald, followed
by a dramatic *Resurrection* by H. S. Palmer—Christ seems to
break forward from the picture plane into the church's interior—and, finally, the *Ascension* by H. S. Stansfield, in which
Jesus lifts upward from a thick Canadian forest, past a range
of snowcapped mountains wreathed in cloud.

The story of how St. Anne's came to be such a storehouse
of decoration begins in 1907, when the rapidly expanding
working-class parish decided to demolish its first home—built
on the present site in 1862—to make way for a new and larger
structure. It was then that Lawrence Skey, rector of St. Anne's,
proposed that the design of the new building be imperial
Byzantine. The choice was unusual in Gothic Toronto, but
not eccentric. Anglicanism, which has always believed itself to
be non-papal Catholicism, has been fascinated by the similarly
non-papal Orthodox Church since the century of
Reformation. (The fascination has not been reciprocated.)
Other Toronto Anglicans might worship in stone or brick
churches in the High Medieval-Revival style—but, if Father
Skey got his way, the people of St. Anne's would pray under a

great dome, surrounded by splendid art recalling the glorious triumph of Constantinian Christianity over Roman paganism, and affirming the ecumenical character of Anglican faith and practice.

As matters turned out, the rector got half his way. His Byzantine church, designed by architect Ford Howland to seat 1,400 people, was built in 1908. But Skey had to wait for fifteen years, and the death of a generous parishioner, who left St. Anne's $5,000, to start on the decorations. In 1923, the rector assigned the work of finishing the church's interior to painter and friend J. E. H. MacDonald, and the architect William Rae. When the massive job was finished (in the same year), St. Anne's people found themselves with an stunning interior adorned with paintings by thirteen of Canada's most important contemporary artists.

Some sixty years later, and after an extensive restoration in the 1960s, St. Anne's is still alive, though not exactly thriving. The present rector, like all his recent predecessors, is trying to keep the church going—a worthy task, if only because of the building's unique status as the only project of decoration jointly undertaken by Group of Seven artists. Each artist worked on his own painting in his studio, and each brought to his assigned topic individual treatment—but the colour scheme and the mix of flat, stiff representation borrowed from Byzantium with more subtle modelling taken from the Italian Renaissance was dictated by J. E. H. MacDonald.

The decoration does not always work in harmonious concord. But in a city whose church-folk still clung to the academic English Gothic styles of yesteryear, St. Anne's stands out as a fresh, visually forceful and ecumenical reply to the Edwardian decline of faith into narrow denominationalism, and a restatement of the embracing universality to which Christian witness has always aspired.

CLASSICAL

Dwellers in cities hardly need to be reminded that we live among buildings decorated with ancient Greek and Roman ornament; the things are everywhere. But while so deeply ingrained into the Toronto cityscape that we just take it for granted, Classicism can still turn up in some surprising places.

One need not look for such surprises in the neighbourhoods of the rich and powerful, since they've always deployed the Classical as a power style. In 1807, when Toronto was little more than a squared-off village in the mud, the near-mythic adventurer and merchant Laurent Quetton St. George classed-up his otherwise blockish house with a Palladian front window and a pedimented portico supported by four graceful Ionic columns. From that time through the 1860s in all antiquarian earnestness, and until the Second World War in the solemn name of academic propriety and "tradition," and again, nowadays, in the playful, gaming spirit of post-Modernism, the columns and Classical orders, have not ceased to appear. (An "order," incidentally, is usually defined as a distinctive ensemble of a column with capital, and, on top of that, a load-bearing superstructure of some kind, called an entablature. A base or pedestal for the column is optional.)

While the Classical vocabulary can be readily detected in most city districts built up more than sixty years ago, the most high-profile uses of this visual language are to be found, of course, in large, noble non-residential structures. Union Station, which opened for business in 1927, exudes businesslike formality from every sweeping, dead-level sculpted line and from every massive Tuscan pillar. As far as the momentous Classical mode is concerned, Toronto's central station cannot be exceeded.

A different tone—still dignified, but less imposing and altogether more celebratory than that of Union Station—is struck by the lofty Ionic columns and elegantly detailed façade of Thomas William Fuller's curving Dominion Public Building (1926–1931), at 1 Front Street West. And speaking of the Ionic Order, my current favourite Toronto building in this traditionally feminine style is the headquarters of the Provincial Ombudsman, formerly the University of Toronto's Lillian Massey Department of Household Science, built between 1908 and 1912 at the intersection of Bloor Street West and Queen's Park. On its west front, this ample, intelligent structure presents an exceptionally fine Ionic portico, whose full columns are echoed by the half-columns in the graceful wings extending from it north and south. No non-monumental Classical building front in Toronto displays such expressive rhythms of light and shade, extrusive ornament and receding aperture, with more harmony or clarity.

But Toronto's noble and important buildings in the classical style have been written up in many textbooks and guide-books, and so need no further comment. Every city has its columns and pediments and such. I should like to know, however, if every city has so much evidence of the Classical on streets like the one I live on, in an unimportant working-class district which appeared on downtown's west side on the eve of the First World War. Though the factory I live in sports no classical elements, many of the other domiciles along the street do. Were one to tell the neighbours that their front porches are really pedimented porticos supported by pedestalled columns of a simplified Doric order, approaching the stolidity of the Tuscan, and inspired by the Renaissance theories (and misunderstandings) of Andrea Palladio, they might find this news amusing, if not very interesting. But, in any case, that's what they've got, *mutatis mutandis.*

The occurrence of Classical columns and pediments is certainly not limited to my street. That's just where I started noticing it. Nor do I want to suggest that any of the Roman porticos on my street, or those on dozens of other ordinary Toronto steets built up between late Victorian times and the

Depression, are remotely comparable, in beauty, grace or nobility, to those you find adorning the city's great public buildings, or mansions in Rosedale.

But if the builders who gave us working-class and middle-class Toronto around the turn of the century rarely observed the precise proportions and ornamental canons of Classicism, they nevertheless left the city of Toronto haunted by echoes of the ancient architecture of virtue, empire and the gods.

On many old Toronto porches, we find each column resting on a pedestal, just as illustrated in the standard Renaissance manuals on ancient Greek and Roman building—though the Toronto style is, more often than not, to make the pedestal of brick, cap it with a slab of concrete, and run it as much as halfway up to the horizontal superstructure supported by pedestal and column together, i.e. the entablature. Occasionally the column extends the whole distance between floor and entablature, as it sometimes did in Roman times.

The typical Toronto column, standing on the pedestal or on the floor, is almost always of wood, machine turned and mass produced. It is shaped into a semblance of a round, unfluted Doric (or Tuscan) column, with a simple square base and simple square capital, usually devoid of either the volutes of the Ionic style, or the pretty vegetative flourish of the Corinthian. Columns here often present a white shaft with capital and base in some other colour, green and brown being preferred hues on my street.

The entablature is customarily just a horizontal wooden beam, innocent of ornamentation. Supported by the columns and resting on the entablature is the pediment, its triangle defined by the two sloping eaves of the gabled porch. The whole affair is attached to the front of a blocky brick house. On spring and summer evenings, families across old Toronto may be seen visiting in these porticos, whose ancient prototype once served as the ceremonial portal to a god's dwelling place.

Something of this archaeological information may have

been known by those who quickly developed the west-side Toronto neighbourhoods some ninety years ago, though it has long since been lost, if such data ever had a life in popular knowledge. But the history of a culture is largely that of people living in structures whose meanings they do not know, or have forgotten. This amnesia doesn't change the fact that ordinary Toronto neighbourhoods are graced with the most venerable symbolism, which can be brought to life and recollected, given a bit of background and a little imagination.

An ordinary street of working-class houses with plain Doric columns out front then becomes a row of temples sheltering the sacred flame of marriage. As we are told by Vitruvius, the Roman who wrote the only ancient architectural manual that has survived from antiquity, the first Doric temple was dedicated to Hera, the jealous first lady of the gods and guardian of marriage. The Doric columns themselves advertise the sort of people who build such things, as totem poles do. From Vitruvius we learn that the sons of Dorus, a Greek, were a race of conquerors; the writer compares the unadorned Doric columns to naked men—that is, to soldiers (since the Greeks fought naked).

Subsequent architectural authors have always seen in the Doric style a certain militant, plain manliness, as distinct, for example, from the supposed femininity of the more slender, prettier Ionic columns. The Doric ornamentation on countless old Toronto houses thus proclaims them to be the family homes of conquerors, returned from the wars of empire, and the shrines of our era's new hero, the working man.

This is all fantasy, I admit. But it's probably not too different from what many loyal, turn-of-the-century Torontonians of British descent would have found a noble vision indeed. Was this proliferation of innumerable Doric-like columns in Toronto really just happenstance, a matter of builders mindlessly copying out of architectural pattern books? Or were those porticos meant consciously to recall the greatness of

Rome, and the British imperial ideology of the hour?

Such questions inevitably lead not to answers, but to other questions. For while this architectural usage gives a touch of charm and aesthetic elevation to many an otherwise shabby house front, why did it really happen? Why does the spirit of the Classical haunt urban imagination like an unlaid ghost, sometimes quiescent, but always ready to start dispensing plans and comfort again whenever the faith in the current anti-Classicism (Gothic, International Style, or whatever) becomes shaky?

I have no answer, nor a satisfactory explanation. It does little good to say that the business of columns, entablature and so on we have inherited from Classical antiquity is safely beautiful and noble, since Classicism certainly does not ensure the beauty and nobility of the outcome. Though straining mightily for great Classical effect, both the White House and the United States Capitol building, for instance, today look like Victorian mantel clocks with quaint Classical airs.

It would be absurd to argue that New World structures crafted in other revived Old World architectural styles cannot turn out as lovely and high-minded as anything in the Classical mode. One thinks of Toronto's Old City Hall, or the numerous gems of Victorian Gothicizing that dot our downtown; or, for that matter, the Toronto-Dominion Centre, splendid in heroic anti-historicism.

But the fact remains that, for the moment anyway, Modernism is *démodé*. And nobody builds in the palatial Romanesque style of Old City Hall any more, and probably nobody ever will again—while the Classical is with us in new hotels out at the airport, in new condo towers on the lakeshore, in monster houses, and on the columned porches of turn-of-the-century workers' housing that is not monstrous at all. Classicism has never been out of fashion for very long in Europe, or in lands touched by European civilization, since attaining its typical form upwards of 2,500 years or so ago.

The riddle of this durability has exercised the imaginations of many architectural writers, though none has come up with a really convincing solution to it. Quinlan Terry, a contemporary British architect best known for his devotion to reviving architectural antiquity, has written that the Classical Orders—the "timeless and universal recipe for architecture"—were given to Moses by God on Mount Sinai, along with the Ten Commandments. From Moses, this truth of architecture was passed by reverential architects from the Hebrews to the Greeks, and thence to us, via Rome. I do not personally find this explanation particularly helpful, though I do find it colourful.

Less colourful, more modern-sounding and provocative, and conceivably more helpful, is the idea set out by the British art historian John Onians in his 1988 book *Bearers of Meaning*. In this study of the development of the Orders, Onians theorizes that our evolution into humans has predisposed us to like columns, entablature and so on. The forest-dwelling early members of the human family most likely to survive were those primates, naturally, who could discriminate which tracks signalled "food up ahead," and which ones meant danger. Later, the alertness born in the forests was applied to architecture. "Trees and woods have always had a special importance as at once the best source of food and the favourite haunt of enemies," Onians writes. Hence, the persistence in architecture of "tree-like columns and columnar shafts and...capitals which occupied a position similar to that...of the tops of tree trunks."

Are deeply-buried memories of forest hopes and fears, then, the sources of the special pleasure we feel when beholding some contrivance of Classical pillars?

Or could it be, simply, that the Classical style is the most versatile, expressive, desirable and pleasing building style ever devised? The latter view is faintly distasteful, because believing it would mean that a final answer to a problem of the first order—shelter—had been discovered very early in our history; and we Western people *do* like thinking of ourselves as just about the best improvers, inventors and developers humanity has yet produced. But much of this intelligence is inherent in the very stuff of Classical architecture, in the stock ornaments and elements architects receive from the past. Even in the hands of so-so architects, these elements still contain unmistakable sophistication and cultural resonance. Perhaps our admiration for this vocabulary of building has to do with its deeply reassuring unoriginality, and its dependable usefulness as a means of bestowing dignity on the urban streetscape.

CYBERNETIC

Most city folk who attend a traditional church or synagogue, as well as all outright pagans, would probably have found themselves as bemused as I was by a news release I received early in 1993, promoting an impending trade fair of ecclesiastical technology. The event would be known as Inspiration '93.

Described as "four dynamic days of cutting-edge seminars, workshops, choral readings and exhibits," Inspiration '93 was to feature displays of such unusual worship aids as bass guitars, backdrops, lighting dimmers, "sound effect libraries," "video walls" and audio carts.

Several workshops were advertised. Participants could learn about the "latest in Bible Software," "how to prepare sermons for overhead projectors," and "how a MIDI-equipped organ can become a dynamic instrument in your ministry." (MIDI is the acronym for Musical Instrument Digital Interface. I cannot explain how this device works, or how it could give "unlimited musical possibilities" to an old two-manual wheezer, as promised in the brochure.)

The prospective visitor is also invited to "learn the applications of dimmers, mixers and...gels to change moods and set atmosphere to cover all events," and to discover the mind-expanding possibilities of teleconferencing, "a coming technology that major Religious organizations are looking into."

The most intriguing seminar on the roster, as far as I was concerned, was the one called "Using Pyrotechnics in Religious Settings." That's because, like most people, I love fireworks. But I like thinking up bizarre scenarios even better. Now I have always thought Anglican Evensong—especially when complete with glorious anthems and choral psalms and chants and perhaps a whiff of incense , but even when stripped of all that carry-on—is a quite satisfactory way to pay one's homage to the Most High. It *is* fun to think, however, of what an Evensong would be like with some spark-spitting pinwheels on the chancel railing, air-burst bombs going off high in the nave, criss-crossing fire-trails of sparkle rockets, and a few Red Devil sizzlers in the chancel. While it's hard to see how fireworks would make Evensong better than it is, you can bet your last Roman candle that a service with sizzlers would mark a major break with traditional Anglican practice.

But that appears to be exactly what all the gizmos and workshops at Inspiration '93 were about: inventing popular, eye-catching alternatives to the admittedly low-tech, bookish, predictable services standard in the mainstream churches of North America.

It's no secret that an increasing number of people are finding spiritual bliss, or something akin to it, in the kind of "worship environment" proffered by Inspiration '93, and that old-fashioned religious observance is finding fewer and fewer takers. Hence, the equipment and know-how offered by this fair: all you need to create an iridescent, ever-shifting and enveloping architecture of light, moods and sonorities, insubstantial and pretty as dawn cloud, appropriate to the popular new-time religion.

This evanescent architecture of spectacle is as new as the light and sound technology that's making it possible, and it represents one of the most stunning breaks ever to take place in the history of settings for Christian worship. The architects of these environments, from the earliest days of church-building, have long assumed solidity and a certain grandeur appropriate in their products, to make them identifiable as churches, if for no more important reason. These values have been embodied in many forms, from the Classical and basilical arrangement, through the Romanesque temple and high medieval Gothic cathedral and parish church, to the extraordinary pilgrimage chapel of Notre-Dame-du-Haut, designed by Le Corbusier in 1950 and finished in 1955.

Even in something so "modern" as Notre-Dame-du-Haut, the heavy stuff of churches has traditionally been orchestrated to remind us of the enduring greatness of God, and of those times when the Divinity was, we presume, worshipped more loyally than at the present time. This impulse can go horribly wrong, as it certainly did in my home-town Episcopal church in the American South, put up in the 1950s in English-Gothic style. Among its more amusing features are stained-glass depictions of people receiving baptism, Communion and other ministries of the Church, with all the men uniformed as Confederate officers, and all the women dressed like Scarlett O'Hara—heroes and heroines of the Lost Cause, here turned into saints.

But even if the treatment was newfangled and ultimately silly, the urgency to invoke comfortable cultural traditions was something the architects of this parish church shared with church architects of almost every Christian age and area.

Heavy architecture connects the believer to heavy virtues and loyalties: the electronically produced mood-buildings link the believer with those appliances common in the home, hence to the one haven in a heartless world—television, the family-room, the spectacle of plenty, luxury and leisure. The medieval churches (which dot Toronto) expressed beliefs that were at least as much a part of common English culture as the holly and the ivy. Built in the heyday of Victorian doubt, as optimistic, secular progressivism was becoming the true religion of Toronto's democratic masses, the Victorian Gothic is a visual argument for the traditional mix of piety, duty and Englishry already coming unstuck in Canada. Similarly, the light-and-sound shows of the newer Christian piety appear to be reaching out to a perishing class—this time, a devout middle class, watching its prayers for prosperity and mobility of traditional suburban life go unanswered, as the recession deepens and jobs vanish. I suspect the techniques promoted by Inspiration '93 will do no more to keep religion alive in mass culture, in the long term, than the techniques of the Victorian master-builders.

Be all that as it may, I've decided to go down with the neo-Gothic Anglican ship, clutching my antique Book of Common Prayer to my bosom, and praying that, in the Heaven that surely awaits me, there will be no gels, MIDIs, sound-effect tapes, or teleconferencing devices—though a small fireworks display now and again might prove a nice enhancement of our heavenly work of praise.

Houses and Home

Ravine house, Weston

AT HOME

These thoughts about home, and its disappearance, began to form in my mind one bright afternoon a few summers ago, after a visit to Toronto's Power Plant gallery, when I decided to hoof the couple of kilometres back to *The Globe and Mail's* headquarters on Front Street West.

Anyone familiar with Hogtown's deep-downtown geography can visualize the northwesterly route. It went from the waters of the harbour, across high-speed roads and almost unused railway tracks, under the elevated belly of the Gardiner Expressway, up over the Spadina Avenue bridge, and thence to work.

It wouldn't have been a walk worth writing up, had it not been for the discovery of a kind of campsite in the last place one might expect to find one: the dark, sharp angle between the ground and an exit ramp descending from the expressway.

No one was there. To prevent discovery, the occupant had stashed his or her belongings up under the ramp, on a narrow steel ledge. From the looks of the stuff, its owner had been sleeping in this place for some time and was no travelling rich kid, playing poor. The rolled sleeping bag was filthy and ripped, the cooking gear dented, the cans full of the cheapest noodles and stew you can buy. In a nook rendered invisible from the busy roadway nearby by concrete pylons, there were signs of a cooking fire and a scatter of empty cans, plastic bags, booze bottles.

It was, of course, the digs of a homeless person, and merely one among many such sites in downtown Toronto's empty buildings, ravines and dead zones. Nevertheless, it gave me the sense of being an inadvertent trespasser in someone's house—embarrassed, curious, nervous. That feeling of trespass has haunted me since then. How very little it took—just a sleeping bag, a smudge of burnt wood and paper, a couple of battered aluminum pans—to mark that spot as a dwelling place, and to mark me out as an intruder!

The site under the expressway was no home. Home is where the heart is, or, more precisely, the bed. And the human mate we share the bed with, the surly cat who insists on snuggling between us under the duvet every night, the boring religious tome on the bedside table. Home is the familiar faces, objects, personal problems and intimacies to which we return at the end of a day spent in the mutable, impersonal world outside. It need not be always, or altogether, comfortable; it does have to be usual, and as secure as possible against the chops, cheats and changes of contemporary urban reality.

Or so, at least, goes the common wisdom about homes and home-making, and about the things the homeless presumably do not have.

While I have no wish to trivialize the plight and misery of homeless people, the experience under the expressway did leave doubts about the way we speak of "home" and "homelessness," and about just where the line between the two should be drawn.

Just what part, for instance, does *permanence* play in our idea of home? The person who slept under the Gardiner could claim no permanent place of his own. But is anyone cheeky enough to do so today? The idea of endless migration and transition, permanent homelessness—the "starter home," the second home, the third, ad infinitum—is now on hold in Toronto, due to the decline in real-estate prices during the early 1990s. The suspension is temporary; and, before long, city-dwellers will once again be caught up in the churn of selling and buying, moving and moving again.

This is not a phenomenon I can take an unqualified stand against. Virtually every good thing that's come to me in my life has been accompanied by uprooting, temporary rerooting, and moving on. As I grow older, however, home has become intrinsically bound up for me in the specific built entity I now inhabit—one certainly open to architectural experiment and revision, but the place in which I now intend to live, experience the intimacies of human existence, change and learn and, *Deus volens,* die. Everything else is *house*—mere architecture, real estate—whether a Rosedale mansion or a sleeping bag under an expressway ramp.

My convictions about home have not been changed by the gradual awareness that, while I have a dwelling place that is much more secure and comfortable than a cold, muddy spot under a bridge, something of homeness is still lacking in the house I live in. This sense is bound up, somehow, with my house's gradual occupation by impersonal, non-intimate technologies of pleasure. The homeless person's dwelling is vulnerable to the environment, both natural and industrial; my walls no longer stand impermeable, defining the line between private intimacies and the public world of mass-distributed images and noises. The once-solid definitions of home have now got holes in them—a television cable network, pumping in images from the outside world. No family, perhaps, sits together for hours in the dark unspeaking, transfixed by glowing images, as they did in TV's early days. There is a TV for every person's taste, and VCRs, so when the family tires of the various othernesses brought to us by TV, it's easy for us all to rent a video and enjoy exotic mass-produced fantasies, distractions and amusements as we will.

Telephones and the answering machine are always there, dissolving the walls between us and voices in what we used to confidently call the "outside world." With a compact disc player, it is now possible to "stay home" forever, but locked into solitary pleasures created wholly in an outside world. But what exactly are those iridescent discs, if not just more holes in the walls of home, openings to those welcome invaders, the merchandisers of mass-produced simulated experiences?

I do not advocate the course taken by David McDermott and Peter McGough, New York painters who live together in a Manhattan brownstone according to an archaeologically exact Victorianism, with wall-to-wall period decor, only candles and oil lamps for light, and coal for heat. If technology and the rapid spins and displacements of capitalist culture have taken away home forever, turning us all into transients and wanderers, contaminating human intimacy—perhaps even the very idea of the human—with myriad impersonal pleasures, those pleasures are still marvellous, and the well-spring of every urban person's most continual and evocative challenge. "To be modern," literary critic Marshall Berman has written, "is to experience personal and social life as a maelstrom, to find one's world and oneself in perpetual disintegration and renewal, trouble and anguish, ambiguity and contradiction: to be part of a universe in which all that is solid melts into air. To be modernist is to make oneself somehow at home in the maelstrom."

CASA LOMA

In the last century, many a city too great for such nonsense allowed a monstrosity to be built on its best urban elevation. Paris let Sacré Coeur ruin the summit of Montmartre; Montreal let the pious raise St. Joseph's Oratory on the slopes of Mount Royal—while Toronto gave over a splendid escarpment site to, of all things, the apallingly vast curiosity known as Casa Loma. Built between 1911 and 1914 by Sir Henry Pellatt, an industrialist, at a cost of $3.5 million in Edwardian dollars, this story-book castle is of no historical significance. Nothing important ever happened there, and nobody important ever slept there, though Sir Henry did set aside a suitably opulent guest-room fit for a British royal, should one ever drop by. None ever did.

Leased from the city and operated for charity by the Kiwanis Club since 1936, Casa Loma is more nearly finished than it was in 1923, when Sir Henry's business belly-flopped, his wife died, and the knight moved out. It is a confection of looming ramparts and soaring chimneys out of *Boy's Own* illustrations, a pastiche of gaudily carved wood-panelling and marble and stone veneer, plaster gargoyles and useless towers, among myriad other architectural pomps and fusses and flushes. But despite the romantic insistence of its Walter Scottishness, there is little dour or gloomy or overbearingly *northern* about Casa Loma. Viewed from a standpoint in its cramped grounds, in fact, this busy architectural pastry seems almost too light and frothy—a cardboard fantasia, a set thrown up in a day to serve as the backdrop for a B-movie about knights of yore.

Yet despite its seeming flimsiness, this Moloch required coal costing $25,000 a year to keep it warm. Sir Henry employed a squadron of forty servants to run it, and kept busy nobody knows how many European dealers in baronial junk and pseudo-antiques, the suppliers of the armour and stag-horns and carved, ponderous oaken furniture with which its rooms were once stuffed. These chambers were denuded during the 1923 bankruptcy auction, which lasted five days. Today, in their far more sparsely furnished state, the compartments of Casa Loma seem decently proportioned and detailed—which is what we would expect of E. J. Lennox, its distinguished architect—though they would be just as acceptable in less sensational wrapping.

I think Torontonians make a mistake in dismissing Casa Loma. Whatever else it may be, the grand folly is surely the most conspicuous evidence in our midst of a tendency that's been present in European building a long time—the English *nouveaux-riches* were putting up medieval storybook castles as early as the seventeenth century—but which mutated into a kind of mass neurosis among the North American rich in the decades leading up to the First World War.

Modern military technology had made them obsolete as defences, of course, and advanced communications and industrial rationalization made them look silly as residences. Nevertheless, vast castles proliferated in North Carolina, Ontario, New Jersey, California, all of them staggeringly expensive antiquarian rhapsodies pastiched from a dozen historical sources. Unity of composition was avoided at all costs. Wanted, instead, was what architectural critic Robert Harbison has called "bogus historical scenarios," such as the "neo-Classical" room inserted into the presumably much earlier "Norman" tower at Casa Loma.

But why did this extraordinary event in the history of building take place? While we may just be dealing with the durable human proclivity to pompous nonsense, one senses behind this construction boom in ponderous architectural fictions the quite specifically late nineteenth-century anxiety over loss—the loss of historical bearings, the disintegration of everything once certain and solid and reliable about life. Everywhere in Europe and America, this dissolving of received wisdom and fact by the acid of technological Modernity was proceeding apace. One remedy popular among the very rich, the one chosen by Sir Henry and Lady Pellatt, was to sally forth on an impossible retreat from reality into fantasy, from modern politics into story-hour scenarios of chivalry, from the discomforts of truth into an almost unbelievably expensive prison of pleasing obsolescence. Casa Loma deserves to be considered as a late freak in an era during which the very rich, the only people who could afford such freakishness, made our century's first stand against advancing Modernity. (There would be others; they would not be so harmless.) It deserves membership in that club of such fellow-oddities as Balmoral (1853–1855), King Ludwig II's Neuschwanstein (1868–1886) and Castle Drogo, the monstrous make-believe fort built before and after the First World War by Sir Edwin Lutyens for Julius Drew, a mad tea trader.

The people who operate Casa Loma bristle at the common opinion that Sir Henry was just a colonial Edwardian fat cat with baronial delusions. Their offence is at least partly justified. He was an outstanding and inventive businessman, for at least the first years of his career. He was a lavish philanthropist, a loyal soldier and officer, still fondly remembered by his regiment, the Queen's Own Rifles, which today maintains a display of its memorabilia at Casa Loma. And if he was a pompous man, he was also a tragic one, eventually felled by his doomed attempt to escape the present-day world.

Before his death in 1939, Sir Henry lost just about everything a man can lose, including two wives, his fortune—estimated, at its peak, to have reached $200 million in today's dollars—his farm in nearby King City, his successively smaller houses and apartments. He ended his days in a room in the smudged industrial lakeshore suburb of Mimico. In addition to everything else, he lost his place in Toronto's historical imagination as a pioneer of the city's electrification, and one of Toronto's most fabulously successful industrialists; and is now remembered almost exclusively as the obese, crazy tycoon who built the silly castle high atop the Lake Iroquois escarpment.

But Casa Loma is not the only North American fantasy-castle haunted by a sad story of financial ruin, or untimely death, or lost love. Such miserable tales seem to haunt such structures, which now are vanished, or lying in pathetic ruins, or operated as tourist draws across this continent—as befits their absurd hugeness and their studious anachronism.

But let the soul of Casa Loma's builder rest in peace, and our memory of him be the compound of civic contribution and lunacy he deserves. And let his fantasy be a reminder to us not to condemn those whose extravangance is all that sets them apart from the rest of us, sharers in the same struggle. For there are few of us who do not yearn for safe harbour, a castle of small or great scale, in which the forces of technology, mass communications and mass culture continually dismantling and reorienting life are kept at bay.

MONSTER HOUSES

If I long avoided writing about the monster houses, all instances of which I hate, it was because I never felt I understood exactly what distinguished this newish architectural type from the ample, eclectic houses I admire. It has mostly to do with television and air-conditioning. Of that, more presently. Right now, some words about the oft-repeated criteria which are said to enable us to recognize the differences between monster and non-monster houses.

One is the aggressive occupation of site, a billowing-out to the property lines on all four sides, dwarfing the older houses on the street. The flagrant violation of street scale is a charge that, I believe, can be made to stick, though the most horrible monster houses I have seen do not stand on little Toronto lots, but on spacious grounds outside town. This very isolation only seems to accentuate the "if you got it, flaunt it" vulgarity, sometimes noted as the most despicable thing about the monster house craze of the 1980s.

Moneyed people, especially those not yet comfortable with wealth, have always gone for the big effect in home-building, with often memorable results. One of the richest spectacles of this sort to be found in Toronto is the George Gooderham house, now the York Club, located just north of the University of Toronto, at the busy corner of St. George and Bloor streets. Put up by a booze tycoon around 1890 in the most deluxe architectural style of the day—the massive sandstone and terracotta neo-Romanesque associated with the name of American architect H. H. Richardson—this house proclaims its occupant's big bucks in every ponderous detail, from its grand corner tower and soaring chimneys to its deeply shaded porches and ancient-looking carved arches and windows.

The house is shameless in its advertisement, and wonderful for its operatic inclination and insistence on effect. While the contemporary monster house and the Gooderham mansion share a purely philistine spirit of ostentation, the older building shows off the wealth of its owner with much flair. There is none of the vulgar archaism of the monster house, the almost hernia-producing drive to cobble together "period elegance" with the casual. Perhaps the thing that makes monster houses really monstrous is their *seriousness*—grandiose scale, and the reckless deployment of ornaments filched from this or that despotism in the past, and inflated—pediments and Classical columns from Imperial Rome, boldly rusticated exterior walls composed of real or fake massive blocks of stone separated by deep-cut grooves, from the Florentine Renaissance or Georgian England.

What makes it possible to be so serious (hence irresponsible) about mere appearances is that the traditional structures and ornaments in noble architecture—the ponderous walls, deeply inset windows, porches and so on—have, with startling speed, lost all practical usefulness. I suggested earlier that monster-house building has something to do with television and air-conditioning. That these two technological innovations, prewar in origin but common everywhere only after the Second World War, have had a drastic effect on urban life is not an idea I can take credit for. Marshall Berman, in his book *All That is Solid Melts into Air,* speaks at some length of the abrupt postwar emptying of the summer streets in his Jewish neighbourhood in New York, as everyone abandoned the porch and the neighbourly evening stroll to watch TV in cool comfort inside.

This wasn't just a New York phenomenon, of course. I still remember the summer, in the mid-1950s, when TV sets and air-conditioners started arriving in my small home town. Suddenly, nobody was dropping by after dinner for some cool lemonade and leisurely conversation on the south veranda any more. A certain kind of evening talk which traditional Southern families had always enjoyed—rambling, casual, apparently to do with weather, cotton and politics, but really a recital of conversational art for its own sake—came abruptly to an end, and a peculiarly blank silence fell upon the house, never to lift again.

It has taken a while, but, if I'm right, the silence precipitated by the introduction of new technologies has finally found its typical expression in a new architectural style, that of the monster house. It is not the archaism of these buildings that is hateful. Nor is it the mindless exacerbation of old-fashioned ornament I resent—use of the thick walls and heavy overhangs, deep-set windows and so on all for show, without the least attention to the problems these features were designed to address: the bright sun at noon and the long shadows of evening, the direction of breezes in the various seasons, and other weather phenomena. Air-conditioning made this contempt for context possible, even as bad taste, it appears, makes it mandatory.

Perhaps George Gooderham would have built a dull monster-house box, if he'd had the chance. But if he controlled the fates of myriad distillery workers in Toronto, he did not control the Toronto weather. It, not he, dictated the way his house must be built, and accounted for the rich play of exterior variation we find so pleasing in his house, and usually absent in the monster house, except as bits of showoff. The rhythm of porches and cornices, the arrangements of windows, the orientation of portals, and the interplay of house with trees and grounds—all moulded and mandated by the natural elements, by the need for circulating fresh air in the dog days of August, and the corresponding need for shielding from the damp, icy February blasts.

I have less to say about television, since I do not watch it, know little about what people see, and have no interest in finding out. But I do know that watching TV is preferred by many people to having drop-in company after dinner. Hence the tendency of *all* new housing, not just monster housing, to reduce the veranda or porch to the status of disused ornament, degrade the living room to an expensive, purely ceremonial area—a process begun in nineteenth-century domestic architecture with the introduction of the "parlour"—and exalt that zone of sloppy manners and incivility known as the family room.

To my mind, the curious blankness of the monster house is the visual analogue to the barren silence and social blankness created by television, which gathers the family around itself each evening, then forbids conversation, discourages evening callers and regiments life by the clock with a rigidity hitherto known in few places outside Trappist monasteries.

CASTLE HILL

While not exactly hating Castle Hill, my newspaper has always officially disdained this exclusive Georgian-style residential development at the south-west corner of Spadina and Davenport roads. The paper's dislike, by the way, appears to be casual and habitual, merely a quirk in our corporate culture, not active or vicious. In editorial pronouncements and asides, the phrase Castle Hill often appears with the word kitsch stuck to it, as though the two belonged together, like pea and pod.

I herewith dissent from the prevailing view. In visible contradiction to *The Globe*'s neglect during the half-decade or so since developer Murray Goldman put his ninety-one suites on the market (almost losing his shirt thereby), and in defiance of much else that's made it a hard sell—the slump in Toronto real-estate prices, dismal architectural neighbours (high-tension hydro lines, public parking lots), few fine shops nearby—Castle Hill stands today as one of Toronto's most suggestive recent experiments in high-density, low-rise residential planning.

The design, by the Toronto firm of Gabor & Popper, is a miniaturized paraphrase of the grand-manner Georgian terrace, its long rows of faceted luxury townhouses crisply defining the treed and sidewalked street between them. Gabor & Popper have obeyed the sound Georgian rule for the building line: generally uniform and continuous, but open to quiet differentiation among the various horizontal strata.

Thus, the lowest course in the Castle Hill façade scheme consists of a single storey faced with warm limestone blocks, quarried on the Bruce Peninsula and cut in Cambridge, Ontario, and punctuated by deeply inset entries topped by flat lintels accented by keystones. Above the first storey are the next two storeys, both set behind an uninflected plastered wall tinted light beige. The third, uppermost course is a low, almost white filled balustrade. Again in keeping with the British model, all the railings and fencings on the site are painted black. The insensitive use of the same standardized windows used throughout the project does tend to cast a pall of prefab monotony over Castle Hill. But, in general, the humane visual message created and maintained by the standardized front elevation is one of dignity, stability without ostentation, and sturdy, affluent domesticity.

Unlike an eighteenth-century terraced street, flat on the London mud, however, the Castle Hill site lies on the sloping shoreline of glacial Lake Iroquois, where it drops off below Casa Loma. The architects could have timidly avoided the potential risk to serene Georgian horizontality by orienting their buildings parallel to Davenport. Instead, they forthrightly broke the rules, and stepped their large freehold suites up the hill, emphasizing each "step" with a conspicuous high pilaster, its parts echoing the grey-beige-white upward sequence of the façade.

Among Gabor & Popper's most interesting challenges involved the short north-south stretch of Walmer Road bisecting the site. As the only public street available for architectural definition on both sides—the other two principal boundary roads, Davenport on the north and Spadina on the east, provided no such opportunity—Walmer was their one chance to invoke the distinctive atmosphere of grandeur and intimacy, vista and urbanity which is the glory of the best Georgian thoroughfare. And indeed, they just about did it, albeit on a doll-house scale. The resulting streetscape is stately without being stuffy, exclusive without being forcefully or obviously exclusionary. And perfectly framed by the pleasant ellipse allowed by a concave backup of the building line halfway up Walmer Road, dear and awful old Casa Loma never looked so gracious.

To those observers who think there's something wrong with Castle Hill's old-fashioned styling, because it was developed a couple of centuries ago, I can only say: so what? If you want to kick some architecture, Toronto's full of ugly Tudoroid mansionettes fully deserving a whack. But lay off buildings and street schemes that work. The Georgian town plan worked—for people, for their vehicles, for the eye and for the soul—as well as anything devised in Europe since antiquity.

The problem with Castle Hill, however, is that it's not a town, but only a whisp of alternative urbanism. It's a fragment, a vignette, and not the large, multifarious area implied by its urbane forms, ornament and abbreviated street definitions. There are no ground-floor shops in the complex, none of the variegated amenities that would make the settlement a haven. Castle Hill requires the use of automobiles as much as any Modernist high-rise-in-the-park. Because it is a zone of homogeneously upmarket residences, sharply demarcated from the surroundings, this little, open development is unfortunately pervaded by a whiff of that incompleteness and sterility we sense in suburban American compounds where the rich huddle in their alarmed monster houses behind electrified fences patrolled by armed guards.

Well, Castle Hill, if diminutive, incomplete and flawed, is still not that bad; and it is a notably pleasant, sheltered place to stroll through on a crisp November morning, or windy one in spring. It is also a spot that deserves study by young architectural dreamers able to keep alive the vision of large-scale planning through the current, pitiful design-build drought which developers have visited upon the city fringes. Of all Toronto housing developments I know, Castle Hill is the one I would like to see become bigger, and finally grow up to be the many-streeted town it will not and unfortunately cannot become.

SONIC CITY

It would never have occurred to me to write this had it not been for a peculiar thing that happened a couple of years ago, while I recovered at home from minor surgery.

I had lived in cities for the better part of my life, and had taken the urban cacophony for granted, rarely giving much mind to the city's noise makers, other than the particularly terrible jackhammer or siren or cement truck. But during those weeks of convalescence—for reasons nobody has been able to explain satisfactorily—the walls of my house seemed to evaporate, and my hearing briefly became astonishingly, painfully acute.

As long as I was in that state, the most disturbing sounds were not the occasional gruntings of garbage trucks or shrieks of kids playing in the alley, but ambient noises with permanence and dense opacity, like that of brick and cement—the

Night, February

steady whine, never louder and never softer, emitted day and night by the toilet factory a block away, for instance; and the great unceasing whirr of innumerable automobile tires rolling on the city's streets.

The noises made by certain household appliances—the high-pitched wheeze of the microwave oven, the buzzing purr of the refrigerator—have something of the same monotonous quality, and were similarly unbearable. Briefly, I took to wearing those hard ear-muffs used by people who guide aircraft to their parking spots; but to little avail. The sounds of the city, and the monotone appliances, came right through their shields of plastic and fibre battings.

Fortunately, this phenomenon was short-lived. Today my hearing and aural tolerance are not appreciably worse or better than anyone else's. But the experience will always be memorable, inasmuch as it made the invisible architecture of urban sounds present to me for the first time—as tangible and complex as the solid architecture of streets and buildings that comprise the central visible fact of our urban dwelling.

Unlike the built city, the sonic city is diffuse, lacking the walls and fences or other strict demarcations that make visible the abstract, modern notion of "real estate." Sound structures blend perversely, puddling and congealing and spreading with no respect for firm boundaries, or for gravity. In the darkness just before dawn, the never-stopping whirr of car tires on pavement washes through streets and lanes and seeps through the windows of my study, being no respecter of legal bounds, such as property lines or walls.

This pervasive noise is joined by the sharper crunching and motor-driven grind of local street work, which in turn mixes with the first rattling of a mechanical operation beginning its day over by the railroad tracks. An early-bird jet on final approach to Toronto's international airport—its business travellers readying themselves for meetings in rooms sonically broadloomed with the grey, muffled stir of air-conditioning, its engines and down-thrust flaps violently stirring the air outside, showering down its roar, which muddles with the eternal toilet-factory whine, the incessant gasp of industrial air-intakes and the slapping flutter of fume exhaust systems. And somewhere far beyond my windows, in a direction I cannot identify, a mechanized chatter of unknown origin has begun its familiar daily utterance, contributing its strange stutter to the ramshackle construction of the invisible city of sound.

One of the lessons of my curious convalescence was that, like the solid, visible city, the unseen, aural one has its distinctive historical architectures, styles, meanings. Lounging on a deck or patio on a sunny day, or sailing, or sitting near a window on a rainy afternoon, one can hear the only two sounds of energy exchange which had any economic significance to urban humankind from the beginning of time until the mid-eighteenth century: wind, and the rush of water down the drain spout. Wind power is without importance nowadays, and falling water's power to drive engines and mills is much less significant than it used to be; hence their sounds are now thought by city dwellers to be restful, restorative—something that would never have occurred to a miller with wheat to grind under his water-driven wheel, or a farmer with a crop waiting for rain that does not come.

Most of the non-human sounds we hear come from sites of energy exchange that are new in the long history of technology—motors, engines, industrial and domestic appliances. The racket they produce, people still sometimes find strange and stressful: especially the absolutely steady, unrhythmic drones, whines, drizzles, grinds and wheezes of such systems and instruments as electric heating and cooling plants, pumps, power saws and drills. These are the bricks, so to speak, in the sound-structure of the late-modern city. But the sonic bricks of the post-Modern city are with us already, in the form of eerily banal absences of sound.

Gone from many a workplace are the typewriters that once

saturated the air with the battering of paper by tiny metal bits. And gone from our "industrial parks" are the smoky heaving and pounding that were once conspicuous shapers of the urban soundscape. Computers will be doing more and more of our work, as information becomes the central commodity of the postindustrial age; but we may well find the airless silence in which computers work to be as difficult to live with as the still-new whines of the current era.

The sounds of late industrial modernity and the curious silences of post-Modernism have almost displaced old-fashioned sonic factors in Toronto culture—but not quite. If you happen to be fortunate enough to live near a railway track, you can still hear the clatter of metal wheels hitting the rail-joints, which is perhaps just another sound (like the tick-tock of the Depression-era clock in my kitchen) that is only endearing because of its firmly dated quality. Virtually gone, however, is the hectic, jubilant pulse of car-honks once characteristic of every city's financial district—silenced, partly, by anti-noise laws, but mostly by the tight sealing of recent office buildings against everything outside, including sound.

No writer on urbanism, to my knowledge, has bothered trying to sort out the historical types and forms of urban sounds and their evolution, in the way building types were long ago arranged into a coherent history. Architectural guides to cities concentrate on built artifacts, rarely referring to the network of human use in which these artifacts exist, and never to the unseen net of historical sounds within which we live and work. But if your hearing ever becomes abnormally acute—a fate I wish on no one—or even if you just sit still right now and pay attention to the tableau of sound around you, you will find yourself in the midst of noises as rich in cultural heritage and history as any factory, soaring skyscraper or noble antique edifice.

At the Edge

Swansea Asphalt

HEARTLAND

On the outskirts of every substantial North American city, often near the airport, you find a version of Mississauga's Heartland Business Community, with the same uncanny, new and weightless experiences which technological modernity has brought us.

Laid out on flat Ontario farm country in this far-western Toronto suburb by Orlando Corporation, one of Canada's largest commercial real-estate developers, Heartland is a 1,200-acre zone of broad and curving streets with names that often recall venerable or romantic sites in other countries (Avebury, Venice, Rodeo Drive), and have nothing to do with any local fact or history. The street names appear to have been picked for general effect, as contributions to the scenography or atmospherics of the place, and with the intention of suggesting "tradition." This, in contrast to the names given boulevards of another commercial park in the vicinity, such as Satellite, Orbiter and Shuttle, which long ago lost their their newish-ness and pizazz, becoming tawdry. In Heartland's avenues, no battles are commemorated, no famous people recalled. History, as the collective weight of memory embodied in the names of things, is absent.

The streets in Heartland's developed portions are bordered by enormous vacant lots—one of them the site of a proposed public "sports complex"—and many manicured and primly planted lawns and low, mostly new office buildings and warehouses, curtained with mirrored glass, slabs of concrete or vast sheets of metal. Orlando has put up each of these buildings on more or less the same plan, and with the same internal structurings, but has given each a frosting of architectural stylistics. Some sport delicate green-glazed facades topped by hints of gable and pediment, while others are faced by long horizontal walls of reflective midnight-blue glass, banded with strips of stainless steel. Some are glass all round, while others are blocks of windowless concrete with a glistening glass portico or portal out front.

There is much sameness in this apparent variety, even as the buildings are alike in mystery—their reticence to give away their function—and their air of disuse. In the earlier stages of our industrial era, warehouses looked like warehouses, and factories like factories. Architects believed, as a matter of faith and principle, that this should be the case. And there used to be evidence of *work* everywhere: trash and dirt, smoke and noise, and patches of bare, poisoned dirt. In Heartland, the buildings are immaculate, and very quiet. After several passes around the immense Pepsi Cola building, with its mirrored front and escarpments of windowless concrete, I gave up trying to figure out what happens in it. A process of production, or perhaps of bottling? Or merely of storage and distribution, of marketing and administration? Like everything else in Heartland, it sits, incommunicative, on its clipped, flowered lawn, telling no stories, revealing nothing.

The sky is wide in much of the great semi-circle of postindustrial emplacements surrounding Metropolitan Toronto; but it is nowhere wider than in Heartland. The horizon is low and distant, and the landline would be as dead-flat as a corn field were it not for the huge ziggurats and tumuli of landfill, rising like the sand-covered graves of ancient Mesopotamian cities against the sky. The streets and parking lots are virtually empty on weekends, making it perfect for summer Sunday bicycling and quiet walks. No one lives there, of course. It is thus the exact opposite of the thing represented by Toronto's famously untidy Kensington Market, and an excellent example of a still-strange new urban form created by the car: a market-place emptied and filled, used and abandoned, according to the abstract, arbitrary rhythm of the nine-to-five, Monday to Friday work-week. Heartland seems to rest very, very lightly on its site. A high wind, it seems, could blow away its fragile curtains of glass and new-laid turf and marigold beds. Even the marble-lined, post-Modern *film noir* lobby of Britannia Place, the centrepiece office block of Heartland, seems tentative, theatrical.

But curiously, the postindustrial fantasia of Heartland does recall one historical reality, which is the tale of the very ground it stands on. On weekends, it is as quiet as that former Ontario farmland; on weekdays, as busy. It is a clear space, with the sense of clearing about it, and of keeping things tidy and visually clarified, hence like the land on which it has been installed. I was interested to learn from a commemorative plaque—the surest evidence that true history has passed from popular memory into extinction—that Heartland stands on what was once called Gardner's Clearing, an agricultural community settled in 1821. In late Victorian times, patriotic villagers changed their town's name to Britannia, then left their farms and vanished into the mass of industrial workers clustered around the mills and factories in downtown Toronto. The only reminders of their village is a lovely neo-Gothic United Church, built in 1843, an 1876 schoolhouse and a farm building or two, now used to give Mississauga children a picture-postcard view of Life Before Now.

Someday, as surely as Britannia, Heartland itself will vanish, in the endless process of making and remaking, destroying and emptying and rebuilding and repopulating that have always been the motive forces in urban existence. Will anything of fragile, light Heartland remain, to give schoolkids of the future a taste of life in the age of information, pure surface, mere sheen?

NOWHERE

Like many people, I often engage in what's laughably called "travel" on behalf of my employer—laughably, because nothing could be more absurdly unlike real travel.

The real thing is moving smoothly, rapidly in a comfortable train coach through Canadian farm country on a wintry afternoon, bewitched by the fleeting tints of last sunlight glancing across the snow. Or it's driving around a bend in Arkansas—hot, sick of being behind the wheel, tired of rap music thumping away on the kid's boom-box in the back seat—and finding the answer to a prayer: Joe Bob's Reptile Ranch and BBQ Pit, with a beehive-hairdoed waitress named Billie, a sleepy old alligator, and a tall glass of iced tea.

Travel, at least of the sort I do, usually by air, is devoid of virtually everything deserving the name. Excitement in the air is not, by the way, what's being asked for. I got enough of that to last a lifetime some twenty years ago over Detroit, when my jet carrier barely escaped mid-air collision with another by performing a disconcerting and unforgettable flip. But getting from place to place by air should have something to do with motion. Paradoxically, this most high-speed means of transport imposes on its users utter immobility, a sense of paralyzed waiting, of being passively processed.

It starts at home, in the dull interval between the panic of packing and the moment the taxi-driver buzzes; and it continues the moment the taxi heads onto the airport-bound expressway, a trip too short to read anything, too familiar to make looking out of the window a revelation.

The descent into boredom continues, quickening, on the airport grounds. Upon entering the precincts of Toronto's

Pearson International, for example, one is immediately disoriented by the tangle of broad highways, forking off in improbable directions, and by a horizon defined by uniformly flat-roofed structures. Even the passenger becomes alert to the omnipresent signage, for fear of being whisked away by the driver to the wrong terminal. Here, in this labyrinth of signs, begins the true onset of that helplessness and routinization typical of every moment of air travel.

That's not to say the airport is an architectural object unworthy of attention. In fact, it is a crucial, focal site of economic and social activity in any city that's got one. Because of this prominence, it has attracted notable interest and homage from writers on urbanism. In his recent book *The 100 Mile City,* British architect Deyan Sudjic hails the contemporary airport as "one of the most intricately interwoven spatial hierarchies to be found anywhere in the world city—as complex as the Forbidden City of Beijing."

Sudjic then describes its four "domains," each one more "forbidden" than the one before. First comes the public concourse, devoted to shops and check-in counters, where anyone can go. Second comes the waiting-room, restricted to passengers who have passed the inspection of their bodies and carry-on luggage for weapons and bombs. Next is the expanse of concrete on which the mighty planes move grandly and park, attended by the scurry of little service vehicles, all of which may be witnessed, but only through windows, by passengers who've been certified weapon-free. Finally, there are the many wholly hidden zones ordinary folk will never see, where baggage is sorted, radar signals are monitored and the movements of those huge machines on the ground and in the air are managed.

As intellectually engaging as I find this description, which I include here for your edification, my actual experience of airports, as a user, remains numbness and impotence, even a kind of degradation. Emotions must be suppressed. When a guard is frisking your crotch for guns with his electronic wand, and the device squeals hysterically because it's sensed a dollar-coin in your pocket, you dare not show your disgust, for fear of being forbidden entry to the yet-more inner sanctum. One must take pains to act and look "normal," by reason of the same fear—a reasonable one, incidentally, since there's a sign suggesting that the guards are on the lookout for people who are not behaving normally. Nothing spells out what "normal" is, and you dare not ask. The important thing is to catch one's plane. To do so requires the adoption of an attitude as sedate, standardized and abject as the chrome and plastic seating, the broadloom, the soporific colour scheme—the adoption of the visible pose of a person who is patient, resigned, passive and stupidly happy with the prospect of air travel.

This prospect is not what it was. Once deluxe, streamlined and "special," the contemporary passenger aircraft has become merely an extension of the airport lounge: a cylinder, instead of a box, for waiting. One moves in an airplane, but does not travel, since there is virtually nothing to see (except worthless movies) and nothing to do, except try to read, between inter-ruptions by service personnel at drink time, feeding time, hot-towel time, and the time it takes the captain to tell us it's going to be smooth flying all the way.

The passenger waits as the plane taxis and takes off; of course, one is supposed to read, or look at the pictures in, the brochure indicating what to do should your B-767M aircraft dive into the water or crash-land. I don't read these things, since I am convinced that they are meant merely to make me feel better. I prefer realism. In case of accident, I'm doomed.

Then one waits a certain period aloft and waits to land, waits to be taxied to the terminal, waits to exit the plane and to enter a terminal identical in almost every respect to the one he or she has left. Terminal, plane, terminal become a single and continuous site of waiting.

This is not the alert, wholly human waiting one learns in the spiritual practices of all great religions. It's not the exhilarating wait for a lover to arrive at the front door, nor the expectant waiting for death's release and peace after a grave illness. The waiting that air transit enforces is a training in blandness, anonymity, insignificance, in putting up with bad food and cramped quarters. How did it happen that the one conspicuous moment of public immobility our high-speed technical culture allows us has to be so stupid? Why can't every airport, and at least one or two take-offs in life, be just a little more like the final scene in *Casablanca*?

PROSPECTS FOR THE MIDDLE

Young architects hardly need reminding that they face a dim future. They know better than anyone about the decline of the three great forces that have traditionally provided marvellous opportunities for young architects to recast great European and American cities boldly and memorably.

One is hostile aerial bombardment, which, in a few hours, can level many square kilometres and a half-dozen centuries of pompous building, as it did in Germany a half-century ago. But perhaps it's just as well this sort of radical urban renewal is currently out of fashion in most parts of Europe, and everywhere in North America. As the banal rebuilt downtowns of Frankfurt, Kassel and Cologne abundantly testify, few imaginative planners or architects got the chance to build on the empty lots opened by such sweeping demolition.

Too, megalomaniacal visionaries—with armies of workgangs and strong sympathetic governments behind them—are in notably short supply in our mass-democratic times. Where are the enlightened, autocratic vandals who, every hundred years or so, used to sweep away old Berlin, and raise a new city in its place? And just when smart young architects need him, where is the Toronto or Vancouver rendition of Georges-Eugene Haussmann (1809–1891), the remorseless destroyer and reshaper of Napoleon III's Paris?

Last, and perhaps most important for their future as changers of the cityscape, architecture grads must discount earthquakes, fires and windstorms, all traditional allies in their quest for interesting commissions. The tall building, glory of urban culture in our time, would quite possibly have never arisen and flourished in Chicago had not the city's central clutter been cleared by the fire of 1871. And, long before the human-inspired devastation of central Toronto in the 1960s and 1970s, important fires in 1849 and again in 1904 gave local architects extraordinary chances to ply their trade, try their ideas, make over the city centre in the current image of the New.

I had harboured hopes that such an opportunity for Toronto architects might arise again one day. Not in the wake of earthquake or fire, mind you, but in the natural, slow course of things. By reducing the threat of widespread devastation by meteorological or geological powers, structural engineering during the past century has proved to be a mixed blessing for architects at best, and, at worst, a grounds for their elimination. But even the best-engineered tall buildings (I told myself) grow old, become weather-beaten and worn, and must be taken down, to make way for something new.

Or so I thought, until a visit to the Scarborough headquarters of Dowdell Pal Ellis Shim Consulting Engineers. Founder and senior partner Gordon Dowdell—whose projects include Ontario Place and the Sherway Gardens shopping plaza in Etobicoke—quickly put one of these thoughts out of my head. If maintenance is kept up, the steel-framed tall buildings that define our skyline, says Dowdell, should last. "We don't have a problem of deterioration with time. There are wind and earthquake stresses, and a building moves all the time. But if the whole exterior cladding is designed to withstand such forces, you have no problem."

If the load-bearing metal cage of a tall building is indefinitely reliable, in theory anyway, little else about it is. At the end of a particularly severe test of nineteenth-century construction techniques—the San Francisco earthquake and fire of April 18, 1906—the steel-framed, curtain-walled edifices were among the few built things to survive. But as the panoramic photos recently shown at Montreal's Canadian Centre for Architecture reveal, that's about all that did survive: gone was the glass, the furniture, the people—everything, in fact, *but* the frame.

Given the frailty of cladding commonly in use these days, Dowdell is concerned about what will happen to Toronto's tall glass-walled structures in the entirely possible (through statistically improbable) event of a significant earthquake. His more immediate worries, however, have to do with a long-term failure to maintain buildings against more subtle threats to their existence. "Many of our tall buildings have precast concrete panels tied onto the surface, anchored back into the building. The anchorage is supposed to be corrosion-resistant, but I question this. Our weather is becoming more and more corrosive. Too, glass ages and becomes more brittle with time. The other snag is that glass can tolerate very little distortion." The problem will always be one of maintenance—of just how long, that is, the owners can afford to pay for upkeep against the slowly declining structural reliability of a given high building. Abandon this pricey, permanently necessary maintenance, and, even if it won't fall down, a tall building becomes a dangerous proposition.

The precast concrete apartment buildings built all over Metro since the 1950s are another story, says Dowdell. Long before the high office towers are done for, atmospheric corrosion and crumble, salt damage (especially in the underground parking garages), and the natural shrinkage of all poured stone, known as "creep," will necessitate the demolition of apartments, which were produced faster and with less structural resistance to the elements than their commercial cousins. Current life expectancy of these buildings: about twenty-five years.

So here's my advice to architecture students who hope to build something on the Toronto city site before the middle of the next century. Forget about 150 storeys of deluxe office space, glistening towers piercing the golden mists of a summer morning, monumental spectacles of steel and glass punctuating the city's low horizon. Start thinking up something new to do with the family room.

EDGE CITIES

Lest this seems an unnecessarily gloomy and apocalyptic vision to lay on a young architect, I hasten to add that there is an alternative scenario on the horizon; and, if it comes into play on the urban stage, future designers will have their hands full indeed.

This more hopeful prognosis first began to form in my head in early 1993, when several birds of distinctly unlike feather—vulture developers, gentle spotted environmentalists, red-crested socialists and unionists, and wattled grey entrepreneurs—hopped on the same snowy branch and together began twittering the praises of the Ontario government's new $2.5-billion mass-transit plan for the greater Toronto area.

Not joining the noisy perch-in, however, was *The Globe and Mail's* resident Raven of Doom, Colin Vaughan. He took one look at the plan and its proposed funding formula for the big projects—a pooling of provincial and municipal tax money with $500 million from private pockets—rustled his black feathers on his lofty rafter, and solemnly quoth: Nevermore!

On the off-chance you haven't heard what lies in the future for us mobile Torontonians, here's a quick summary. A new 5.4-km subway line, burrowed eastward from the north-south Yonge line into the settled, upmarket suburbia around Sheppard Avenue East and Don Mills Road. A 4.4-km subway, run out from Eglinton West station to a currently unpopulated wasteland, on whose dank flats is destined to rise the commercial, residential and governmental towers of the new York City Centre.

The Spadina Subway—always called that, despite the fact that most of its track lies above ground (and not under or along Spadina Avenue)—will be pushed north-west 5.2 km to York University, at last hard-wiring the farflung campus into the urban transportation grid.

The Scarborough Rapid Transit line will be extended 3.4 km farther into the largely unexplored forests of north-east Metro, where wolves and a few residual pioneers have been sighted in recent years. Finally, dedicated express lanes will be built on the margins of the west-suburban superhighway 403 to accommodate fast bus travel for the huge numbers of people always in a great rush, I've noticed, getting from nowhere to nowhere in northern Mississauga.

"The outlook is grim," croaked Vaughan. The biggest bone sticking in his craw was the predicted private-sector involvement. Vaughan was aflutter, specifically, at the thought of developer Murray Frum dropping a few million into the pot for the Scarborough Rapid Transit extension, which happens to be going out where he has land, presumably in hopes of turning a profit.

Now for those, like this writer, who would like to see all public utilities and most government services, including the streets and mass-transit system, sold off to private investors, the Frum involvement will probably seem like a very tiny step indeed in the right direction. But for Vaughan, it would be the first step on the slippery slope to perdition. "The cheque book has become the keystone of the new urban order in Ontario," he lamented. "To hell with the planning priorities....Sadly, we can only wait for the horror stories to emerge."

But why wait? From Vaughan's perspective, the shape of things to come is already plain, and plainly unacceptable.

Magnified urban sprawl. The hastened flight of populations from urbia to suburbia. The final quenching of Toronto's downtown seethe, and the surrender of our once-busy city to drug fiends and prostitutes, and desertion by the kindly and decent.

Missing from this dark view is, first, the acknowledgement that subways run two ways. If everybody's hurtling downtown to work every morning now, who's to say that they'll be hurtling in the same direction forever? In my crystal ball, I see the future toiling masses of Information City—for so Toronto has already become—living right downtown, in increasingly compact, enjoyable high-rise and low-rise circumstances, and scattering on outbound subways to the several neo-urban centres of commerce, industry and high-tech work that will then stand, like beacons, on the city's periphery. These neo-urbanisms will perhaps (or perhaps not) re-create, on a little scale, the traditional metropolitan mountain, with skyscrapers at the centre, lower industrial/commercial blocks beyond, and, beyond them, yet lower residential neighbourhoods. (After all, there's no law that says every city must have the same sort of skyline forever.)

A jet traveller landing at Pearson International in fifty years, for instance, may well see a rumpled carpet of high and low housing and amenities where the old high city core of Toronto used to be, before we took a few skyscrapers down. On the far eastern, western and northern horizons, the same traveller glimpses things that seem to belong to a former time: a scatter of widely separated vertical points composed (on closer view) of numerous compact tall and low buildings. These neo-urban units would house our workplaces, our workshops, factories and offices. Though broadly spaced, each would be connected to each other and to the residential downtown by even more marvellous electronic gadgetry than we have today, and by the branching mass-transit system already in place, and about to be given a nudge by the Ontario government and its capitalist allies.

Only after this fantasy came to me did I discover, not to my surprise, that architects and urbanists had already begun drawing up plans for the diffused city. In 1991, the University of Texas's Center for American Architecture and Design held a symposium in Austin called "New Centers on the Periphery: the Case of Four Texas Metropolitan Areas." Several talking-points, later published in the Center's journal, struck me as provocative, but none more than the assertion that we've always had it wrong about North American cities. They've never been uniformly centred, with lessening densities from the inside out. "In fact," said Texas architectural historian Robert Bruegmann, "the American city has been decentralizing and re-centering virtually as long as there has been an American city." While acknowledging the near-universal hatred of edge-towns by academics and upholders of the high-density downtown ideal, Bruegmann urged his colleagues to think *diffusion,* since so little is really known about this phenomenon. "It may be daunting, but what a challenge: to try to understand urban systems at the moment that they are fundamentally transforming themselves before our very eyes."

And, anyway, what have we got to lose? Pick a centre on the edge—any one will do—and try to understand how it works, *if* it works, how people use and move and respond to each other within it. Your spot may be far removed from downtown Toronto, but it need not be. I'm interested in the mixed-use York City Centre, the new focus of the constituent Metro city, and a proposed terminus of a new spur off the Spadina line. From the imaginary pictures made by its planners, it appears that this place will feature rather old-fashioned skyscrapers, showing more stone than steel and glass, hence more intimate than the distinct "tall building" of classic postwar Modernism. Moreover, as if intent on declaring the centre's theme to be proximity rather of purity, the principal York skyscrapers will be linked near the top by dramatic flying bridges and, near the bottom, by what appears to be a lofty atrium or mall.

When completed, of course, York City Centre may not function as planned. But if it doesn't, it can be written off as yet another experiment in the long attempts of urban populations to adjust to the inevitable arising of concentrated but widely spaced urbanism. Some day, it's to be hoped, the planners and developers will get the formula right, and start consciously creating fringe-cities architecturally responsive to the opportunities for scatter opened by contemporary communications technology, transport systems, and social desire.

YORK

In the waning days of 1993, I suddenly recalled that Toronto's bicentennial was rapidly slipping away, and that I'd not done so much as raise a glass of Diet Coke to my adopted home town. Whereupon I resolved to pay a visit to pay an honorific visit to the spot my city got its start.

First, I had to find it. Of the civilian town called York—the town site's name until its incorporation in 1834—nothing has survived except a patch of rigid street-plan. To the best of my knowledge, no persisting architectural marker indicates the town's original boundaries. When I sat down with a city map and thought about exactly where York had been—something anyone who'd lived in Toronto a quarter-century *should* know—I was surprised to find the whole business a bafflement. All I knew was that York was somewhere quite downtown, and east of Yonge Street, but north of Front Street, simply because that was the shoreline of Lake Ontario until nineteenth-century landfill pushed the docklands far south.

Getting this bit of information was easy. Eric Arthur's *Toronto No Mean City*, shows the civilian town as a tiny rectangle bounded by present-day George, Berkeley, Adelaide and Front streets. Even if one knows where the borders are, however, they are bound to seem arbitrary. The territory they

enclose is an especially undistinguished swatch of east downtown, with the usual gaps of parking lots, a few tottering late
Victorians somebody forgot to demolish, and many Vic-but-
renovated and twentieth-century buildings. But this very non-
descriptness has a venerable history. Right from 1793, what
Arthur calls the "practical but indescribably mean and
unimaginative" plan of York never had architectural focus or
definition. "Had there been provision for a school, a church,
or, more particularly, a village green, the plan of Toronto today
would have been different," he bewails. "It also lacked direction, so that when expansion was inevitable, the town grew
merely by adding more squares."

The tight plan Arthur so disliked was the brainchild of one
Alexander Aitkin, who drew it off for Lieutenant-Governor
Simcoe in 1793. Aitkin hardly deserves such harsh condem-
nation. He was, after all, a military engineer, not a visionary;
and his checkerboard of streets was, and is, the most common
ever devised to lay out a new city intended to serve as a trading, industrial, administrative or military centre. It was introduced early into Britain's American colonies, and thereafter
never lost its utility, or its appeal to planners.

The grid proposed by Thomas Hulme for Philadelphia in
1682, for example, provided no more common space than the
1793 plan for York. Following the example of Hulme and
new-town planners like him, Aitkin simply ignored the bumps
and dips in the land which, in Toronto's case, included
swampy beaver ponds, deep hardwood forests, meandering
creeks and gullies cut in the clay. Beginning with Aristotle and
finding especially fertile ground among radical philosophers
of habitation in our own century, this imposition of rectilinear
order on wiggly Nature has always been a persistently popular
idea in serious Western urbanism. Unless Western civic idealism is coming permanently unstuck, the grid will return to
favour—though it should be said that, for some time, it has
been in deep disfavour indeed. Visiting Hulme's Philadelphia

in 1842, Charles Dickens complained of the city's "distract-ingly regular" layout: "After walking about it for an hour or two, I felt that I would have given the world for a crooked street." Eric Arthur, Dickens's latter-day soul mate, is then voicing no new gripe against the old gridiron, when he con-demns the Toronto street-plan "with which we have had to cope for almost two hundred years and with which posterity will have to deal till the end of time."

Sure enough, addition of new city blocks became quickly necessary after 1793, since Aitken had plunked York too close to the great, foggy swamp that, 125 years later, would be filled and become the Port Industrial District. Within only a few years of tiny York's creation, the more important colonial citi-zenry had begun to move their homes and businesses, and hence the centre of town, westward, in the direction of the financial district focusing the financial and corporate city today. (As late as 1834, the middle of the city still was no far-ther west than the corner of King and Frederick, and Queen Street had not been pushed as far west as Yonge Street.) Old York, which could have been our Boston Common or Wall Street, had it presented a suitable focus, was swiftly abandoned to an ignominious fate as a Victorian warehousing and factory zone, of no further immediate importance to the growing metropolis.

After reading Arthur's dismissive account, it seemed unfit-ting to visit this ill-starred place only by day, when it would be swarming with trucks and haulers and office workers, busy with their metropolitan day-jobs. Alexander Aitkin's York *is* the cemetery of a noble, unpopular historical idea in town planning, applied here with the same nonchalant disregard for topographical reality as elsewhere. So I decided to pace off York's emptied lanes and streets in the hours long before dawn one winter's night, when even the unsleeping city nods off just a little, and a certain pensive quiet dawns in the downtown lanes.

On the principal streets, of course, darkness had long been vanquished by unblinking lamps—though in the dark alleys hidden among tall buildings, the chilling dank gloom that once shrouded the little wooden houses of eighteenth-century York could still be felt. I love downtown night laneways such as these—very still, littered with the detritus of industry, at times shadowed thick as ink, at other times turned into a mad fantasia of black and bright white by a glaring light-bulb installed on a rear wall. Due to the clay on which it was built, "Muddy York" deserved its early nickname—though the point is driven home with special force when, making your way over the icy, broken pavement of an alley and trying not to seem snoopy about the deal going down between the hooker and john a few metres down the lane, you put your foot into a deep pot-hole filled with that oily, never-draining water that gave York its odd handle.

If the site lacks a central square, or anything else that would have kept the burgeoning city anchored there, it does have one accidental peculiarity that I'd never noticed before my wintry night prowl. An outsized number of the businesses installed in the old warehouses and new, short towers have to do with communications and transportation; and all are bathed in the faint electric buzz and sublimnal vibratory haze characteristic of such structures. The large office building and printing plant of *The Toronto Sun*—its lunchroom empty, its ground-floor editorial nooks flooding the dark, damp sidewalk outside with glaring fluorescent light—stands upon the site of old York, along with the headquarters of *Saturday Night* magazine, and Greyhound's parcel-handling facility. There are innumerable copy shops, their machinery hooded in plastic at night, and countless design firms, ad agencies, little publishing outfits and printing companies.

If old York did not become our geographical urban centre, it did become the forum of Toronto's key industry in the late twentieth century, which is information. In this place are monuments to Toronto's essential existence, though hidden from sight: a thick web of wires wriggling through walls and under the streets, telecommunications and data-processing equipment glowing and pulsing, receiving and sending; and whole buildings whose human staff is organized by its computers and cybernetic gadgetry. Without meaning to, Old York today typifies what Toronto seems always destined to become, and which many students of cities now appear deeply uneasy about: an urban complex much poorer in focused, stately, famous architectural space and construction than Eric Arthur would have liked, without greatness as a manufacturing centre or port, yet dense and busy with the electric impulses, data, words, texts, stories, movements, service exchanges and flying numbers which together constitute whatever significance it has, and whatever its claim to the name *cosmopolis*. I am not downplaying the importance of our architectural monuments in brick, stone and glass; I do not need to do so. Long before I sat down to write the columns on which this book is based, the solvent of high technology had already dissolved the historic substantiality of the modern city, leaving behind only the Cheshire Cat's mercurial smile—the information and ornamentation, the evanescent visual and sensuous codes which are now the central facts of contemporary urban experience and, by an interesting historical irony, the principal products being made today in Old York.

But fear not: I am not about to launch into a happy-faced hymn to the "information highway"—the most vulgar buzzword so far invented by the futurologists of the 1990s—or claim Old York to be a particularly key crossroads in it. Though I believe the city as we have known it is vanishing, and a new city as we have *not* known it is coming to be—which makes it utterly beyond words, hence beyond thought—I find much joy in contemplating the evidence of the historic Western idea of urbanism as stolid, choreographic, disciplined. A number of buildings in the Old York district,

Near Dupont Street, Evening

for example, have what is known as "historic interest," a polite way to say they have become ideologically obsolete—dear, discarded shells left as the city became something new. Among the oldest extant buildings is also the most lovely: the august structure at 252 Adelaide Street East, originally the Bank of Upper Canada and now a restored "period" post office, built in the late 1820s. There are also a few other brick Victorian churches and commercial and industrial buildings, and a nice ultra-*moderne* one or two, standing among the parking lots, pocks and pot-holes, now decidedly beyond the eastern edge of the gleam-zone of our financial and banking skyscrapers.

Because I find "architectural tours" too tightly aesthetic and sentimental, and invariably unconscious of urban context and the dissolving process that turns discreet buildings into city sites—from *things* into fields over which meaning flickers and dances—I've made the acquaintance of these buildings on my own over the years. Visiting them should be, and is, like dropping in on elderly friends, retired actors in a long-running play, now closed. My gradual learning has usually come after admiring, by chance, the grace of a porch or column, the polite or haughty or faintly overbearing ways the façades address the street, the way they silently request the passer-by to realign her posture or straighten his tie.

Most of the buildings and occurrences in the district do not remind us of terribly serious matters, only interesting and intellectually vivifying ones. Not so—at least for this walker in the city—the sight of St. James' Cathedral, and the sounds of its bells, pealing over the city. I know as well as any Torontonian that St. James is destined forever to be one of the top ten stops on any architectural guide's tour of Ye Olde Toronto. Surely because I am a Christian believer, this church is an abrupt comeuppance to thoughts adrift on the intellectual draughts and alluring currents I'm habitually drawn to. For if the city is changing into something new, and seems to be tending in directions hardly thinkable in traditional categories, I am not.

This point came home to me on a Saturday just before I finished this book, when walking the streets of Old York, I heard the bells of St. James. Almost without thinking, I dropped in on the noontime Eucharist, which, to my surprise and pleasure, was being celebrated by a priest I had not seen in some time. As she offered the bread and wine to God, using the ancient words of thanksgiving, I was reminded of the men and women who had been bringing to this place, for two centuries, the little-changing sorrows, dreads and joys and exaltations of human existence. These experiences know no ethnic boundaries, no limits of time. Perhaps at the very heart of our historic need to live in cities is this longing not to be alone—to be continually aware that we are surrounded by a cloud of witnesses, living and dead and yet unborn, to the immense richness and endless discovery of urban experience.

"Heritage" concentrates on the dead merely, making them seem remote in their quaint costumes and antique surroundings. The service at St. James, however, called to mind the persistent issues and needs, and the nearness of the long-dead to our own condition, and our enduring community with them. Frankly, that plain Saturday celebration of the Holy Communion was the first occasion on which I had felt good about the Toronto bicentennial, simply because it was the first time I felt proper gratitude for Toronto, this experiment in urbanity begun two hundred years before. It was also a moment of thanksgiving for those who will come after this book and its author, and all those now living, have been forgotten—our children and newcomers, who will be drawn by the lights of the Emerald City, and decide to inhabit this place, pace it off and consider it anew, learn from and build it again, forever.

SOURCES
AND
RESOURCES

Thinking Places

Banham, Reyner. *A Concrete Atlantis: U.S. Industrial Building and European Modern Architecture 1900–1925*. Cambridge, Mass.: The MIT Press, 1986.

Canadian Auto Workers Local 303 Heritage Committee. *You Can't Bring Back Yesterday*. Toronto: CAW Local 303, 1993.

Charles, Prince of Wales, quoted by Jonathan Bate in *Times Literary Supplement*, June 12, 1992.

Le Corbusier. "Three Reminders to Architects, (I): Mass," in *Towards a New Architecture*. New York: Dover Publications, 1986.

Macaulay, Rose. *Pleasure of Ruins*. London: Thames and Hudson, 1984.

Roberts, V.M. "Toronto Harbour," *Canadian Geographical Journal*, Vol. XV, No. 2 (August, 1937).

Ruskin, John. *The Seven Lamps of Architecture*. London: George Routledge & Sons, 1907.

Sobel, David. *The Moving Past: A Presentation of Archival Work Films*. Toronto: Labour History Imagers Group, n.d.

Sobel, David, and Susan Meurer. *Working at Inglis: The Story of a Toronto Factory*. Toronto: James Lorimer and Co., 1994.

Stinson, Jeffrey. *The Heritage of the Port Industrial District*, Vol. I. Toronto: The Toronto Harbour Commissioners, 1990. See also Suzanne Barrett and Joanna Kidd, *East Bayfront and Port Industrial Area: Pathways: Towards an Ecosystem Approach, A Report to the Royal Commission on the Future of the Toronto Waterfront*. Ottawa: Minister of Supply and Services Canada, 1991.

On the Land

Gregory, Dan, and Roderick MacKenzie. *Toronto's Backyard: A Guide to Selected Nature Walks.* Vancouver and Toronto: Douglas & McIntyre, 1986.

Moore, Charles W., William J. Mitchell, William Turnbull, Jr. *The Poetics of Gardens.* Cambridge, Mass.: The MIT Press, 1988.

Pielou, E.C. *After the Ice Age: The Return of Life to Glaciated North America.* Chicago and London: The University of Chicago Press, 1991.

Solomon, Barbara Stauffacher. *Green Architecture: Notes on the Common Ground (Design Quarterly 120).* Minneapolis: Walker Art Center, 1982.

Binding and Loosing the Waters

Hough, Michael. *City Form and Natural Process.* London and New York: Routledge, 1991.

McGlade, Terry, James Brown, Whitney Smith. "The Garrison Creek Community Project." Grant application, n.d.

The Task Force to Bring Back the Don. *Bringing Back the Don.* Toronto: City of Toronto Planning and Development Department, 1991.

Tales of the Pioneers

Beck, Julia, and Alec Keefer, eds. *Vernacular Architecture in Ontario.* Toronto: The Architectural Conservancy of Ontario, 1993.

Benn, Carl. *Historic Fort York 1793–1993.* Toronto: Natural Heritage/Natural History, Inc., 1993.

Brunskill, R.W. *Illustrated Handbook of Vernacular Architecture.* London and Boston: Faber and Faber, 1987.

Duncan, Dorothy. *Life in the Past Lane.* Toronto: Metropolitan Toronto & Region Conservation Foundation, n.d.

Mika, Nick, Helma Mika and Gary Thomson. *Black Creek Pioneer Village.* Belleville: Mika Publishing Co., 1988.

A Pictorial History of Weston. Toronto: The Weston Historical Society, 1981.

Rempel, John I. *Building with Wood.* Toronto: University of Toronto Press, 1980.

Stacey, C.P. *The Undefended Border: The Myth and the Reality.* Ottawa: The Canadian Historical Association, 1967.

The Town Project. "Railways in Weston," "Churches in Weston" and other publications, n.d.

Pleasures in Places

Colvin, Howard. *Architecture and the After-Life.* New Haven: Yale University Press, 1992.

Etlin, Richard A. *The Architecture of Death: The Transformation of the Cemetery in Eighteenth-Century Paris.* Cambridge, Mass.: The MIT Press, 1987.

Filey, Mike. *Mount Pleasant Cemetery: An Illustrated Guide.* Toronto: Firefly Books, 1990.

Francis, Mark, and Randolph T. Hester, Jr. *The Meaning of Gardens: Idea, Place and Action.* Cambridge, Mass.: The MIT Press, 1990.

Harris, Neil, and Benjamin Portis. *Civic Visions, World's Fairs.* Exhibition catalogue. Montreal: The Canadian Centre for Architecture, 1993.

Ring, Dan, Guy Vanderhaeghe and George Melnyk. *The Urban Prairie.* Exhibition catalogue. Saskatoon: Mendel Art Gallery and Fifth House Publishers, 1993.

Robinson, John. *Once Upon a Century: 100 Year History of the 'Ex.'* Toronto: privately printed, 1978.

Rybczynski, Witold. "Building the City Beautiful." *Times Literary Supplement*, November 20, 1992.

Toronto Historical Board. "Colborne Lodge, 1837." Mimeographed handout, n.d.

Webster, Donald. "Colborne Lodge Furnishings." Mimeographed training text. Toronto Historical Board, n.d.

Modern

Dal Co, Francesco. *Figures of Architecture and Thought: German Architecture Culture 1880–1920.* New York: Rizzoli, 1990.

Fenton, Joseph. *Hybrid Buildings (Pamphlet Architecture 11).* 2nd ed. Princeton: Princeton Architectural Press, 1985.

Ferriss, Hugh. *The Metropolis of Tomorrow.* New York: Ives Wasburn, 1929.

Frith, Valerie, ed. *Toronto Modern: Architecture 1945–1965.* Toronto: Coach House Press and the Bureau of Architecture and Urbanism, 1987.

Huxtable, Ada Louise. *Architecture, Anyone? Cautionary Tales of the Building Art.* Berkeley and Los Angeles: University of California Press, 1986.

Jacobs, Jane. *The Death and Life of Great American Cities.* New York: Random House, 1961.

Jencks, Charles. *Modern Movements in Architecture.* Harmondsworth: Penguin Books, 1982.

The Royal Commission on the Future of Toronto. *Regeneration.* Ottawa: Minister of Supply and Services Canada, 1992.

Shopping

Cappe, Lorne, *Window on Toronto.* City of Toronto Planning and Development Department, 1990.

Laycock, Margaret, and Barbara Myrvold. *Parkdale in Pictures: Its Development to 1889.* Toronto Public Library Board, 1991.

Little, Bruce. "Retail Sales? It's not our department." *The Globe and Mail,* November 1, 1993.

Pevsner, Nikolaus. *A History of Building Types.* Princeton: Princeton University Press, 1989.

Sorkin, Michael, ed. *Variations on a Theme Park: The New American City and the End of Public Space.* New York: The Noonday Press, 1992.

Wills, Garry. "Chicago Underground." *The New York Review of Books,* October 21, 1993.

Suburban Idylls

Adorno, Theodor. "Culture and Administration." Trans. by Wes Blomster, in Dennis Crow, ed. *Philosophical Streets: New Approaches to Urbanism*. Washington, D.C.: Maisonneuve Press, 1990.

Bonis, Robert R., ed. *A History of Scarborough*. Toronto: Scarborough Public Library, 1968.

Le Corbusier. "The Hours of Repose," in *The City of To-Morrow and its Planning*. Mineola, N.Y.: Dover Publications, 1987.

Sewell, John. *The Shape of the City: Toronto Struggles with Modern Planning*. Toronto: University of Toronto Press, 1993.

Spain, Daphne. *Gendered Spaces*. Chapel Hill and London: The University of North Carolina Press, 1992.

Teshima, Ted, et al. "Moriyama & Teshima: Architecture as a Work of Life." *Process: Architecture* No. 107 (December, 1992).

Wilson Alexander. *The Culture of Nature: North American Landscape from Disney to the Exxon Valdez*. Toronto: Between the Lines, 1991.

Concrete Dreams

Billington, David P. *The Tower and the Bridge: The New Art of Structural Engineering*. Princeton: Princeton University Press, 1983.

Brown, David J. *Bridges: Three Thousand Years of Defying Nature*. London: Mitchell Beazley, 1993.

McKillop, David. *The Motel Strip Study*. Toronto: City of Etobicoke Planning Department, 1986.

Zwarts, Moshé. "Why Are Car Parks So Ugly?" In Maarten Kloos, ed. *Architecture Now: A Compilation of Comments on the State of Contemporary Architecture*. Amsterdam: Architectura & Natura, 1991.

Streets

Cameron, C.J. *Foreigners or Canadians?*, quoted in text accompanying "The Magic Assembling: Metropolitan Toronto Storefronts and Street Scenes," an exhibition organized by Michael McMahon and Lillian Petroff for the Metropolitan Toronto Archives, March, 1993.

Dendy, William. *Lost Toronto: Images of the City's Past.* rev. ed. Toronto: McClelland & Stewart, 1993.

Donegan, Rosemary, introduction by Rick Salutin. *Spadina Avenue.* Vancouver: Douglas and McIntyre, 1985.

Kostof, Spiro. *The City Assembled: The Elements of Urban Form Through History.* Boston: Little, Brown and Company, 1992.

Myrvold, Barbara. *Historical Walking Tour of Kensington Market and College Street.* Toronto: Toronto Public Library Board, 1993.

Gardens

Baraness, Marc, and Larry Richards. *Toronto Places: A Context for Urban Design.* Toronto: University of Toronto Press, 1992.

Brown, Jane. *Gardens of a Golden Afternoon: The Story of a Partnership, Edwin Luytens and Gertrude Jekyll.* New York: Van Nostrand Reinhold, 1982.

Ellacombe, Canon Henry N. *In a Gloucestershire Garden.* London: Century Hutchinson, 1986.

Holl, Stephen. "Double House." *Rural and Urban House Types in North America (Pamphlet Architecture 9).* Princeton: Princeton Architectural Press, 1983.

Pavord, Anna. "Back to the fuchsia." *Times Literary Supplement,* July 17, 1992.

Pollan, Michael. "Why Mow? The Case Against Lawns." *The New York Times Magazine,* May 28, 1989.

Sackville-West, Vita. *V. Sackville-West's Garden Book.* Edited by Philippa Nicolson. London: Michael Joseph, 1989.

Moral Management

Campbell, Mary, and Barbara Myrvold. *The Beach in Pictures 1793–1932.* Toronto: Toronto Public Library Board, 1988.

Conway, Hazel. *People's Parks: The Design and Development of Victorian Parks in Britain.* Cambridge: Cambridge University Press, 1992.

Department of Parks and Recreation, City of Toronto. "High Park: Past to Present." Toronto, n.d.

Foucault, Michel. *Discipline and Punish: The Birth of the Prison.* Trans. by Alan Sheridan. New York, 1977. See also Joseph Masheck. *Building-Art: Modern Architecture Under Cultural Construction.* Cambridge: Cambridge University Press, 1993.

Rosenzweig, Roy and Elizabeth Blackmar. *The Park and the People: A History of Central Park.* Ithaca: Cornell University Press, 1993.

Moderne *Variations*

Bliss, Michael. "The Historical Significance of Maple Leaf Gardens." Memorandum submitted to the Toronto Historical Board, November 26, 1989.

Fleming, Keith R. *Power at Cost: Ontario Hydro and Rural Electrification 1911–1958.* Montreal and Kingston: McGill-Queen's University Press, 1992.

Greif, Martin. *Depression Modern: The Thirties Style in America.* New York: Universe Books, 1988.

Hawes, Elizabeth. *New York, New York: How the Apartment House Transformed the Life of the City (1869–1930).* New York: Alfred A. Knopf, 1993.

Holl, Stephen. *The Alphabetical City (Pamphlet Architecture 5),* 2nd ed. Princeton: Princeton Architectural Press, 1980.

Huxtable, Ada Louise. *The Tall Building Artistically Considered: The Search for a Skyscraper Style.* Berkeley and Los Angeles: University of California Press, 1992.

Toronto Historical Board. "Heritage Property Report: Maple Leaf Gardens, 438 Church Street." December 1989.

High Styles

Hersey, George. *The Lost Meaning of Classical Architecture: Speculations on Ornament from Vitruvius to Venturi.* Cambridge, Mass.: The MIT Press, 1988.

Hersey, George, and Richard Freedman. *Possible Palladian Villas.* Cambridge, Mass.: The MIT Press, 1992.

Hitchcock, Henry-Russell. *The Architecture of H.H. Richardson and his Times.* rev. ed. Cambridge, Mass.: The MIT Press, 1986.

Onians, John. *Bearers of Meaning: The Classical Orders in Antiquity, the Middle Ages and the Renaissance.* Princeton: Princeton University Press, 1988.

Pelt, Robert Jan van, and Carroll William Westfall. *Architectural Principles in the Age of Historicism.* New Haven: Yale University Press, 1993.

Summerson, John. *The Classical Language of Architecture.* rev. ed. London: Thames and Hudson, 1980.

Houses and Home

Berman, Marshall. *All That is Solid Melts into Air: The Experience of Modernity.* New York: Simon and Shuster, 1982.

Denison, John. *Casa Loma and the Man who Built it.* Erin, Ont.: The Boston Mills Press, 1982.

Harbison, Robert. *The Built, the Unbuilt and the Unbuildable: In Pursuit of Architectural Meaning.* Cambridge, Mass.: The MIT Press, 1991.

Kolb, David. *Postmodern Sophistications: Philosophy, Architecture and Tradition.* Chicago: The University of Chicago Press, 1992.

Rybczynski, Witold. *Looking Around: A Journey Through Architecture.* Toronto: HarperCollins, 1993.

At the Edge

Adams, James L. *Flying Buttresses, Entropy, and O-Rings: The World of an Engineer.* Cambridge, Mass.: Harvard University Press, 1991.

Bergh, Wim van den. "Mental Transparency." In Maarten Kloos, ed. *Architecture Now: A Compilation of Comments on the State of Contemporary Architecture.* Amsterdam: Architecture & Natura, 1991.

Bruegmann, Robert, and Tim Davis, "New Centers on the Periphery." *Center*, Vol. 7, 1992. See also Mildred Friedman. *Edge of a City (Pamphlet Architecture 13)*, Princeton: Princeton Architectural Press, 1991.

Elliot, Cecil D. *Technics and Architecture: The Development of Materials and Systems for Buildings.* Cambridge, Mass.: The MIT Press, 1992.

Isin, Engin F. *Cities without Citizens: Modernity of the City as a Corporation.* Montreal and New York: Black Rose Books, 1992.

Kelly, Colleen. *Cabbagetown in Pictures.* Toronto: Toronto Public Library Board, 1984.

Kostof, Spiro. *The City Shaped: Urban Patterns and Meanings Through History.* Boston, Toronto and London: Little, Brown and Company, 1991.

Macrae-Gibson, Gavin. *The Secret Life of Buildings: An American Mythology for Modern Architecture.* Cambridge, Mass.: The MIT Press, 1988.

Sudjic, Deyan. *The 100 Mile City.* London: Andre Deutsch, 1992.